STABILITY AND PERFORMANCE OF SLOPES AND EMBANKMENTS-II

Volume 2

Proceedings of a Specialty Conference
sponsored by the
Geotechnical Engineering Division of the
American Society of Civil Engineers

co-sponsored by the
San Francisco Section of ASCE, and
The University of California, Berkeley

in cooperation with the
Association of Engineering Geologists
U.S. Committee on Large Dams
ASFE/The Association of Engineering Firms
Practicing in the Geosciences
Association of State Dam Safety Officials

Berkeley, California
June 29 – July 1, 1992

Edited by Raymond B. Seed
Ross W. Boulanger

Geotechnical Special Publication No. 31

Published by the
American Society of Civil Engineers
345 East 47th Street
New York, New York 10017–2398

ABSTRACT

This proceedings, *Stability and Performance of Slopes and Embankments II*, consists of papers presented at the Specialty Conference sponsored by the Geotechnical Engineering Division of the American Society of Civil Engineers held in Berkeley, California, June 29—July 1, 1992. The conference provided a forum for the discussion of advances in the analysis and design of slopes and embankments, review of new data on performance, presentation of recent developments in remediation, and forecasting of future developments. Some of the topics covered by this proceedings are: 1) Soil property evaluation; 2) static stability and deformation; 3) seismic stability and deformation; 4) stability of landfills; 5) residual strength evaluation; and 6) site improvement or remediation. This proceedings provides the engineer with the opportunity to assimilate new information and cope with the joint challenge and trauma that change inevitably brings.

Library of Congress Cataloging-in-Publication Data

Stability and performance of slopes and embankments II: proceedings of a specialty conference / sponsored by the Geotechnical Engineering Division of the American Society of Civil Engineers; co-sponsored by the San Francisco Section of ASCE, and the University of California, Berkeley; in cooperation with the Association of Engineering Geologists ... [et al.]: edited by Raymond B. Seed, Ross W. Boulanger.

p. cm.—(Geotechnical special publication: no. 31)

Includes indexes.

ISBN 0-87262-872-8

1. Embankments—Congresses. 2. Slopes (Soil mechanics) — Congresses. 3. Soil stabilization — Congresses. I. Seed, Raymond. II. Boulanger, Ross W. III. American Society of Civil Engineers, Geotechnical Engineering Division. IV. American Society of Civil Engineers, San Francisco Section. V. Association of Engineering Geologists. VI. Series.

TA760.S73 1992

624.1′62—dc20 92-14707

CIP

Library of Congress Catalog Card No: 92-14707

ISBN 0-87262-872-8

Manufactured in the United States of America.

Session 4
SEISMIC STABILITY AND DEFORMATION ANALYSIS
Moderator: R. B. Seed

Session 5
FILLS AND EMBANKMENTS OVER SOFT CLAY
Moderator: J. M. Duncan

Session 6
NATURAL SLOPES AND RESIDUAL STRENGTH EVALUATION
Moderator: J. D. Smart

Session 7
CASE HISTORIES - VARIOUS
Moderator: C. N. Baker

Session 8
WORKSHOP SESSION: BENCHMARK PROBLEMS FOR SLOPE STABILITY ANALYSIS
Moderator: D. F. Munger

Session 9
PROBABILITY AND ENGINEERING JUDGEMENT
Moderator: M. E. Hynes

Session 10
WASTE FILLS AND REPOSITORIES
Moderator: J. K. Mitchell

Session 11
SETTLEMENT/COMPRESSION OF DEEP FILLS
Moderator: H. E. Wahls

Session 12
SITE IMPROVEMENT/REMEDIATION
Moderator: J. P. Welsh

GEOTECHNICAL SPECIAL PUBLICATIONS

1) TERZAGHI LECTURES
2) GEOTECHNICAL ASPECTS OF STIFF AND HARD CLAYS
3) LANDSLIDE DAMS: PROCESSES RISK, AND MITIGATION
4) TIEBACKS FOR BULKHEADS
5) SETTLEMENT OF SHALLOW FOUNDATION AND COHESIONLESS SOILS: DESIGN AND PERFORMANCE
6) USE OF IN SITU TESTS IN GEOTECHNICAL ENGINEERING
7) TIMBER BULKHEADS
8) FOUNDATIONS FOR TRANSMISSION LINE TOWERS
9) FOUNDATIONS AND EXCAVATIONS IN DECOMPOSED ROCK OF THE PIEDMONT PROVINCE
10) ENGINEERING ASPECTS OF SOIL EROSION DISPERSIVE CLAYS AND LOESS
11) DYNAMIC RESPONSE OF PILE FOUNDATIONS—EXPERIMENT, ANALYSIS AND OBSERVATION
12) SOIL IMPROVEMENT—A TEN YEAR UPDATE
13) GEOTECHNICAL PRACTICE FOR SOLID WASTE DISPOSAL '87
14) GEOTECHNICAL ASPECTS OF KARST TERRAINS
15) MEASURED PERFORMANCE SHALLOW FOUNDATIONS
16) SPECIAL TOPICS IN FOUNDATIONS
17) SOIL PROPERTIES EVALUATION FROM CENTRIFUGAL MODELS
18) GEOSYNTHETICS FOR SOIL IMPROVEMENT
19) MINE INDUCED SUBSIDENCE: EFFECTS ON ENGINEERED STRUCTURES
20) EARTHQUAKE ENGINEERING & SOIL DYNAMICS (II)
21) HYDRAULIC FILL STRUCTURES
22) FOUNDATION ENGINEERING: CURRENT PRINCIPLES AND PRACTICES
23) PREDICTED AND OBSERVED AXIAL BEHAVIOR OF PILES
24) RESILIENT MODULI OF SOILS: LABORATORY CONDITIONS
25) DESIGN AND PERFORMANCE OF EARTH RETAINING STRUCTURES
26) WASTE CONTAINMENT SYSTEMS: CONSTRUCTION, REGULATION, AND PERFORMANCE
27) GEOTECHNICAL ENGINEERING CONGRESS
28) DETECTION OF AND CONSTRUCTION AT THE SOIL/ROCK INTERFACE
29) RECENT ADVANCES IN INSTRUMENTATION, DATA ACQUISITION AND TESTING IN SOIL DYNAMICS
30) GROUTING, SOIL IMPROVEMENT AND GEOSYNTHETICS
31) STABILITY AND PERFORMANCE OF SLOPES AND EMBANKMENTS II (A 25-YEAR PERSPECTIVE)
32) EMBANKMENT DAMS - JAMES L. SHERARD CONTRIBUTIONS

SLOPE STABILIZATION AT THE FORKS OF BUTTE PROJECT

by Stephen J. Klein, Member, ASCE[1]
David K. Hughes, Associate Member, ASCE[2]

Abstract

This paper describes a case history involving the evaluation, design, and construction of remedial measures required to stabilize a tunnel portal excavation at a hydroelectric power project in Northern California. Slope movements were large enough that tunneling work had to be suspended until the slope was stabilized. The slope was successfully stabilized by installing multi-strand cable tiebacks anchored in rock. Monitoring data concerning the performance of the stabilized slope after tunneling operations resumed is also summarized and discussed.

Introduction

The Forks of Butte Hydroelectric Project is an 11.6 MW run-of-the-river hydroelectric power project located in Butte County about 20 miles (32 km) east of Chico in northeastern California. The project involves diverting water in Butte Creek, with a diversion dam and intake structure, into a 560-foot (170 m) deep shaft, where it flows through an 11,800-foot-long (3600 m) power tunnel and a 140-foot-long (43 m) steel penstock section outside the tunnel to a powerhouse on the banks of Butte Creek. The project was constructed by a private developer under the jurisdiction of the Federal Energy Regulatory Commission (FERC).

The powerhouse and tunnel portal are located in a large landslide deposit. The landslide, which covers an area of about 13 acres, is generally considered an ancient dormant feature. In late 1989 slope movements were observed in an area of the slope above the tunnel portal. Construction

---

[1]Associate and [2]Senior Staff Engineer, Woodward-Clyde Consultants, 500 12th Street, Suite 100, Oakland, CA 94607

work up until the time of this movement consisted of grading activities and slope excavation to develop a staging area for tunnel construction. Cut slopes up to 65 feet (20 m) high had been excavated above the tunnel portal. Slope movements were detected by slope inclinometers installed above the tunnel portal. These movements accelerated in April and May of 1990 when tunneling activities began and reached a critical point when safety concerns required construction activities to be suspended until the slope could be stabilized. Woodward-Clyde Consultants (WCC) was retained to evaluate the slope and develop recommendations for stabilizing the slope so construction activities could safely resume. This paper focuses on the evaluation and implementation of the required slope stabilization measures.

Site Conditions

The Forks of Butte site is located in the California foothills in an area exhibiting rugged mountainous terrain with high relief. Elevations range from 1,200 to 2,500 feet (365 to 760 m) above mean sea level (MSL). Butte Creek crosses the site from north to south and constitutes the main drainage in the site vicinity.

Grading operations to develop the tunnel portal and staging area involved slope excavation and construction of a Tensar wall in order to provide a level bench for staging tunneling operations (Figure 1). The bench is about 85 (26 m) feet wide and over 250 feet (76 m) long. Cut slopes up to about 65 feet (20 m) high were made above this bench. These slopes were excavated at 0.5:1 (horizontal to vertical) with an intermediate bench and were reinforced with shotcrete and multi-strand cable tiebacks of various lengths. In addition development of the powerhouse site required a 65-foot (20 m) high cut slope below the Tensar wall. This excavation was completed after stabilizing the upper slope excavation.

Geologic Conditions

The Forks of Butte project area is underlain by a complex assemblage of Mesozoic metamorphic rocks including argillite, serpentine, and meta-volcanics. Rocks in the site vicinity are generally oriented along northwest-southeast trends with very steep dips.

The slopes above the powerhouse and tunnel portal comprise landslide debris deposits underlain by bedrock. Exploratory boring data indicate that the landslide debris deposits range in thickness from about 40 to 70 feet (12 to 21 m) in the project area. The landslide debris com-

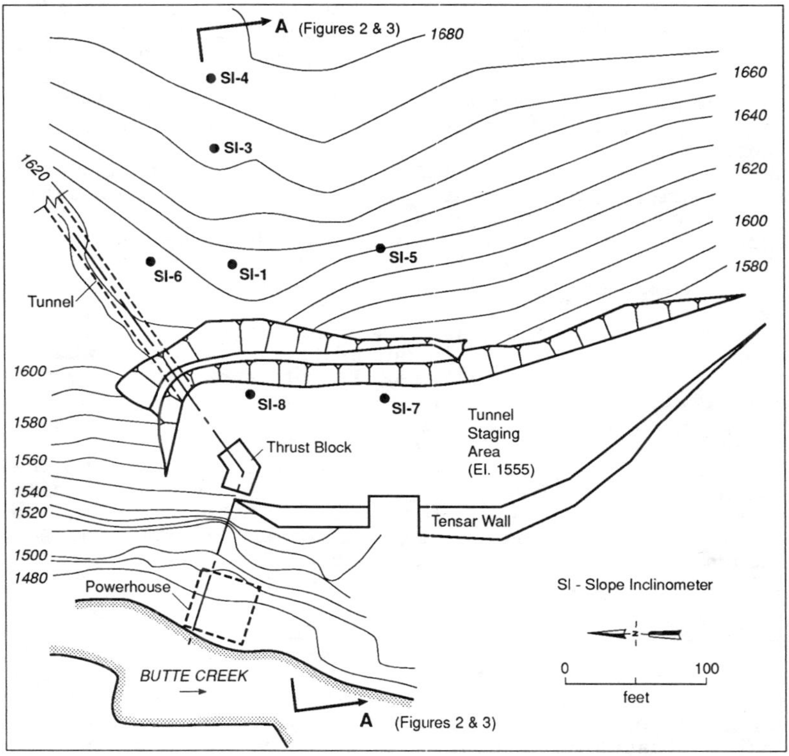

Figure 1 - Tunnel and Powerhouse Plan

prises a heterogeneous unconsolidated, poorly sorted, mixture of gravel, cobbles, and boulders embedded in a matrix that varies in texture from gravelly or sandy to clayey. The cobbles and boulders made it virtually impossible to obtain undisturbed samples for laboratory strength testing. Laboratory classification tests on samples of the matrix material indicate that the matrix varies in composition from a poorly graded silty gravel to a clayey sand. Grain size analyses indicate that the fines fraction (fraction finer than the #200 sieve) ranges from about 8 to 43%. Atterberg limit data for the fines fraction showed liquid limit values ranging from 33 to 52 and plasticity index values ranging from 14 to 26.

Groundwater level data below the portal slope were obtained from monitoring two of the slope inclinometer installations. Based on this data the depth to groundwater was observed to have been fairly stable at a depth of about 65 feet in SI-1 and about 74 to 75 feet in SI-3 with fluctuations of less than about 1 foot during the period of January to June, 1990. These relatively small fluctuations are interesting considering that the area received about 30 inches (750 mm) of rain during this time period. It should be noted that the groundwater levels in these inclinometers was 8 to 15 feet (2.4 to 4.6 m) below the bedrock surface, indicating that the groundwater levels are apparently below the landslide debris deposits. Groundwater was encountered at shallow depths (less than about 10 feet) in the borings completed at the powerhouse site. This depth generally corresponds to the water level in the adjacent Butte Creek.

The seismic history of the site vicinity is characterized by activity along discontinuous northwest trending faults. About 14 historic earthquakes larger than Magnitude 5.0 have occurred within 50 miles (80 km) of the Forks of Butte powerhouse site (California Department of Water Resources, 1989). The Magalia Fault, part of the Foothills Fault System, is the closest active fault to the Forks of Butte powerhouse, approaching within 1 mile (1.6 km). This fault may be capable of a Magnitude 6.0 to 6.5 earthquake, based on work carried out for the proposed Auburn Dam (WCC, 1977) and on unpublished work by the California Division of Mines and Geology.

Slope Movements Prior to Stabilization

Slope inclinometer data were reviewed to evaluate the slope movements which occurred prior to the suspension of construction activities in May 1990. The inclinometer data indicated a pattern of deformation consistent with a general slope movement. The slope movement appeared to be taking place on a relatively planar surface that cuts through the landslide debris deposits above the bedrock surface. The location of this surface is shown on Figure 2, which is a cross-section at the tunnel portal excavation in May 1990 drawn through inclinometers SI-1, SI-3, SI-4, and SI-8. Slope movements accelerated after tunneling work started and movements ranging from about 0.2 to 1 inch (5 to 25 mm) occurred in the month after the start of tunnel construction. Total slope deformations ranging up to 4 inches (100 mm) had been observed since the start of construction. Slope movement was most likely triggered by the excavation of the cut slope to construct the tunnel portal and tunnel staging area. Although 6 rows of tiebacks were installed on or above the cut slopes to stabilize this excavation, only 2 of the rows appear to have been anchored substantially beyond the plane of the slope movement (Figure 2).

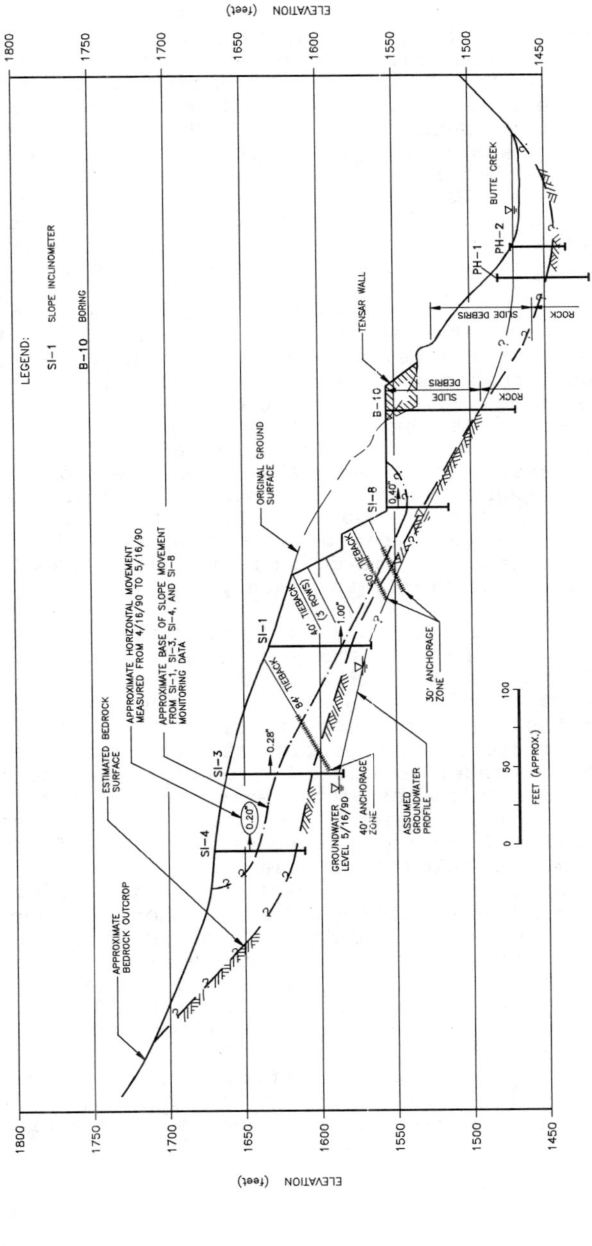

Figure 2 - Existing Conditions, Section A-A

Slope Stability Analyses

Existing Conditions - The tunnel portal cut slope was analyzed to evaluate the stability of slope and to identify slope stabilization measures that could be implemented to raise the factor of safety of the slope. The slope stability analyses were made using the computer program PCSTABL5M (Achilleos, 1988). This version of the PCSTABL program was developed at Purdue University and was chosen principally because it is a general purpose slope stability program that also has the ability to incorporate the influence of tiebacks into the analysis. The factors of safety were calculated using the program's version of Modified Janbu's Generalized Procedure of Slices.

No reliable strength parameters were available for the landslide debris materials from the previous subsurface explorations at the site, nor was it considered feasible to obtain undisturbed samples of the landslide debris for physical testing. However, the observed slope movements provided an excellent opportunity for the back-calculation of strength parameters for the slope materials based on a qualitative assessment of the stability of the slope. The section used in this analysis is shown on Figure 2 and the location is indicated on Figure 1. This section (A-A) was selected for these analyses because four slope inclinometers are located along it defining the geometry of the failure surface relatively accurately.

The back-calculation was done in three steps because the effectiveness of the originally installed tiebacks was unknown. First, the friction angle of the failure surface was back-calculated based on an assumed factor of safety of one (for the excavated slope prior to stabilization), assuming a failure surface configuration as indicated by the slope inclinometer data (see Figure 2), zero cohesion, and ignoring the influence of the existing tiebacks. The results of this calculation (i.e., 24.5 degrees) essentially established the upper bound value for the friction angle. A second calculation was made using the same configuration described above except that the effective tiebacks (i.e., tiebacks thought to be anchored beyond the failure surface) were included in the analysis. The result of this analysis (i.e., 22.5 degrees) is considered to be representative of a lower bound value for the friction angle. An average friction angle of 23.5 degrees was determined from the results of these two calculations.

A third stability analysis was performed using the average friction angle (i.e., 23.5 degrees) and the same configuration as the second case described above. The estimated factor of safety with this average friction angle was found to be about 1.05, suggesting that the actual factor of safety of the upper slope shown on Figure 2 was probably in the range of 1.00 to 1.05 prior to implementing additional measures to increase the stability of the slope. These results appeared to be consistent with general observations regarding behavior of slope in that small destabilizing factors such as blasting in the tunnel tended to produce slope movements which gradually subsided when construction activities were halted.

Slope Stabilization Analyses - Slope stabilization studies focused on the use of tiebacks to stabilize the existing cut slopes and the planned cut slopes for the powerhouse excavation, primarily because grading operations to flatten or buttress these slopes were not feasible. There are several reasons for this including: the necessity of maintaining the full bench width for staging tunnel construction operations, space limitations in the relatively narrow valley, the unavailability of a place to dispose of excavation spoil, and also the desire to avoid damaging the existing tiebacks that were contributing to the stability of the slope.

To evaluate possible stabilization schemes, stability analyses were performed for Section A-A. This section was drawn through the powerhouse site and the highest part of the existing cut slope and was considered to be the critical section. Other sections through the slope were checked to confirm the results of the analyses for Section A-A but are not discussed in this paper due to space limitations.

The stability of the slope section shown in Figure 2 was analyzed using the average back-calculated friction angle discussed above. Three potential failure modes of the slope were checked in order to evaluate the stability of the section. These modes include sliding of: the existing cut slopes above the tunnel portal excavation, the cut slopes for the powerhouse excavation, and the overall slope. The powerhouse excavation cut slopes are about 65 feet (20 m) high and were required to be 0.5:1 slopes with two intermediate benches as shown on Figure 3. The failure surface geometry for the portal cut slopes was reasonably well defined by the inclinometer data. In evaluating the powerhouse cut slope and overall slope stability it was necessary to supplement the inclinometer data with a computer search to establish the probable critical failure surface (i.e., surface giving the lowest calculated factor of safety). The geometry of slope and the critical surfaces are shown on Figure 3 for the operating condition. For the construction condition it was assumed that two additional rows of permanent tiebacks (i.e., Rows AP and BP) would be installed on the upper slope and five rows of tiebacks (i.e., Rows EP through IP) would be installed for the powerhouse cut slope as shown on Figure 3. For the operating condition it was assumed that the original tiebacks (which are not corrosion protected) do not provide any long-term slope stabilization benefit, that three additional rows of tiebacks (i.e., Rows CP, DP, and JP, in addition to the tieback Rows AP and BP installed for construction) would be installed in the upper slope, and that a buttress fill would be added to buttress the cut slope following construction (Figure 3). It was assumed that each row of tiebacks would have a tested capacity of at least 35 kips (156 kN) per lineal foot across the slope face (i.e., acting as a line load) and that all of the new tiebacks would be anchored in rock below the landslide debris deposits. Factors of safety calculated for the construction and operating conditions described above are summarized below in Table 1.

TABLE 1

Summary of Factors of Safety
From Results of Stability Analysis For Section A-A

Location	Construction Condition	Operating Condition
Portal Cut Slope	1.20	1.60
Powerhouse Cut Slope	1.30	1.55
Overall Slope	1.35	1.50

The groundwater levels used in the analyses were based on the limited available information. Although groundwater level data indicated that the landslide debris appeared to be free draining, the influence of higher groundwater levels were checked by adding 20 feet (6 m) to the estimated levels at SI-1 and SI-3 and tapering this increase down to match the water level in the creek as indicated in Figure 3. This did not result in a significant reduction in stability, which is attributable mainly to the depth of the groundwater level relative to the failure surface. Similarly, the effect of a flood (up to Elevation 1489 feet [454 m]) raising the groundwater levels at the base of the slope was analyzed and its effect on the factor of safety was found to be minimal.

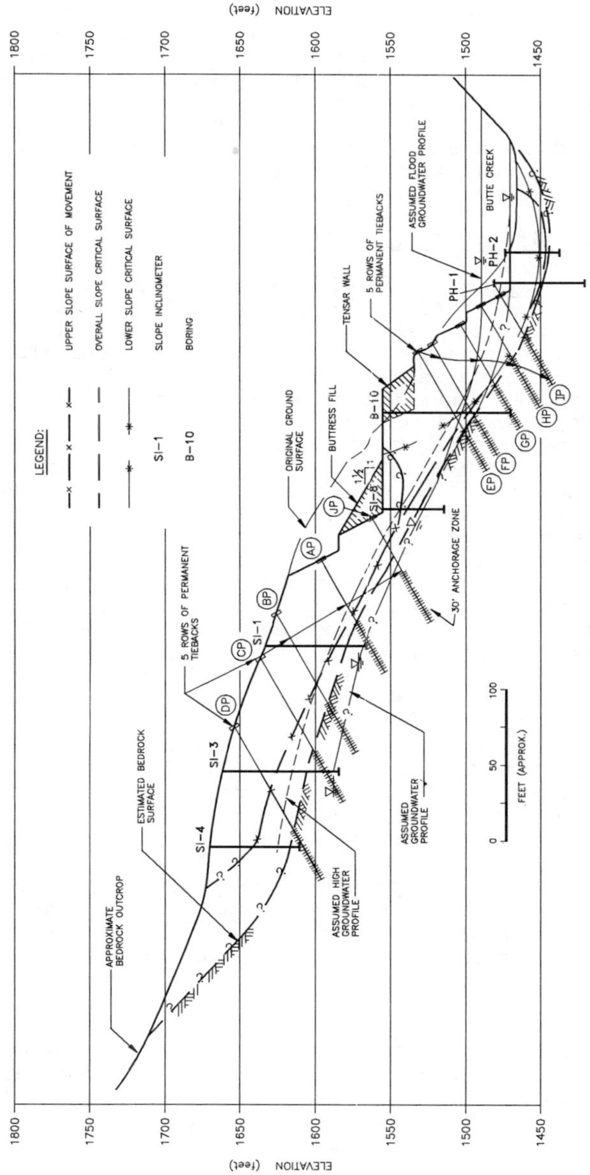

Figure 3 - New Tieback Locations, Section A-A

Based on the results of these analyses, tiebacks for stabilizing the slope were specified to be permanent, double-corrosion protected rock anchors with an anchorage length of at least 30 feet (9 m) in fresh, competent rock. For a tieback spacing of 10 feet (3 m) along the slope, face tiebacks were required to have a demonstrated capacity of 350 kips (1557 kN). Approximately 80 tiebacks, ranging in length from 80 to 150 feet (24 to 46 m), were required to stabilize the 200-foot-long (61 m) cut slopes for the tunnel portal, and another 73 tiebacks were installed to stabilize the powerhouse cut slopes.

Seismic Analysis of Slope

Method of Evaluation - The seismic stability of the slope was evaluated by estimating the amount of permanent deformation that the slope would experience during a Magnitude 6.0 earthquake event occurring at the closest distance of the Magalia Fault to the site. Deformations of the slope due to seismic loading were estimated using the simplified procedure developed by Makdisi and Seed (1978). The method uses the yield acceleration concept of Newmark (1965) to evaluate the permanent deformations. The method assumes that deformation occurs on a well-defined slip surface with elastic material behavior at stress levels below the yield strength and plastic behavior above yield. The yield acceleration is assumed to be the pseudo-static seismic coefficient that produces a factor of safety of one on the slip surface. Movement of the slope can occur when the acceleration of the sliding mass exceeds the yield acceleration. The amount of movement is a function of the amplitude and duration of the differences in the earthquake-induced accelerations and the yield acceleration.

Potential Ground Motions - For the seismic stability evaluation of this project, the potential earthquake motions can be characterized by the peak acceleration. Response spectra are not necessary for the simplified analyses used to estimate the potential seismic deformations of the slope. The peak acceleration was estimated using selected published empirical attenuation relationships. These relationships describe the variation of peak acceleration with earthquake magnitude and distance between the source of the earthquake and the site.

Attenuation relationships developed by the following investigations were used to estimate the peak acceleration at a rock site: Seed and Schnabel, 1980 (presented in Seed and Idriss, 1982); Campbell, 1988; and Idriss, 1985. The median peak horizontal acceleration in the free field at the project site for a Magnitude 6.0 earthquake on the Magalia fault was estimated using each of the attenuation relationships. Although the maximum magnitude of an event on the Foothills Fault System is estimated to be in the range of 6.0 to 6.5, use of a Magnitude 6.0 event 1 mile from the site is considered appropriately conservative as the recurrence interval for an event of this size is estimated to be about 30,000 years. The average of the median peak accelerations from the three attenuation relationships is 0.51 g. This value was used for estimating the seismically induced slope deformations.

Results of Seismic Stability Analyses

Estimated Deformations - The yield accelerations were calculated from a pseudo-static analysis using the PCSTABL computer program. Because much of the potential slope movement is along an existing plane of movement above the groundwater level, use of the back-calculated strengths are considered appropriate for this analysis. The lowest calculated yield acceleration is approximately 0.17 g for the overall slope failure surface shown on Section A-A (Figure 3), assuming that the tiebacks develop a capacity of 40 kips (178 kN) per lineal foot at the yield point.

The maximum average accelerations induced by the maximum credible earthquakes were estimated to be equal to the estimated bedrock motions of about 0.51 g. The seismically induced deformations of the slope during this event are estimated to range from about 1 to 8 inches (25 to 200 mm) in a horizontal direction with a best estimate of the most likely deformation ranging from about 3 to 4 inches (75 to 100 mm). The uncertainties in these estimates are not insignificant, but considering the close distance assumed between the earthquake and the site (i.e., only about 1 mile), it was concluded that these estimates are reasonably conservative.

The main consideration with slope deformations of this magnitude is the ability of the tiebacks to accommodate these movements. Horizontal slope movements at the upper end of the estimated range are about 8 inches (200 mm), which corresponds to a deformation along the tieback axis of about 7 inches (175 mm). These deformations would tend to increase the tension in the tiebacks. Deformations transverse to the tiebacks are not considered to be of consequence because the tendons are flexible and will not generate significant bending moments due to these deformations.

The tiebacks used in stabilizing the slopes have an ultimate capacity of 468 kips (2083 kN) but will be locked off at 265 kips (1180 kN). Permanent tieback lengths were expected to range from about 80 to over 110 feet (24 to 34 m), which provides for a 30-foot (9 m) rock anchorage and a stressing length varying from 50 to 80 feet (15 to 24 m). For an 80-foot (24 m) stressing length the tieback could deform or stretch about 4.5 inches (113 mm) before reaching yield (based on a yield strength of 90% of ultimate for low relaxation tendons). The strain level if the maximum axial tieback deformation of 7 inches (175 mm) occurred would be about 1.3%, which slightly exceeds the yield strain of 1.0% but is significantly less than the 3.5% minimum strain at rupture specified by ASTM (ASTM A-416). At this strain level the maximum load in the tiebacks would be about 91% of the ultimate strength. It should be noted that test data submitted by the tieback supplier indicates that the actual strain at rupture for the permanent tiebacks is about 7%, with a standard deviation of about 0.5%.

For a similar capacity tieback with a 50-foot (15 m) stressing length, the tieback can accommodate a deformation of about 3 inches (75 mm) before yielding. The strain level for the 7-inch (175 mm) maximum axial tieback deformation is about 1.7%, which exceeds the yield limit by 0.7% but is still a factor of 2 less than the 3.5% rupture strain. At this strain level the maximum load in the tiebacks would be about 93% of the ultimate strength. Therefore, tiebacks were required to have a 50-foot (15 m) minimum free length to provide enough strain capacity to accommodate the maximum calculated seismic slope deformations. Considering the conservative assumptions made in selecting the seismic event used in this evaluation (i.e., a Magnitude 6.0 event 1 mile from the site), this 50-foot (15 m) minimum free length appears to provide a reasonably conservative safety margin during this extreme event.

Slope Monitoring and Instrumentation Program

A slope monitoring and instrumentation program was considered essential for checking the validity of the assumptions made in the analyses summarized above and for monitoring the behavior of the slopes during construction and later during operation of the facility. The monitoring and instrumentation program has four basic components:

1. Slope inclinometers for detecting and measuring slope movements.

2. Load cells for monitoring tieback tension.

3. Groundwater observation wells for monitoring groundwater levels.

4. Survey control points to check for ground movements by surveying methods.

At the completion of construction the results of the construction monitoring program will be evaluated to determine the long-term instrumentation and monitoring requirements.

Implementation of Slope Stabilization Measures

It was important to stabilize the slope as quickly as possible so that the delay in tunnel construction activities could be minimized. Tunneling resumed after installing about 30 tiebacks, which raised the factor of safety of portal cut slope to about 1.2. To minimize any potential influence that blasting vibrations could have on the slope movements, the amount of explosive was limited to 5 lbs (2.3 kg) per blast with no more than 1 lb (0.5 kg) of explosive per delay of each blast. These careful tunnel excavation procedures were successful in controlling slope movements. Tieback installation continued concurrently with tunneling operations so that the stability of the slope was being increased as the tunneling work progressed.

Tieback Construction - Initially, bearing pads were constructed against the slope face to transfer the tieback load to support the slope. The bearing pads were constructed using shotcrete. These pads are 2 feet (0.60 m) thick and 9 feet (2.7 m) square and are reinforced with two mats of reinforcing steel.

The tiebacks were supplied to the site in a variety of lengths ranging from 90 to 150 feet (27 to 46 m) to accommodate the variability in the depth to bedrock. Each tieback contained 8 strands with a 30 foot (9 m) bond length. The strands consisted of seven stress relieved wires conforming to ASTM A-416. In the free length, each of the eight strands were individually encapsulated with corrosion inhibiting grease and polyethylene tubing. All eight strands were in turn protected by an outer 3-inch (75 mm) diameter PVC plastic sleeve. The bare strands in the bond length were protected with a corrugated PVC sheath and fully encapsulated with neat cement grout.

A Klemm 806 Hydraulic Crawler rig equipped with double-head-overburden drilling unit was selected for drilling the tieback holes because of its ability to continuously case the hole through the unstable landslide debris deposits. The holes were advanced efficiently through the boulders and other landslide debris with a downhole hammer (DHH) fitted with an inner bit and a disposable crown bit. Cuttings were expelled up the annulus between the casing and drill string using moist air pressure. The casing and drill string were advanced simultaneously, although the rotation and advance pressure of each were independent. The casing was advanced through the landslide debris and generally well into the bedrock. After drilling the required length into competent rock, the disposable crown bit on the DHH was pulled off against the casing and the drill string withdrawn. In several instances, particularly in the vicinity of the tunnel portal, the drill became jammed while advancing through the landslide debris. The hammer tended to deflect off the side of large boulders in the landslide and the casing was unable to follow. Several holes and considerable time was lost in the initial stages of installing the tiebacks due to this problem.

The tiebacks were installed through the casing to the bottom of the hole. Any excess length was cut off. The tiebacks had a pre-installed interior tremie grout line inside the PVC sleeve and sheath, and a second line was taped to the exterior of the sheath during installation. A neat cement grout was used to obtain the rock bond in the anchorage zone and for encapsulating the strands to protect them from corrosion. Due to the pervious nature of the landslide debris deposits, it was not generally possible to fully encapsulate the tiebacks above the bedrock with grout. Instead, in order to assure double corrosion protection, the interior of the sheath was fully filled with grout above the bond zone by alternately pumping the two grout lines. Immediately after the tieback had been grouted, the casings were jacked out of the hole. After the casings were removed, grout was pumped through the interior grout line until grout return was observed flowing from the PVC sleeve.

Each tieback was proof tested to 133% of design load; i.e., 350 kips (1557 kN). The tiebacks were tensioned to the maximum test load in five increments and the maximum load was held for 10 minutes. Acceptance criteria for the tiebacks were less than 0.04 inches (1 mm) creep during the time period at the constant test load. All of the tiebacks satisfied the proof test requirements. Performance tests were done on approximately 10% of the tiebacks to check the design assumptions regarding the rock bond strength.

Slope Monitoring Data - Slope inclinometer data were collected daily after resuming tunneling. Figure 4 summarizes slope inclinometer data for Slope Inclinometer 5 (SI-5) and is typical of the observed slope movements. Almost 1 inch (25 mm) of slope movement occurred prior to beginning tunnel excavation work. The movements appeared to accelerate until tunneling work was halted, and then the rate of movement gradually subsided. The rate of movement decreased further following the installation of tiebacks to stabilize the slope. Tunneling resumed after the factor of safety of the slope had been increased to about 1.2 and only small slope movements, less than 0.1 to 0.2 inches (2.5 to 5 mm), were observed subsequently.

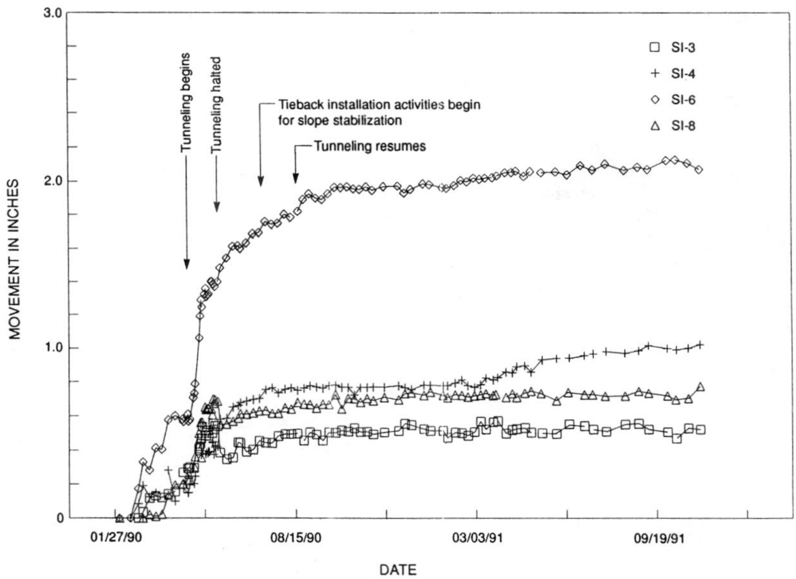

Figure 4 - Slope Inclinometer Monitoring Data

Typical load cell data for the tiebacks is summarized in Figure 5. The load cell data indicates load decreases of varying magnitudes for the instrumented tiebacks. The load decreases are attributed to compression or settlement of the soils beneath the tieback bearing pads.

Decreases in the tieback loads do not necessarily indicate a concern regarding the stability of the slope. Slope inclinometer monitoring data does not show any signs of significant slope movements corresponding with the decreased tieback loads. The reduced tieback loads are not considered to jeopardize the stability of the slope because, if slope movements were to occur, the tieback tension would increase significantly within several tenths of an inch of slope movement, providing additional restraining forces to stabilize the slope movements. Nonetheless, selected tieback groups will be retensioned before the completion of construction.

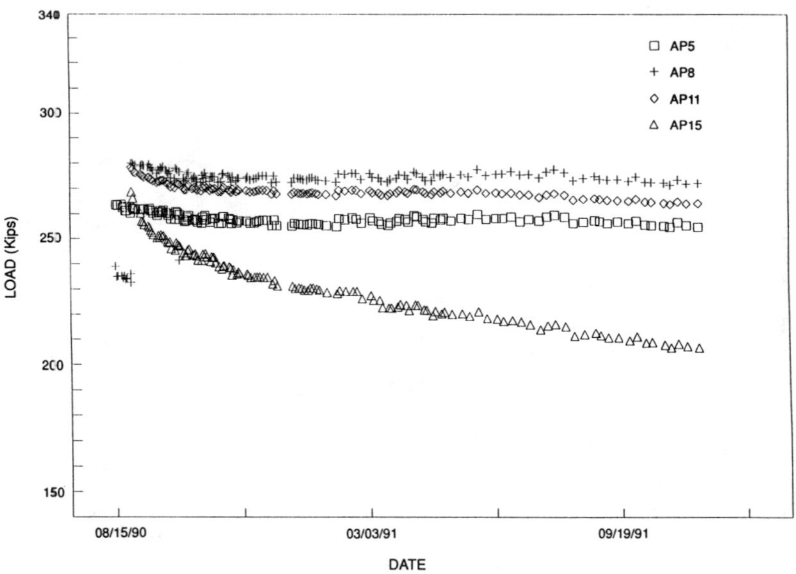

Figure 5 - Typical Load Cell Data

Discussion

This paper describes a case history of a slope failure that resulted in a 3-1/2 month delay in construction of the project and an expenditure of several million dollars to stabilize. The site was recognized to be underlain by a large ancient landslide prior to beginning construction, but tunnel portal excavations re-activated a portion of the landslide leading to slope movements that halted construction activities at the site. This project underscores the importance of treating recognized landslides conservatively to avoid the unfortunate circumstances that hampered this project.

Acknowledgements

The authors would like to thank the owner, Energy Growth Partnership, Alameda, CA for the permission to publish this paper. The efforts of the project team H&M Engineering, Inc., Columbia, MD (project manager); Sverdrup Corporation, San Francisco, CA (civil and structural engineer); Ray Toney and Associates, Redding, CA (powerhouse contractor); Malcolm Drilling Co., Inc., Kent, WA (tieback contractor); and Obayashi Corporation, South San Francisco, CA (tunnel contractor) are gratefully acknowledged. The following members of WCC's Oakland office made important contributions to the project; J. Bischoff, J. O'Rourke, L. Mejia, G. Ford, R. Green, and K. Weaver. The authors also acknowledge the support of WCC's Professional Development Program for sponsoring preparation of this paper.

REFERENCES

Achilleos, E. (1988). "PCSTABL5M USER GUIDE". Informational report prepared by Purdue University, School of Civil Engineering for the Joint Highway Research Project.

ASTM (1987). "Standard Specification for Steel Strand, Uncoated Seven-Wire Stress Relieved for Prestressed Concrete". Designation A 416-86, 1987 Annual Book of Standards, p. 339-341.

California Department of Water Resources (1989). "The August 1, 1975 Earthquake Investigation". Supplement to Bulletin 203-78. Bulletin 203-88, p. 617.

Campbell, K.W. (1988). "Preliminary report on empirical studies of horizontal strong ground motion for the Diablo Canyon site, California". Report to the U.S. Nuclear Regulatory Commission, October.

Idriss, I.M (1985). "Evaluating seismic risk in engineering practice". Proceedings, Eleventh International Conference in Soil Mechanics and Foundation Engineering, San Francisco, California, Vol. 4, pp. 255-320, August.

Makdisi, F.I., and Seed, H.B. (1978). "Simplified procedure for estimating dam and embankment earthquake-induced deformations". Journal of Geotechnical Engineering Division, American Society of Civil Engineers, Vol. 104, No. GT7, pp. 849-867, July.

Newmark, N.M. (1965). "Effects of earthquakes on dams and embankments". Geotechnique, Vol. 15, No. 2, pp. 139-160, June.

Seed, H.B., and Idriss, I.M. (1982). "Ground motions and soil liquefaction during earthquakes". Earthquake Engineering Research Institute, Berkeley, California, Vol. 5 of a series entitled: Engineering Monographs on Earthquake Criteria, Structural Design, and Strong Motion Records.

Woodward-Clyde Consultants (1977). "Earthquake Evaluation Studies of the Auburn Dam Area, Regional Geology and Tectonics, Vol. 3". Prepared for the U.S. Bureau of Reclamation.

ANALYSIS OF SLOPE FAILURE AND REMEDIAL DESIGN OF AN EARTH DAM

By: Michael J. Mann[1] and Robert E. Snow[2] Members, ASCE

ABSTRACT

A 45-foot (13.7 m) high earth fill dam was constructed in eastern Ohio from the summer of 1983 through the fall of 1984. Immediately upon completion of construction and prior to reservoir filling, a crack was reported along much of the length of the crest with subsequent progressive movement of the upstream slope. Exploration and monitoring of the upstream slope in 1986 indicated continued creep movements on the order of 0.1 inches (2.5 mm) per month (horizontal) or less with some evidence of localized, elevated pore pressures.

A remedial design was prepared to buttress and stabilize the upstream slope. During design of the buttress a value of shear strength had to be selected to represent the portion of the slide zone that would remain in place. Laboratory testing and back-analysis of the slope stability were performed to determine strength parameters which reflected the existing creep behavior and to estimate at what stress levels the creep movements could be expected to stop. A remedial design was prepared consisting of a buttress, reconstruction of the upstream slope and crest, and internal drainage control. Construction of the remedial design was completed in the summer of 1988 and the reservoir is filled.

[1]Senior Project Manager, GZA GeoEnvironmental of New York, Buffalo, New York 14225

[2]Principal, D'Appolonia Engineering Division of Ground Technology, Monroeville, Pennsylvania 15146

INTRODUCTION

The Scout Reservation Dam, located in Carroll County, Ohio is an earth fill structure approximately 45 feet (13.7 m) high, as measured from the upstream toe and 800 feet (243.8 m) long. The dam creates an approximate 30-acre (121,410 m^2) recreational lake for the Buckeye Council, Boy Scouts of America. The original design called for a homogeneous earth fill dam with a 3 horizontal to 1 vertical (3H:1V) downstream slope and a 2H:1V upstream slope. The dam was constructed of residual silty clay soils beginning in the summer of 1983 and was completed in the fall of 1984. The principal spillway consisted of a concrete decant tower and discharge pipe near the west abutment to control the normal pool of the reservoir. An emergency open channel spillway is located on the east abutment.

Gradual movement of the upstream slope was reported immediately after construction was completed in the fall of 1984. Initially, a crack was noted along most of the length of the crest. As the upstream slope continued to move downward and away from the rest of the dam, a scarp developed, which was 1 to 2 feet (.30 to .61 m) high in January 1985, 3 to 4 feet (.91 to 1.2 m) high in July 1985 and 5 to 6 feet (1.5 to 1.8 m) high by the summer of 1986. Shortly after the slope movements were noted, a soil and rock fill buttress was placed upstream of the primary spillway tower and the crest was lowered in this area to slow movements and minimize damage to the structure. Figure 1 presents this general configuration of the dam.

SUBSURFACE EXPLORATION

Two subsurface exploration programs were performed at the site. An initial program, which consisted of borings (B-series), was performed shortly after failure of the slope was detected in 1984. Subsequently, nine exploratory borings were drilled in 1986 at the locations designated as D-1 through D-9 in Figure 1. The discussions herein primarily relate to the 1986 exploration program.

The exploratory borings were advanced using hollow stem augers. Standard Penetration Tests (SPTs) were performed at 3-foot (.91 m) intervals through the embankment and continuously through the suspected failure zone(s). Offset borings were drilled east or west of the original boring (i.e., D5E) and Shelby tube samples were obtained in the suspected failure zone.

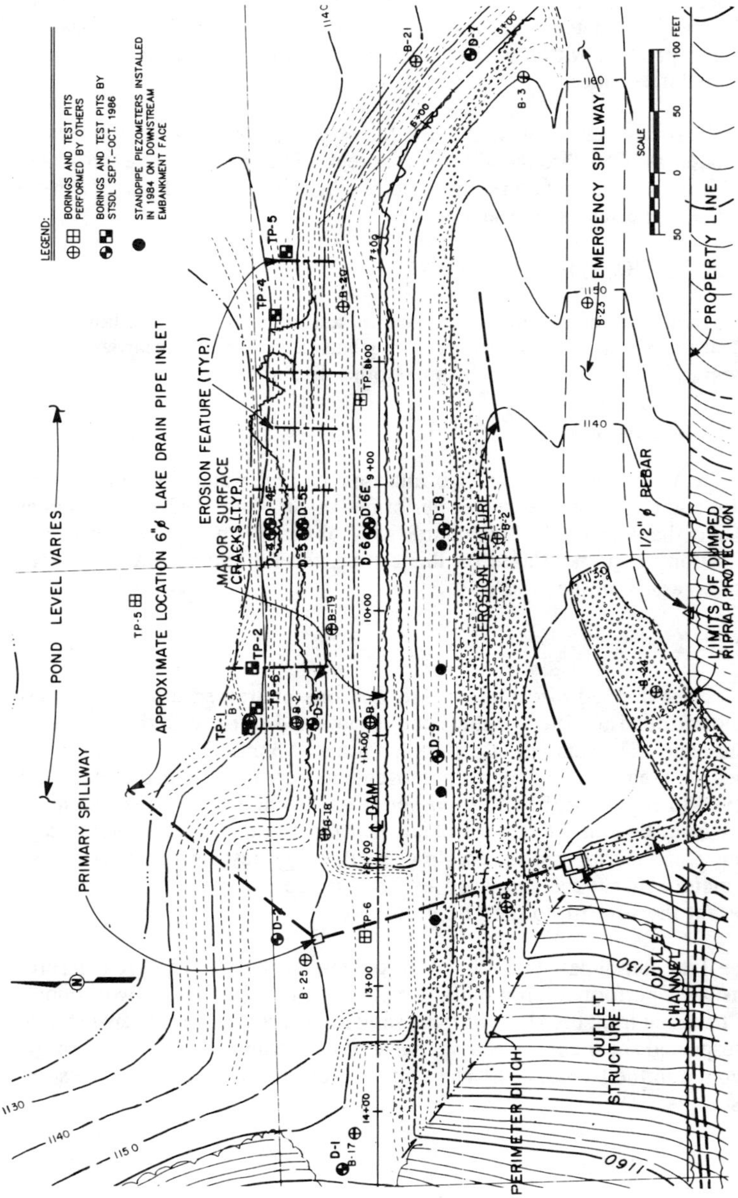

FIGURE 1 - SITE PLAN OF EXISTING CONDITIONS

Inclinometer casings were installed in borings D-2 through D-9 to measure movement of the upstream and downstream slopes (if any). Pneumatic piezometers were installed in borings D-5E and D-6 to measure the water pressure within the embankment and, in particular, in the vicinity of the failure surface.

Test pits were excavated at the toe of the upstream slope, at locations designated as TP-1 through TP-6 in Figure 1. The test pits were excavated through the suspected failure zone and Shelby tube and block samples were obtained in the suspected shear zones.

INITIAL LABORATORY TESTING

Initial testing included classification and direct shear (constant strain rate) strength tests. All laboratory testing was performed in accordance with the applicable ASTM standards unless otherwise noted.

Classification Testing

Atterberg limits, grain size analyses, natural water content, specific gravity, and unit weight tests were performed on SPT and Shelby tube samples of the embankment fill and natural soils obtained during the subsurface exploration program. Based on this testing, the fill and natural soils were classified as CL in the USCS.

Strength Testing

Four drained direct shear tests and one undrained direct shear test were performed on Shelby tube samples obtained in or near the suspected failure zone. The samples tested from borings D-3 and D-4E were from the immediate area where inclinometer movements were noted and, therefore, are believed to represent the soil in or very close to the failure zone. Samples tested from test pit TP-6 were obtained from the zones where significant evidence of shearing was noted. In general, staged tests were performed so that material from the suspected existing failure surface could be used to represent the strength to the extent possible.

Two consolidated undrained triaxial tests with pore pressure measurements were performed on a block sample from test pit TP-1. Two samples were trimmed from the block with the samples oriented in a manner such that the suspected shear zone was at an angle of approximately 60 degrees to the horizontal in the triaxial cell. The samples were sheared and pore pressures measured.

The results of initial strength testing presented in Table 1 indicate that for soil samples taken from near the existing failure zone, the peak friction angle is between approximately 23 degrees and 28 degrees and the residual angle of internal friction is between 17 and 30 degrees.

Consolidated undrained triaxial tests with pore pressure measurements were also performed on samples from borrow areas for use in buttress construction for the remedial design. These data are also presented in Table 1.

Table 1: Summary of Strength Test Results

	Sample No.	Elev.	Test Type (1)	Angle of Internal Friction (degrees)		Peak Undrained Shear Strength (KPa)	Residual Undrained Shear Strength (KPa)
				Peak	Residual		
TP-6	ST-3	1122.6	DDS	--	23.8		
TP-6	ST-2	1122.4	DDS	23.8	--		
D3	ST-4	1118.2	DDS	27.8	27.8		
D4E	ST-1	1122.5	DDS	24.8	17.4		
TP-1	BL-1	1122.5	CU		30.6(2)		
TP-1	BL-1	1122.5	CU		28.2(2)		
TP-6	ST-3	1122.6	UDS	--	--	53.0(3)	32.3(3)
TP7 &8(4)	T1-T3	---	CU	34.6	--	--	--
TP7 &8(4)	T4-T6	---	CU	34.1	--	--	--

(1) DDS = drained direct shear test; CU = consolidated undrained triaxial test with pore pressure measurements; UDS = undrained direct shear test.
(2) Sample failed along pre-existing shear plane.
(3) Confining stress = 49 KPa
(4) Tests on proposed buttress material.

EXISTING SITE CONDITIONS

Geology

The site lies in the unglaciated portion of the Appalachian Plateau physiographic province. The terrain is hilly with stream channels 200 to 300 feet (61.0 to 91.4 m) beneath the tops of hills. Bedrock strata in the site vicinity are generally horizontal and comprised of beds of the lower portion of the Pennsylvanian age Conemaugh Group. The rock consists of interbedded gray shale, gray and red claystone and gray sandstone.

Surficial Conditions

A surficial reconnaissance of the site in September 1986 disclosed numerous cracks and erosion features on the upstream embankment face, the more significant of which are shown in Figure 1. Although the majority of the cracks were typically 0.5 to 1 inch (12.7 to 25.4 mm) wide and appear to extend to depths of a few feet, some cracks were several inches wide and several feet deep. The cracking east of Station 7+00, was most likely associated with desiccation of the surficial soils and some slumping of the crest soils into the upstream riprap slope protection.

The spillway tower located on the upstream slope was tilted approximately 19 inches (.48 mm) toward the northeast. Inside the spillway tower cracks were noted radiating upward from the existing 6-inch (152.4 mm) diameter drain pipe along the east wall and extending around the south wall of the tower.

Subsurface Conditions

The existing subsurface conditions are depicted in the cross section on Figure 2. The borings disclosed predominately medium stiff to very stiff silty clay soils with traces of coal, roots, and some wood fragments in the fill for the dam, underlain by medium stiff to stiff silty clay and subsequently weathered claystone and hard gray siltstone and sandstone. Test pits excavated near the upstream toe of slope, disclosed distinct planes with evidence of movement within the fill for the dam.

Figure 3 presents the boring log for boring D-5, which illustrates the embankment conditions at midslope including SPT, water content and field estimates of unconfined compressive strength based on a hand penetrometer.

In general, the embankment and foundation soils were classified as silty clays of low to medium plasticity. Typically, 70 to 90 percent by weight was finer than the No. 200 sieve, liquid limits ranged from 30 to 42 and the plasticity index varied from 8 to 21 with all soils being classified as CL in the USCS. Water contents were typically in the range of 16 to 20 percent.

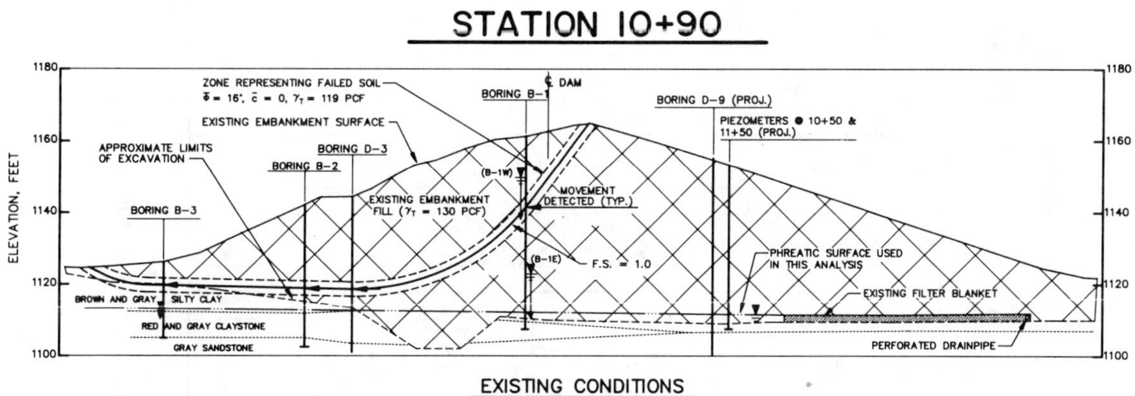

FIGURE 2 - CROSS SECTION OF EXISTING CONDITIONS

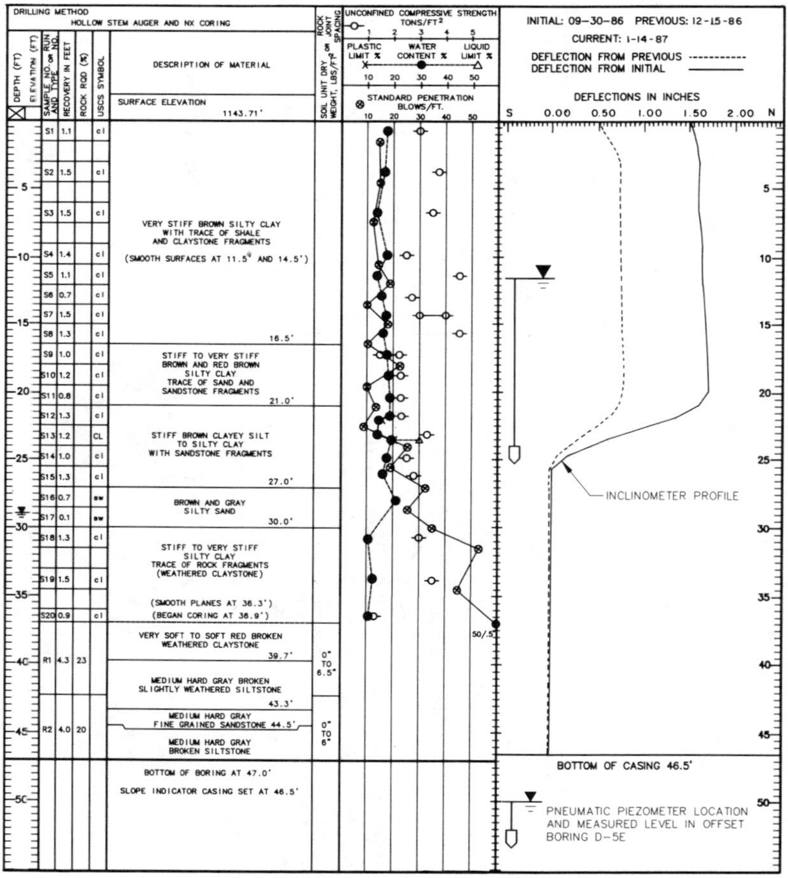

FIGURE 3 - BORING D-5 LOG

The specific gravity and unit weight testing indicated that the soils, in general, appeared to be unsaturated in most of the embankment, and into the foundation with some indications of randomly located saturated zones. Samples from boring D-4E between elevations 1118.5 to 1122.6 feet appear close to saturation, while a sample at elevation 1118.2 feet was not. In boring D-6W samples tested from elevation 1135.3 to 1141.6 feet, well above the pool level, appear to be close to saturation.

Monitoring Data

Figure 3 illustrates the typical inclinometer profile results at boring D-5 and shows the general location and behavior of the slope movement. A similar distinct failure zone was observed at each of the inclinometers in the upstream portion of the embankment at the locations indicated on the cross section in Figure 2. The lower portion of the failure zone is nearly horizontal and is between elevation 1119 and 1120 at Station 10+90. A similarly shaped failure zone was served at Station 9+30. The upper portion of the failure zone can be hypothesized as either a wedge or curved surface through the upper inclinometer monitoring location to the scarp at the top of the slope.

Monitoring of the inclinometer and piezometer installations was performed from October 1986 until the dam was repaired in the summer of 1988. The data are presented in Figure 4 for the period of monitoring. Based on this data, the embankment appeared to be creeping in the upstream direction at a rate of approximately .1 inches (2.5 mm) horizontally per month. However, minor grading of the slope (such as during subsurface exploration activities) and other factors, such as heavy rainfall, appear to increase the rate of movement.

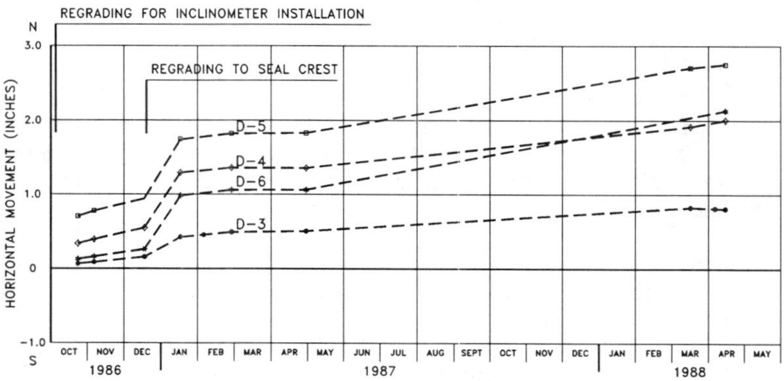

FIGURE 4 - INCLINOMETER MEASUREMENTS

The pneumatic piezometers installed in 1984 and 1986, and the standpipe piezometers installed upon completion of the embankment construction were monitored. The location of some of the sensing zones and typical piezometer measurements are shown on Figure 2. In general, the groundwater level lies within the natural soil, close to the top of rock and tends to fluctuate a foot or two depending on rainfall. The pneumatic piezometers installed within the lower rock zones also indicate this trend, however, those installed in the embankment indicate some fluctuations and some localized higher water levels (B-1E, B-1W), which may be associated with perched water resulting from runoff into open cracks.

SLOPE STABILITY OF EXISTING CONDITIONS

The existing condition cross section shown on Figure 2 at Station 10+90 was analyzed for slope stability. The objective of the stability analyses was to define the existing shear stress along the failure surface. A cross section at Station 9+30 was also analyzed and similar results to those described below were obtained.

Slope stability analyses were performed for the existing conditions utilizing the groundwater levels and the failure surface and material properties shown in Figure 2. The analyses were performed utilizing a version of the Janbu Method employed by the computer program STABL2 developed by Purdue University.

The results indicate that for a factor of safety of unity, the existing shear stress on the failure surface in terms of angle of internal friction (friction angle) is 16 degrees at Station 10+90. The actual failure surface or zone may be shaped slightly differently than the hypothetical surface used in this analysis, however, little difference was observed in the results by minor modifications to the shape of the failure surface.

The backcalculated friction angle represents the existing stress condition on the failure surface at which progressive failure continued to occur. The backcalculated value (16 degrees) was considerably less than the residual friction angles measured in the laboratory on soil samples from in or near the slide zone (17 to 30 degrees).

Additional analyses were made of the existing conditions using higher pore pressures in the slide zone to attempt to explain this inconsistency. It was found that when using the lowest measured residual friction angle (17 degrees) along the horizontal portion of the failure surface and a typical value of 24 degrees for the inclined portion, the pore pressure along the entire failure surface would have to be at elevation 1145 for a factor of safety of unity. This condition was considered unlikely because piezometric measurements did not indicate pore pressures of this magnitude.

Therefore, it was hypothesized that the soils along the failure surface were deforming at levels of stress below their peak residual shear strength. The concept of time dependent behavior of clayey soils and the fact that some soils fail under sustained levels of stress significantly below the peak stress (measured in a short duration test) has been documented by many researchers as summarized by Mitchell (1976) and Tavenas and Leroueil (1981). Additional laboratory testing was performed in an attempt to characterize this behavior.

CONSTANT STRESS DIRECT SHEAR TESTS

Constant stress direct shear tests were performed to characterize the creep behavior of the soils along the failure surface. Samples were tested from test pit TP-6 (B1), and from boring D-4E (ST1). Index properties of these samples are shown below:

Boring/ Test Pit No.	Sample	Elev. (ft) Top of Sample	Natural Water Content %	Liquid Limit %	Plastic Limit %	Dry Unit Weight (PCF)	Dry Unit Weight (KN/M^3)
D4E	ST1	1122.4	17.8	30	11	111.8	17.6
TP6	B1	1118.0	19.6	35	15	98.2	15.4

The samples were prepared and installed in the direct shear machine under a normal pressure of approximately one kg/cm^2 (98.1 KPa), which represents the effective vertical stress at a point slightly below the mid height of the embankment. Peak and residual strengths were initially measured with the load applied at a constant strain rate. Subsequently, the proving ring was removed and a wire and pulley arrangement was installed so that a constant load could be applied to the sample. Loading was initiated at approximately 25 percent of the maximum residual shear load that the sample had withstood in the constant strain rate test, and then increased in 10 to 15 percent increments to failure. Each constant stress increment was generally allowed to remain in place for one (minimum) to several days.

The deformation rate was calculated throughout the test. Typically the deformation rate decreased with time (linearly on a log-log scale) and increased with increasing applied stress. Plots of stress versus deformation rate at 1,000 minutes and 10,000 minutes for the samples tested are shown on Figure 5. Friction angle values corresponding to the applied shear stress (considering 1 kg/cm² confining pressure and a cohesion intercept of zero) are also shown on the plots. For each sample, values of the maximum residual shear strength obtained from the constant strain rate tests (at the same confining pressure) are shown at the top of each plot.

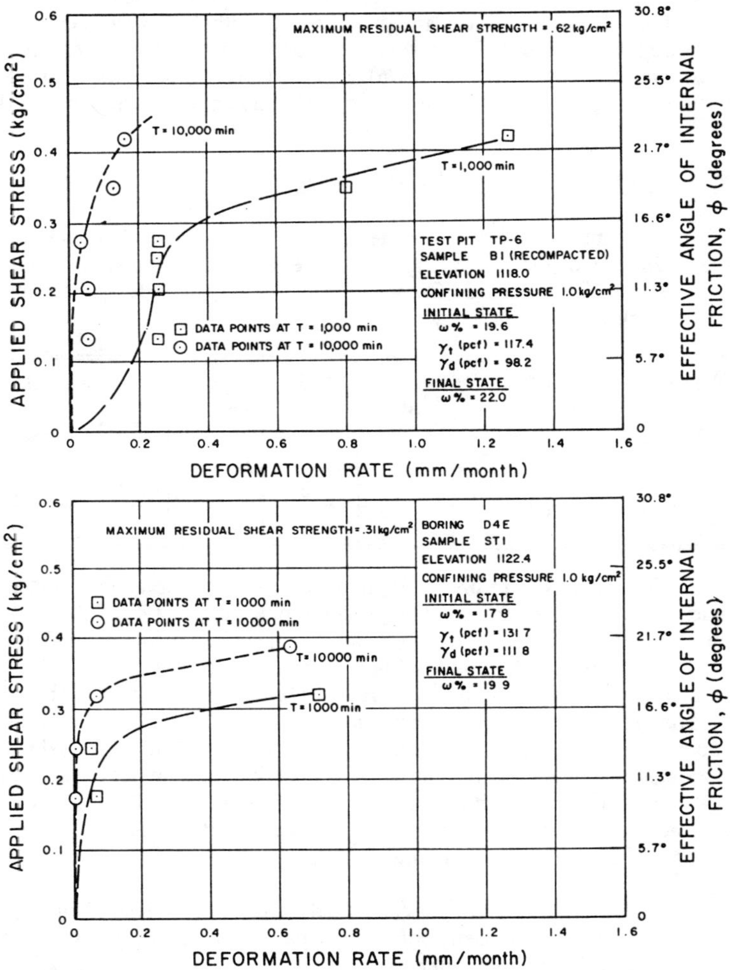

FIGURE 5 - STRESS CONTROLLED DIRECT SHEAR TESTS

For all of the samples tested, a significant increase in the deformation rate was noted when the applied shear stress exceeded about 0.3 kg/cm^2 (29.4 KPa) (corresponding to a friction angle of 16 to 17 degrees). For one of the samples (TP-6 B1), the increase in deformation rate occurred at a stress level of about 50 percent of the maximum residual shear strength. For the other sample (D4E, ST1), the increase occurred at a stress level only slightly less than the maximum residual value.

Considering the close agreement between the backcalculated friction angle (16 degrees) and the values at which creep movement initiated in the constant stress tests (16 to 17 degrees), it appeared that the soils along the failure surface were continuing to deform slowly at stress levels below the residual shear strength. Undoubtedly the higher pore pressures as recorded in some of the piezometers were contributing to the slope movement, however, this effect was considered to be small. Therefore, the existing shear stress along the failure surface was considered to be represented by a friction angle of between 16 and 17 degrees. This was reduced for design, as described below.

REMEDIAL DESIGN AND STABILITY ANALYSIS

Remedial Design

Figure 6 illustrates in cross section the remedial design at Station 10+90. Key features of the design include:

- Removal and replacement of the upper 10 to 15 feet (3.0 to 4.6 m) of the embankment crest and upstream slope to repair cracking in this zone, and facilitate construction of other remedial design features;

- Construction of a buttress against the upstream slope to stabilize the embankment and provide for a factor of safety against continued unacceptable slope movements;

- Additional seepage control through potentially cracked zones in the existing embankment; and

- Removal of the existing fill around the spillway tower and repair of the tower and lake drain pipe.

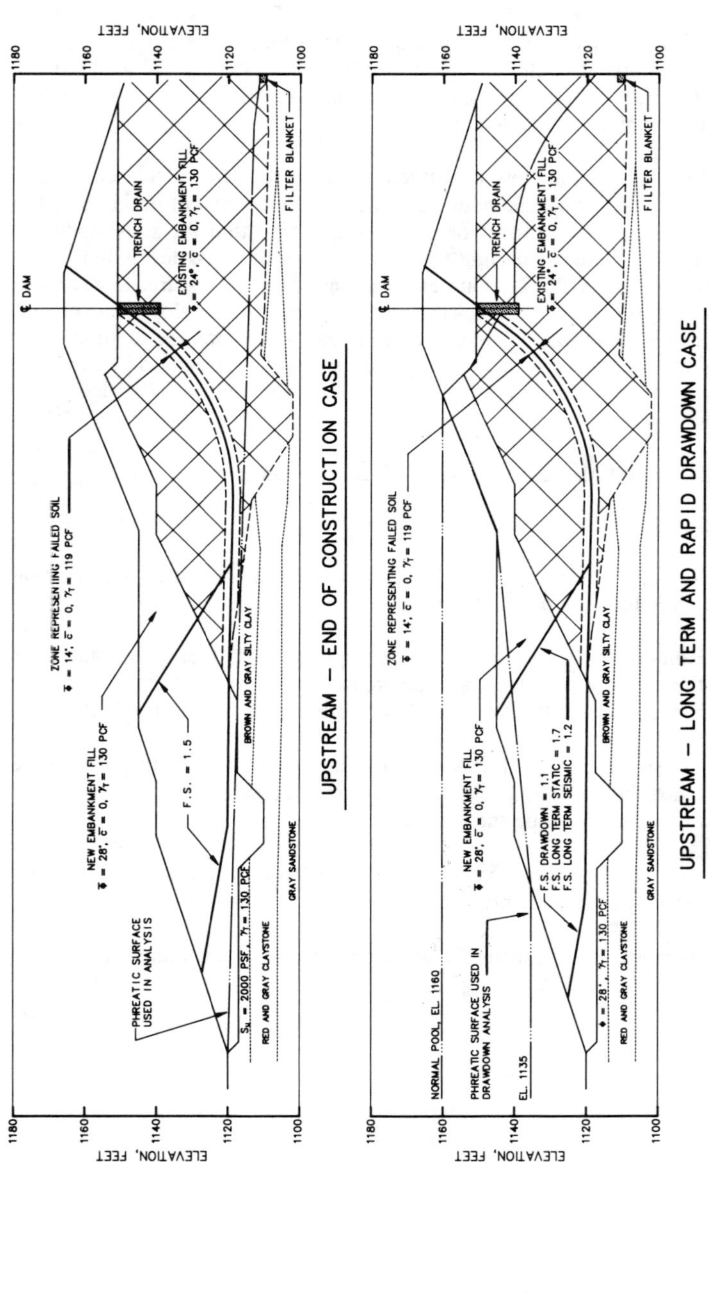

FIGURE 6 - STATION 10+90 STABILITY ANALYSES CROSS SECTION FOR REMEDIAL DESIGN

Stability Analyses

Analyses were performed to model several conditions during the embankment life including buttress construction, end of construction, rapid drawdown from full reservoir level, and long term steady seepage under static and seismic conditions. The analyses were completed using a computerized version of the Janbu Method employed by the computer program STABL2. The results of the upstream design analyses are presented on Figure 6. Safety factors recommended by the U.S. Army Corps of Engineers (1970) were used for design. Downstream stability analyses were also completed, however, are not presented with this paper.

Engineering parameters assigned to the various soil zones are shown on the figures and were based on judgement and the results of laboratory testing. The value of shear stress assigned to the slide zone for design was reduced from the estimated existing value (i.e. represented by a friction angle of 16 to 17 degrees) so that the design would be based on stress levels associated with a stable condition (i.e., no movement). As shown on Figure 5, essentially no creep was noted at stress levels corresponding to a friction angle of 14 degrees or less. Furthermore, when test samples that were exhibiting creep deformation were unloaded to a stress level corresponding to a friction angle of 14 degrees, movement stopped. Therefore, a friction angle of 14 degrees was selected to represent the slide zone for design.

A friction angle of 28 degrees was assigned to the buttress material. This is less than the value measured in the laboratory (34 degrees) and was reduced so that the design would be based on a strength value that would develop at low strain.

For the end of construction condition, a value of 2,000 psf (95.8 KPa) was selected for the undrained strength of the natural silty clay soil. For this case no increase to the shear strength along the slide zone was allowed due to the weight of the overlying buttress material. The shear strength of the buttress material was considered both in terms of undrained shear strength and effective stress parameters and the most conservative values used for the analyses.

The rapid drawdown analyses controlled the proportioning of the buttress. For these analyses, the reservoir level was dropped from normal pool at elevation 1160 to elevation 1135 with the water level maintained at the ground surface within the embankment. The buttress was proportioned so that failure surfaces extending the entire length of the buttress and those shearing through the buttress would have similar calculated factors of safety.

Analyses performed for the long-term steady state seepage condition utilized the phreatic surface shown in Figure 6. Earthquake loading was represented by applying a horizontal acceleration equivalent to 0.05 times the acceleration due to gravity (g) to the embankment mass as recommended by the Applied Technology Council (1978) for this area, and performing a static slope stability analysis.

SUMMARY AND CONCLUSIONS

A remedial design was prepared to buttress and stabilize the failed upstream slope of the Scout Reservation Dam. During design of the buttress, a value of shear strength had to be selected to represent the slide zone. Constant strain rate direct shear tests were made on Shelby tube samples obtained from near the slide zone to define the residual strength of these soils. The residual shear strength values, however, were significantly greater than those calculated for a factor of safety of unity, if no excess pore pressures were included in the analyses. It was concluded that either significant excess pore pressures continued to exist along the failure surface or creep movements were occurring at stress levels below the residual shear strength.

In an effort to determine the stress levels at which creep movements would occur, constant stress direct shear strength tests were made on Shelby tube soil samples from near the slide zone. Creep movements were detected at constant stress levels significantly less than the residual strength measurements from the constant strain rate tests and approximately equal to the calculated stress level for a FS = 1 with no excess pore pressure. While the potential for excess pore pressure on the failure surface could not be completely ruled out, the testing suggested that to restrict creep movement, the stress level used for design should be less than the residual shear strength and less than that backcalculated for the existing conditions.

A shear stress level for the slide zone corresponding to no creep deformation was selected for design based on judgement and the results of the constant stress tests. The size and configuration of the buttress was then determined, based on stability analyses performed to evaluate conditions during and after construction.

In conclusion, it appears that residual shear strength values from constant strain rate laboratory tests may be significantly greater than the levels of stress at which low rates of deformation will occur along an existing failure surface. Backcalculation of the stress level for a factor of safety of unity provides a reasonable estimate of the existing average stress conditions. However, if movement of the slope is ongoing, then the backcalculated stress level should be reduced for design to correspond to a condition of no movement. Constant stress tests provide a useful aid to predict the stress levels that low rates of deformation will be initiated.

The remedial design includes a monitoring program to measure slope movement and piezometric levels during and post construction. These data are presently being collected and will provide information regarding the stress-strain-time behavior of the remedial design.

REFERENCES

Applied Technology Council (1978). "Tentative Provisions for the Development of Seismic Regulations for Buildings", NBS Special Bulletin 510, June.

Department of the Army (1970). Office of the Chief of Engineers, "Engineering and Design Stability of Earth and Rockfill Dams", EM 1110-2-1902, April.

Mitchell, J.K. (1976). "Fundamentals of Soil Behavior", John Wiley and Sons, Inc., New York, New York, 291-339.

Tavenas, F. and Leroueil, S. (1981). "Creep and Failure of Slopes in Clays", Canadian Geotechnical Journal, Vol, 18, 106-120.

UNDERWATER SLOPE FAILURE, PORT HUENEME

By W.H. Roth[1], Member, ASCE, D.T. Liu[1], Member, ASCE, M. Tischuk[1], Member, ASCE, and T. Hjort[2]

ABSTRACT

In late 1987 construction of the Wharf-2 extension project at Port Hueneme, California was halted by underwater slope failures which occurred during excavation. A preliminary assessment of the possible failure causes resulted in a fast-track re-design of the pile supported wharf which enabled construction to be resumed within 4 weeks. By installing a 5-foot high cantilever steel bulkhead at the berth line, it was possible to flatten the slope without changing the original wharf design. More importantly, this arrangement avoided the exposure of a cohesive layer which is believed to have set the conditions for hydraulic ground failure during excavation of the toe of the slope. With respect to liability, the owner insisted that the costs for additional investigations, design, and construction delays be borne by his consultants and contractor. This position was rejected with the argument that this would be tantamount to the owner's contract with his consultants being an insurance policy against unexpected conditions during construction.

INTRODUCTION

To accommodate the growing business demands at the Port of Hueneme, California, a 700-foot wharf was recently constructed as part of the Port's Wharf 2 Extension project. Subsurface investigations were initiated in 1984, and foundation recommendations for the design of the wharf were provided in 1985. In late 1987 wharf

---

(1) Dames & Moore, 911 Wilshire Boulevard, Suite 700, Los Angeles, California 90017.

(2) URS/SW Engineering, 500 N.E. Multnomah Street, Portland, Oregon 97232.

construction was begun and, after a brief interruption caused by the underwater slope failure which is the subject of this paper, was successfully completed in the fall of 1988.

This paper describes the findings and recommendations of the initial subsurface investigation, and the chain of events triggered by the underwater slope failures experienced at the beginning of construction. In addition to the technical aspects concerning the failure investigation and the implementation of fast-track design revisions to keep construction delays at a minimum, the paper also discusses the question of professional liability.

INITIAL SITE INVESTIGATION

SUBSURFACE CONDITIONS

At the time of the initial subsurface investigation in 1984, the entire wharf site was on dry ground at Elevation +15 feet Mean Lower Low Water (MLLW). A total of 10 borings were drilled, with 5 borings located in the backlands, and 5 along the proposed wharf alignment as shown in Figure 1. The five wharf borings were drilled with truck mounted 5-inch diameter rotary wash equipment, to depths ranging from 51 to 144 feet. Soil samples were obtained with 2.4-inch (I.D.) diameter Dames & Moore Type U drive samplers at 5-foot intervals. Samplers were driven with a 400-pound hammer dropping a height of 18 inches.

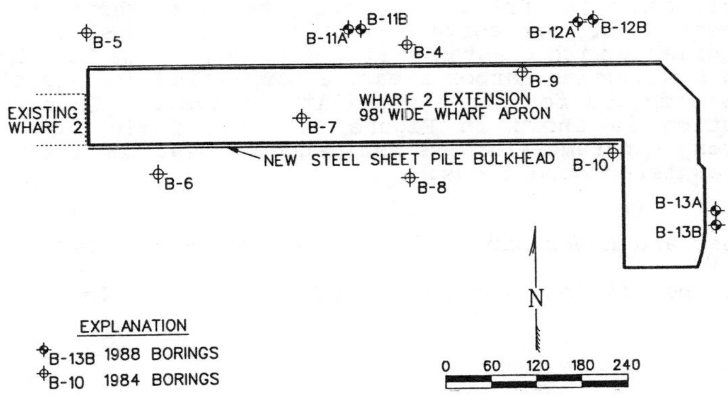

Figure 1. Plot Plan with Boring Locations

A zone of interbedded clays, silts and fine sands was encountered, roughly between Elevations -25 and -36 feet MLLW, sandwiched between upper medium dense to dense sands and lower dense sands. Typical soils in this zone were identified as "loose fine sandy silt" in Boring B-6, "medium stiff silty clay" in B-7 and B-9, and "dense sandy silt" in B-10. Boring B-8 did not encounter this cohesive zone, but indicated continuous sands at the elevations in question.

ORIGINAL DESIGN RECOMMENDATIONS

Because of the interbedded cohesive soils encountered near the proposed harbor bottom at the face of the wharf, preliminary design recommendations included both an anchored bulkhead structure and a pile-supported wharf with underwater slope. Specifically, the bulkhead alternative was characterized as "eliminating potential problems with the underwater slope configuration." Given the fact that the entire site was on dry land at the time of the initial site investigation, this alternative appeared particularly desirable, as it would have allowed installation of the bulkhead with land based construction equipment. Nevertheless, upon the recommendation of the engineering consultant, the owner opted to pursue the more conventional design of a pile-supported wharf. Accordingly, recommendations were developed for the underwater slope configuration.

Based on visual examinations of the soil samples, recorded sample blow counts, laboratory testing (direct shear tests for frictional soils, triaxial shear, pocket penetrometer, and torvane tests for cohesive soils), and experience with coastal soil conditions in the Long Beach and Los Angeles Harbor areas, a simplified cross section was composed for slope stability analysis. This cross section is shown in Figure 2. The following shear strength parameters were estimated to best represent the encountered conditions:

Upper and Lower Sands	PHI=38 degrees, C0=0	
Interbedded Cohesive Soils	PHI=0	C_u=900 psf
		($C_u/\sigma_v' \approx 0.3$)

Based on these soil parameters, an underwater slope configuration of 2(h) to 1(v) was recommended for wharf design.

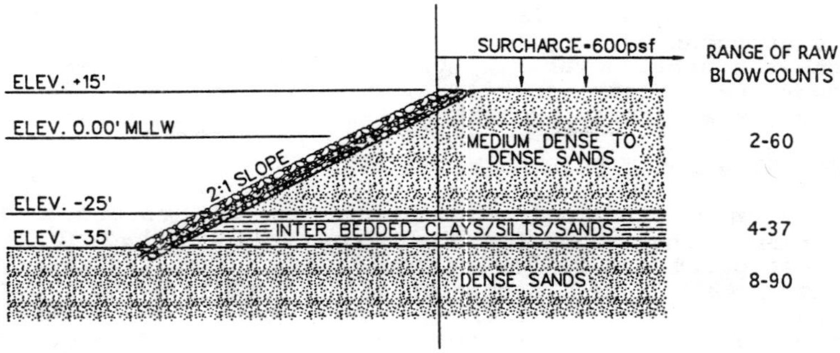

Figure 2. Wharf Slope with Simplified Soil Conditions

UNDERWATER SLOPE FAILURE

CASE HISTORY

The first construction activity after the 1984 site investigation occurred in early 1987, as part of a general harbor dredging program performed by the Corps of Engineers. At that time the wharf site was pre-excavated from dry land at approximate Elevation +15 feet MLLW, down to roughly Elevation -20 feet, with slopes of 3(h) to 1(v).

Actual construction of the wharf began with the final excavation of the underwater slope in late November 1987, in a section adjacent to the existing Wharf 2. Local sloughing at the toe reportedly began on December 1, 1987, when the large, 10-cubic-yard dredge bucket reached a critical elevation of approximately -35 feet, roughly at, or just below, the sandwiched cohesive layer described above. In an attempt to reach Elevation -39 feet for the placement of a rip-rap key at the toe of the slope, dredging was continued without gaining additional depth, until a slope failure occurred which reached above the water line.

Until the slope failure incident, the geotechnical engineer who had provided the original foundation recommendations was not involved in any construction related services. By the time he arrived at the site, on December 6, 1987, the contractor had already attempted excavating another 50-foot section of the slope in the vicinity of Boring B-8. Even though this boring did not indicate any cohesive material at the critical elevation, a similar situation developed as before. After again encountering sloughing near Elevation -35 feet, the contractor had continued excavating without gaining additional depth for about 7 hours. When this resulted in yet another slope failure, construction was halted and the geotechnical engineer was requested to re-evaluate both the constructability of the wharf, as well as its long-term safety in the light of the conditions encountered during excavation.

SEARCHING FOR A FAILURE MECHANISM

In order to find a solution for the encountered problems, the cause for the slope failures had to be investigated. A review of the boring logs, laboratory test results, and slope stability analyses did not produce any obvious clues. As such, this case appeared to be a classical example for the important role of geotechnical construction monitoring as the "final investigation phase" which may well result in design adjustments during construction. The list of likely failure causes quickly focused on strength loss associated with reduced effective stresses - a behavior typical for soils with frictional-type strength characteristics, such as sands and silts. Specifically, the following scenarios/conditions were considered:

1. Excessive disturbance during construction. The pounding, and suction caused by the heavy, 10-cubic-yard clam shell of the dredge may have induced localized liquefaction of loose silty/sandy soil lenses near the toe.

2. Time lag between tidal movement and ground water. Excavation during low tide could have resulted in an outward pressure gradient (from the slope towards the free water surface) which, upon approaching the final toe elevation, may have initiated hydraulic ground failure in the silty/sandy soils of the sandwiched cohesive layer, or the underlying sands.

Indeed, the first incident on December 1, 1987 developed in a period of descending tide (Elevation +6.0 feet at 6:00 am, to +0.1 feet at 1:00 pm), at approximately 1 pm. However, the second incident was first noticed on December 5, at 7:00 am, near the end of a rising tide (Elevation -3.8 feet on December 4, at 9:45 pm, to +6.4 feet on December 5, 8:30 am).

3. Artesian ground water conditions. Regional artesian aquifers of the Oxnard Plain, or local groundwater pools, may have pressurized the sands which were capped by the cohesive layer near Elevation -30 feet. Similar to Scenario 2, the failure could have been caused by hydraulic ground failure at the time of breaking through this "lid" and exposing the sands to the lower pressure environment of the free water of the harbor basin.

RE-EVALUATION OF WHARF DESIGN

While the constructability of the wharf was a pressing immediate issue, due to the high cost of construction delays, re-evaluation of the wharf design with respect to long-term stability was also a major concern. In the original investigation, the sandwiched mixed-soil layer near the toe of the slope had been interpreted as consisting mainly of clay-type soils with cohesional rather than frictional strength. This interpretation had been based on laboratory test results on predominantly clayey soil samples from that layer. On the other hand, the circumstances of the experienced slope failure pointed towards the predominance of more sandy materials. The prospect that the behavior of these sandwiched soils could actually be governed by the frictional shear strength of intermixed medium dense to loose silty/sandy soils, rather than clay as originally assumed, was of some concern. Hence, the stability of the proposed underwater slope was re-analyzed with revised shear strength parameters reflecting this possibility. Specifically, the cohesion of C=900 psf was replaced with a friction angle of PHI=23 degrees without cohesion.

Based on slope stability analyses with the new strength parameters, the underwater slope configuration was modified. The new configuration involved a flatter slope (2.5 : 1) and more importantly, avoided exposure of the suspect soil layers near the toe. By installing a 5-foot high cantilever bulkhead at the berth line, changes to the structural design of the wharf itself were avoided, and construction was able to resume within about four weeks after the slope failure incidents. Figure 3 shows the wharf structure with the toe bulkhead consisting of a RZ10 sheet pile wall. The bulkhead was installed after excavating the slope to a toe elevation of -30 feet, and excavation to the required channel depth at Elevation -35 feet proceeded thereafter.

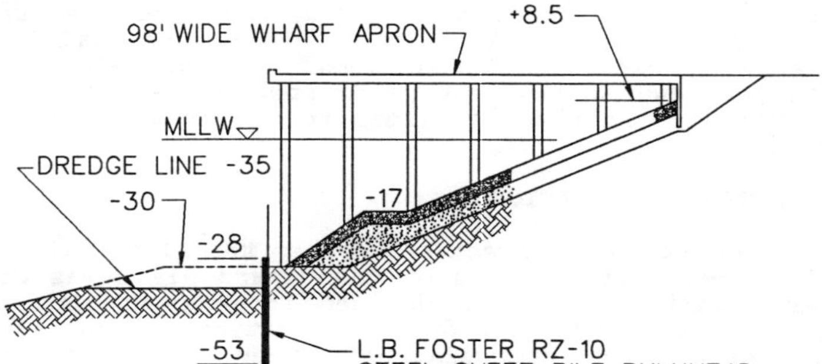

Figure 3. Wharf Structure and Toe Bulkhead

The toe-bulkhead solution was considered to be the most appropriate remedy, independent of the cause of slope failure. However, it was recognized that this design may have to be backed up with a pressure relief system if significant artesian conditions (Scenario 3, above) were to exist at the site. Artesian pressures confined by the upper cohesive layer and the bulkhead could adversely affect the long-term stability of the slope above the bulkhead. Relief wells or wick drains connecting the lower and upper sands, were considered as a possible remedy for this situation. In order to evaluate the need for such measures, a field investigation was launched to explore the groundwater conditions at the site.

However, because the installation of a pressure relief system was not on the critical path of the wharf completion schedule, construction was able to resume without regard to the timing of the groundwater investigation.

INVESTIGATION OF GROUNDWATER CONDITIONS

REGIONAL GROUNDWATER REGIME

At the outset of the groundwater study the question arose whether the Oxnard Plain's historic artesian conditions could have some connection with the problems encountered during wharf construction. According to California Department of Water Resources Bulletin No. 63-1, October 1965, fresh water is contained below the Oxnard Plain in several aquifers charged by the Santa Clara river flood plain several miles north from Port Hueneme. Historically, groundwater used to flow to the ocean through offshore outcrops. Wells drilled into these aquifers had enough artesian pressure to supply the water needs throughout a two-story house. However, these artesian pressures started to decline in the 1920s. Eventually, the hydraulic gradients were even reversed, as agricultural activities expanded and annual well extractions exceeded replenishment. Since about the 1950s groundwater levels have dropped below sea level, resulting in the upper aquifers being intruded by salt water.

PIEZOMETER INSTALLATION

Even though a connection between the slope failures and the regional aquifers of the Oxnard Plain was considered unlikely, it was decided to proceed with the monitoring of local groundwater conditions at the site. In addition to the interrelation between tide fluctuation and piezometric levels in the excavated soils, the possible existence of localized, perhaps seasonally recharged, groundwater pools was to be investigated. To this end, pneumatic pore pressure gages were installed in three pairs of closely spaced twin borings at locations B-11, B-12, and B-13 shown in Figure 1.

The new borings were drilled with a truck mounted rotary wash rig, and 2.4-inch diameter soil samples were taken with drive sampler, in 5-foot intervals in the upper 35 feet, and continuously from there to about 60 feet. They essentially confirmed the originally interpreted subsurface profile consisting of medium dense to dense Upper and Lower Sands, separated by an approximately 5 to 10-foot zone of interbedded cohesive soils near the proposed channel bottom at -35 feet MLLW. As can be seen from the boring logs presented in Figure 4, the detailed makeup of this zone varied considerably from boring to boring.

The adjacent shallow and deep borings at each location contain the pressure gages targeted at the Upper and Lower Sands, respectively. The positions of the gages with respect to the encountered soil materials are marked in the boring logs of Figure 4. Each gage was embedded in a 4-foot sand backfill at the bottoms of the separated twin holes, sealed with a 3-foot bentonite plug, and backfilled to the ground surface with bentonite-cement grout. The pore pressure gages were monitored on an hourly basis during a full 24-hour cycle in May 1988, and again in July 1989. The May 1988 and July 1989 data are presented in the upper (a) and lower (b) portions of Figures 5 through 7, respectively.

INTERPRETATION OF MONITORING RESULTS

The hypothesis to be tested was that the excavated underwater slope may have been undermined by hydraulic ground failure at the toe. Movement of water through soil is associated with seepage forces which affect the grain-to-grain contact stresses and, hence, the shear strength of sandy soils. The amount of strength loss depends on the seepage velocity, which is related to the hydraulic gradient. Because the latter is proportional to the ratio of hydraulic head over the length of the seepage path, even a small head can result in a large gradient, if the seepage path is short. Hence, the sudden removal of the cohesive "lid" on top of the Lower Sands with locked-in elevated piezometric pressures, could have created transient "quicksand" conditions.

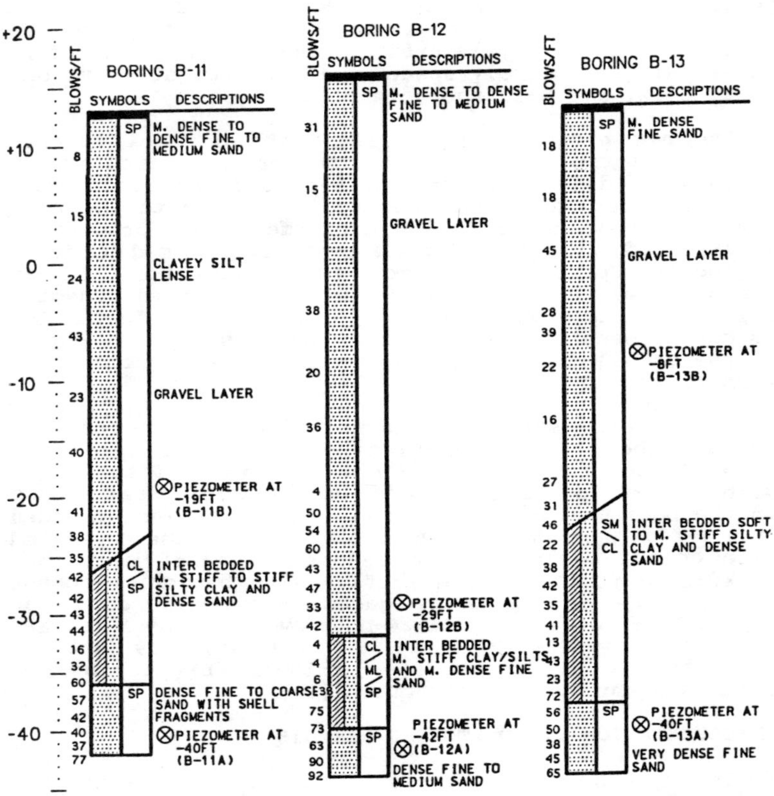

Figure 4. Boring Logs at Piezometer Locations

In an attempt to identify the overriding effects from onshore groundwater sources or sinks, "median" elevations were defined for each of the plotted curves, which mark the midpoint between the maximum and minimum elevations during the recorded 24-hour cycles. In the absence of any source or sink other than the tide, these median elevations should be identical for all curves. Conversely, any observed differences between the median elevations of the tide and piezometric curves could be interpreted as evidence for the existence of an external (e.g. onshore) source or sink.

Figures 5a through 7a present the data recorded during the first monitoring cycle in May 1988. The plots show relatively short, 5 to 8-hour windows with excess pore pressures in the Lower Sands. The maximum excess pressures in these windows range from 1 to 2 feet above tide level, whereas the median piezometric levels are consistently below the median tide. The latter would suggest the existence of onshore groundwater sinks, rather than recharge sources. While these results may be consistent with the aforementioned saltwater intrusion taking place in the Oxnard Plain since the 1950s, they also are inconclusive with respect to explaining the encountered slope failures.

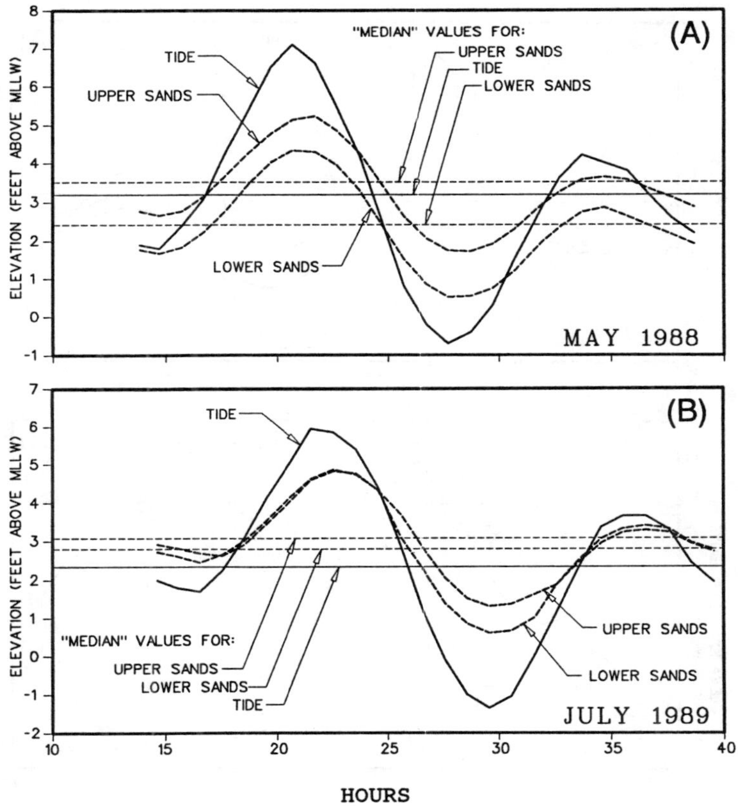

Figure 5. B-11, Tidal and Piezometric Levels vs. Time

Considering the timing of the May 1988 monitoring cycle, several months after the cohesive "lid" had been broken through at several locations, it was recognized that the hydraulic conditions in the Lower Sands may have been substantially different at the time of the slope failures. Therefore, it was decided to conduct an additional monitoring cycle after the Lower Sands had been "sealed off" by the sheetpile wall installed at the toe. This situation was believed to be more representative of the conditions at the time of initial excavation through the cohesive layers covering the Lower Sands.

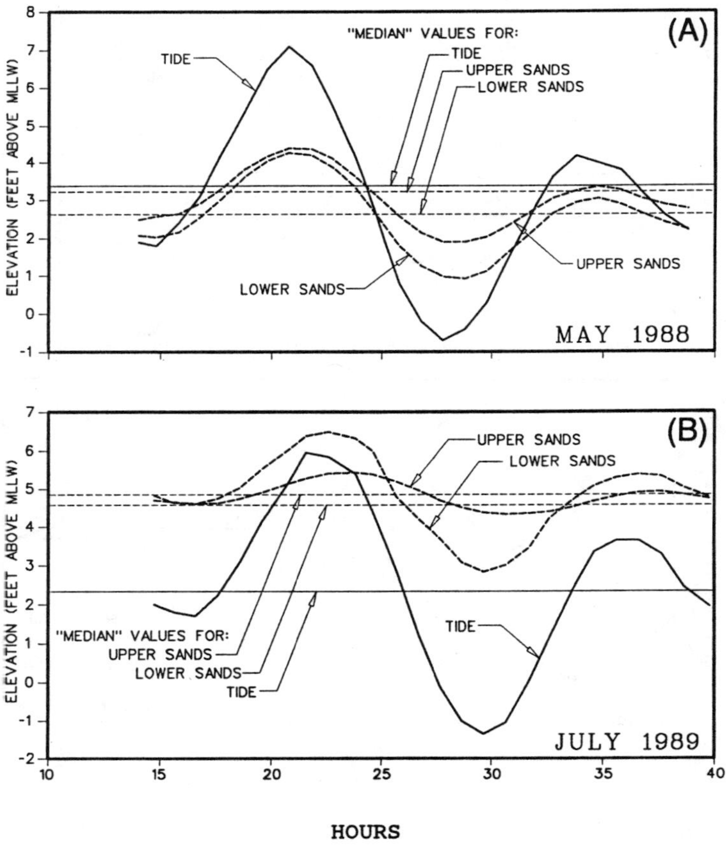

Figure 6. B-12, Tidal and Piezometric Levels vs. Time

Figures 5b through 7b present the results of the second monitoring cycle in July 1989. In contrast to the May 1988 data set, the median elevations derived from the later data reveal a consistent hydraulic gradient from the Lower Sands towards the free water level. For B-11 and B-13 the measured pressure differentials ranged from about 1/2 to 1 foot, and for B-12 it even exceeded 2 feet. The resulting shift of the piezometric curves caused an increase in maximum head difference from about 2 feet to 4 feet. Even more dramatically, the excess pore pressure at B-12 never reversed throughout the entire second monitoring cycle.

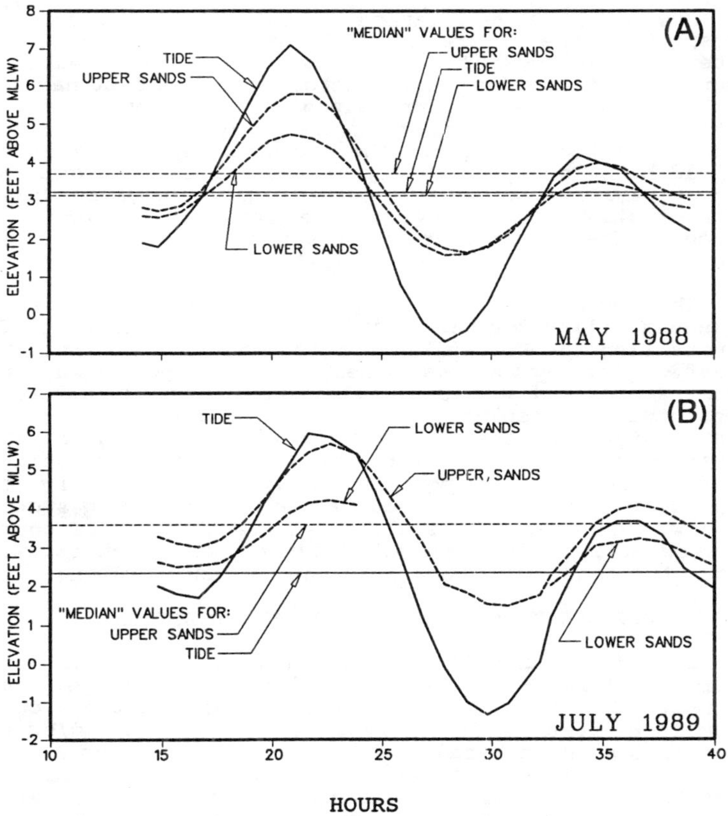

Figure 7. B-13, Tidal and Piezometric Levels vs. Time

Based on the results of the second monitoring cycle, it was concluded that the slope failures may indeed have been triggered by hydraulic ground failure as described earlier in this paper under Scenarios 2 and 3. While the observed excess pore pressures were not considered large enough to pose a threat to the long-term stability of the wharf, it was nevertheless recommended that the pore pressure gages be maintained for future monitoring during extreme rainy seasons. If significant pressure increases were to occur, the need for a pressure relief system would have to be re-evaluated.

QUESTIONS OF LIABILITY

Any soils investigation, short of excavating and inspecting the entire area of interest, can only be a "best estimate" of soil conditions which must be confirmed by construction monitoring. This fact has been acknowledged and emphasized by many renowned practitioners and teachers in the field of soil mechanics since its inception. If the design of earth structures would have to depend solely on the information gathered from small-diameter borings, unreasonable design conservatism would have to be employed in order to accommodate any soil-related surprises without design changes.

The owner in the subject case readily assumed responsibility for the additional construction costs for installing the toe bulkhead. In addition, all involved parties, including the owner, engineering consultant, contractor, and geotechnical engineer mutually agreed to postpone the resolve of any remaining liability aspects until after the design revisions were implemented, such that construction could resume without further delays. The arguments and counter arguments concerning the unresolved issues are discussed in the following section below. The dispute was ultimately settled out of court. The owner's demands were partially met with contributions from the contractor, designer and geotechnical engineer.

COSTS FOR ADDITIONAL INVESTIGATION AND DESIGN

One of the unresolved arguments concerned the costs for additional site investigation, design, and construction delay. The owner insisted that these costs should be born by the engineering consultant and/or the latter's subcontracted geotechnical engineer, as well as the construction contractor. To examine this position, one should consider the following hypothetical situation at a time when neither design nor construction had even begun:

> The geotechnical engineer completes the initially agreed upon scope of work, but finds that additional work is needed due to unusual and unexpected soil conditions. He proposes to drill additional borings and investigate the possibility of artesian pressure and its effect on the proposed structure, etc. Would anybody have seriously suggested that the costs for investigating these unexpected conditions are the consultant's responsibility?

The answer must be **clearly "no"!** Similarly, after learning about the unexpected soil problems, the engineering consultant should have been able to negotiate an additional fee for the increased scope associated with a non-conventional wharf design.

COSTS OF CONSTRUCTION DELAY

More than in any other field of civil engineering, judgement and experience are the main factors guiding the recommendations of the geotechnical engineer. Even though uncertainties can be reduced (not eliminated) by increasing the scope of the pre-design soils investigation, economic constraints usually require a compromise between the consultant's quest for more information and the owner's desire to keep these ("unproductive") costs down.

In any case, it is the construction phase which must be regarded as the "final investigation phase" in foundation engineering. It hopefully confirms the conclusions based on the initial investigation; but sometimes it is necessary to adjust the original design in the construction phase. A rigid design philosophy, which does not allow for such changes, would require excessive conservatism (and construction costs) in order to be "ready for any surprise." Since most midstream design changes, short of starting from scratch and terminating the construction contract, entail far less additional costs than executing such a rigid design, it seems only logical for the owner to expect his consultants to be flexible. The risk associated with this flexibility, however, must be accepted by the owner, since he is the only beneficiary. The consultants' fees do not include allowances for insuring against reasonably unforeseeable circumstances.

In closing, the authors believe that there is no rational basis for an owner to look at his contract with consultants as an insurance policy against unexpected soil conditions discovered during construction. Such an attitude would create a clear conflict of interest for the consultant who recognizes changed conditions during construction. Should he acknowledge the need for a design change (and hence be blamed for associated construction delays), or should he take the risk of "getting by" without a change. Clearly, it is in the interest of the owner to take such constraints off the consultant by accepting the financial responsibility for mid-construction design changes - unless, of course, they are caused by professional negligence.

SUMMARY AND CONCLUSIONS

The construction of the 700-foot long Wharf 2 extension at Port Hueneme involved the excavation of an underwater slope in medium dense sands with intermittent cohesive layers. During dredging of the first two 50-foot sections, slope failures occurred which eventually forced construction to be halted and the wharf design to be revised. When the search for the cause of these failures focused on the possible existence of elevated pore pressures in the sands below the intermittent cohesive layers, a groundwater investigation was launched to test this hypothesis.

Based on the results from two 24-hour monitoring cycles involving hourly measurements of piezometric pressures and tide levels, it was concluded that the slope failures were most likely triggered by hydraulic ground failure at the toe of the slope. The conditions for such failures to occur consisted of elevated pore pressures in the Lower Sands, and the fact that these sands had been hydraulically isolated by a "lid" of cohesive soils. When the excavation bucket broke through this lid, the Lower Sands were abruptly exposed to a lower pressure environment, thereby causing temporarily large hydraulic gradients and related "quicksand"-type, piping ground and failure.

In addition to the transient, tide related, excess piezometric pressures in the Lower Sands, the data suggested hydraulic interaction with other sources, possibly local groundwater pools recharged by rainfall. The measured piezometric heads from these sources, however, did not warrant the installation of a pressure relief system. Rather, it was recommended to continue monitoring during future extreme rainy seasons, and to re-evaluate the need for a pressure relief system if significant increases in piezometric levels were to occur.

The high costs of construction delays necessitated fast-track design revisions regarding both constructability and long-term stability of the wharf in light of the subsurface conditions encountered. This was made possible by exemplary cooperation between the owner, contractor, engineering consultant, and the geotechnical engineer, who mutually agreed to postpone liability arguments until after the technical problems were resolved. The design revisions consisted of flattening the slope to 2.5:1, installing an underwater bulkhead at the toe of the slope after excavating to Elevation -30 feet, and proceeding to excavate to -35 feet thereafter.

With respect to liability, the owner insisted that the costs for additional soils investigations, design revisions, and construction delays be ultimately born by his consultants and contractor. This position was rejected with the argument that this would be tantamount to the owner's contract with his consultants being an insurance policy against unexpected conditions during construction. This would present a great conflict of interest for the consultants, essentially forcing them to provide all but the most conservative design recommendations. Since the owner is the only beneficiary of a more reasonable approach, it is he who should accept the risk of unexpected ground conditions.

APPENDIX

1 in = 0.0254 m;
1 ft = 0.3048 m;
1 cubic yard = 0.7646 m^3
1 lb/ft^2 = 47.880 N/m^2

INVESTIGATION OF MACKAY DAM FOLLOWING THE 1983 BORAH PEAK EARTHQUAKE

By

Leslie F. Harder, Jr., M. ASCE[1]

ABSTRACT: Mackay Dam is an embankment in central Idaho composed primarily of dumped sandy gravel. This dam was originally designed to be 120 feet high with the gravelly fill to be placed by dumping from 25-foot-high timber railroad trestles. However, excessive seepage detected during construction and questionable construction practices led to a suspension of construction in 1910 when the dam was only partially complete. The dam was completed to a reduced height in 1918. Mackay Dam experienced moderate shaking from the 1983 Borah Peak earthquake (M_s=7.3), estimated to have resulted in a peak ground acceleration of 0.22g at the damsite. Despite its questionable construction and the fact that the reservoir was almost full at the time of the earthquake, the dam performed excellently. This paper presents the results of an investigation employing the Becker Penetration Test to evaluate and document the performance of the dam. The results of an analysis using the Becker data in conjunction with the Seed-Lee-Idriss method were in accord with the good performance of the dam.

CONSTRUCTION HISTORY OF MACKAY DAM

General Description

Mackay Dam is a sandy gravel embankment owned and operated by the Big Lost River Irrigation District (BLRID). The dam is situated across the Big Lost River approximately 4 miles northwest of the town of Mackay in central Idaho. The purpose of the dam is to store

---

[1]Supervising Engineer, California Dept. of Water Resources, P.O. Box 942836, Sacramento, CA 94236-0001.

spring runoff for irrigation downstream of the dam during the summer months.

The damsite is located where the alluvial fan of Cedar Creek, sloping down to the southwest, meets an outcrop of White Knob Limestone. The limestone outcrop forms the right abutment of the dam and the Cedar Creek fan forms both the foundation and the left abutment. The dam is approximately 1400 feet long with a maximum height of about 80 feet near the limestone right abutment.

The dam has a small parapet crest fill, about 15 feet wide, on the upstream edge of the embankment. Approximately 5 to 7 feet below this parapet crest, the surface of the embankment widens to approximately 160 feet. This sudden increase in width is due to the fact that construction began when the total design height was to be 120 feet. Due to the safety concerns that developed during construction, the dam was only completed to a height of about 70 feet in 1918. In 1956, the approximately 6-foot-high parapet crest fill was added to the upstream edge of the embankment in order to increase storage within the reservoir (see Figures 1 and 2).

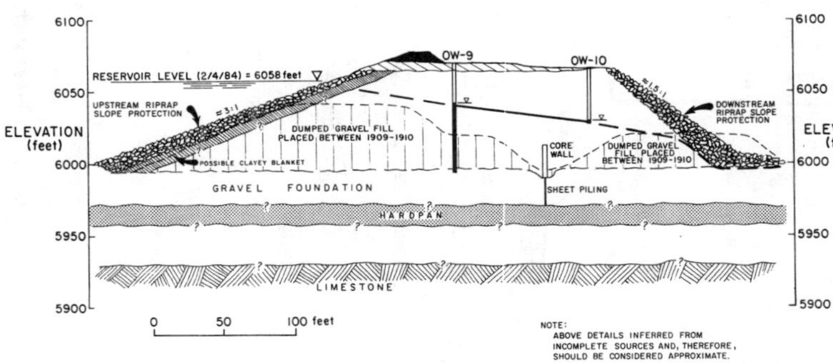

Figure 1: Cross Section of Mackay Dam at Station 17

The upstream side of the dam has a slope of about 3:1 and is covered with riprap up to the pre-1956 elevation. The downstream slope is covered with limestone rubble presumably placed as slope protection. It has an approximate slope of 1.5:1.

The embankment surface is irregular in shape and is covered in many places with vegetation, notably sagebrush. There are several seeps exiting from the downstream slope near the toe of the dam. This has led to the development of a semi-marshy area at the

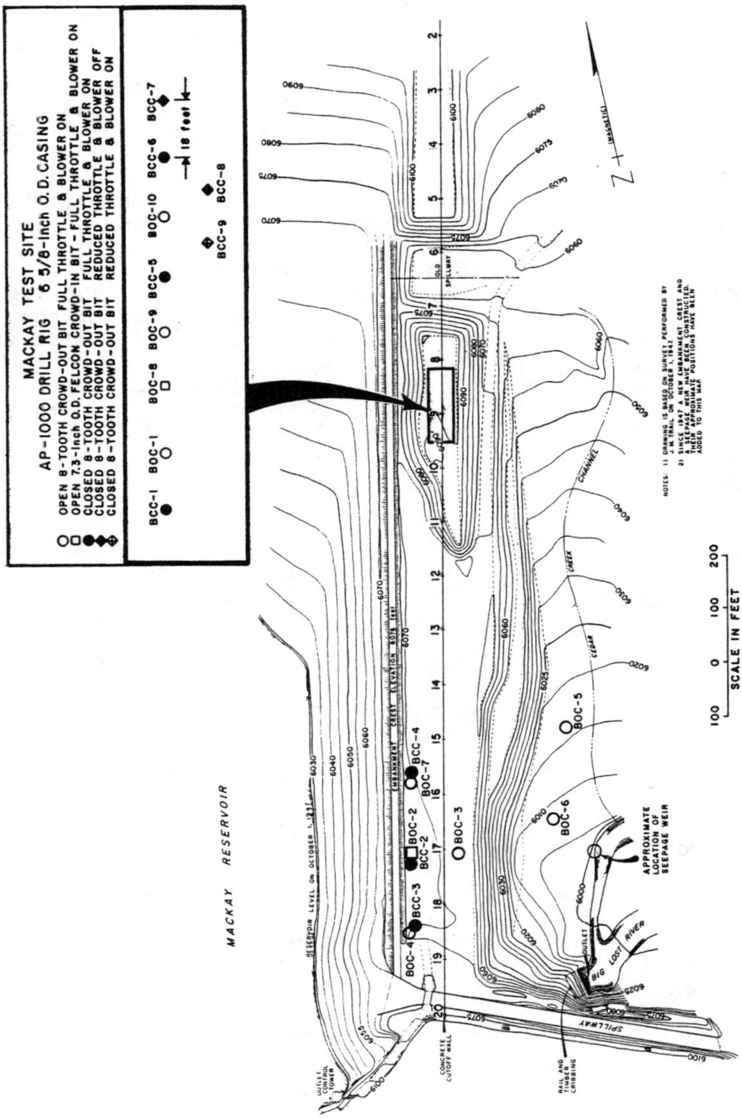

Figure 2: Plan View of Mackay Dam Illustrating Locations of Becker Boreholes

downstream toe where thick brush and trees grow. The individual seepage flows meet in a pool near the outlet release where a 6-foot weir has been established to measure flows. Measurements made by BLRID staff indicate that this seepage flow generally ranges between 10 and 25 cfs, depending on the reservoir level.

Site Geology

Preconstruction geologic studies are not available, although some subsequent reports refer to a number of preconstruction churn borings which apparently gave conflicting information regarding foundation conditions. After excessive seepage was detected and construction was halted, three churn boreholes were drilled in 1911 through approximately 25 feet of fill into the dam foundation near the limestone right abutment. The logs from these borings indicated that the foundation near the limestone abutment consisted of sand and gravel together with zones of non-clay hardpan. All three boreholes encountered an apparently impervious non-clay hardpan layer lying about 20 feet below the surface of original ground. Two of the borings found this hardpan layer to be about 6 to 12 feet thick with more gravel and sand lying beneath it. In both of these boreholes, artesian water pressures were found to exist beneath the hardpan layer. The third boring was located closest to the limestone right abutment and it indicated that the hardpan layer was about 30 feet thick with limestone lying beneath it.

A geologic examination of the damsite was made by L. D. Jarrard of the United States Bureau of Reclamation (USBR) in 1949. In addition to visual inspection of the area, 9 diamond drill holes were made in the limestone right abutment and in the Cedar Creek fan downstream of the dam. Jarrard concluded that the Cedar Creek fan consisted of fanglomerate that is tightly cemented with particles ranging from a fraction of an inch up to 6 inches or more in size. He also concluded that the dam rests on younger reworked gravels of this fan. The USBR boreholes placed downstream of the dam generally encountered sand and gravel. Based on the materials and water levels encountered in the different boreholes and a review of previous explorations, Jarrard (1950) concluded that a significant amount of seepage was passing beneath and through Mackay Dam.

Construction History

Many details concerning the original design specifications, materials, as-built drawings, and subsequent additions to the dam are apparently no longer available. Much of the construction history that is known about Mackay Dam has been obtained from consultant's reports regarding the safety of the dam, interpretations of a few limited photographs and drawings, and visual inspections of the dam. Details of the interpretations made to develop an understanding of the construction history of Mackay Dam are presented in the studies by Harder (1988). The construction history of Mackay Dam apparently involved four phases:

- Phase 1 (1909-1910)
- Phase 2 (1917-1918)
- Phase 3 (1932)
- Phase 4 (1956)

<u>1909-1910 Construction Period</u>

Construction began in June 1909 with the goal of a 120-foot-high dam with an upstream slope of 2.5:1 and a downstream slope of 2:1. The dam was apparently intended to have a concrete core wall 2-feet-thick rising from a 6-foot-deep foundation trench to Elevation 6040 feet (70 feet below the planned crest of the dam). Beneath the core wall near the limestone right abutment, sheet piling was to be placed in the foundation down to "Firm Rock."

Placement of the fill consisted of dumping embankment material from railroad cars on trestles approximately 25 feet high. Embankment fill came from borrow pits located upstream and above the dam on the Cedar Creek fan forming the left abutment. The placement method entailed construction of railroad trestles on both sides of the dam axis and allowed a gap for construction of the core wall. By June 1910, the second lift on the upstream side and the first lift on the downstream side had been completed. In addition, portions of the third and fourth lifts on the left abutment had been completed. These higher lifts on the left abutment can be seen today and rise above the current crest of the dam.

Water began to rise behind the partially built dam in May 1910 and flow through the lower part of the fill. The seepage through the dam caused concern for the dam's safety and rumors began to circulate that the dam was unsafe. A consultant retained by the citizens of Mackay, Mr. J. B. Lippincott, produced a report in August 1910. Shortly thereafter, the State Land Board formed an Investigating Board of Engineers to investigate the safety of Mackay Dam and a report from this board was issued in September 1910. Both of these reports were extremely critical of the design and construction of the dam and questioned its safety. As a result of these reports, construction of the dam was halted in late 1910.

1917-1918 Construction Period

Between 1910 and 1917, political and financial difficulties prevented completion of the dam. Under new ownership, construction was restarted in 1917. This construction apparently consisted of placing new gravelly fill in layers to complete the embankment to an elevation of 6065 feet (approximately 70 feet above the river channel). In addition, an upstream sloping core of clayey material was placed against previously-constructed fill. However, the exact limits of the clayey material are somewhat uncertain. The sloping core was protected by the placement of a thick upstream covering zone of riprap made of selected boulders and cobbles. A downstream facing composed of broken limestone excavated from the right abutment was also to be added to the finished dam.

The core wall was generally left with a height terminating between Elevation 6013 and 6040 feet. Only an extension on the right abutment up to the crest was added to the core wall.

By 1918, this second phase of construction was complete and most of the current dam was in place. This apparently left the dam between Stations 11 and 20 with a crest elevation of approximately 6065 feet and a crest width of about 200 feet. The old third and fourth trestle lifts on the left abutment were apparently left largely in place up to Elevation 6100 feet.

1932 and 1956 Construction Periods

In 1932 a new spillway was built into the limestone right abutment. This new spillway was an ungated structure with concrete walls and floor. Limestone rubble from the spillway excavation was apparently spread on top of the upstream portion of the wide crest between Stations 11 and 20 thus raising the fill in this area by approximately 6 feet to about Elevation 6071 feet.

In 1956, the BLRID was able to win legal challenges and increase the height of the dam by approximately 6 feet. This was accomplished by taking embankment material from the third and fourth trestle lifts on the left abutment and building a small parapet fill on the upstream side of the existing dam. The resulting crest is at elevation 6076 feet and is about 15 feet wide. During this period, a five-foot-high crest weir was placed in the spillway thus raising the potential maximum reservoir operating level up to Elevation 6067 feet.

PERFORMANCE DURING THE 1983 BORAH PEAK EARTHQUAKE

1983 Borah Peak Earthquake

The Borah Peak Earthquake occurred at approximately 8:06 a.m. local time on October 28, 1983. The earthquake was located in a rural area between the towns of Challis and Mackay, Idaho. The National Earthquake Information Service (NEIS) reported a surface wave magnitude (M_s) of 7.3 for the main shock. According to Stover (1985), the earthquake resulted in the deaths of two children from falling building rubble in the town of Challis, and an estimated damage to buildings and property of approximately 12.5 million dollars.

The earthquake occurred on the Lost River Fault and propagated northwesterly from the epicenter in the Thousand Springs Valley. Investigations by Crone et al. (1985) showed that the surface expression of the fault rupture formed a Y-shaped pattern and extended up to 24 miles in length. Surface expressions of fault displacement were primarily normal slip accompanied by a minor left lateral component.

Performance of Mackay Dam

Mackay Dam is located approximately 10 miles southeast of the epicenter and in the opposite direction taken by the propagation of the fault rupture. Based in part on the damage intensities in the area, the earthquake was estimated to have produced a peak ground acceleration of approximately 0.22g in the area of Mackay Dam and a period of strong shaking that lasted between 20 to 30 seconds (Harder, 1988).

On the day of the earthquake, Mackay Reservoir was at Elevation 6058.8 feet, approximately 17 feet below the crest of the dam. Despite the moderate level of shaking, the dam performed very well with damage consisting principally of two longitudinal cracks along the crest between Stations 15 and 18. The first crack extended about 100 feet along the downstream edge of the 6-foot parapet crest fill added in 1956. The second crack was very similar and was found about 12 feet from the downstream edge of the 1932 crest. This second crack had a length of approximately 200 feet. Both cracks were less than about 1/4 of an inch in width.

Immediately after the earthquake, the pool at the downstream toe where the seepage is measured was observed to have become cloudy for a few hours. Reports generally indicated that the individual seepage flows leading from the dam to the pool had muddied up. Along with the development of muddy water, post-earthquake measurements at the seepage weir indicated a temporary and slight increase (5 percent) in flow. Most of the cloudiness in the water cleared within a few hours after the earthquake and after a few days, the water was completely clear. There were no functioning piezometers, movement devices, or survey monuments at the dam during the time of the earthquake.

Overall, the dam performed surprisingly well considering that through much of its history its safety was so much in question.

FIELD AND LABORATORY INVESTIGATIONS

1984 Churn Drill Boreholes

Although eight observation wells had been installed in the dam in 1947, these wells had long been abandoned and covered over. At the time of the investigation, none of the readings from these old wells were available. Accordingly, to determine what portion of the embankment was saturated, two churn drill boreholes were drilled at Station 17 in January and February 1984 and well casing was installed to create Observation Wells OW-9 and OW-10. Measurements made in February 1984 when the reservoir was at an elevation comparable to that during the earthquake showed that the lower half of the embankment was saturated (see Figure 1). Records later obtained from the State of Idaho showed that the old observation well measurements gave comparable readings for similar reservoir levels.

1984-1985 Becker Penetration Tests

In August 1984 and in July 1985, Becker Penetration Tests were made within the dam and foundation. During these two periods, nineteen Becker boreholes were made with both open and closed bits. Figure 2 illustrates the locations of the Becker boreholes. All of the Becker boreholes were drilled using 6.6-inch O.D. casing with an AP-1000 drill rig owned by Becker Drills, Inc. In general, the open-bit and plugged-bit Becker soundings produced similar blowcounts.

Penetration resistance measured with the Becker equipment was converted into equivalent Standard Penetration Test (SPT) blowcounts using the procedures developed by Harder and Seed (1986) and Harder (1988). Figure 3 presents the equivalent SPT N_{60} blowcounts from the soundings made in the third and fourth lifts of the dumped trestle fill on the left abutment. Using a C_N correction curve developed by Harder (1988) to normalize blowcounts in gravelly soils to 1 tsf overburden pressure, the equivalent corrected SPT $(N_1)_{60}$ for the left abutment dumped fill would be approximately 10 blows per foot.

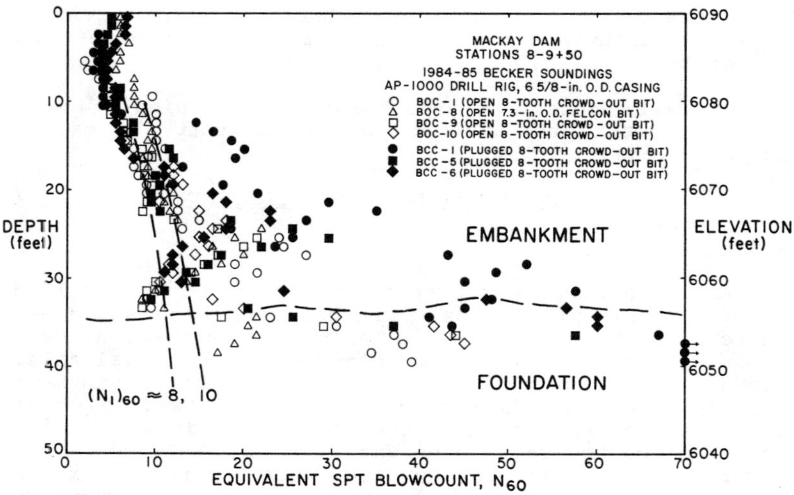

Figure 3: Equivalent SPT Blowcounts Determined in Mackay Dam Dumped Fill Between Stations 8 and 10

Figure 4 presents the equivalent SPT N_{60} blowcounts from the soundings made in the main part of the embankment between Stations 16 and 19. Below about Elevation 5995 feet lies foundation material. Between approximately Elevations 5995 and 6030 feet lies the dumped trestle fill. Above about Elevation 6030 feet lies fill placed subsequent to the dumped trestle fill. As may be observed, the fill placed subsequent to the dumped trestle fill appears to be actually slightly looser than the dumped fill. However, the foundation penetration resistance is shown to be substantially stronger than either embankment material. Using the same gravel C_N curve, the equivalent corrected SPT $(N_1)_{60}$ for the dumped fill below Elevation 6030 feet would average approximately 8 to 9 blows per foot.

1984 Ring Density Tests

Four ring-density tests were performed in August 1984 within the dumped trestle fill on the left abutment. For each ring-density test, a back-hoe was used to excavate a pit approximately 5 to 6 feet deep in order to gain access to gravel that was relatively free from

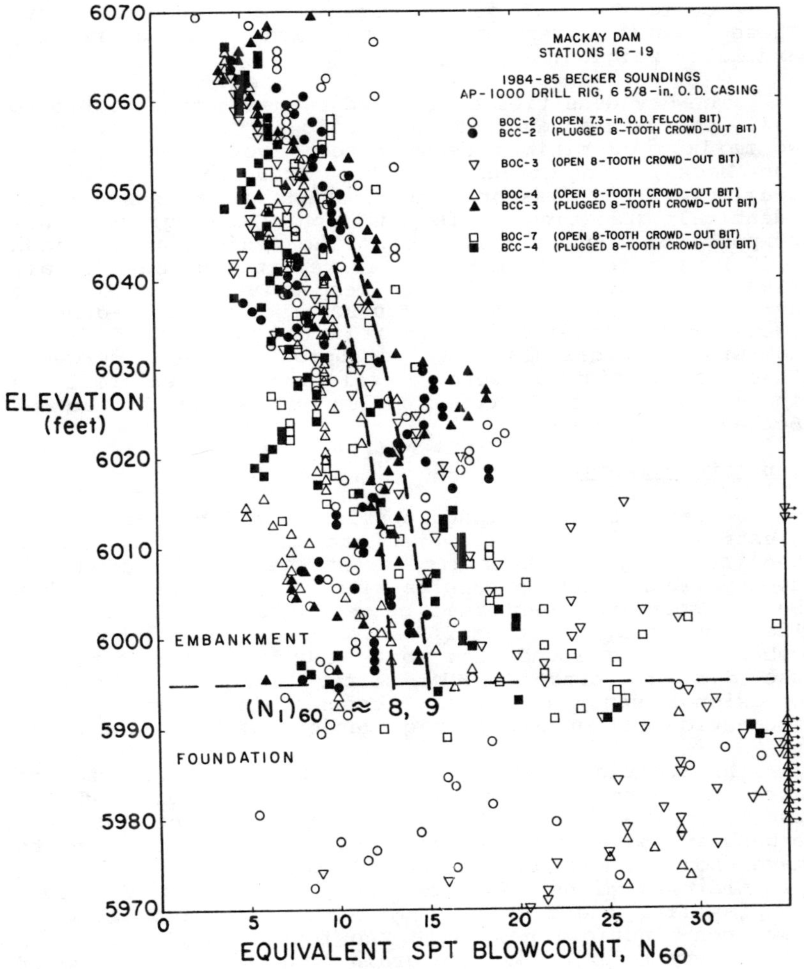

Figure 4: Equivalent SPT Blowcounts Determined in Mackay Dam Between Stations 16 and 19

surface disturbances over the years. Using shovels and hand labor, an additional 1 to 2 feet of gravel was cleared off to remove gravel disturbed by the back-hoe and to create a level surface. A four-foot ring was used together with four-mil plastic to make volume measurements with water. The samples removed for each test consisted of silty, sandy gravel. Approximately 3 to 5 cubic feet of gravel was removed for each test. These materials were collected, dried in a laboratory, and later weighed.

The dry densities determined ranged between 125.5 to 134.0 pcf with an average void ratio of 0.328. Although no maximum or minimum density tests were performed for the Mackay Dam Gravel samples, void ratio data was available from Livermore Gravel which had an essentially identical gradation. In addition, both gravels are composed of rounded particles. The maximum and minimum void ratios for Livermore Gravel reported by Seed et al. (1984) are 0.455 and 0.166, respectively. Using the average in situ void ratio of 0.328 from the ring-density tests, the estimated average relative density of Mackay Dam Gravel is calculated to be approximately 45 percent. This is consistent with a relatively loose fill as indicated by both the construction process and by the low Becker blowcounts.

Laboratory Testing

Laboratory testing consisted of gradation and plasticity tests of recovered samples. In addition to the four ring-density tests, twenty-seven embankment and twenty-two foundation samples were obtained from open-bit Becker soundings performed between Stations 15 and 19. The results of the classification tests showed that the embankment and foundation soils found in this reach of the dam are predominantly silty, sandy gravels with occasional sandy or clayey lenses being found in the foundation and in the upper elevations of the embankment.

Within the saturated, dumped, trestle fill, samples were found to have little to no plasticity. Of the twenty-one samples obtained from portions of the embankment beneath the phreatic surface, 10 were found to have non-plastic fines with 7 other samples having plasticity indices of 4 or less. Three of the four samples from the ring density tests were also found to have non-plastic fines, with the fourth sample having a plasticity index of 4. The grain size characteristics of the samples are summarized in Table 1. The gravel content (percent coarser than the No. 4 sieve) of the dumped fill both at the left abutment and in the main part of the dam averaged approximately 65 percent. With fines contents averaging about 7 percent, the dumped fill was assigned a Unified Soil Classification System symbol of GW-GM.

TABLE 1: GRAIN SIZE SUMMARY FOR MACKAY DAM SAMPLES

Material	Gravel Content (%)	Sand Content (%)	Fines Content (%)
4 Ring Density Samples of Dumped Mackay Dam Fill on Left Abutment	Range = 56-75 Mean = 64	Range = 18-36 Mean = 29	Range = 6- 8 Mean = 7
21 Becker Samples of Mackay Dam Embankment Material Obtained Beneath Phreatic Surface	Range = 42-83 Mean = 69	Range = 13-54 Mean = 26	Range = 3-10 Mean = 6
16 Becker Samples of Mackay Dam Foundation Material Obtained Beneath Phreatic Surface	Range = 2-89 Mean = 57	Range = 8-78 Mean = 32	Range = 1-73 Mean = 11

SEISMIC STABILITY ANALYSES

Static Finite Element Analysis

A static stress finite element analysis was made to estimate the static stresses within Mackay Dam prior to the earthquake. Because the maximum section, generally found between Stations 15 and 19, was considered critical, it was this section that was chosen for analysis. Accordingly, a finite element mesh was constructed to represent the geometry and conditions existing within the embankment and the upper layers of the foundation in this reach. A linear-elastic gravity turn-on plane strain analysis was used to determine the static stresses. Seepage and buoyancy forces were simulated by equivalent nodal forces.

One-Dimensional Dynamic Response Analyses

Earthquake-induced dynamic shear stresses were estimated by conducting a set of one-dimensional dynamic response analyses using Program SHAKE (Schnabel et al., 1972). Analyses were carried out using three soil columns through the embankment and foundation (see Figure 5). Based on studies by Seed et al. (1984), a $(K_2)_{max}$ value of 50 was selected for determining the maximum dynamic modulus of the embankment and a $(K_2)_{max}$ value of 70 was selected for the foundation. The average gravel modulus reduction curve together with the average sand and gravel damping curve from Seed et al., (1984) were used to represent strain-dependent properties.

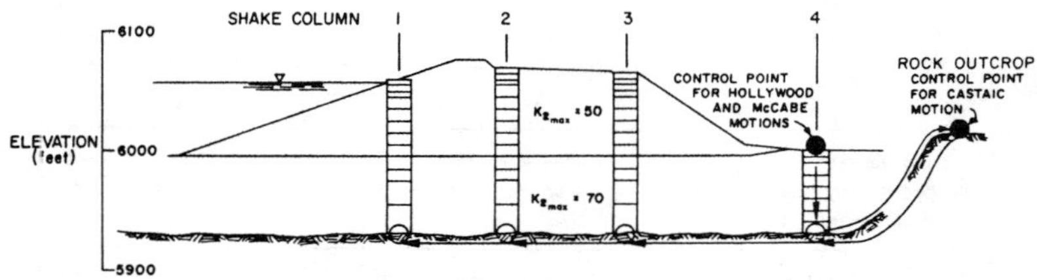

Figure 5: SHAKE Columns Used to Analyze Mackay Dam

Because there were no seismographs near Mackay Dam during the 1983 main shock, it was necessary to select accelerograms recorded in previous earthquakes. It was judged that, by using three different accelerograms and taking the average of the calculated dynamic stresses, much of the potential variability in frequency content could be accounted for. All three records were scaled to have a peak acceleration of 0.22g. Two of the accelerograms were recorded in previous earthquakes on soil sites and were input at the surface of a foundation soil column to determine an equivalent rock outcrop motion for use in analyzing the three embankment soil columns. The third accelerogram was a rock record and was directly input as a rock outcrop motion (see Figure 5 and Table 2).

TABLE 2: ACCELEROGRAMS USED TO ANALYZE MACKAY DAM

Earthquake	Station Location	Instrument Component	Approximate Source Distance (mi.)	Original Peak Acceleration (g)
San Fernando (2/09/71) M_L=6.6	Castaic Old Ridge Route [ROCK]	N-21-E	13	0.316
San Fernando (2/09/71) M_L=6.6	Hollywood Storage PE Lot [SOIL]	N-90-E	22	0.211
Imperial Valley (10/15/79) M_L=6.4	McCabe School Array 11 [SOIL]	S-50-W	10	0.382

Cyclic Load Resistance

Estimates of the cyclic load resistance of the dumped fill were made using the results of the Becker penetration data in conjunction with the correlation developed by Seed et al., (1985) between liquefaction and

SPT resistance in sands. By averaging the Becker data obtained in the dumped fill on both the left abutment and the main embankment, the equivalent SPT $(N_1)_{60}$ corrected blowcount would be about 9 blows per foot. The gravels have an overall fines content of about 7 percent, but it is believed that the liquefaction characteristics are controlled more by the sandy matrix. In addition, the liquefaction resistance curves in the Seed et al. (1985) correlation are based on fines contents of sandy soils. Accordingly, the fines content used in this analysis was based on the sand portion of the Mackay Dam gravels. Based on the gradation tests, this value was set at 12 percent. For a Magnitude 7.5 earthquake, the Seed et al. (1985) SPT correlation would predict a liquefaction resistance (expressed as a normalized cyclic resistance ratio) of 0.13.

The SPT correlation with liquefaction, however, is limited to level ground conditions with effective overburden pressures of less than 1 tsf. For different effective stress conditions and magnitude values different than Magnitude 7.5, use is made of Equation 1:

$$\tau_L = (\tau_L/\sigma_O')_{M=7.5} \times \sigma_O' \times C_M \times K_\alpha \times K_\sigma \quad (1)$$

where:

τ_L = Cyclic load resistance.
$\tau_L/\sigma_O')_{M=7.5}$ = Cyclic resistance ratio from SPT correlation = 0.13.
σ_O' = Effective stress at depth.
C_M = Magnitude correction (1.03 for M_S=7.3).
K_α = Correction factor for sloping ground.
K_σ = Correction factor for confining pressure.

Values of K_α and K_σ were obtained from those suggested by Harder (1988) and Seed and Harder (1990) for different effective stress conditions. Typical K_α values ranged between 1.00 and 1.15. Typical K_σ values ranged between 1.00 and 0.68.

Liquefaction Triggering Analysis

The factor of safety against triggering liquefaction was calculated for each element of the saturated gravel fill using Equation 2:

$$F_L = \tau_L / \tau_{avg} \quad (2)$$

where:

F_L = Factor of safety against triggering liquefaction.
τ_L = Cyclic load resistance (from Equation 1).
τ_{avg} = Average earthquake-induced shear stress.

The average earthquake-induced shear stress was obtained by averaging the peak dynamic shear stresses from the three SHAKE analyses for each soil column, and then using a normalizing factor of 0.65 to account for the irregular nature of the shear stress time histories. For elements located between and/or beyond the SHAKE columns, dynamic stresses were estimated by interpolation and extrapolation.

Figure 6 presents the results of the computations of the factor of safety against triggering liquefaction. The calculations indicate a limited zone in the central portion of the embankment immediately below the phreatic surface where the factor of safety is only between 1.1 and 1.2. In addition to this zone, there is a somewhat larger central zone where the factor of safety is calculated to be between 1.2 and 1.3. However, the greater parts of both the unsaturated and saturated fill were predicted to have factors of safety against liquefaction in excess of 1.3. These results would indicate that the dam would develop some excess pore pressures; but that significant zones of completely liquefied soil would not be expected to occur. This prediction is in excellent accord with the observed good performance of the dam during the 1983 earthquake.

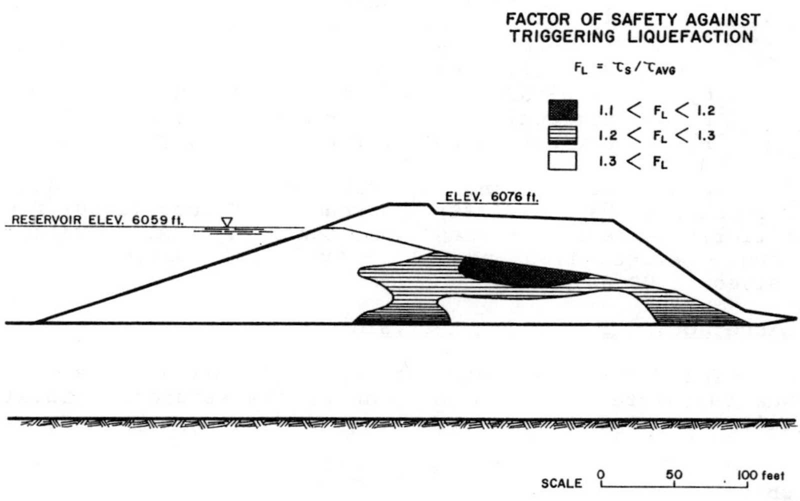

Figure 6: Factors of Safety Against Triggering Liquefaction Calculated for Mackay Dam

CONCLUSIONS

The main conclusions resulting from this study may be summarized as follows:

1. Mackay Dam is an embankment dam composed of silty, sandy gravel with a relative density of approximately 45 percent. It experienced an approximate peak ground acceleration of 0.22g during the 1983 Borah Peak Earthquake, but performed excellently during and after the event.

2. The Becker Penetration Test was an effective tool in quantifying the penetration resistance of the dumped gravelly fill and in obtaining samples for classification testing. The equivalent SPT $(N_1)_{60}$ value of 9 determined for the dumped gravel fill is in good accord with the method of construction and the results obtained from the ring-density tests.

3. Analyses of liquefaction potential using the Seed-Lee-Idriss approach for applying the results of the Becker soundings resulted in factors of safety against liquefaction of approximately 1.3. Thus, the analyses provide results in excellent accord with observed behavior. However, since muddy water was observed coming from some of the seeps at the downstream toe of the dam soon after the earthquake, it would appear that some excess pore pressures were generated in the embankment by the earthquake shaking. This latter behavior is also in good accord with the marginal safety factors calculated for liquefaction triggering.

ACKNOWLEDGEMENTS

The studies described in the proceeding pages were supported in part by Grant No. CEE-84-12366 from the National Science Foundation. The work was conducted while the author was a student under the late Dr. H. Bolton Seed. Dr. Seed's leadership, review, and guidance during the phases of this study were instrumental in their successful completion. The consent granted by the Big Lost River Irrigation District and the special assistance provided by Ms. Rita Lundy and Mr. Doyle Jensen of the District are gratefully acknowledged.

APPENDIX I: REFERENCES

Crone, A., Machette, M., Bonilla, M., Lienkaemper, J., Pierce, K., Scott, W., and Bucknam, R. (1985). "Characteristics of Surface Faulting Accompanying the Borah Peak Earthquake, Central Idaho," Proceedings of

Workshop XXVIII on the Borah Peak, Idaho, Earthquake, convened under the Auspices of National Earthquake Prediction and Hazard Programs, October 3-6, 1984, United States Geological Survey Open-File Report 85-290.

Harder, Jr. Leslie F. (1988). "Use of Penetration Tests to Determine the Cyclic Load Resistance of Gravelly Soils," Dissertation submitted as partial satisfaction for the degree of Doctor of Philosophy, University of California, Berkeley.

Harder, Jr. Leslie F. and Seed, H. Bolton (1986). "Determination of Penetration Resistance For Coarse-Grained Soils Using the Becker Hammer-Drill," Report No. UCB/EERC 86/06, Earthquake Engineering Research Center, University of California, Berkeley, May.

Jarrard, L. D. (1950). "Geologic Report on Mackay Dam, Big Lost River Project, Idaho," Bureau of Reclamation, United States Department of the Interior, September.

Schnabel, P. B., Lysmer, John, Seed, H. Bolton (1972). "SHAKE, A Computer Program for Earthquake Response Analysis of Horizontally Layered Sites," Report No. EERC 72-12, Earthquake Engineering Research Center, University of California, Berkeley, December.

Seed, H. Bolton, Wong, Robert T., Idriss, I. M., and Tokimatsu, K. (1984). "Moduli and Damping Factors for Dynamic Analyses of Cohesionless Soils," Earthquake Engineering Research Center, Report No. EERC 84/13, University of California, Berkeley, September.

Seed, H. Bolton, Tokimatsu, K., Harder, L. F., and Chung, Riley M. (1985). "Influence of SPT Procedures in Soil Liquefaction Resistance Evaluations," Journal of the Geotechnical Engineering Division, ASCE, Vol. 111, No. GT12, December.

Seed, Raymond B. and Harder, Jr. Leslie F. (1990). "SPT-Based Analysis of Cyclic Pore Pressure Generation and Undrained Residual Strength," Proceedings of the Memorial Symposium for H. Bolton Seed, BiTech Publications, Ltd.

Stover, Carl W. (1985). "The Borah Peak, Idaho Earthquake of October 28, 1983--Isoseimal Map and Intensity Distribution," Earthquake Spectra, Earthquake Engineering Research Institute, Vol. 2, No. 1, November.

APPENDIX II: UNITS

1 foot = 0.305 meters	1 tsf = 95.7 kilopascals
1 mile = 1.61 kilometers	1 pcf = 0.0160 grams/ $cm.^3$

Seismic Response of Landfill Slopes

D. G. Anderson[1], B. Hushmand[2], and G. R. Martin[3]
Members, ASCE

Abstract

The seismic stability of slopes at a landfill in the Los Angeles area is discussed in this paper. Slopes that form the perimeter of the landfill are as steep as 1.3H:1V and have heights in excess of 200 feet. Acceleration levels at the base of the landfill during the design seismic event are predicted to be as high as 0.5 g. As part of the seismic stability assessment, seismic stations were installed at the base and top of the landfill. Results from seismic monitoring during four earthquakes are used to make conclusions regarding stability of slopes at the landfill during the design seismic event.

Introduction

The design of landfill slopes has drawn considerable attention over the past decade. This attention results primarily because of the potential consequences if landfill slopes fail, such as pollution to groundwater or emission of toxic gases.

Numerous uncertainties must be addressed during the design of landfill slopes. One of the most significant involves landfill material characterization. Landfill materials include a variety of solids and liquids. The solid portion

---

[1]Senior Geotechnical Engineer, CH2M HILL, 777 108th Ave. NE, Bellevue, WA 98004.

[2]Principal, Hushmand and Associates, 3831 Magnolia Street, Irvine, CA 92714; Visiting Professor, California Institute of Technology, Pasadena, CA 91125.

[3]Research Professor, University of Southern California, Los Angeles, CA 90089.

ranges from fine-grained soil to parts of automobiles and larger; liquids vary from petroleum-based products to water. Given the range in material composition, establishing engineering properties with any confidence is a major task.

Predicting performance of landfill slopes during seismic events introduces even more uncertainty. In addition to uncertainties associated with the static strength of the waste material, there is the added uncertainty of how these materials respond dynamically. The size and heterogeneity of the landfill material make conventional dynamic testing extremely difficult, usually forcing the engineer to rely on judgment rather than documented behavior.

This paper describes dynamic response measurements and predictions made for a landfill. During the remedial investigation/feasibility study (RI/FS) for this landfill, questions were raised regarding the potential for seismically induced slope failure. In response to these questions, a program involving landfill characterization, seismic ground motion monitoring, and seismic slope stability assessments was initiated.

Landfill Characteristics

The seismic stability evaluation was carried out for the Operating Industries, Inc. (OII) landfill. This landfill is located in Monterey Park, California, approximately 7 miles east of downtown Los Angeles. It is surrounded by residential and commercial development; the heavily traveled Pomona Freeway is at the base of one side of the landfill (Figure 1).

The landfill is approximately 190 acres in size. It was formed by filling what was once a gravel pit. The landfill was closed for operations in 1984. Presently, the landfill is nearly 330 feet thick at the center. The top of the landfill ranges from 70 to 225 feet above the adjacent ground surface. Side slopes vary from 3H:1V (horizontal to vertical) to as steep as 1.3H:1V.

Contents of the OII landfill include a variety of hazardous municipal and industrial waste. No special provision was taken during disposal to contain these wastes, other than the placement of interim cover materials.

Site Characterization and Seismic Monitoring

A number of site characterization studies have been conducted at the landfill. These have included drilling and sampling, cone penetrometer testing, and installation of settlement and slope inclinometer monitoring systems. Laboratory studies involving material classification and strength testing have also been conducted. Siegel et al. (1990) provides a summary of these efforts.

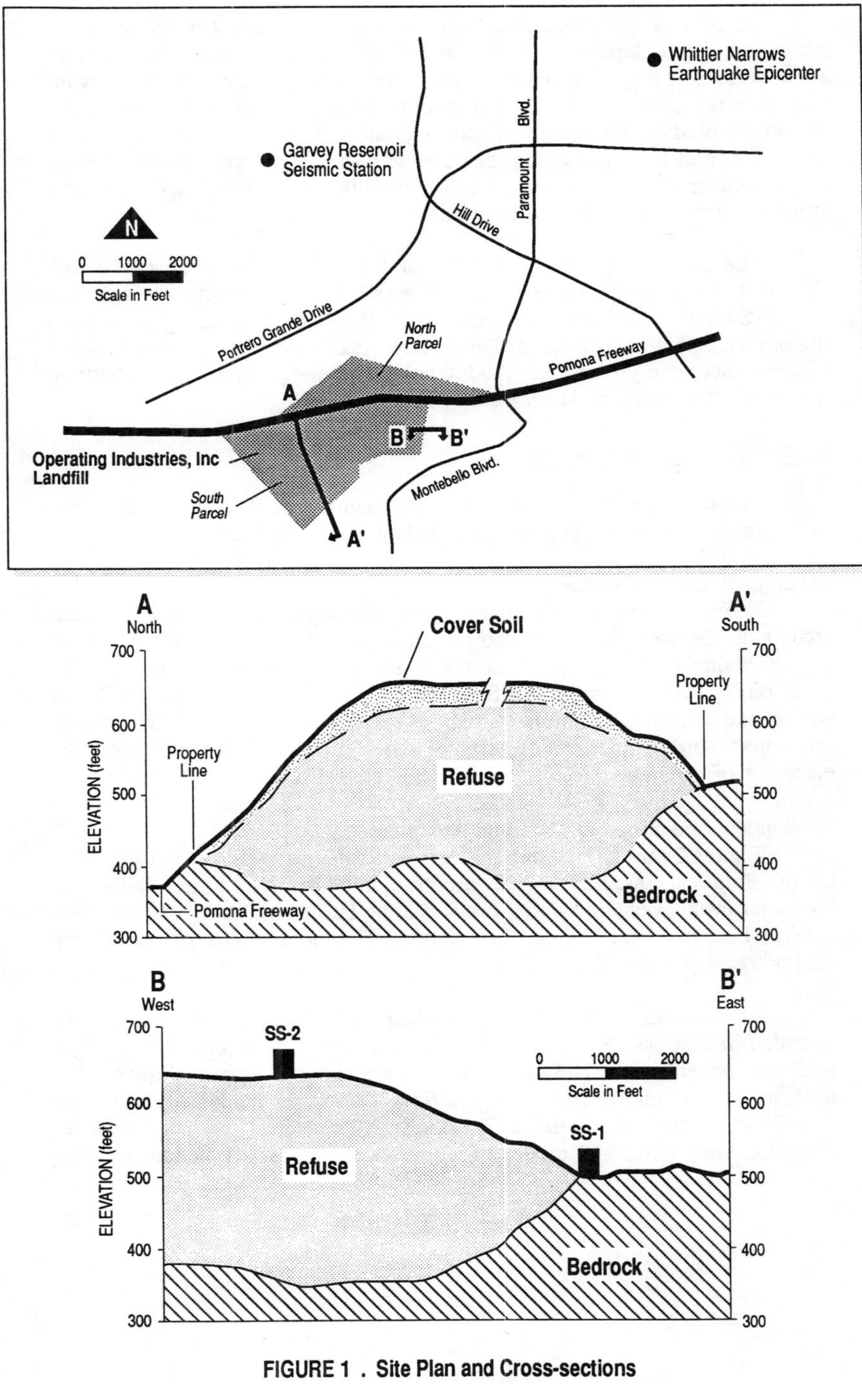

FIGURE 1 . Site Plan and Cross-sections

As part of the characterization studies, two seismic stations were installed on the eastern side of the landfill (Figure 1) to obtain information about seismic energy propagation through the landfill. The need for this information resulted in part from the inability to properly determine either the magnitude of shear modulus and material damping of the landfill material or the variation of these dynamic properties with shearing strain. In the absence of this information, it was difficult to say with confidence whether or not ground motions would amplify or attenuate through the landfill.

The station at the base of the landfill (SS-1) is located on native soil. This material varies from a very hard clayey silt-silty clay to siltstone and claystone. Station SS-2 is located over nearly 250 feet of refuse. The seismic stations consist of Kinemetrics digital accelerographs with three orthogonal force-balance accelerometers. Additional information about the monitoring system is summarized by Hushmand et al. (1990).

Seismic Response Monitoring

The seismic stations were activated in September 1988. Since that date four complete sets of earthquake records have been obtained.

Earthquake Characteristics

Table 1 provides a summary of five earthquakes that have occurred recently in the area, the last four of which were recorded. The Whittier Narrows earthquake occurred just before the seismic stations were installed. The acceleration level on firm ground from the main shock of the Whittier Narrows earthquake was estimated to be as high as 0.47 g based on acceleration measurements recorded at Garvey Reservoir located approximately 1 mile from the epicenter (Figure 1).

Earthquake Accelerogram Data and Analyses

Figures 2 and 3 show time histories recorded at Stations SS-1 and SS-2 for the Brea and East Los Angeles events. Time histories for the Malibu and Pasadena events were presented previously by Hushmand et al. (1990). The data in Figures 2 and 3 show that peak accelerations decreased as seismic energy traveled from SS-1 to SS-2.

Figures 4 and 5 show the Fourier Amplitude Spectra (FAS) for motions recorded at Stations SS-1 and SS-2 during the two most recent events. [The FAS is a convenient method of showing the amount of energy at different frequencies for a given record.] In Figure 6 the FAS at the landfill top is divided by the FAS at the base to define Fourier amplitude ratios, or energy transfer functions. For horizontal modes of vibration, the transfer functions are less

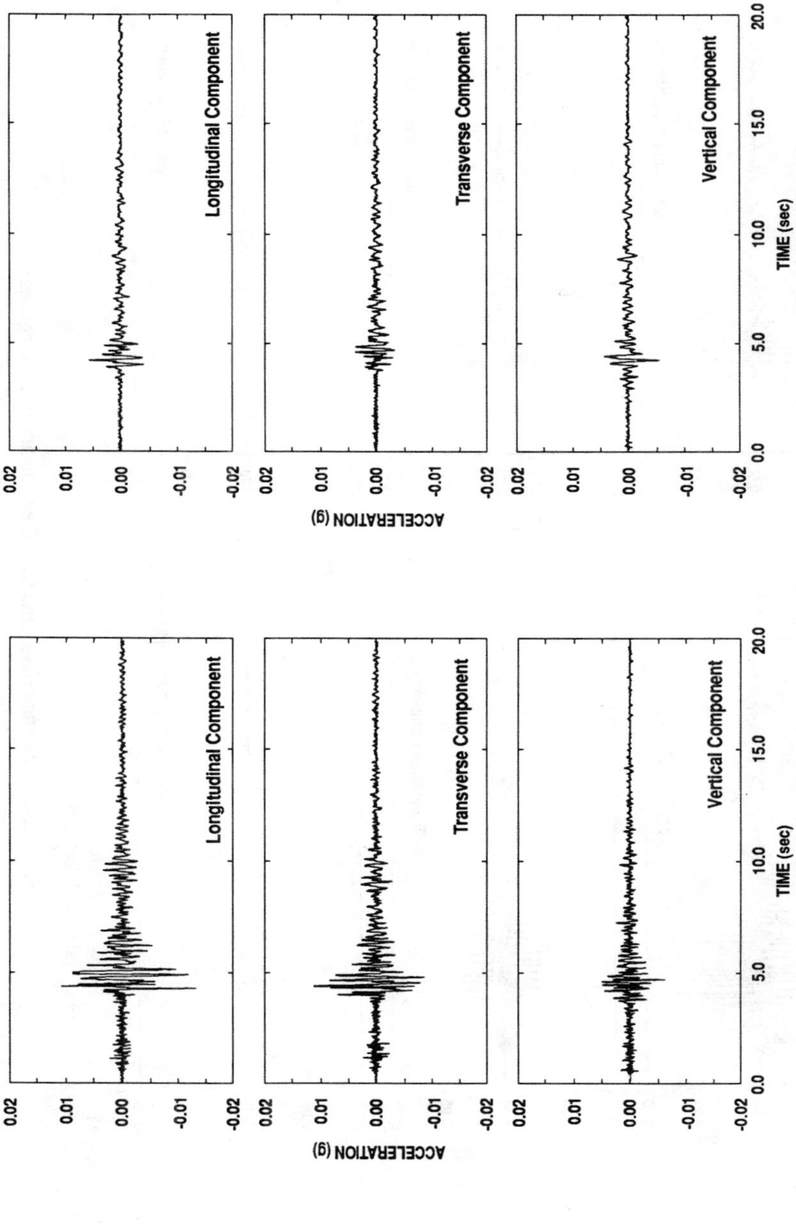

(a) Base Seismic Station

(a) Top Seismic Station

FIGURE 2. Acceleration Records from Brea Earthquake

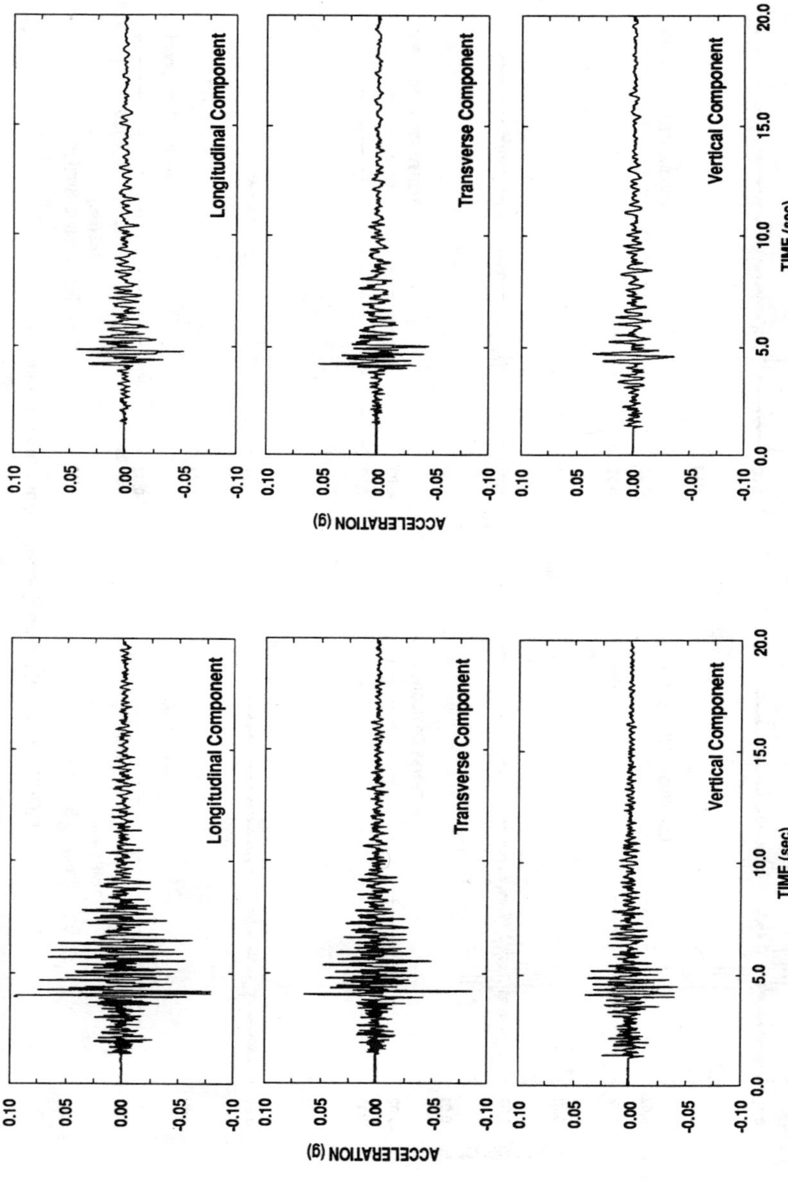

(a) Base Seismic Station

(a) Top Seismic Station

FIGURE 3. Acceleration Records from East Los Angeles Earthquake

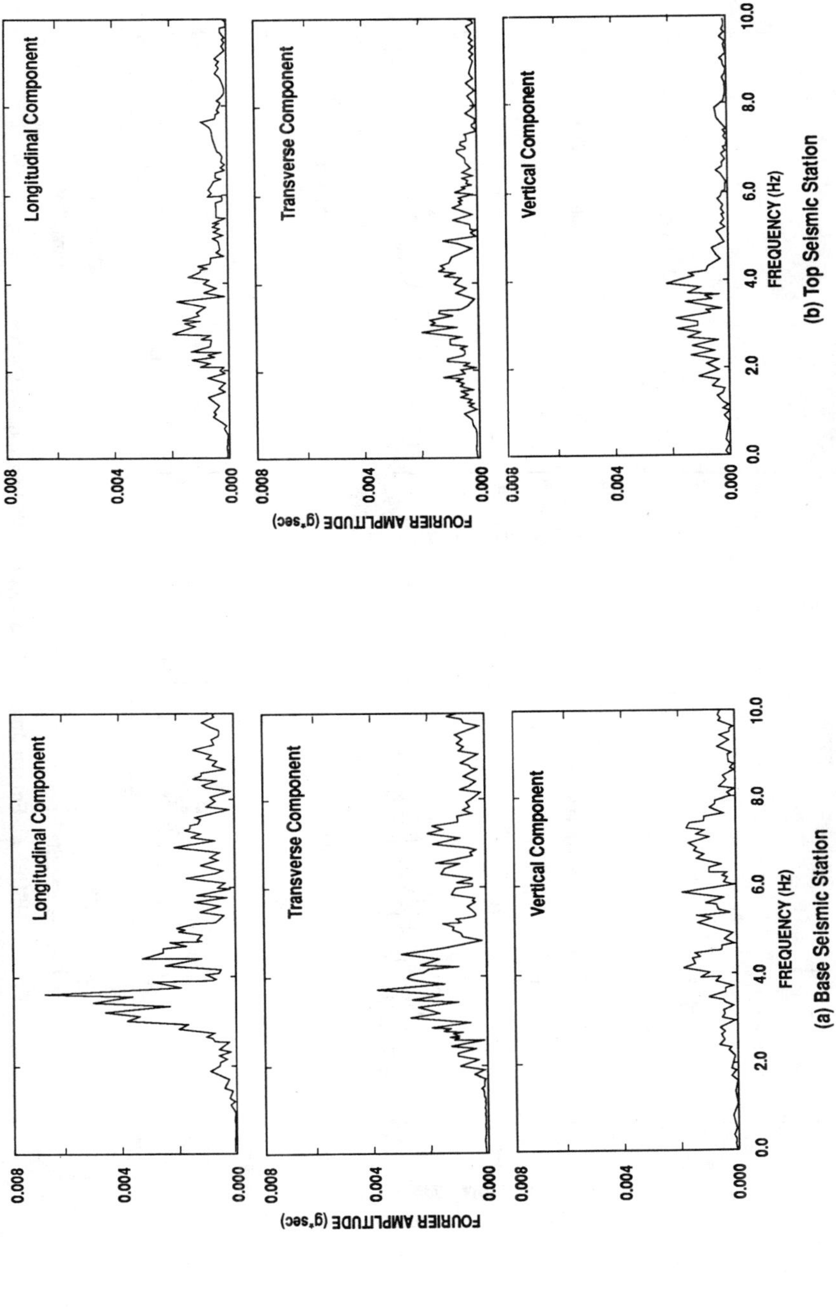

FIGURE 4 . Fourier Amplitude Spectra for Brea Earthquake

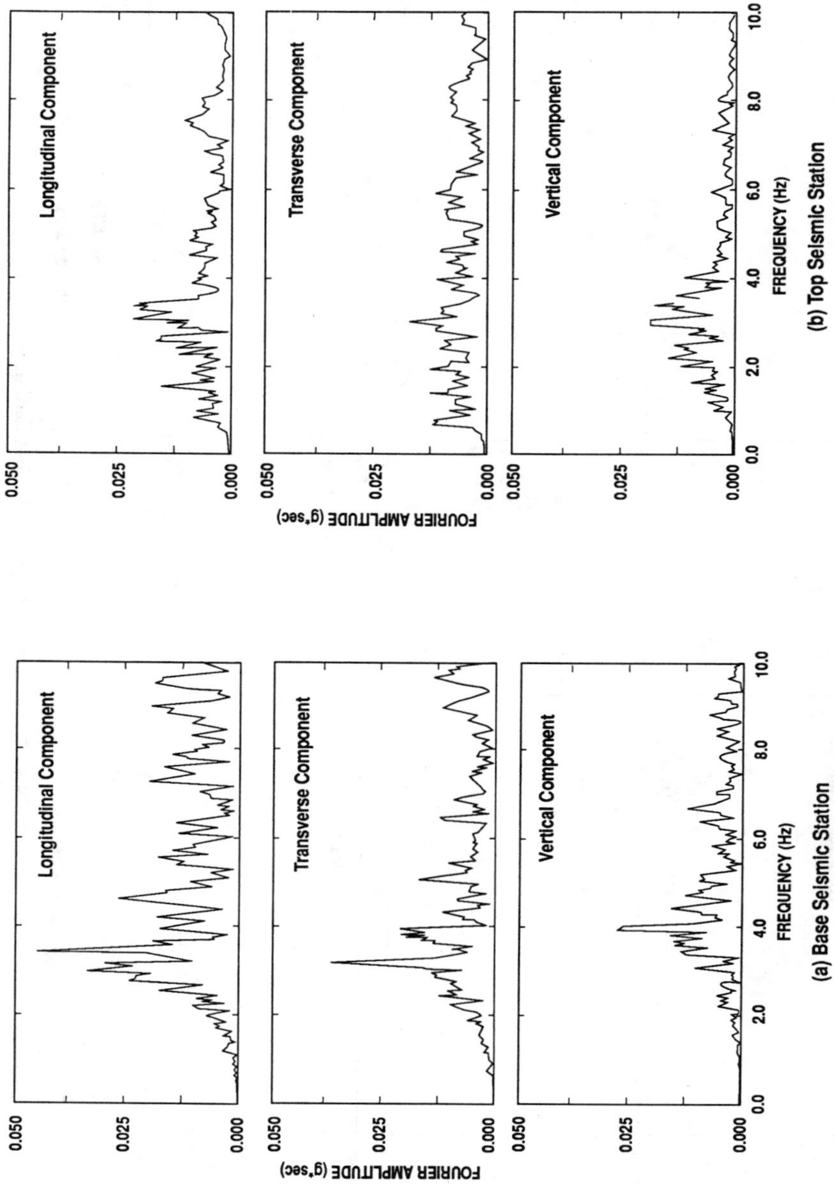

FIGURE 5 . Fourier Amplitude Spectra for East Los Angeles Earthquake

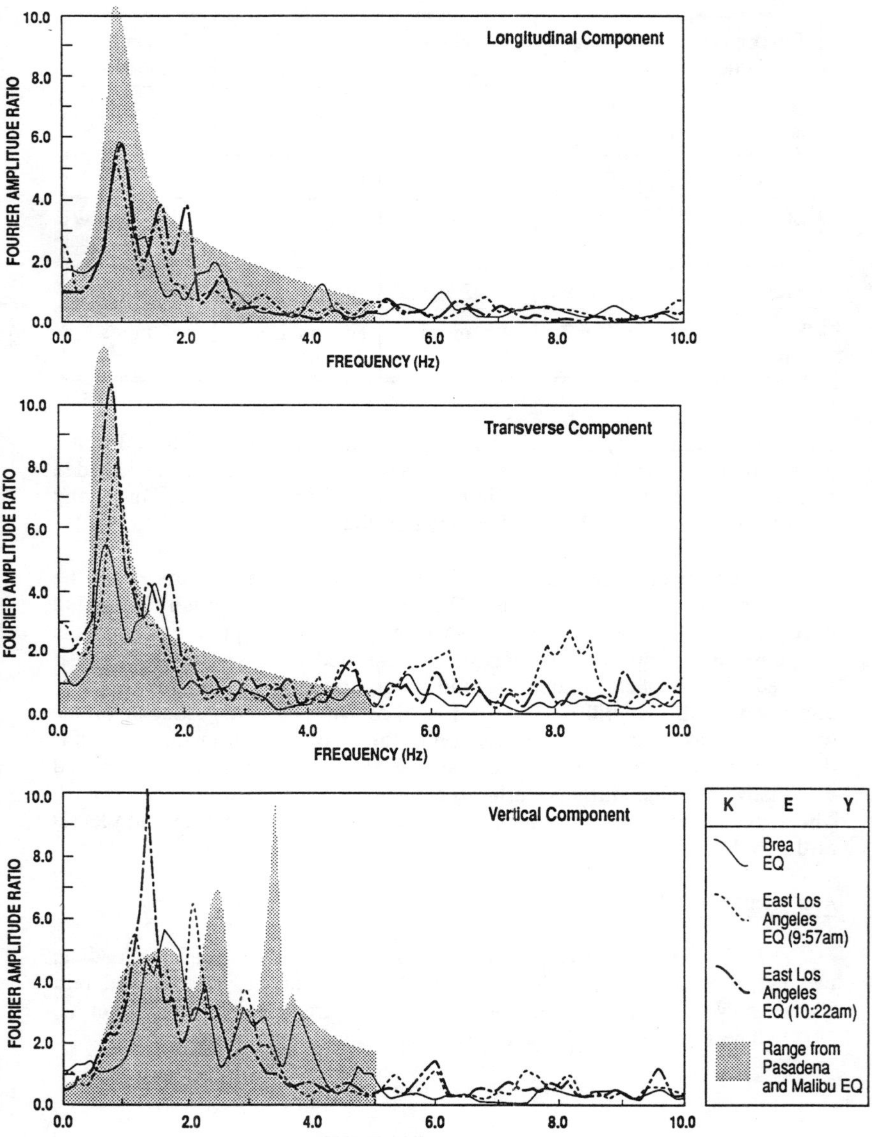

FIGURE 6. Transfer Function between Base and Top of Landfill

Table 1: Earthquake Characteristics

Earthquake Name	Event Date	Richter Magnitude	Epicenter Location	Epicenter Distance
Whittier Narrows	10/01/87	5.9	See Figure 1	2.3 miles
Pasadena	12/03/88	5.0	Pasadena	9 miles
Malibu	01/18/89	5.0	South of Malibu	31 miles
Brea	05/06/89	3.7	Brea	15 miles
East Los Angeles	06/12/89	4.5 and 4.1	East of Los Angeles	5 miles

than 1.0 (attenuation) above about 2 Hz, while the transfer functions are greater than 1.0 (amplification) at frequencies less than 1 to 2 Hz. The shaded zone in Figure 6 shows the range in transfer functions reported by Hushmand et al. (1990) for the Malibu and Pasadena earthquakes.

The fact that the transfer functions for the four events appear to be similar has significant implications. The peak accelerations within the landfill during the Malibu and Brea events were very low, e.g., approximately 0.01 g. Response under this level of excitation is expected to be linear, i.e., little, if any, reduction in shear modulus should occur. In contrast, the peak accelerations within the landfill during the East Los Angeles and Pasadena events were as high as 0.2 g. Since the location of the peak of the transfer function is determined by the stiffness of the material, the landfill apparently exhibited essentially the same stiffness during the small- and intermediate-sized events. This suggests that landfill behavior was reasonably linear to acceleration levels of 0.1 to 0.2 g.

Analytical Modeling

To better understand how the landfill behaved during seismic loading, two-dimensional, equivalent-linear finite element analyses were carried out. These analyses involved use of the computer program QUAD4 (Idriss et al., 1973). The Pasadena earthquake record served as the input motion.

For the finite element analyses, the landfill material was characterized by a shear wave velocity (V_s) of 800 feet per second (fps). This velocity is similar to values determined by Woodward Clyde (1987) during seismic refraction surveys at the site. The variation in shear modulus with shearing strain was represented by the curve developed by Seed and Idriss (1970) for peat; damping was assumed to be similar to that for clay. Linear QUAD4 analyses with V_s = 800 fps were also performed to evaluate response for the case where no modulus reduction occurs.

Figures 7 and 8 show comparisons between predicted and recorded ground motions, FAS, and transfer functions for the transverse component of the Pasadena record. Of the two predictions (linear and equivalent linear), the linear prediction appears to be much more similar to recorded response. This observation is consistent with previous comments regarding the similarity in transfer functions between the small- and intermediate-sized seismic events. It also suggests that the modulus curve for peat, which is commonly used for representing landfill materials, is inappropriate for the OII landfill.

As a check on the QUAD 4 predictions, an estimate of the fundamental frequency of vibration was made using the simple, one-dimensional relationship $f_n = V_s/4H$ where V_s is the shear wave velocity and H is the landfill thickness. This check gave a natural frequency of approximately 1 Hz when V_s was equal to the low-strain amplitude shear wave velocity (i.e., 800 fps). This fundamental frequency is similar to the locations of the transfer function peaks in Figure 8. Although one would expect the actual fundamental frequency to be slightly greater than the value determined on the basis of the one-dimensional relationship because of two- and three-dimensional effects, the similarity between the fundamental frequency from the one-dimensional equation and the locations of the peaks in the transfer functions in Figures 6 and 8, again suggest that the landfill was responding nearly elastically.

Slope Stability During Seismic Loading

The potential for slope movement during seismic loading was estimated using the Newmark method (Newmark, 1965). It was recognized that the Newmark method is a simplified representation of the likely failure mechanism during an earthquake. However, it was also reasoned that more rigorous analyses would require material properties that would be very difficult to establish.

Predicted Deformation

Studies by Woodward Clyde (1987) determined that the accelerations at the base of the landfill on firm ground could range from 0.25 g for a design

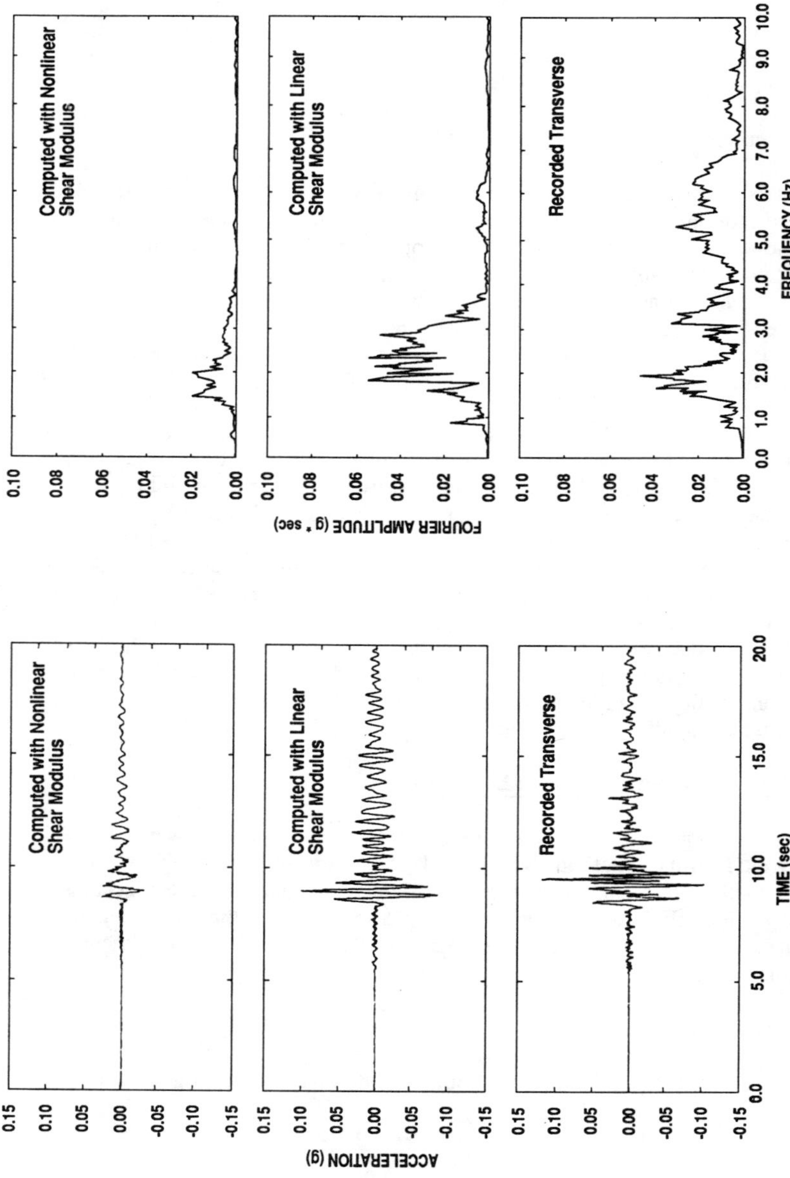

(a) Base Seismic Station

(a) Top Seismic Station

FIGURE 7. Comparison of Computed and Recorded Transverse Acceleration and FAS at SS-2

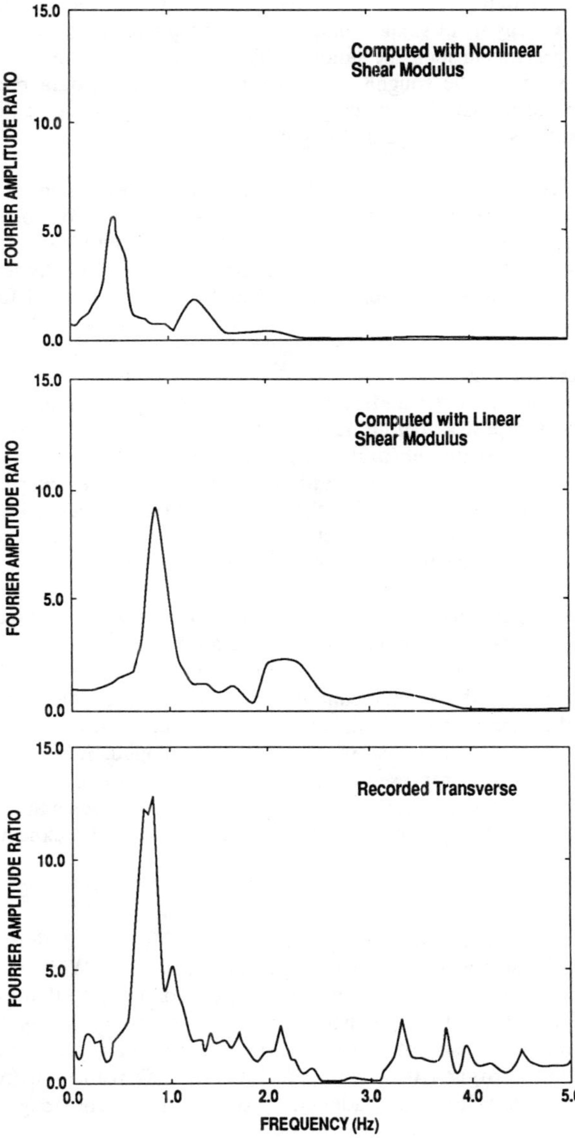

FIGURE 8 . Comparison of Computed and Recorded Transverse Transfer Functions

event on the San Andreas fault to 0.5 g for a design event on the Whittier-Elsinore fault system. Results of seismic monitoring and the QUAD4 study indicate that energy from smaller-magnitude earthquakes (e.g., M<5) will attenuate as it passes through the landfill. The amount of attenuation in peak acceleration seems to be roughly 50 percent, based on the data recorded to date. This suggests that for a peak ground acceleration of 0.5 g at the landfill base, accelerations within the slope might be 0.25 to 0.3 g.

The Newmark analysis required that the yield acceleration of the slope be determined before deformations could be estimated. For lower bound strength property assumptions (ϕ = 30°, c = 500 psf), the yield acceleration was less than 0.2 g. Since the yield acceleration was only slightly lower than the predicted ground acceleration, only nominal deformations (<1 foot) were predicted.

Observed Deformations

Slope inclinometer and settlement monitoring instruments were installed on the steepest slopes in 1988 (CH2M HILL, 1989). These systems have been monitored for the past three years. Results from the monitoring program have revealed continuous settlement and downslope creep. The annualized downslope creep has approached 0.5 foot, while annualized settlement has been as high as 1.5 feet (Siegel et al., 1990). No dramatic changes in settlement or downslope creep were recorded after any of the earthquakes.

Visual inspections of the landfill were made after the larger earthquakes. Some cracks were noted after the Pasadena earthquake; the sizes of these cracks were typically less than 1/2 to 1 inch in width and less than 150 feet in length. More significant cracking occurred during the Whittier Narrows earthquake. Cracks from this event were up to several inches wide and 300 feet long. In locations where cracks developed, it was difficult to determine if the cracks represented the head of scarps or differential settlement. These observations seem to be consistent with the Newmark deformation estimates; i.e., little if any downslope movement would be expected.

Discussion

Results of these analyses suggest that the OII landfill slopes should remain relatively stable during a large seismic event. The Whittier Narrows earthquake likely produced a peak ground acceleration close to that associated with the design event of 0.5 g. Inasmuch as the time history and FAS for this event are generally consistent with what has occurred during the monitoring period, it could be argued that the landfill has already withstood the design event and therefore will respond adequately during any future design event.

However, there are several uncertainties that could lead to higher than predicted movements. The first uncertainty is related to the tendency of the landfill to amplify energy in the 1-Hz range. This tendency suggests that large, distant seismic events could be more serious than close events. Another source of uncertainty is the strength characteristics of the landfill material under static and seismic loading. Larger deformations could occur if strength decreases with time due to material decomposition. The final uncertainty is associated with the actual slope failure mechanism and how it will develop with time. As noted previously, the Newmark method provides a very simple representation of movement. Under larger, longer duration ground motions, more rigorous modeling methods may be necessary to obtain reasonable estimates of deformation.

These uncertainties will be difficult to resolve by ordinary exploration, laboratory testing, and analytical methods. The most prudent approach seems to be to continue monitoring slope deformations, settlements, and seismic response. By collecting an adequate data base for landfill behavior, confidence in the long-term performance will increase.

Conclusions

The seismic monitoring program at the OII landfill has provided valuable information regarding landfill performance during seismic events. Based on the seismic records obtained to date, the following conclusions can be made:

- Peak accelerations from small- to intermediate-sized earthquakes attenuate through the OII landfill. Results of transfer function calculations suggest that this is due to attenuation of energy above 2 Hz.

- Finite element analyses can provide a reasonable estimate of the transfer function if a linear soil model is used. At higher acceleration levels, nonlinear behavior is expected. The response at higher strains is expected to resemble more conventional soil modulus-shearing strain properties.

- Newmark deformation predictions seem to be compatible with observed slope response during seismic events. The greatest uncertainty in the deformation prediction is the estimate of strength properties.

In closing, seismic monitoring programs such as the one initiated at OII provide valuable information regarding the response of landfill materials during seismic events. Similar monitoring programs must be encouraged.

Acknowledgments

The work discussed in this paper was conducted by CH2M HILL for the United States Environmental Protection Agency (EPA) as part of the Remedial Investigation/Feasibility Study. The Earth Technology Corporation of Long Beach, California, installed the seismic monitoring station and carried out the QUAD4 analyses. Seismic monitoring and analysis are currently being performed by Hushmand and Associates under subcontract to Sciences Application International Corporation (SAIC), who are performing work under contract to EPA. Assistance of Dr. John Zellmer of SAIC in providing data from the last two seismic events is gratefully acknowledged. This paper does not necessarily reflect the EPA's view, and no official endorsement should be inferred.

Appendix I. References

CH2M HILL (1989). "Technical Memorandum, Slope Stability and Seismic Response Studies, Operating Industries, Inc., Landfill," prepared for the Environmental Protection Agency, March.

Hushmand, B., Anderson, D.G., Crouse, C.B., and Robertson, R.J. (1990). "Seismic Monitoring and Evaluation of a Solid Waste Landfill," *Proceedings, 4th U.S. National Conference on Earthquake Engineering*, Palm Springs, May.

Idriss, I. M., Lysmer, J., Hwang, R.N., and Seed, H.B. (1973). "QUAD4, A Computer Program for Evaluating the Seismic Response of Soil Structures by Variable Damping Finite Element Procedures." Report No. EERC 73-16, Earthquake Engineering Research Center, University of California, Berkeley.

Newmark, N. M. (1965). "Effects of Earthquakes on Dams and Embankments." *Geotechnique*, Vol. 15, No. 2, pp. 139-160.

Seed, H.B., and I.M. Idriss (1970). "Soil Moduli and Damping Factors for Dynamic Response Analyses," Report No. EERC 70-10, Earthquake Engineering Research Center, University of California, Berkeley.

Siegel, R.A., R.J. Robertson, and D.G. Anderson (1990). "Slope Stability Investigation at a Landfill in Southern California," *Geotechnic of Waste Fills—Theory and Practice*, ASTM STP 1070, pp. 259-284.

Woodward Clyde Consultants (1987). "Seismic Stability Evaluations, Operating Industries, Inc. Landfill, Monterey Park, California," prepared for Camp Dresser McKee, May.

Appendix II. SI Equivalents

1 inch	=	25.4 mm
1 foot	=	0.305 m
1 mile	=	1.61 km
1 fps	=	0.305 mps
1 psf	=	0.048 kPa

POST-EARTHQUAKE SLOPE STABILITY OF TWO DAMS WITH LIQUEFIED GRAVEL FOUNDATIONS

D. W. Sykora,[1] M.ASCE, J. P. Koester,[1] M.ASCE, R. E. Wahl,[1] M.ASCE, and M. E. Hynes,[2] M.ASCE

ABSTRACT

This paper presents a brief summary of the procedures used to conduct the post-earthquake stability analyses and specifics of the analyses and results for two federally-owned dams in the western United States. The seismic stability evaluations for these two dams have been completed by the U.S. Army Corps of Engineers. The liquefaction potential of gravel in foundation and embankment zones was evaluated using the Becker Hammer penetration test. The results of liquefaction analyses at these two locations indicate that if the maximum credible earthquake (MCE) is experienced, excess pore water pressures are expected to develop within the embankment and foundation soils and extensive zones of gravel in the foundations of both these dams are expected to liquefy. Consequent analysis of post-earthquake (slope) stability using limit equilibrium methods was then performed. One dam with a MCE of moderate magnitude was found to be unstable. The other dam is expected to remain despite having a much larger MCE. The comparison of the results for these two projects supports the need for thorough exploration and analyses when evaluating the seismic stability of an earth dam and has implications for future design. Other general conclusions are also presented.

---

1 - Research Civil Engineer, and
2 - Supervisory Civil Engineer, Geotechnical Laboratory, USAE Waterways Experiment Station, Vicksburg, MS.

INTRODUCTION

A seismic stability evaluation of a dam by the COE involves the consideration of a number of potential modes of deleterious response and instability. One potential mode of instability is the alteration of material properties and the generation of excess pore water pressure resulting from the cyclic straining produced by an earthquake. The stability of the slopes of the dam and foundation system are analyzed with these predicted changes in strength and pore water pressure increase using traditional equilibrium methods of analysis with trial circles and wedges. The overall procedure is called a post-earthquake stability analysis. More sophisticated methods of deformation analysis are used to design remedial treatments and to estimate the extent of deformation for cases that are not well characterized by rigid-body sliding assumptions. These methods of analysis provide the basis for engineers to judge whether a dam is safe.

Brief summaries of the post-earthquake stability evaluation of two dams are presented in this paper. The two dams are located in the western United States and are presently owned and operated by the U.S. Department of the Interior, Bureau of Reclamation. The COE performed the seismic stability evaluation of these two dams since it was responsible for the original design and construction.

The first dam considered is Mormon Island Auxiliary Dam (MIAD), part of the Folsom Dam and Reservoir Project located near Sacramento, California, which was constructed by COE, Sacramento District, in the 1950's. Mormon Island Auxiliary Dam has been determined to be unstable against sliding following a MCE with a moderate magnitude and remedial actions are in progress. The other dam is Ririe Dam located near Idaho Falls, Idaho. This dam was constructed by COE, Walla Walla District, in the 1970's. Ririe Dam has been determined to be stable against sliding for post-earthquake conditions following a large, near-field design earthquake event.

METHODOLOGY

General. The method used by the U.S. Army Engineer Waterways Experiment Station (WES) to calculate factors of safety against sliding for post-earthquake conditions, particularly for dams with gravel zones or foundations, has been presented in stages by Hynes-Griffin (1987), Hynes (1988), Marcuson, Hynes, and Franklin (1990), and Marcuson, Hynes, and Franklin (1992) and in reports for the two subject studies. This analysis is subsequent to dynamic response calculations and the comparison of earthquake-induced shear stresses with cyclic strengths.

The Becker Penetration Test (BPT) is an integral tool for site characterization, liquefaction analysis, and post-earthquake stability studies at sites with gravel (Harder and Seed 1986). The BPT is performed using a Becker Hammer drill to drive a modified steel pipe pile with a diesel pile hammer. The steel pipe pile is double-walled with an OD of 6.6 in. (16.8 cm) and may be open or closed at the tip. Open tip soundings are used to obtain samples whereas the closed tip soundings are more reliable for correlation with SPT blowcounts (Harder and Seed 1986).

In order to use the present methods of analysis, BPT blowcounts must be converted into equivalent Standard Penetration Test (SPT) blowcounts, corrected to an effective stress of 1 tsf (9.0 kPa), and adjusted to correspond to values that would be produced by a hammer system capable of imparting 60 percent of theoretical maximum free-fall energy to the SPT split spoon sampler, $(N_1)_{60}$. Next, the equivalent corrected blowcounts are used to estimate cyclic shear strength by the Seed performance-based procedure (Seed et al. 1985). Appropriate corrections to cyclic strength are made for earthquake magnitude (K_M), confining stress (K_σ), initial shear stress (K_α), and fines content of the material (Marcuson, Hynes, and Franklin 1990).

These cyclic strengths are then compared to the cyclic shear stresses imposed by the earthquake. If the ratio of shear strength to shear stresses approaches unity for a particular region, post-liquefaction residual strengths are assigned. Residual strengths corresponding to liquefied materials are estimated using the empirical relationship proposed by Seed and Harder (1990) as shown in Figure 1. The relationship between residual excess pore water pressures, r_u, and factor of safety against liquefaction, FS_L, are derived from results of laboratory testing of representative samples (Hynes 1988).

If the materials are expected to be non-liquefiable, adjusted Mohr-Coulomb effective strength parameters are used along with effective stresses reduced by excess pore water pressures. Average values of r_u are used to adjust static strength parameters to account for the distribution of excess pore water pressures. The effect of these earthquake-induced excess pore pressures is to cause a reduction in available frictional strength (by means of a reduced effective angle of internal friction, ϕ') for input to slope stability studies. An artificially reduced ϕ' was calculated for non-liquefied materials using the following expression (after Hynes-Griffin, 1987):

$$\sin \phi'_{r_{u \neq 0}} = (1-r_u) \sin \phi'_{r_{u=0}} \qquad (1)$$

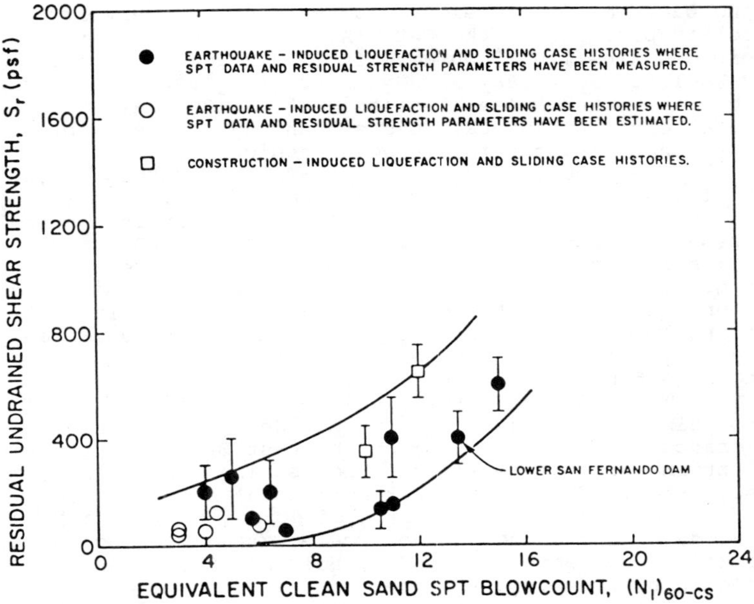

Figure 1. Empirical relation used to derive residual strengths (Seed and Harder 1990)

Limit Equilibrium calculations. The slope stability studies described in this paper were performed using the computer program UTEXAS2 (developed by Prof. S. G. Wright at the University of Texas at Austin) adapted for microcomputer use as documented by Edris and Wright (1987). This program may be used to determine factors of safety against sliding along circular or non-circular surfaces using one or more computational schemes. The limit equilibrium technique by Spencer (1967) was used for the two studies described in this paper. Other comparable programs may also be used.

Factors of safety against sliding for post-earthquake conditions, FS_{PES}, are calculated for effective stress conditions using limit equilibrium methods. If the factors of safety are greater than unity, the dam or embankment is considered to be safe for this aspect of seismic stability. Additional studies are then performed to estimate displacements, such as Newmark or non-linear finite element approaches.

MORMON ISLAND AUXILIARY DAM, CALIFORNIA

General. The Folsom Dam and Reservoir Project consists of a Concrete Gravity Dam (CGD), Right and Left Wing Dams, MIAD, and eight saddle dikes. Reservoir control is provided by the CGD. At gross pool, the reservoir contains 1,010,000 acre-ft (1.2×10^9 m^3). Reports by Wahl et al. (1988) and Hynes et al. (1990) document the Phase I and Phase II seismic stability evaluations and a report by Sykora, Koester, and Hynes (1990) presents considerations for remedial design.

The MIAD is a zoned, earth-fill embankment, 4,820 ft (1,470 m) long with a maximum height of 165 ft (50 m) from the base of the core trench. The impervious core extends to amphibolite bedrock. For about 900 ft (270 m) of its length, the shells of the embankment are founded on gravelly dredged tailings remnant from gold mining at around the turn of the century. For about 700 ft (215 m) of its length, the shells of the dam rest on (undredged) gravelly alluvium. The remainder of the dam (about 3,170 ft, or 965 m) rests on bedrock.

The MCE used for the seismic stability study of the Folsom Dam Project is a magnitude 6.5 event at a distance of 15 km, located on the East Branch of the Bear Mountains Fault Zone (Tierra Engineering Consultants, Inc. 1983). The design peak horizontal acceleration, a_{max}, is 0.35 g with a duration of strong motion ($a \geq 0.05$ g) of 16 seconds. Two accelerograms meeting these design values were provided by Bolt and Seed (1983) and used in the response analyses.

The site was extensively investigated prior to and during construction with test shafts, borings, and core excavation. During this study, additional investigations included surface and crosshole geophysical tests, test pit excavations to measure in-situ densities, in-situ permeability measurements, and several series of BPT measurements. The distribution of average $(N_1)_{60}$'s estimated from all BPT sounding programs conducted for Phases I and II and remedial action studies are shown in Figure 2.

Results of liquefaction assessment. The gravelly dredged tailings portion of the foundation beneath MIAD was determined to be susceptible to liquefaction when subjected to the MCE in the Phase I seismic safety evaluation. It was concluded from a Phase II study that the portions of the dam founded on rock or on undredged alluvium would be stable against design earthquake shaking. Additional analyses were performed to re-evaluate post-earthquake slope stability using additional BPT measurements obtained after completion of the Phase I study. The updated stability analysis results were intended to assist in the development of remedial

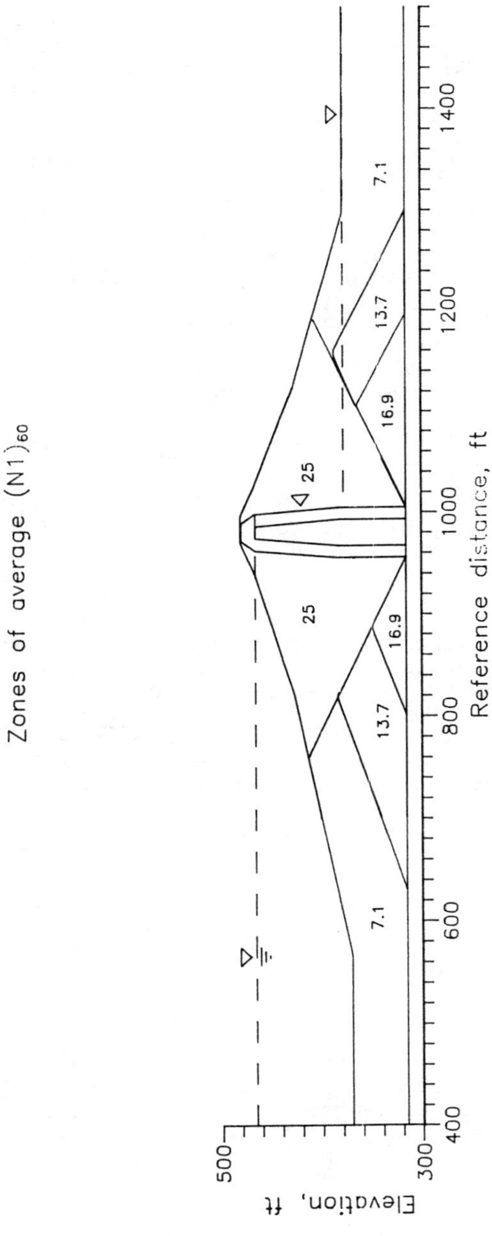

Figure 2. Idealized distribution of $(N_1)_{60}$ blowcounts in cross section at MIAD

measures against liquefaction within and beneath those segments of MIAD founded on dredged tailings.

Evaluation of post-earthquake stability. Material properties assigned to the cross section of MIAD developed for post-earthquake stability evaluation are shown in Figure 3. Two slip surfaces are shown in Figure 4 through the upstream slope for post-earthquake conditions of MIAD for loading conditions relevant to the design pool level (i.e., gross pool at el 466) and MCE. Surfaces passing through materials above and upstream of the surface labeled F_{PES} = 1.05 in Figure 4 were computed to have factors of safety against sliding of less than one. The minimum overall upstream safety factor was computed to be 0.24 for the smaller surface shown in Figure 4. Factors of safety against post-earthquake sliding increase rapidly for surfaces passing through materials deeper than and downstream of the surface with FS_{PES} = 1.05. This latter surface is similar to one for which FS_{PES} was computed to be 0.96 in the Phase I study.

Post-earthquake sliding surfaces were evaluated for reservoir levels lower than the design pool. Material strengths were increased as a result of the increased effective overburden pressure in portions of the slope stability section that were submerged by the design pool. Critical slip surfaces (i.e., those having $F_{PES} \approx 1.0$) were found to be very similar to those determined for design pool conditions. The increase in strengths of affected materials were apparently offset by additional driving force caused by embankment gravel not submerged at lower pools.

Subsequent sensitivity analyses revealed that slip surfaces deep enough to threaten the core of MIAD would have $F_{PES} \leq 1.0$ if only the free-field gravel liquefied. Figure 5 depicts a critical slip surface (F_{PES} = 1.0) for these conditions, as well as the slip surface deemed critical for the design earthquake case (F_{PES} = 1.73 for surface that intersects the core below the pool level). The studies indicated that the material zone having $(N_1)_{60}$ = 13.7 must become unstable before deeper slip surfaces would develop. (This is consistent with the results of more sophisticated deformation analyses in progress at the time of this writing; these studies also indicated that remedial measures are needed beneath the slopes of the dam, not just in the free field areas upstream and downstream.)

RIRIE DAM, IDAHO

General. Ririe Dam was constructed as a federal, multipurpose, water resource development project. It consists of a zoned earth- and rock-fill embankment with a central impervious core, a gated outlet works, and an emergency spillway. At maximum pool, the reservoir

<u>Zoning for UTEXAS2:</u>

1	Gravel shell, moist, γ_m = 146 pcf, ϕ' = 43°, c = 0
2	Transition SM, moist, γ_m = 136 pcf, ϕ' = 37°, c = 0
3	Dredge tailings, moist, γ_m = 136 pcf, ϕ' = 41°, c = 0
4, 5, 6, 7	Gravel shell, γ_t = 152 pcf, ϕ' = 30°, c = 0*
8, 9	Dredge tailings, saturated, γ_t = 136 pcf, ϕ' = 19°, c = 0*
10, 17	Dredge tailings, liquefied, S_r = 100 psf, γ_t = 136 pcf
11, 16	Dredge tailings, liquefied, S_r = 400 psf, γ_t = 136 pcf
12, 15	Dredge tailings, liquefied, S_r = 650 psf, γ_t = 136 pcf
13, 14	Gravel shell, liquefied, S_r = 1200 psf, γ_t = 152 pcf
18	Transition SM, saturated γ_t = 142 pcf, ϕ' = 37°, c = 0
19	Clayey core, γ_t = 135 pcf, ϕ' = 29°, c = 0
20	Bedrock

*Note: ϕ' in these materials was reduced to simulate strength reduction caused by earthquake-induced excess pore water pressure, according to $(\sin\phi')_{R_u \neq 0} = (1-R_u)(\sin\phi')_{R_u=0}$ (see text)

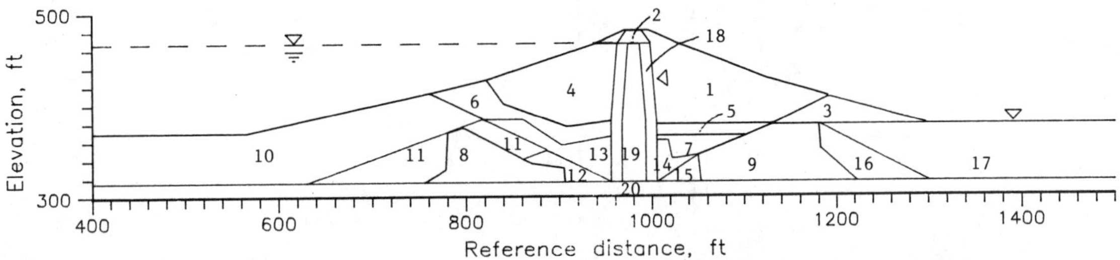

Figure 3. Material properties used for post-earthquake stability analysis of MIAD

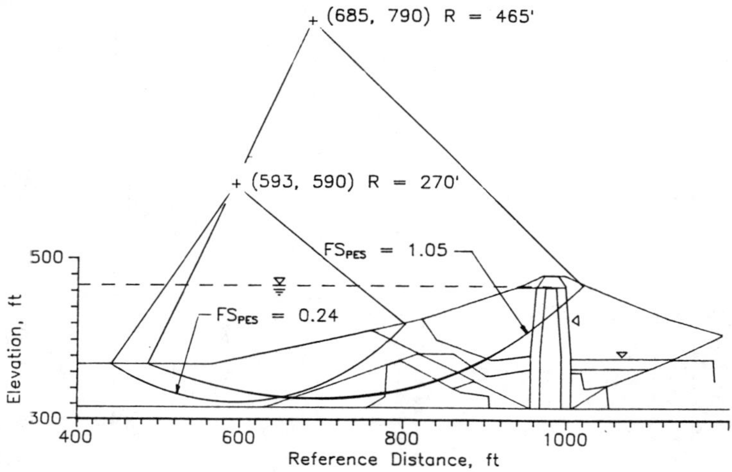

Figure 4. Potential post-earthquake slip surfaces through the upstream slope of MIAD for loading conditions relevant to the design pool level

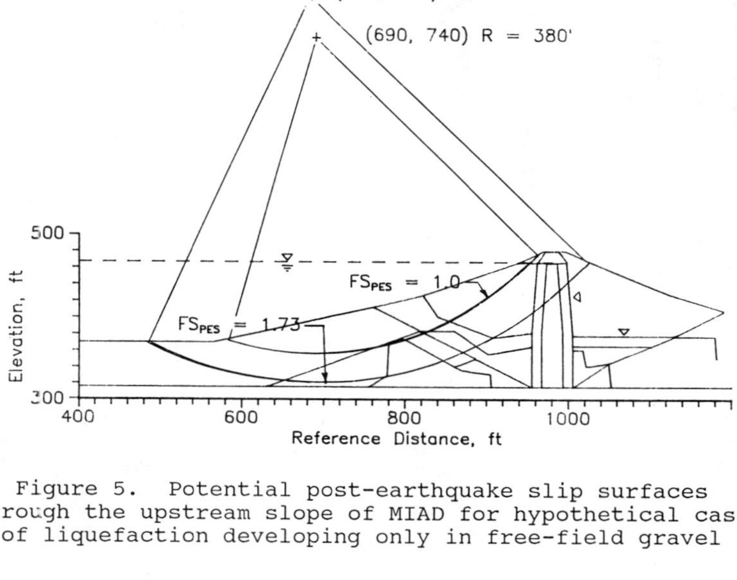

Figure 5. Potential post-earthquake slip surfaces through the upstream slope of MIAD for hypothetical case of liquefaction developing only in free-field gravel

contains 100,000 acre-ft (1.2×10^8 m^3) and there is a freeboard of 9 ft (2.7 m). Reports by Sykora et al. (1992) and Sykora, Koester, and Hynes (1992) document the seismic stability evaluation.

The dam rests at a bend of a steep, narrow canyon and is curved slightly upstream with a crest length of 840 ft (256 m) and a maximum height of 250 ft (76 m), 185 ft (56 m) above the top of the valley alluvium. The core extends through alluvium to basalt bedrock. The shells are founded on gravel alluvium that has a maximum thickness of about 65 ft (20 m). The idealized cross section for analysis of slope stability is shown in Figure 6.

The entire near-field region, specifically the Grand Valley Graben, was interpreted as active and capable of producing earthquakes with $M_s = 7.5$ (Krinitzsky and Dunbar 1991). The nearest segment of the graben is about 8 km from the site. The MCE has a site intensity of MM X, $a_{max} = 1.22$ g, and a duration of strong motion ($a \geq 0.05$ g) of 33 sec at the site. The hypothetical earthquake record representing this event were synthesized by Seed (1987) by scaling the S 19° E component of the Pacoima Dam, California, record during the 1971 San Fernando Earthquake and a portion of the N 21° E component of the Taft record from the 1952 Kern County Earthquake.

A suite of geotechnical and seismic geophysical exploration techniques were used to complement measurements made during construction and readings from existing instrumentation to characterize the materials and subsurface conditions at Ririe Dam. Explorations for the seismic stability study included rotary drilling, Becker Hammer drilling, cone penetration testing with pore pressure and downhole seismic measurements, standard crosshole and downhole seismic measurements, and the construction of an eight-foot-diameter (2.4 m) test shaft for sampling.

The cyclic strengths expressed as stress ratios used for the liquefaction analysis are summarized in Table 1 for $\sigma_o' = 1$ tsf (9.0 kPa), and level ground conditions ($K_\alpha = 1.0$). Values of $((N_1)_{60})_{avg}$ were obtained using the results of BPTs. Laboratory studies were performed to determine the pore water generation characteristics of the various materials, including cyclic triaxial undrained shear tests on samples of gravelly alluvium.

Results of liquefaction assessment. The results of a performance-based liquefaction evaluation using the results of a two-dimensional dynamic response analysis including three-dimensional valley effects indicated that submerged random fill and large portions of the upper alluvium will liquefy as a consequence of the MCE. The actual extent of zones of alluvium that will liquefy are confined to upper layers of loose gravel and medium dense gravel that were idealized to exist in the upper portion of alluvium as shown in Figure 7. The results of the field investigations indicated that such relatively loose material is present, but it is unlikely that continuous zones with large lateral extent actually exist at the site. Consequently, it was felt that this idealization was conservative. The excess pore water pressures developed in the remaining portions of the alluvium are very high. The rock fill, gravel fill, core, very dense gravel alluvium (at depth), and dry silt alluvium are not expected to develop significant excess pore water pressures.

Table 1. Material strengths used for liquefaction analysis at Ririe Dam

Material type	$((N_1)_{60})_{avg}$	% Fines	$\frac{\tau_{avg}}{\sigma'_o}$
Random fill	25	7	0.31
Dry alluvial silt	11	45-50	NL*
Loose gravel	16	4	0.18
Med. dense gravel	26	4-5	0.31
Very dense gravel	51	7	NL*

* Non-liquefiable

Evaluation of post-earthquake stability. The cross section used for the slope stability analysis corresponds to a smooth curve in plan that generally follows the centerline of the river valley. This cross section was selected because it is believed to best represent the conditions acting on potential failure surfaces. The geometric effect of using a curve to develop the cross section is to lengthen horizontal distances and flatten slopes; vertical distances are unaffected. The differences between the normal cross section (Section AA) and one for a curve (Section CC) may be seen by comparing Figures 6 and 7.

The material strengths used for the slope stability analysis are summarized in Table 2. These strengths were derived from an evaluation of data used in design, measurements made during construction, and the results of in-situ investigations performed for this study. Residual strengths, S_r, were assigned to materials expected to liquefy. The residual strengths of liquefied gravel at

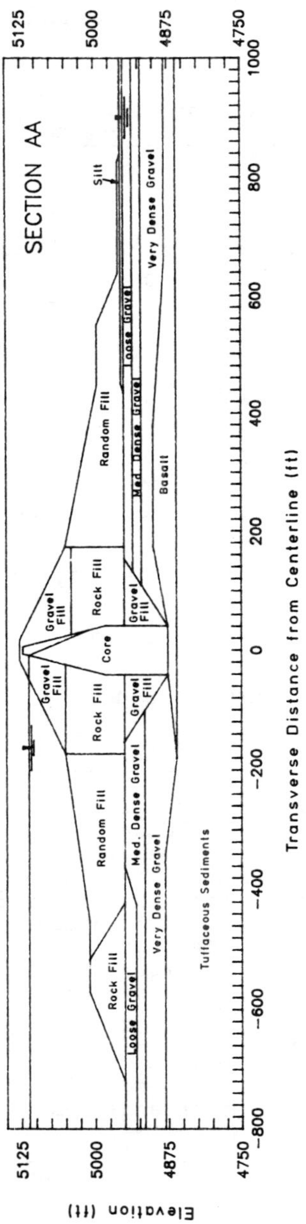

Figure 6. Idealized cross section of Ririe Dam used for evaluation of post-earthquake stability

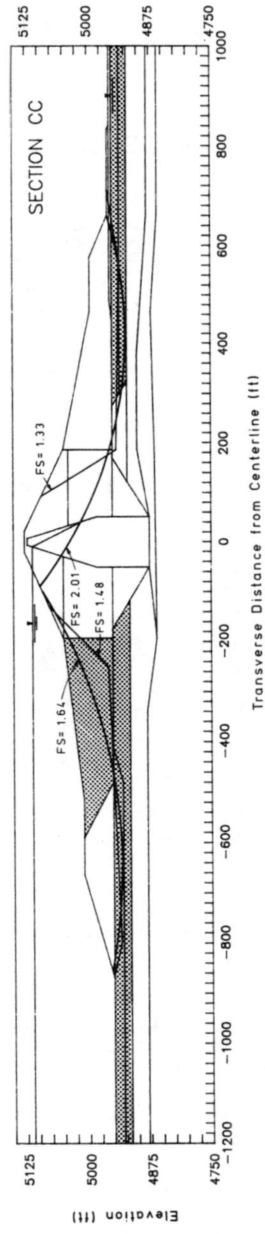

Figure 7. Locations of critical surfaces and factors of safety for Ririe Dam

Ririe Dam are expected to be relatively high because the values of $(N_1)_{60}$ are relatively large (the driving stresses caused by the large design event produced liquefaction of medium dense soils).

Table 2. Material strengths used for limit equilibrium analysis of slope stability at Ririe Dam

Material	Phreatic Condition	Total Unit Wt. (pcf)	ϕ' (deg)	c' (psf)	S_r (psf)
Gravel fill	Unsubmerged	135	48	0	--
	Submerged	145	48	0	--
Silt Core	Unsubmerged	120	35	0	--
	Submerged	125	35	0	--
Rock fill	Unsubmerged	135	48	0	--
	Submerged	145	48	0	--
Random fill	Unsubmerged	130	45	0	--
	Submerged	140	--	--	1,200
Loose gravel	Submerged	135	--	--	500
Med. dense gravel	Submerged	144	48	0	1,200
Very dense gravel	Submerged	147	50	0	--
Tuffaceous sediments	Submerged	155	30	1,500	--
Basalt	Submerged	150	33	5,000	--

The minimum factors of safety against sliding for the upstream slope were 1.64 and 1.48 for the critical circular and non-circular surfaces, respectively. The minimum factors of safety against sliding for the downstream slope were 2.01 and 1.33 for the critical circular and non-circular surfaces, respectively. The locations of these surfaces are shown in Figure 7 along with the zones expected to liquefy. The factor of safety for all four critical surfaces is greater than unity, representing safe conditions. The location of the critical surface on the upstream side suggests that even if this failure surface were to develop, slope failure would not affect the ability of the dam to retain the reservoir. Subsequent Newmark sliding block deformation analyses also contributed to this conclusion.

Sensitivity analyses were conducted and indicate that the results are moderately sensitive to the residual strengths of liquefied materials. The results appear to be insensitive to minor changes in several other parameters. It can be concluded using the performance-based approach that the mode of failure corresponding to loss of freeboard due to slope failure is not a potential threat to Ririe Dam for the MCE.

CONCLUSIONS

Seismic stability evaluation is a difficult undertaking and requires careful analysis of data obtained from a careful site investigation. This paper illustrates that a dam predicted to experience liquefaction in the foundation if subjected to very strong earthquake motions can still be predicted to remain stable. This is a consequence of the locations of liquefiable materials, their thickness and lateral extent, and the sufficiently high residual strengths. Conversely, it was predicted that a dam constructed on poor materials and subjected to moderate ground motions would not remain stable. These projects should be useful examples to other engineers considering the seismic stability performance of gravelly deposits. The designer may be able to consider residual strengths for earthquake resistance without automatically removing potentially liquefiable materials. Deformation analyses may be required, however.

The uncertainty in using residual strengths for seismic analysis and design is considered to be offset by the general conservatism in selecting ground motion parameters and appropriate earthquake records inherent with deterministic approaches. Studies involving probabilistic assessment of seismic hazard should probably use more conservative values of S_r.

The BPT was used successfully at MIAD and Ririe Dam to explore, idealize, and evaluate the liquefaction hazard in gravelly deposits in the foundations of both sites. The true value and accuracy of the BPT may only be realized following the study of more case studies of liquefaction of gravel soils, particularly in and beneath embankments.

ACKNOWLEDGEMENTS

The tests described and the resulting data presented herein, unless otherwise noted, were obtained from research conducted for the US Army Engineer Districts, Sacramento, and Walla Walla by the US Army Engineer Waterways Experiment Station. Permission was granted by the Chief of Engineers to publish this information. The authors would like to acknowledge Messrs. John Nickell, SPK, and Fred Miklancic, NPW, for their support in publishing this paper and providing review comments. The authors also wish to thank Leslie F. Harder, Jr. for providing some additional stability results for the Ririe Dam study.

REFERENCES

Bolt, B. and Seed, H. 1983. "Accelerogram Selection Report for Folsom Dam Project, California," Contract Rpt. DACW05-83-Q-0205, US Army Engr District, Sacramento, CA.

Edris, E., Jr. and Wright, S. 1987. "User's Guide: UTEXAS2 Slope Stability Package," Tech. Rpt. GL-81-1, Vol 1, US Army Engr WES, Vicksburg, MS.

Harder, L., Jr. and Seed, H. 1986. "Determination of Penetration Resistance for Coarse-Grained Soils Using the Becker Penetration Test," EERC Report No. UCB/EERC-86-06, Univ. Calif., Berkeley, CA.

Hynes-Griffin, M. 1987. "Seismic Stability Evaluation of Folsom Dam and Reservoir Project, Report 1: Summary Report" Tech. Rpt. GL-87-14, US Army Engr WES, Vicksburg, MS.

Hynes, M. 1988. "Pore Pressure Generation Characteristics of Gravel Under Undrained Cyclic Loading," PhD Thesis, Univ. Calif., Berkeley, CA.

Hynes, M., Wahl, R., Donaghe, R., and Tsuchida, T. 1990. "Seismic Stability Evaluation of Folsom Dam and Reservoir Project, Report 4 Update: Mormon Island Auxiliary Dam - Phase I" Tech. Rpt. GL-87-14, US Army Engr WES, Vicksburg, MS.

Krinitzsky, E. and Dunbar, J. 1991. "Geological and Seismological Evaluation of Earthquake Hazards at Ririe Dam," Tech. Rpt. GL-91-11, US Army Engr WES, Vicksburg, MS.

Marcuson, W., III, Hynes, M., and Franklin, A. 1992. "Seismic Stability and Permanent Deformation Analyses: The Last Twenty Five Years," Proc., Stability and Performance of Slopes and Embankments II, ASCE, Berkeley, CA.

Marcuson, W., III, Hynes, M., and Franklin, A. 1990. "Evaluation and Use of Residual Strength in Seismic Safety Analysis of Embankments," Earthquake Spectra, Vol. 6, No. 3, pp. 529-572.

Seed, H. 1987. "Hypothetical Ground Motion Accelerogram for Ririe Dam," Ltr. Rpt. to US Army Engr WES.

Seed, H., Tokimatsu, K., Harder, L., Jr., and Chung, R. 1985. "Influence of SPT Procedures in Soil Liquefaction Resistance Evaluation," Journ., Geotech. Engr. Div., ASCE, Vol 111, No. 11, pp 1425-1445.

Seed, H., Wong, R., Idriss, I., and Tokimatsu, K. 1989. "Reevaluation of the Lower San Fernando Dam, Report 2: Examination of the Post-Earthquake Slide of February 9, 1971," Contract Rpt. GL-89-2, US Army Engr WES, Vicksburg, MS.

Seed, R. and Harder, L., Jr. 1990. "SPT-Based Analysis of Cyclic Pore Pressure Generation and Undrained Residual Strength," Proc., H. Bolton Seed Memorial Symp., ed. J. Duncan, Vol 2, pp 351-376.

Spencer, E. 1967. "A Method of Analysis of the Stability of Embankments Assuming Parallel Inter-Slice Forces," Geotechnique, Vol 17, No. 1, pp 11-26.

Sykora, D., Koester, J., and Hynes, M. 1990. "Estimation of Seismic Hazard and Slope Stability Assessment of Mormon Island Auxiliary Dam of Folsom Dam and Reservoir Project," Draft Rpt to US Army District, Sacramento by US Army Engineer WES, Vicksburg, MS.

Sykora, D., Koester, J., Hynes, M., Wahl, R., Yule, D., and Stark, T. 1991. "Seismic Stability of Ririe Dam and Reservoir Project, Report 1: Construction History and Field and Laboratory Studies," Tech. Rpt. GL-91-22, 2 Vols, US Army Engr WES, Vicksburg, MS.

Sykora, D., Koester, J., and Hynes, M. 1991. "Seismic Stability of Ririe Dam and Reservoir Project, Report 2: Stability Calculations, Analysis, and Evaluation," Tech. Rpt. GL-91-22, 2 Vols, US Army Engr WES, Vicksburg, MS.

Tierra Engineering Consultants, Inc. 1983. "Geologic and Seismologic Investigations of the Folsom, California, Area," Rpt. to US Army Engr District, Sacramento, contract DACW05-82-C-0042, Sante Fe, NM.

Wahl, R., Crawforth, S., Hynes, M., Comes, G., and Yule, D. 1988. "Seismic Stability Evaluation of Folsom Dam and Reservoir Project, Report 8: Mormon Island Auxiliary Dam - Phase II," Tech. Rpt. GL-87-14, US Army Engr WES, Vicksburg, MS.

Stability of the Olga C Test Embankment

J.G. Lavallée[1]
G. St-Arnaud[1]
R. Gervais[1] M. ASCE
Y. Hammamji[2] M. ASCE

ABSTRACT

Hydro-Québec is planning to build earth structures on soft, sensitive clay deposits, as required for the NBR hydroelectric project. A committee of specialists issued recommendations on design specifications to ensure the stability of the earth structures. The Olga C test embankment was built accordingly to check the performance of wick drains in the clay deposits. The committee's recommendations for ensuring stability were followed for design and construction of the test embankment. No signs of instability have been observed.

---

1. Geotechnical Engineer, Division Géotechnique, Service Géologie et Structures, Hydro-Québec, 855 Sainte-Catherine Street East, Montreal, Quebec H2L 4P5.

2. Chef de division Géotechnique, Hydro-Québec.

INTRODUCTION

To date, Hydro-Québec has been able to build the vast majority of its hydroelectric projects on bedrock or on ground that is firm or not very compressible. The utility is now planning to build a new hydroelectric complex with installed capacity of more than 8000 MW in a flat region where much of the bearing soil is soft, highly compressible clay. The new complex is to be sited southeast of James Bay and will include the basins of the Nottaway, Broadback and Rupert rivers (NBR Complex). See Figure 1.

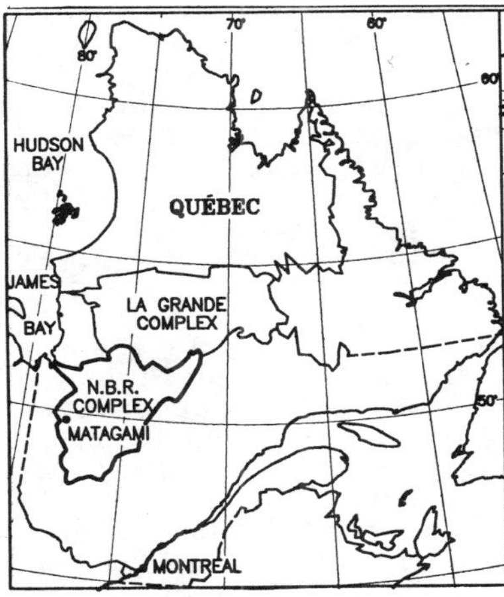

Figure 1 - Location

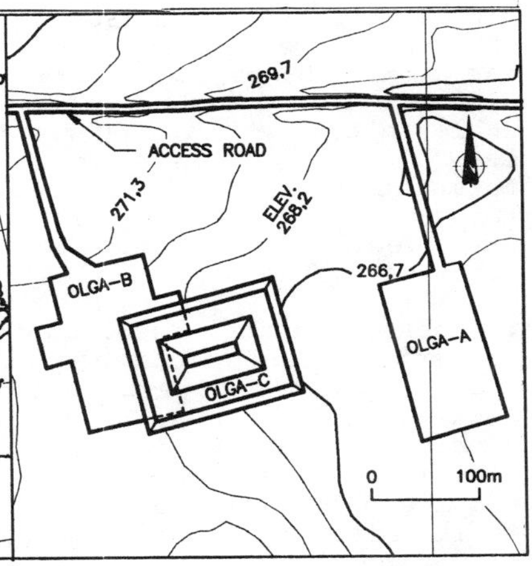

Figure 2 - Olga A, B and C

About 50 km of water retaining structures--most less than 12 m high but some as much as 25 m high--will have to be built on the sensitive clay deposits of the NBR Complex region and there is little precedent for construction of such structures on this clay. The clay is covered with a layer of peat about 3 m thick. The best option for investigating its in situ strength for design purposes is a light, reliable equipment such as a field vane since many of the structure sites are swampy and located far from roads.

To check the reliability of stability analyses based on shear-strength profiles obtained from field vane soundings, Hydro-Québec in the early 1970s constructed two test embankments, building up the structures to failure. One test embankment, Olga A, was in a lacustrine clay zone close to Matagami in the southern part of the NBR

Complex territory; the other, at site R-7, was farther north, close to the Rupert River where the upper part of the soil deposit is marine clay. Failure heights of both test embankments were lower than expected. At Olga A, failure height was about half as high as predicted based on shear-strength profiles derived from field vane tests. The findings of these in-situ failure tests were published by O. Dascal et al. in 1972 and 1973. Two other embankments were built during the period, one at each of the two sites, to determine compressibility of the two types of clay (Dascal et al. 1975).

A Committee of Specialists on the Sensitive Clays of the NBR Complex was formed in the early 1980s by the Société d'énergie de la Baie James (SEBJ), a subsidiary of Hydro-Québec. The committee was tasked with detailed analysis of the clay covering much of the watersheds of the Nottaway, Broadback and Rupert rivers and with formulation of design procedures, according to safety rules of water retaining structures and given the little precedent for constructing embankments on the clay deposits in the region, SEBJ (1983).

To accelerate the consolidation rate of the clay, where necessary, the committee recommended sand drains installed via the "Dutch" jet-bailer method rather than wick drains--because of the lack of documentation on wick drains installed in sensitive, brittle clay such as that found in the NBR Complex region. The committee's recommendation that wick drains be tested was followed in building the Olga C embankment described below. Significant savings are expected with the wick drains used in many projects since the committee issued its report in 1983.

The Olga C test embankment was built in the summer of 1990. It is adjacent to the Olga B embankment built in 1971 to study settlement and is about the same size. The Olga C embankment was designed to study the performance of wick drains installed vertically in the clay foundation (see Figure 2).

Ten months after the 1990 test embankment was built, intact soil was tested in the field and in the laboratory to check for clay-strength gain after much of the final settlement had been achieved through primary consolidation.

This paper discusses potential failure-related behaviour of the Olga C embankment. Performance analysis of the wick drains installed and accuracy of the settlement and strength-gain predictions will be discussed at a later date. The stability of the embankment was analyzed with specific rules set by the committee of specialists, who studied in detail the behaviour of the earlier test embankments.

The committee worked with a substantial amount of data from field and laboratory tests performed during as well as after construction of the embankments. In addition to the Olga and R-7 test embankments, the committee studied the future sites of two large earth dams in the lower basin of the Broadback River, B-2 and B-6, this one being described in Ladd (1991).

CONCLUSIONS AND RECOMMENDATIONS OF THE COMMITTEE OF SPECIALISTS

After several years of intensive discussion and careful analysis of available data, the Committee of Specialists on Sensitive Clays of the NBR Complex concluded that embankments could be built on both types of clay encountered in the NBR territory, that even embankments more than 12 m high were feasible as long as strict rules were followed and specific precautions taken during design and construction of the structures. Given the complexity of the problem, however, the committee insisted that suitable exploratory procedures be developed and that the exploratory work be closely supervised according to the specific nature of the problem in order to satisfy the high degree of safety required for water retaining structures.

A methodology was developed to analyze the stability of embankments resting on the marine clay of the Broadback River. The committee used two different approaches for stability analysis--one for embankments less than 12 m high and one for embankments more than 12 m high--but both methods are based on undrained strength analysis. Effective stress analysis was deemed inappropriate for evaluating embankment stability and unsafe for foundations of highly sensitive clay like that of the James Bay area, since with incipient failure, very high shear-induced pore pressures are recorded in the foundation of high embankments.

RECOMMENDATIONS FOR STABILITY ANALYSIS OF PROPOSED NBR EARTH STRUCTURES

For embankments of less than 12 m, the committee recommended analyzing stability directly from shear-strength profiles determined with field vane soundings. Two adjustments of the profiles are made: Bjerrum's correction factor for anisotropy, progressive failure and related effects is applied (Bjerrum 1972); and a correction is made for the weathered crust--that is, the upper part of the NBR clay deposits--whose strength behaviour differs from that of deeper clay layers.

Frost may in the past have penetrated 2 m to 3 m into the clay deposits in the James Bay area, depending on the site of the deposit. The crust has been subjected to different types of weathering in the past and is still subjected to weathering. The result is desiccation of the clay and horizontal as well as vertical cracking that divides the clay into more or less regular cubes. Desiccation and crack frequency decrease gradually with depth as the intact portion of clay deposit is approached. Cracks are, however, filled with water, since the water table is above the top of the clay.

As a result, measured vane strengths of the weathered crust are considerably higher than those of underlying clay. Suspecting that natural-state crust strengths (that is, before recompression by new fill) measured with the field vane did not reflect the strength of the clay as a whole, the committee requested that a series of large box shear strength tests and plate loading tests be conducted at the Olga site. Undrained shear strength mobilized in the crust did indeed prove much lower than measured vane strength around the tested area and approximately equal to measured vane strengths for intact clay immediately below the crust. The field test programme for the weathered crust was complemented with triaxial compression, triaxial extension and simple shear tests, all of which confirmed the results of the in-situ tests (Lefebvre et al. 1987).

The results of the testing programme indicated that for unconfined conditions, such as at the toe of the embankment, design strength applied on the crust must equal measured vane strength of the intact clay immediately below the crust. Under the embankment, shear strength has to be no more than 0.25 times the vertical stress of the embankment fill at the top of the crust, decreasing gradually to the value of the intact clay at the bottom of the crust. Measured vane strengths for the crust are then used only to determine crust thickness.

Embankments more than 12 m high induce foundation stress that greatly exceeds preconsolidation pressure. Stability analysis with these higher embankments must also be based on undrained shear strength obtained at the end of the primary consolidation. Embankments more than 12 m high require staged construction and vertical drains to accelerate the rate of consolidation.

Stage 1 stability is computed using vane strength as

recommended for embankments of less than 12 m; the influence of vertical drains already installed in the foundation are not considered. After a waiting period, primary consolidation nears completion and authorization can be given to continue construction.

Design shear strength used to evaluate embankment stability after consolidation of the clay underneath stage 1 is determined with a combination of the following laboratory tests: undrained triaxial compression, triaxial extension and simple shear tests. The three types of tests were selected to simulate the different zones of applied stress induced in the foundation by the embankment: compression simulates the active zone where the potential failure surface penetrates the clay foundation underlying the highest part of the embankment; simple shear models the lower part of the failure plane; and extension simulates the passive zone under the toe of the embankment. A design shear strength corresponding to a strain of 2% was selected to account for strength loss caused by progressive failure. This method also is recommended to check the stability of the first stage, but using results of laboratory tests for initial conditions is more adequate for a non circular failure plane analysis.

THE CLAY DEPOSIT

The Olga test embankments rest on a deposit of soft, sensitive varved clay. Deposit thickness varies with the topography of the subjacent moraine layer of dense sandy and silty till encountered 13.6 m to 15.3 m below the Olga C embankment. About 0.5 m of peat covers the clay. Water table is at the surface of the clay; however, there is an artesian head of about 1 m under the clay deposit due to the presence of a hill less than 1 km from the site.

Prior to construction of the embankment, four boreholes were put down, twenty Nilcon field vane soundings were performed and pneumatic piezometers were installed. Figure 3 shows the main findings. Surface clay is oxidized and weathered to a depth of at least 2 m. The clay is brown and varves are not discernable. The weathering takes different forms--changes in water content as well as in Atterberg limits, shear strength or consolidation. The dessication effect is noticeable.

To a depth of about 10.5 m, varves are rather regular, especially between 7 m and 10.5 m. From 4 m to 7 m, however, 20% to 25% of the varves are sloping or deformed, indicating soil shifts during deposit formation. Beyond 10.5 m, varves become thicker (40 mm or more),

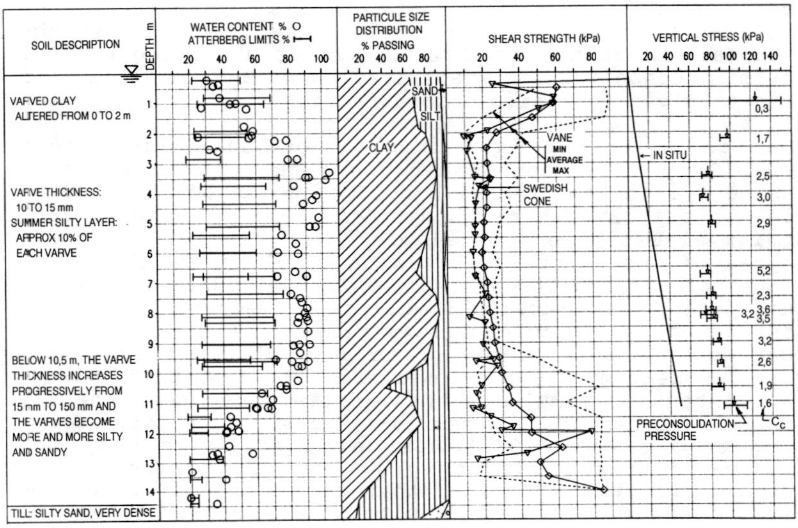

Figure 3 - Geotechnical Parameters

sandier and siltier with depth.

Like water content, the vane strength envelope is tighter between 7 m and 10 m than it is closer to the surface. Beyond 10 m or 11 m, mean shear strength values increase markedly, with significant scatter, a reflection of the stratification of the lower part of the deposit.

Strength values obtained with a Swedish cone are consistently 5 kPa lower than those obtained with the field vane. Measured strength of remoulded clay is slightly less than 1 kPa, which indicates sensitivity of 15 to 20.

The clay is 40 kPa to 60 kPa overconsolidated relative to effective earth pressure; overconsolidation is even greater (about 60 kPa to 80 kPa) in the upper part of the clay deposit, until about 5 m below the surface of the clay. Especially in the upper part of the intact clay, preconsolidation profile of the clay quite closely matches the profile obtained with a ratio of $c_u/\sigma'_p = 0.28$.

THE EMBANKMENT

There are very few case histories of embankments in eastern Canada with a clayey foundation consolidated by installing vertical drains. The committee of specialists expressed some concerns about the performance of such drains--as opposed to sand drains which have a history of good performance. Hydro-Québec has to make sure the drains can operate under the high NBR-Complex embankments requiring staged construction.

The Olga C test embankment is adjacent to the Olga B test embankment built in 1971 to study settlement (see Figure 2). Top length of the Olga C embankment is 50 m and top width is 10 m. With a height of 6 m, the embankment induces a vertical stress of 120 kPa at the base of the embankment, exceeding significantly the 40 kPa to 60 kPa overconsolidation.

Three test sections were built to test two different wick drains, one of these at two different spacings; one control section was built without drains. The undrained section is 15 m long and is located at the end of the top of the embankment. Figure 4 illustrates a cross section of the embankment at this end.

A number of instruments were installed between the drains, under the untreated section, under the berm and at the toe of the embankment to measure pore pressures and movements. Figure 4 shows locations of instruments from which data was taken for stability evaluations, that is, all piezometers outside the drained sections and inclinometers under the berm and at the toe of the embankment.

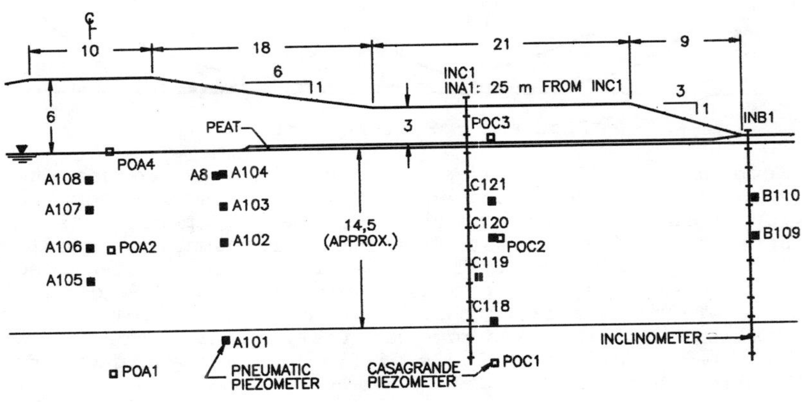

Figure 4 - Embankment Section Showing Instrumentation

The height and the geometry of the upper part of the embankment were designed to promote intense one-dimensional consolidation under the centreline. The geometry of the berm was designed for a safety factor of 1.5 upon completion of construction, as per the committee's recommendations for the embankments of the NBR complex. The method used to analyze embankment stability is described below.

PRECONSTRUCTION STABILITY EVALUATION

If stability were the only consideration, the Olga C embankment could have been built with the same section as the Olga B embankment, which has performed well for over twenty years. It was deemed best, however, to follow the recommendations of the Committee of Specialists on Sensitive Clays of the NBR Complex so the embankment could serve as a test case. Figure 5 shows preconstruction analysis with calculated critical failure surfaces.

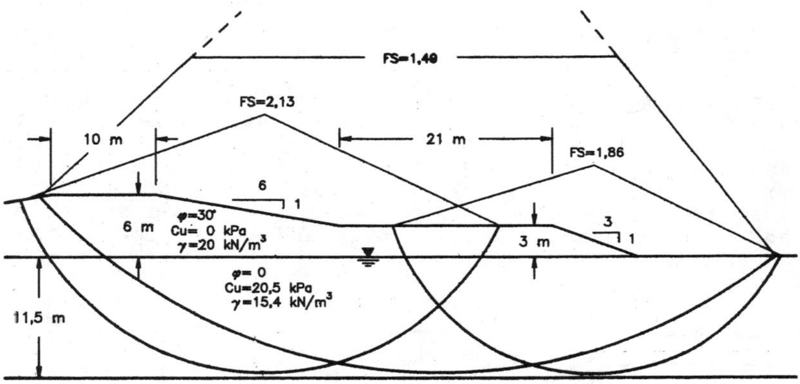

Figure 5 - Failure Surfaces

Methodology. The four-step procedure used to compute stability follows recommendations in Appendix II of the report issued in 1983 by the Committee of Specialists on Sensitive Clays. First, basic site information is collected. Second, a simplified soil profile is developed from shear-strength profiles derived from field vane data. Third, stability charts are used for a preliminary analysis to ascertain project feasibility. Fourth, a detailed stability analysis of final embankment geometry is performed.

Collection of basic site information. No shear strength data were available for the exact site where the Olga C embankment was to be built. However, over the years Nilcon field vane profiles had been obtained for the area around the Olga B embankment. Data used for the preconstruction stability evaluation were issued by SEBJ in 1981. Only profiles for the ground east of Olga B were used, that is, the north toe of Olga C (See Figure 2). These preconstruction vane profiles are similar to those presented in figure 3. The stability analysis was based on mean shear strength values, not minimum values, as recommended by the Committee. The lower limit of the soft clay under Olga C was estimated at 11.5 m based on the vane profiles. Below this depth, mean shear strength values increase, with significant data scatter related to increasingly thicker silty varves.

Simplified soil profile. As per the report issued in 1983 by the Committee on Sensitive Clays, field vane strengths must be multiplied by a correction factor which is a function of soil plasticity. The committee suggested a factor of 0.9 for the Olga site. In addition, shear strength values for the weathered clay crust must be reduced. The method recommended by the committee for crust correction has already been described above.

Preliminary analysis. Preliminary stability computations were performed with a commercial software package that uses the simplified Bishop method of analysis. Several test runs were also made with the Morgenstern-Price method to analyze noncircular failure surfaces.

At this stage, stability charts could have been used. However, since the charts do not cover the geometry of Olga C, a parametric study was conducted to determine the effect of geometrical and geotechnical parameters. Given site conditions, the study suggested slopes of 10H:1V for a safety factor of about 1.5. This short-term safety factor was recommended by the Committee; it was considered adequate for water retaining dykes lying on any type of clay encountered in the NBR territory. Figure 5 shows deep failure circles running as deep as twice the thickness of the fill. With this particular embankment width/clay depth ratio, circular failure surface analysis is still appropriate. However, noncircular failure surface analysis is recommended when the width of the embankment is significantly greater than the thickness of the clay foundation.

OLGA C EMBANKMENT GEOMETRY

Potential mechanisms of earth fill failure were considered when designing the geometry of the Olga C embankment. Design features given special attention were the height and the slope of the berm, the upper embankment slope, the berm width and the unit weight and the shear strength of the fill material.

To date, short-term stability analyses have been conducted as recommended by the Committee on Sensitive Clays. A complementary field and laboratory programme is, however, under way at this site to measure strength gain through the consolidation process. Pore-pressure measurements will be used to determine the long-term safety factor from effective strength parameters. Findings will be published at a later date.

Height and slope of the berm. A berm half as high as the embankment was selected; with embankments more than 6 m high, double berms may be more efficient. Given possible events during construction, a berm slope of 3H:1V was selected to obtain an estimated safety factor of 1.86.

Upper embankment slope. The same slope as for the berm (3H:1V) could have been selected for the upper part of the embankment. To meet stress uniformity requirements, however, upper slopes of 6H:1V were selected instead. This mass in the upper part of the embankment had to be compensated with wider berms; the section is somewhat inefficient volumewise as a result.

The part of the failure surface passing through the granular fill raised questions about mechanical parameters to use for stability analysis. The Committee on Sensitive Clays suggested a 30° friction angle. This value was used for the analysis as well as other assumptions which will not be detailed here. A safety factor of 2.13 was estimated for failure from the top of the embankment through the berm.

Berm width. A berm 21 m wide was selected to provide a safety factor of 1.49 against general failure. Note that berm width is extremely sensitive to the shear strength value assumed in stability computations.

Fill unit weight. A mean unit weight of 20 kN/m3 was assumed for the computations. Two cases assuming different fill unit weights for the berm and the top of the embankment were also considered. It was found that safety factor dropped only 5% with a 6% increase in unit weight.

Failure mode. Safety factor for deep failure mode from the top of the embankment through the clay is 1.49. Noncircular surfaces were analyzed using the Morgenstern-Price method with two various sets of strength assumptions which are not detailed in this paper.

FINAL STABILITY EVALUATION

Twenty Nilcon field vane soundings were performed in the Olga C region right before construction. Figure 3 shows average shear strength profiles developed from the field data. Elevation of surficial peat, of the surface of the clay layer and of penetration refusal, as well as penetration refusal depths, clay thickness and average shear strength in the soft-clay zone were all considered for the evaluation. The most significant factor stability-wise was average Cu in the soft clay zone. Note that the clay is about 2 m deeper under Olga C than Olga B.

Two orthogonal sections passing through the clay foundation were analyzed where lowest average vane strengths were measured. These sections are also where the clay is the deepest. Transverse section shows a lower safety factor than longitudinal profile. Minimum FS is 1.55. The final analysis used shear strength values that are slightly higher than the pre-construction assumptions, explaining the slight difference in computed FS.

DISCUSSION

Figures 6 and 7 show pore pressures and horizontal movement recorded from the beginning of construction of the embankment until several months after the work was completed. Pore pressures are expressed in meters of water above the surface of the clay deposit without taking settlement of the clay into account (only about 10 cm by the end of construction). All horizontal movement was towards the outside of the embankment.

Under the centreline of the embankment, pore pressures measured 11.2 m above the surface of the clay; these pore pressure values were obtained from piezometers A105 and A106 installed at a depth of 10.5 m and 7.5 m respectively. Similar pore pressures were recorded under the slope of the embankment, at piezometer A102 installed at a similar depth. Note that the 1-m artesian head in the till layer subjacent to the clay must be considered in data analysis. Before construction of the embankment, the artesian head persisted almost to the surface. By the end of construction, pore pressures in this layer were 2 m above the surface of the clay.

These high pressures are consistent with the low overconsolidation values (about 40 kPa) in the lower part of the deposit. They suggest that up to about 40 kPa, which corresponds to 2 m of fill, the clay skeleton absorbs about half the load of the fill. After 2 m of fill, additional loading induces an equivalent increase in pore pressure.

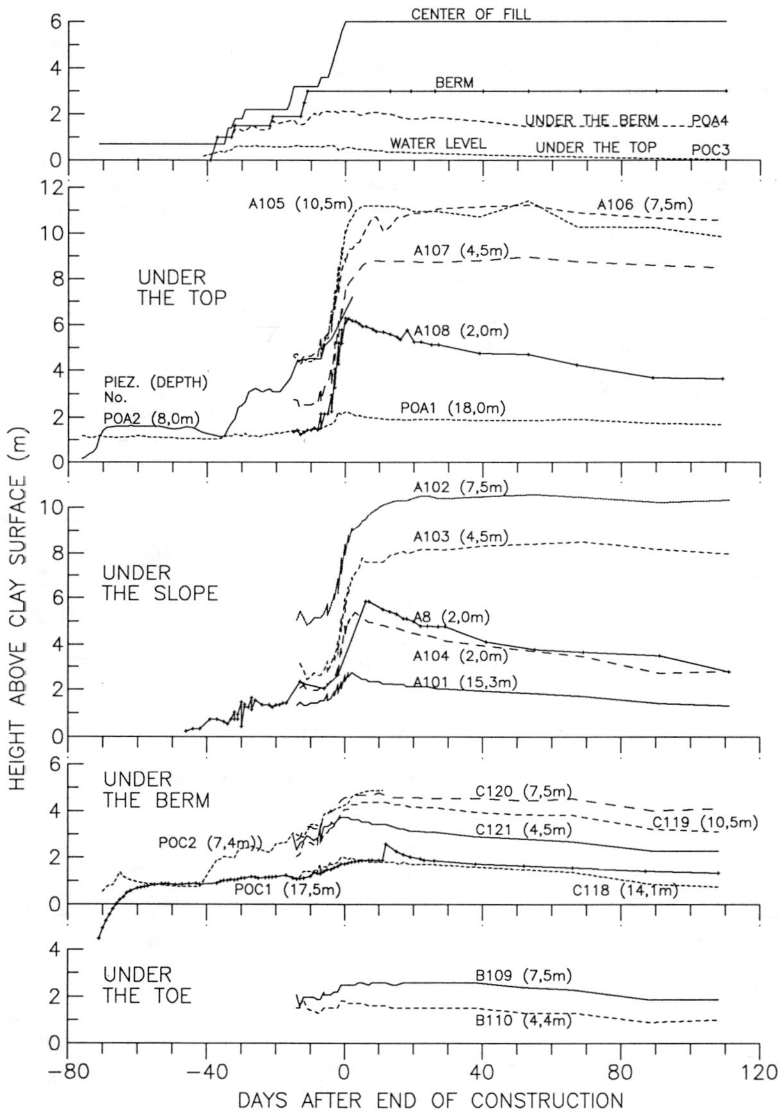

Figure 6 - Piezometric Data

Below the centreline of the embankment, at a depth of 4.5 m below the surface of the clay, pore pressures were about 2 m less than at depths of 7.5 m and 10.5 m. This drop is more or less consistent with the difference in overconsolidation, which is about 20 kPa more at 4.5 m than at greater depths.

Near the bottom of the clay deposit, pressure may have varied with local conditions; this part of the deposit was less well-defined via the exploratory work than the more compressible zones. In the upper part of the deposit, on the other hand, pore pressures at a depth of 2 m pore pressures ran 5 m to 6 m above the surface of the clay, and they began to dissipate quite rapidly.

Given its height (3 m), the berm constituted a load liable to exceed the overconsolidation of the lower part of the deposit. However, water table rose 2 m in the fill of the berm (POC3) and pore pressures under the clay deposit rose to 2 m above the surface of the clay (POA1 and POC1). Given all of the above, the pore pressures near 5 m above the clay surface recorded at piezometers C120 and POC3 reflect a real increase of 3 m, which dissipated quite rapidly after construction was completed.

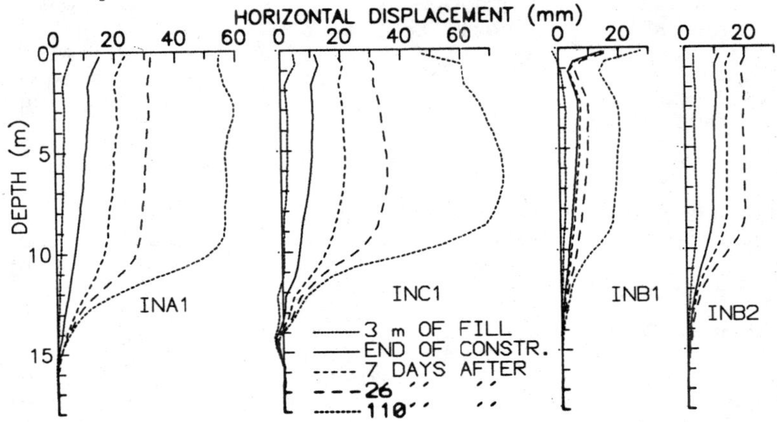

Figure 7 - Inclinometer Data

Two piezometers at the toe of the embankment, installed at depths of 4.4 m and 7.5 m, recorded only a slight increase in pore pressure (pressures ran about 2.5 m above the surface of the clay).

Horizontal movements at the end of construction were negligible, around 1 cm under the centre of the berm and at the foot of the embankment. Eight or nine days after construction was completed, movements had more or less doubled. Movement due to creep was also observed later--up to 7 cm under the berm but very little at the south toe of the embankment. Movement at the north toe could not be measured because the instrument froze.

No cracks were discovered anywhere on the embankment.

CONCLUSIONS

The Olga C test fill constructed on 11 to 12 m soft and sensitive clay deposit at Matagami, Quebec, Canada, has not shown any sign of instability. The maximum height of the fill was 6 m with an average slope of 10H:1V. The minimum safety factor is 1.5 which is considered adequate to insure the safety a water retaining dyke. This minimum safety factor was obtained on a slip circle with the average values of the undrained shear strength measured with the vane Nilcon. The shear strength was modified according to the plasticity index. In the weathered crust near the surface, the shear strength was reduced to values close to those measured in the intact clay immediately below the crust.

At the beginning of construction, the pore pressures were approximately equal to 50% of the load of the fill until the preconsolidation pressure was reached. After this step, the increase of the pore pressures was equal to the addition of the fill. The horizontal displacement were limited to a few centimeters.

ACKNOWLEDGEMENTS

Raymond Morel of Hydro-Québec supervised all work at the embankment site. The National Boring and Sounding Inc. was responsible for geotechnical exploration and instrumentation.

REFERENCES

Bjerrum, L. (1972). "Embankments on soft ground: SOA report." Proc. Specialty Conference on Performance of Earth and Earth-Supported Structures, ASCE, 2, 1-54.

Dascal, O., J. P. Tournier, F. Tavenas and P. La-Rochelle. (1972). "Failure of a Test Embankment on Sensitive Clay." Proc. Specialty Conference on Performance of Earth and Earth-Supported Structures: Part 1, ASCE, 1, 129-158.

Dascal, O. and G. S. Larocque. (1973). "Caractéristiques de compressibilité des argiles du complexe Nottaway-Broadback-Rupert (Baie James)." Canadian Geotechnical Journal, 10, 41-63.

Dascal, O., G. S. Larocque and J. P. Tournier. (1973). "Digue d'essai pour l'étude des tassements d'une fondation d'argile molle et sensible." Canadian Geotechnical Journal, 10, 363-391.

Dascal, O. and J. P. Tournier. (1975). "Embankments on Soft and Sensitive Clay Foundation." J. Geotech. Engrg., ASCE, 101(3), 297-314.

Ladd, C. (1991). "Stability Evaluation During Staged Construction", The twenty-second Terzaghi Lecture, J. Geotech. Engrg., ASCE, 117(4), 537-615,

Lefebvre, G., Paré, J.J., Dascal, O. (1987). "Undrained Shear Strength in the Surficial Weathered Crust", Canadian Geotechnical Journal, Vol. 24, pp. 23-24.

SEBJ, Direction Ingénierie, Centre géotechnique. (1981). Rapport, Programme d'essais 1981, Site de la digue Olga: Volume 1, Aménagement hydroélectrique des rivières Nottaway-Broadback-Rupert.

SEBJ, Vice-Présidence Ingénierie et Développement. (1983). "Committee of Specialists on the Sensitive Clays of the NBR Complexe, Final Report" and Appendix 2 "Embankment Stability Subcommittee Report: Evaluation of Dyke Stability."

PERFORMANCE OF AN EMBANKMENT DAM WITH PARTIAL CUTOFF

Pascual. H. Perazzo[1] and Tauseef. I. Choudry[2], M. ASCE

Abstract

Dike "C" is one of the reservoir rim dikes of the Guri hydroelectric development located on the Caroni river in Venezuela. This dike, a homogeneous earthfill embankment, is founded on decomposed granitic gneiss. A partial cutoff was provided under the dike where the thickness of decomposed gneiss exceeded the corresponding water head. The partial cutoff solution did not prove to be completely effective and seepage increased significantly on raising the reservoir. The seepage occurred mainly along the relic joint structure of the foundation material. Remedial measures were adopted which comprised a series of relief wells together with filter and granular material blanket placed downstream of the dike. These measures have proven to be effective in controlling the seepage flow.

Introduction

The hydroelectric development of the Caroni river, located in the southeastern part of Venezuela, is being undertaken by the state owned company C.V.G.-EDELCA. The lower Caroni development consists of four hydroelectric projects namely Guri, Tocoma, Caruachi and Macagua II, located within a distance of 100 Km before the Caroni river reaches its confluence with the Orinoco river. The Guri project has been completed whereas Macagua II, Caruachi and Tocoma are currently under various stages of design and construction.

The final stage construction of Guri included raising of its main gravity dam and spillway by 52 m. and building of two new gravity dams,a new powerhouse with 10 units, a tailrace channel, two earth and rockfill dams on both river banks and nine closure dikes on the left and right rim of the reservoir.

1.- Section Head, Geotechnical Department, C.V.G-EDELCA. Apartado 62413, Caracas,Venezuela, 1061 A

2.- Associate, Harza Engineering Company, International., Caracas,Venezuela, 1061 A

Dike "C" is one of the main marginal dikes with a total length of 7 Km. and a maximum height of 35 m. This dike overlies decomposed granitic gneiss of variable thickness, which in turn rests on hard weathered rock.

The studies conducted during the investigation stage indicated that the decomposed foundation material varied in depth from a few to 25 m. with a permeability of 1*10^-4 cm/sec. Considering: a) this permeability to be in the semipervious range and b). a positive cutoff would involve excessive excavation, a criteria of partial cutoff was adopted in certain stretches of the dike. The criteria consisted of excavating a partial cutoff trench to a depth of half the water head in those areas where the depth of decomposed gneiss was excessive. In other stretches where the rock was encountered at shallow depths, a positive cutoff was provided. (See typical section in figure1.). A positive cutoff consisted of excavating the trench to the top of hard weathered rock, cleaning the rock surface, and providing a grout curtain.

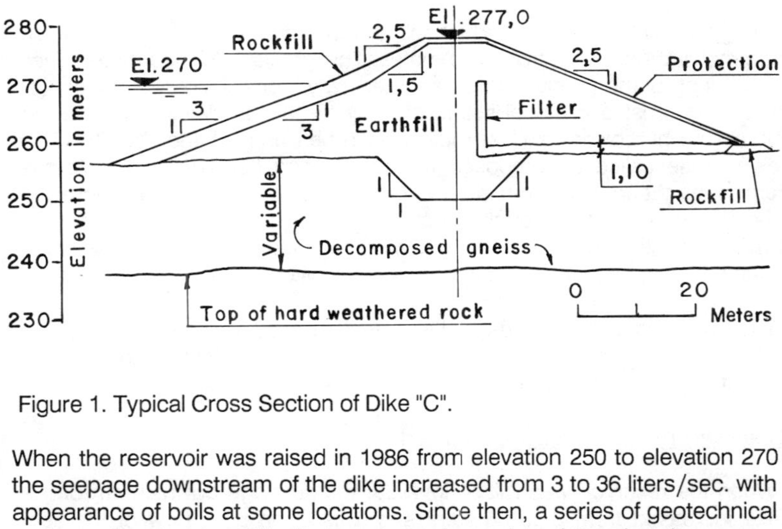

Figure 1. Typical Cross Section of Dike "C".

When the reservoir was raised in 1986 from elevation 250 to elevation 270 the seepage downstream of the dike increased from 3 to 36 liters/sec. with appearance of boils at some locations. Since then, a series of geotechnical and geochemical investigations were performed to determine the causes of the increase in seepage. Based on the studies, certain measures were adopted to control the flow.

The results of the investigations, measures adopted to control seepage, and the performance of those sections of dike "C" with partial cutoff are presented in this paper.

Geological and Mineralogical Characteristics of the Foundation.

The region presents a large structural complexity due to different remobilization periods that have affected the Guayana Shield (De Sola, 1977). The topography at the dam site is generally undulating and is interrupted by many ridges. The drainage areas between these ridges resulted in zones of slope-wash deposits.

The outcropping rocks are metamorphic, belonging to the Imataca Formation (Precambrian age) and are principally characterized by grey and pink granitic gneiss of coarse crystalization and quartz inscrustations (De Sola 1987). The most important structure of the region is the Guri fault, which has developed a sheared zone of a width varying between 500 and 4500 m., covering a length of 1200 m. along the western abutment. This fault, though enormous in size, was determined to be inactive. The fault materials generally consist of highly fractured milonite with tightly closed joints.

The typical profile of the residual soil and the rock is similar to that which commonly develops in tropical climates. This profile is formed by the deep weathering of the granitic gneiss and generally consisits of four zones: (1) a reddish crust of residual soil formed by the accumulation of ferric and aluminium hydroxides and sesquioxides (Eberl and Whitney 1980), (2) a lower zone called decomposed rock containing kaolinite and halloysite formed by the in situ weathering of the parent rock (also called Saprolite). The relic structure of the original rock is frequently preserved in this zone.
(3) The weathered rock and (4) the fresh rock composed of feldspar, quartz, biotite, hypesthene, garnet and muscovite.

The thickness of the residual soil and decomposed rock varies between 3 and 25 m. along the dam axis. The dominant orientation of the joints in the parent rock crosses the dike alignment almost perpendicularly.

Observations and Studies Conducted

When the reservoir was raised in 1986, the seepage downstream of the dike increased from 3 to 36 liters/sec. Based on a review of the field conditions the following observations were made:

1.- At a few locations downstream of the dike, concentrated seepage occurred in the form of small boils (about 5 cm. in diameter), with some signs of soil erosion.

2.- The seepage day-lighted as far as 140 meters downstream from the toe of the dike.

3.- The seepage occurred mainly in those stretches of the dike where it had a partial cutoff section. This was an indication that the seepage was occurring through the foundation i.e., the decomposed gneiss.

4.- Seepage was observed in stretches where partial cutoff existed, even with low gradients i.e., less than two meters of water head.

5.- The seepage occurred mainly in sectors where the joints of the parent rock in the decomposed gneiss are almost perpendicular to the dam axis.

6.- The dissolving of soil mineral (mainly iron content) in water was reflected by some brown spots in locations of incipient seepage.

These observations raised some serious concern for a potential source of piping thereby endangering the stability of the dike. A series of geotechnical investigations were undertaken in order to determine the cause of this problem and to carry out the most appropriate treatment to control the seepage. The investigations consisted of the following:

1.- Review of the as built drawings and test results.

2.- Geophysical studies to establish the depth of soil in problematic areas.

3.- Drilling of exploratory boreholes to obtain SPT blowcounts and soil type.

4.- Performance of triaxial tests on undisturbed foundation samples to confirm or revise the parameters established in the design stage.

Studies and Analysis

From the study of the as built drawings it was concluded that the seepage was mostly concentrated in the stretches where the dike had a partial cutoff section.

The geophysical studies were conducted in areas where the maximun amount of leakage was observed. Geophysical lines were drawn along the dike alignment, confirming the presence of 3 to 25 m. thickness of residual soil.

A series of boreholes were drilled both upstream of the center line and downstream of the dike. Continuous SPT tests were performed to determine if there were any weak zones present within the foundation or embankment materials. These boreholes were converted into piezometers for continuous monitoring. The boreholes confirmed the thickness of decomposed gneiss as determined earlier by the geophysical study.

Some piezometers indicated pressures higher than the ground level near the toe of the dike. However no weak zones were found within the soil layer.

Triaxial Tests and Stability Analysis.

Undisturbed hand-cut samples were taken from the foundation of the dike to perform CIU and CAU triaxial tests. The results of the tests are shown in figure. 2. The tests revealed that the foundation residual soil has a higher shear strength than that originally employed in the design. It was also found that for a low confining pressure (100 KN/m2), the soil presents a highly dilatant characteristic with effective envelopes represented by very low cohesive intercept and friction angles in the range of 40º. For confining pressures greater than 200 KN/m2, the soil is contractive with effective envelopes represented by friction angles in the range of 29º.

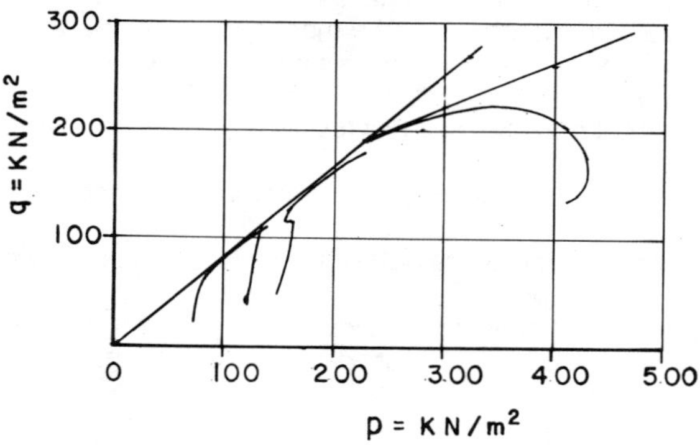

Figure 2. Results of the CAU Tests

To perform the stability analysis, it was necessary to understand the pore pressure distribution within the foundation soil. However, precise record of pore pressure distribution, particularly under the downstream shell, was not available, as most of the piezometers were installed in the bedrock.

Three different conditions or hypothesis were assumed to quantify the pore pressure in the foundation of the dike:

CONDITION 1.- The pore pressure is equal to the ground level at the toe or piezometric pressure in the foundation soil.

CONDITION 2.- The pore pressures measured at the fresh rock remains constant along the full thickness of the foundation soil.

CONDITION 3.- The pore pressure has a linear variation between the pressure at the surface to a maximun value measured at the top of the rock.

Stability analysis was performed for the different conditions at each representative section. The analysis was carried out using the simplified Bishop method. The shear strength for the different materials of the dike and its foundation were characterized by the following parameters:

SOIL	DESCRIPTION	UNIT WEIGHT KN/M3	COHESION KN/M2	FRICTION
1	Filter material	22.0	0	40º
2	Compacted gneiss	20.6	0	31º
3	Rockfill protection	23.0	0	45º
4	Decomposed gneiss (Foundation material)	18.0	0	40º

For conditions Nº 2 and Nº 3, the stability analysis showed some unstable shallow masses near the toe of the dike where the pore pressure was significantly higher than the overburden soil pressure.

The analysis also indicated that if the water table is lowered at the toe of the dam by means of relief wells, the minimum factor of safety for the most unfavorable failure surface, would increase substantially (1.58) thereby approaching the basic requirement of the design i.e., factor of safety of 1.60.

Furthermore, the stability analysis showed that for the most unfavorable interpretation of the foundation water pressure, it is possible to obtain the minimum factor of safety required by the project by means of weighting berms 2,5 m. high, located at the toe of the dike.

Remedial Measures Adopted

Certain remedial measures were adopted to protect the foundation and the embankment against erosion and sliding. These measures included the placement of a granular blanket downstream of the dike with a maximum width of 40 m. and the installation of a series of relief wells, in four different stretches, along the downstream toe of the dike. Additionally, concrete ditches were excavated to channel the leakage to the weirs installed for flow measurement.

Relief wells were installed with a spacing of 10 m, and were constructed introducing a 0.10 m. slotted PVC pipe sorrounded by a geotextile into a hole of 0.30 m. in diameter. The relief wells were carried down to 3 m. into the hard weathered rock. The annular space between the PVC pipe and the hole was filled with sand. The granular blanket consisted of a 1.3 m. thick combined filter of sand and quartzite. This quartzite is a naturally occurring granular material with average size of 3 inches and 10% passing sieve Nº 200. Figure 3, shows the granular blanket and the relief wells in a typical section of the dike.

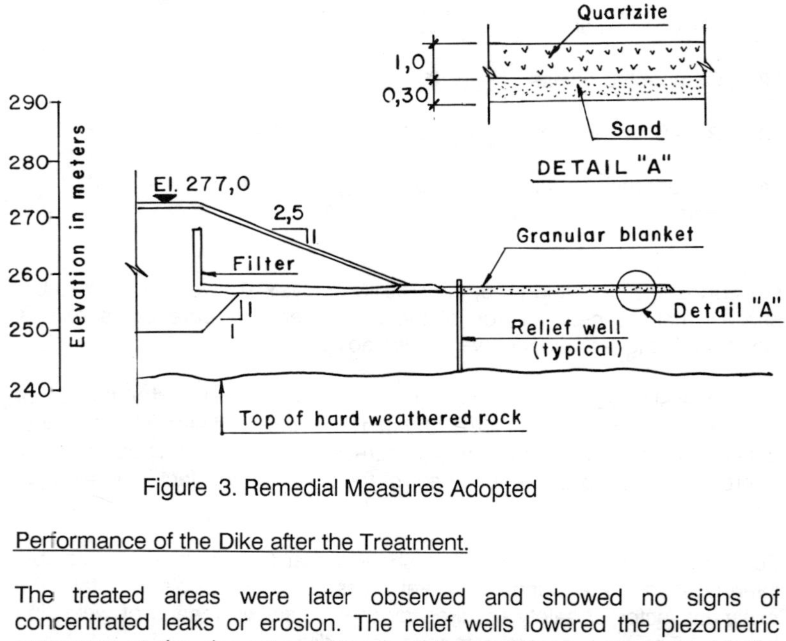

Figure 3. Remedial Measures Adopted

Performance of the Dike after the Treatment.

The treated areas were later observed and showed no signs of concentrated leaks or erosion. The relief wells lowered the piezometric pressures at the downstream toe and the drainage system provided a better control for the seepage collection and measurements.

It was observed that seepage water flowing from some of the relief wells located at topographically higher levels was greater than for those located at lower levels, indicating that the quantity of flow from a given relief well depends more on intercepting relic joints than on the hydraulic head difference. This confirms the hypothesis that the seepage in the foundation is mainly occurring through the relic joints of the decomposed gneiss.

Regular monitoring of the dike did not show any appreciable increase in the seepage flow during the last four years after the aforementioned treatment was carried out. The fluctuations in the reservoir level and the corresponding total seepage for the past six years are shown in figure 4. The increase in seepage flow in 1987 is considered to be mainly due to construction of relief wells.

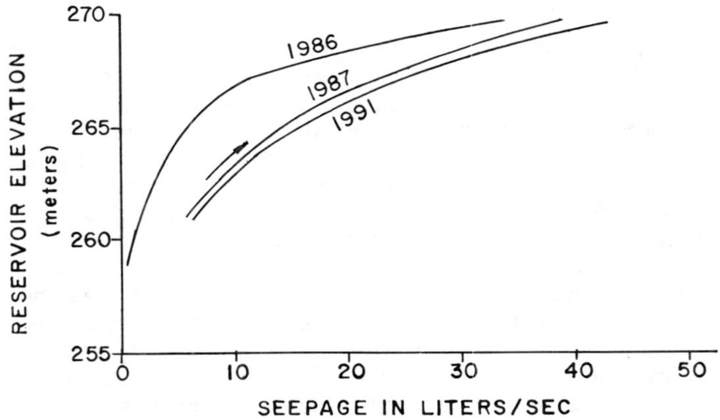

Figure 4. Total Seepage Measured Downstream of Dike "C".

It can also be observed in this figure that there is a marked increase in the seepage flow once the reservoir rises above elevation 262. At this elevation and above, the dike rests on partial cutoff. This provides additional evidence that the seepage flow is mostly coming from the stretches with partial cutoff.

To reanalyze the pore pressure distribution, additional piezometers were installed at the downstream toe of the dike, at variable depths within the decomposed gneiss layer. The readings from these instruments showed maximum piezometric level equivalent to the corresponding ground level.

The stability of the dike was reanalyzed utilizing the information from relief wells and additional piezometers. It showed a factor of safety of 1.58 for the critical section under steady state seepage condition. As this value is close to the design factor of safety of 1.60, the dike is considered safe against sliding. Additional field and laboratory permeability tests were performed on foundation material represented by decomposed gneiss downstream of the dike. Field tests, performed in auger holes of 3 inches in diameter, showed permeabilities with a wide range varying from 3.5 * 10^-5 cm/sec to 2.5 * 10^-3 cm/sec.

This indicates existence of preferential seepage paths within the soil presumably due to the relic joint structure of the parent rock. However, this phenomenon was not evident for the laboratory tests due to a smaller sample size, and the average permeability was found to be of the order of 5.8 * 10^-4 cm/sec.

Chemical Analysis.

Investigations of the geochemical interaction between the foundation material and the reservoir water were performed using long term permeability and triaxial tests. The duration of the tests lasted as much as six months. These tests were performed on hand cut undisturbed samples taken from the first two meters of the decomposed gneiss downstream of the dike. For the triaxial tests, free drainage was permitted and the samples were sheared at different time periods. The test results showed no change in permeability or shear strength with respect to time.

Present information about the geochemical factors that could affect the safety of dams is scant, particularly with respect to the tropical soils. However, Infanti and Kanji (1974) have reported geochemical aspects that affect the horizontal drain. Furthermore the same authors (1981) have presented the data that shows the obstruction of the drainage system by chemical and laterization process in different Brazilian dams. They concluded that in most of the cases the intensity of the problem appears to be related to the parent rock minerals of the soils, including decomposed form of Precambrian granitic rock.

The chemical analysis of the seepage water obtained downstream of the dike "C" showed some increase in total dissolved solids when compared with the reservoir water (from 0.975 in the reservoir to 2.25 mg/liter in the seepage water). Some increase is also found in the cations and anions specially in Calcium, Sodium and Potassium. Inspite of the increase in the dissolved solids, cations and anions, the total quantity is still considered to be very small. There is no appreciable increase in the seepage flow, showing that solutioning is occuring at a very small rate. The chemical studies will be performed on an annual basis to determine any future endangering conditions.

Conclusions.

1.- In the areas where a partial cutoff was used for dike "C", seepage occured with some erosion of foundation soil, which is considered to be mainly due to the orientation of the relic joint structure with respect to the dike alignment.

2.- Seepage increases rapidly when the reservoir rises in the last eight meters of the maximum pool level. This flow is believed to be taking place mostly through the foundation soil under the partial cutoff section.

3.- The remedial measures comprising a series of relief wells together with filter and granular material blanket placed downstream of the dike, have proven to be adequate in controlling the seepage flow.

4.- The chemical analysis performed on the seeping water indicated some gain in dissolved solids as well as in cations and anions. However, this effect has not resulted in any significant change in the quantity of seepage to-date. It is intended to continue with this study to monitor the behavior of the foundation soil.

5.- During the design of foundations for embankment dams on residual soils, in addition to other geotechnical properties consideration must be given to the geological origin of the soil including the orientation of relic joint structure of the parent rock. Also the geochemical interaction of the foundation soil with the reservoir water should be thoroughly studied and analyzed.

Appendix I - References

De Fries, K., and Soto, L.A. (1987). " Reevaluation of the Stability of the Dike "C" at the Guri Hydroelectric Development". De Fries Engineering Office, Caracas, Venezuela.

De Sola, O. (1977). " Geological Report of the "C" area, Final Stage". Miseca Engineering Company, Caracas, Venezuela.

De Sola, O. (1987). ' Evaluation of the Geological Exploration for the Guri Final Stage". Miseca Engineering Company, Caracas, Venezuela.

Eberl, D., and Whitney, G. (1980). " Report on Analyses of Samples from the Guri Dam Site, Venezuela". Harza Engineering Company, Chicago, Illinois.

Infanti, N., and Kanji, M.A. (1974). " Preliminary Considerations on Geochemical Factors affecting the Safety of Earth Dams". Proc. 2nd International Congress of the International Association of Engineering Geology.

Infanti, N., Kanji, M.A., Ferreira, R.C., and Guerra, M.O. (1981). "Geochemistry of the Iron in Engineering Works in Tropical Soils : Summary of main findings". Brazilian Simposium of Engineering in Tropical Soils.

Instrumentation Department. C.V.G-EDELCA. (1987). " Dike C, Foundation Protection, Relief Wells and Seepage Control". Internal Report. Guri, Venezuela.

Vegas,T. (1988). " Preliminary Report on the Chemical Component of the Water from the Seepage and the Reservoir". Internal Report. Guri, Venezuela.

Vegas,T. (1989). " Final Report on the Chemical Component of the Water from the Seepage and the Reservoir". Internal report. C.V.G.-EDELCA. Guri, Venezuela.

STABILITY EVALUATION OF AN OLD DAM WITH A KNOWN HISTORY OF SLIDE

Sukhmander Singh[1], Member ASCE, and
Robert D. Darragh[2], Fellow ASCE

Abstract

The application of an in-situ cone testing technique to define the failure surface of a significant slide in the embankment of an old dam is demonstrated. In performing the seismic evaluation for the dam, the knowledge of the old slide and the location of the slip surface was found to be very helpful in arriving at the proper soil parameters. The seismic stability analysis showed that the seismic displacements were largely controlled by the fling component of the near field design earthquake. A remedial design was completed based on dynamic analyses that required removal and replacement of the old slide plane material.

Introduction

An 35 ft. high earth dam (Fig. 1) that was originally constructed in the period 1920-1922 is presently located in a heavily populated area of the San Francisco Bay Area. An active trace of the Hayward fault is located about 1000 ft. East of the dam. Shortly after construction was completed, a significant slide occurred in the lower portion of the embankment. Since neither the as-built sections of the dam, nor any report on the slide were available, much of the information on the slide was obtained from a 1930 report written by Clyde (1930) and from field memos of the 1950's contained

[1]Chairman, Dept. of Civil Engineering, Santa Clara University, Santa Clara, California, 95053

[2]Partner, Dames and Moore, 221 Main #600, San Francisco, California, 94015

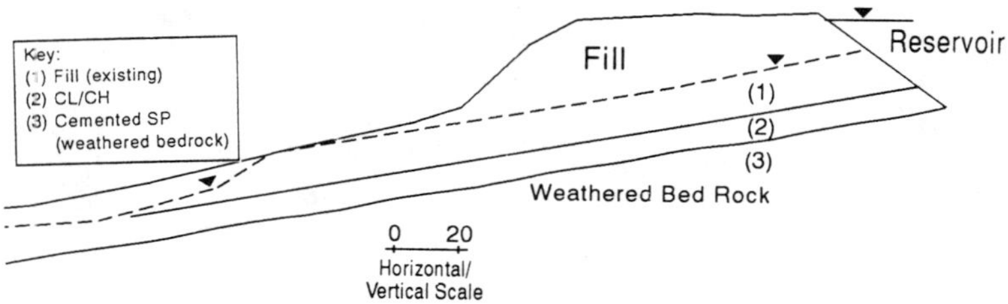

FIG. 1. CROSS SECTION PRIOR TO SLIDE IN 1921 - 1922

in Mr. Clyde's files. The slide area, in plain view, was about 140 by 180 feet (Fig. 2). Dam history as revealed by old notes indicated that between 1922 and 1930, the slide continued moving slowly downhill during rainy seasons. Various remedies to control slide movement were tried. At one time the slide was broken up by shooting with dynamite with the intended purpose of destroying the smooth sliding surface if any. Later, about 1400 cubic yards of large quarry rock was dumped on top of the slide. It was expected that this rock would work downward into the slide mass and take hold to the harder

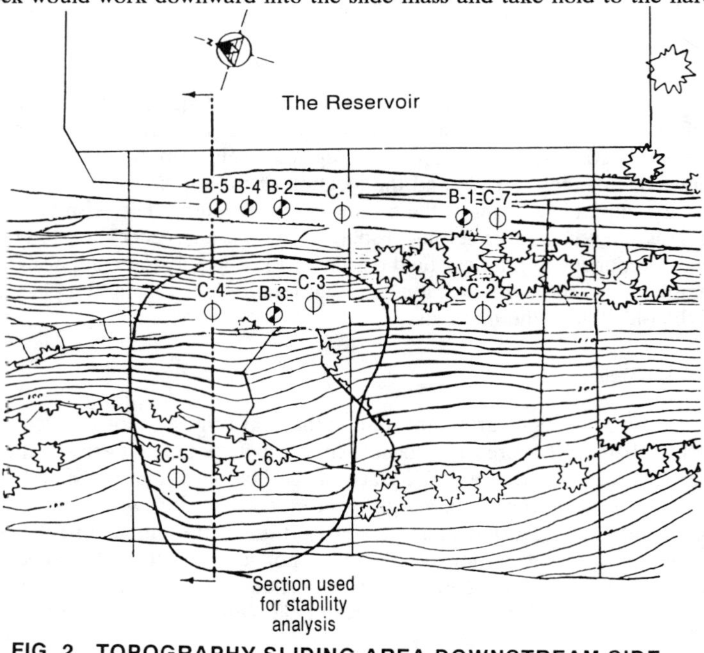

FIG. 2. TOPOGRAPHY SLIDING AREA DOWNSTREAM SIDE OF THE EMBANKMENT

formation. Some of the quarry rock did work its way down into the area leaving a path for water to drain. However, none of the remedial methods proved very successful (Clyde, 1930).

It was the investigation by Clyde in 1930 that indicated the presence of high water table in the slide area and provided a clue as to the cause of the slide. Fig. 3 taken from Clyde's report, shows a cross-section and a transverse section through the slide area. It may be noted in Fig. 3 that the ground-water level at the time of Clyde's 1930 investigation was very near the ground surface within the slide area. Also, there appeared to be a rather abrupt drop in the water level beyond the edges of the slide area.

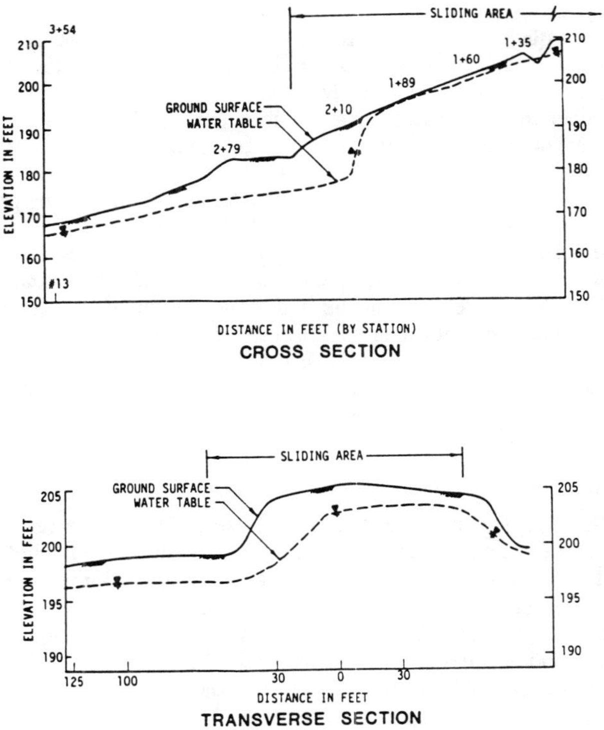

FIG. 3. CROSS SECTION AND TRANSVERSE SECTION OF THE SLIDE AREA IN 1930

Apparently, a high water table persisted the slide area because clays overlying the bedrock prevented drainage. Upon Clyde's recommendation, a 12-inch tile drain was installed in 1930 in the upper third of the slide in an attempt to lower the water table. Although there is no documented evidence regarding the effectiveness of the drain installation, it appears to have been effective gradually reducing in embankment movement. An 18-inch collecting drain system installed in 1954 and improvements in the reservoir lining system also appears to have reduced the high water table conditions. The embankment was remodeled slightly again in 1954, 1956, and 1971. The embankment section in the old slide area remained, however, a matter of concern because preliminary analyses indicated the dam to be unstable in the event of severe earthquake loading conditions.

Accordingly, a comprehensive field and laboratory investigation were performed to establish the cause of the slide, its lateral extent, the location of the existing slip surface, and finally to develop the remedial measures that would render the embankment stable during earthquake loading conditions. This paper describes how the use of cone penetration tests (CPT), in combination with borings and a detailed laboratory investigation, proved to be the most successful means to define the thickness and engineering properties of a thin, weak and highly plastic residual clay layer considered to be the probable surface of sliding for the original slide. Also presented are the results of analytical studies completed for the seismic stability evaluation of the dam. These results clearly show the significance of fling components of near field earthquakes in controlling the seismic displacements. The paper briefly describes the remedial design selected to remove a portion of the embankment in order to replace the clay layer. Finally, a summary of the remedial construction is also presented.

Geology

The dam lies at the base of the Berkeley Hills. The reservoir was made by cutting into a natural slope with a shallow cut on the uphill (easterly) side and constructing an earthen embankment about 35 feet high on the downhill (westerly) side. A previous geologic report (Marliave, 1971), described the Franciscan bedrock formation at the dam site, and in the cut slopes surrounding the reservoir as a predominantly hard sandstone and shale with some metamorphic schistose rocks. Occasional out-crops of the harder bedrock are found along the base of the hills. Most of the surface exposures are highly weathered and some exhibited shearing. Sandstone is at relatively shallow depths over most of the site. One of the primary branches of the Hayward fault passes only about 1000 ft. (300 meters) east of the Dam. The San Andreas fault is located about 19 miles (30 km) west of the Dam.

Field Investigations and Site Conditions

Three borings for soil investigation purposes and two borings for the geophysical study were drilled through the embankment material into underlying bedrock. Four of the borings were located on the crest and one on the downstream slope. A schematic soil profile drawn through the two soil borings on the crest and the boring on the lower road is shown in Fig. 4. After analytical studies showed that a relatively weak clay layer underlying the embankment was the critical soil stratum in terms of slope stability, seven cone penetrometer tests were performed to define the extent and strength of the clay layer. The soil borings and the geophysical borings drilled from the crest encountered brown and gray, sandy, silty, gravely clay (GC-CL) with varying percentages of rock fragments/gravels (max. 3/4 to 1 in size) from

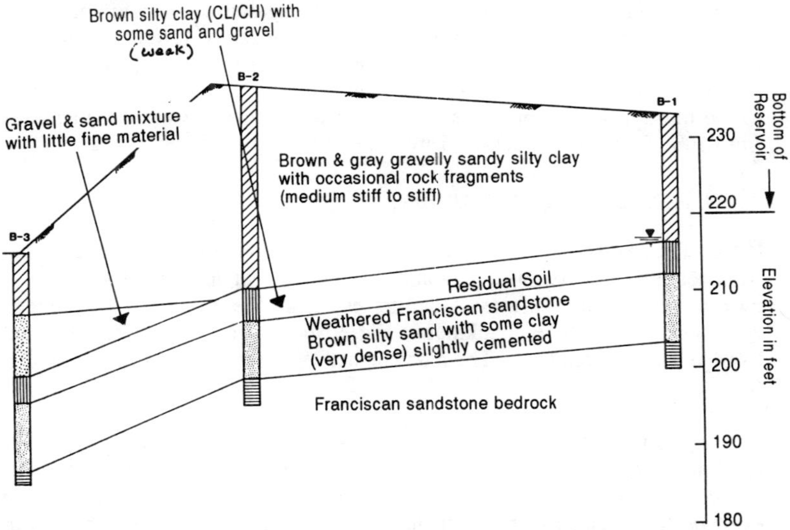

FIG. 4. SCHEMATIC SOIL PROFILE ALONG BORING LOCATIONS
(See Fig. 2 for location of Borings)
(Horizontal not to scale)

ground surface down to about 25 feet depth. This mixture of sands, gravels, silts and clays constitutes the fill used to build the embankment.

Since construction records for the dam were not available, it is difficult to ascertain the manner in which this fill was placed. However, the source of the fill material can be reasonably postulated as having been derived from the cuts for the reservoir area and possibly from some imported fill.

Underlying the embankment material was a brown, stiff to very stiff, highly plastic, silty clay layer about 5 to 8 feet thick. Underlying the clay layer, to a depth of about 10 feet, was the weathered Franciscan sand-stone bedrock which, in turn, grades to unweathered, relatively hard bedrock. Each of the crest borings penetrated about 5 to 10 feet into the sound bedrock.

A sampling interval of about 4 feet was used for the two crest borings (B-1 and B-2) whereas continuous sampling was carried out in the third boring (B-3) drilled from the lower road. Although every effort was made to secure undisturbed samples by use of a 4-inch diameter Pitcher Sampler, the presence of relatively large gravels and rock fragments severely hampered undisturbed sample recovery. However, enough samples of reasonable quality were recovered for the required laboratory testing program. The geophysical investigation to provide in-situ shear wave velocity data was carried out using the "cross hole" method. From the surface to a depth of 15 feet, shear wave velocity increases from 620 fps to 1045 fps. The shear wave velocity remains relatively constant from 15 feet to 30 feet (1030 to 1080 fps). From 30 feet to the bottom of the boreholes at 45 feet, the shear wave velocity increases from 1080 fps to a bedrock velocity of 3660 fps.

Water Levels and Seepage Conditions

The free-standing water levels within the embankment were established from readings of open well piezometers. A ten-year history (1971-1981) of piezometer reading was available to establish the present day seepage conditions within the embankment. Fig. 9 shows the water levels obtained from piezometer readings. The much higher 1930 water level is shown on Fig. 6.

It is important to note that the relatively recent (1971 to 1981) water level in the embankment is generally at the top of the clay layer. Seasonal variations in water level readings are limited to about 5 feet. In order to evaluate the sensitivity of the stability of the embankment to the changes in the water levels, analyses were made for three different water levels: at top of clay layer, 5 ft. and 10 ft. above the clay layer.

Laboratory Test Results

A laboratory testing program consisting of static and dynamic tests was carried out to evaluate the behavior of the embankment fill, underlying clay, and weathered Franciscan sandstone materials under both static and earthquake loading conditions. The results of these tests were used subsequently in the static and dynamic analyses.

The static test program included unconsolidated undrained triaxial compression tests (TX/UU), consolidated undrained multi-stage triaxial compression tests with pore pressure measurements (TX/CU/PP), consolidated drained multistage triaxial compression tests with volume change measurements (TX/CD), and consolidated drained direct shear tests (DS/CD). The dynamic tests, performed primarily on the embankment fill material, included resonant column tests, strain controlled cyclic triaxial tests, and stress controlled cyclic triaxial tests. Since it was not possible to obtain a large number of high quality samples of the clay due to the limited number of borings and the relatively thin nature of the layer, the dynamic strength of the clay was estimated from its static strength.

Embankment Fill Material

The embankment fill material is a brown and gray, gravelly, sandy silty clay with a maximum particle size of 3/4 to 1 inch. The dry density is in the range of 98 to 115 pounds per cubic foot with water contents of 14 to 25 percent. The average liquid limit is 40 with plasticity index of 21. The strength parameters with respect to effective stresses are $\Phi = 29^\circ$ to 32° and C' = 0 to 100 pounds per square foot.

Stress controlled cyclic triaxial tests were also performed to evaluate the cyclic strength and post cyclic strength loss characteristics. The cyclic strength test specimens were anisotropically consolidated in order to represent the initial stress conditions. The test results shown in Fig. 5 indicate that the fill material is not susceptible to post cyclic strength loss or does not lose strength as a result of pore pressure buildup during cyclic loading.

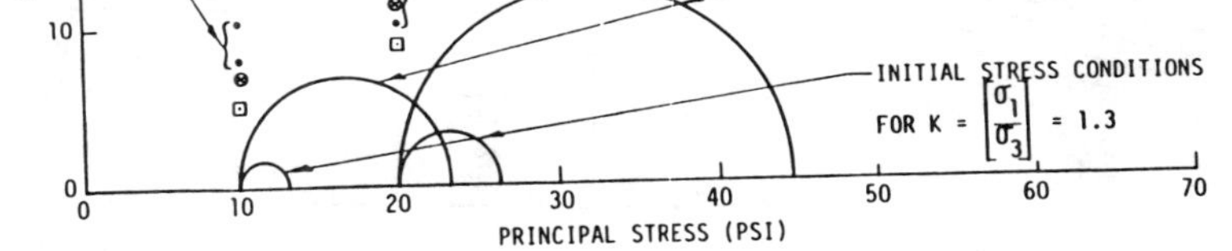

FIG. 5. STRESS CONDITIONS DURING AND AFTER CYCLIC LOADINGS (Total Stresses)

Stability Analyses - Old Slide Area

In order to better understand the stability of the dam, an evaluation of the old slide, which occurred immediately following construction, was performed.

The high ground water table conditions and the weak clay layer underlying the embankment fill were considered to be the probable causes of the original slide. Preliminary analyses indicated that the critical failure surface passes through the weak highly plastic clay layer roughly parallel to the sloping bedrock topography.

Since the exact embankment configuration prior to the slide could not be established due to the lack of construction records, possible embankment sections were estimated from available (1930 and 1953) topographic maps. Fig. 1 shows a typical section of the embankment prior to the slide. However, the pre-failure ground water level could not be established.

When Clyde made his investigation in 1930, he reported that the slide was still moving. Based on this critical piece of evidence and the fact that documented topographic maps and ground water elevations existed for the 1930 conditions, analyses of the 1930 typical cross-section shown in Fig. 6 were performed. The presence of a relatively weak, highly plastic clay layer (approximately 5 to 8 ft. thick) overlying the bedrock was considered to be highly significant.

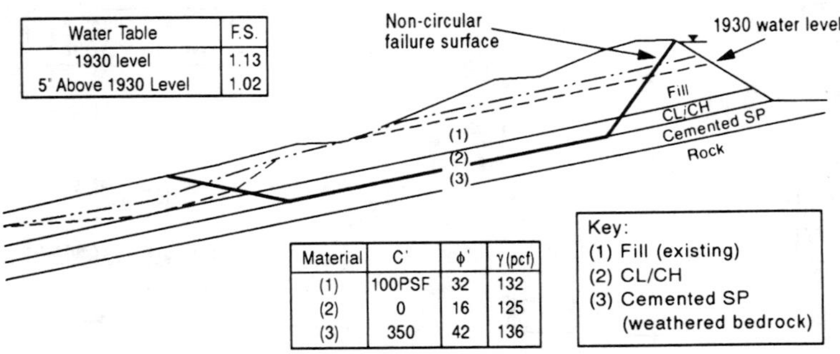

Water Table	F.S.
1930 level	1.13
5' Above 1930 Level	1.02

Material	C'	ϕ'	γ (pcf)
(1)	100PSF	32	132
(2)	0	16	125
(3)	350	42	136

FIG. 6. STATIC STABILITY ANALYSES

Because of the geometry of the weak clay layer, the critical mode of failure was considered to be a planar block slide passing through the clay. Stability analysis for non-circular failure surfaces were carried out for the 1921-1922 and 1930 cross-sections using computer program developed in 1980 at U.C. Berkeley (Celestino and Duncan, 1980). The computer program performs an automatic search for critical noncircular surfaces.

Strength Parameters for the Analyses of 1930 Cross-Section

Since the slide was known to have been moving between 1922 and 1930, and since significant portions of the slip surface passes through the clay layer, the selection of representative long term shear strength parameters for the clay layer was extremely important. The clay material underlying the embankment fill is a brown silty clay residual soil with occasional sand and gravel. The dry density varies from 104 to 108 pounds per cubic foot with water contents ranging from 19 to 25 percent. The average liquid limit is 54 percent with plasticity index of 25. The results of consolidated drained direct shear texts indicated that the effective strength parameters are $\Phi = 22^0$ and C' = 600 pounds per square foot. Using the slope configuration along with the water table conditions by Clyde in 1930, and the strength parameters for the embankment material and the clay layer, the static stability analysis for non-circular surfaces were made and the factor of safety was much higher than unity, which was contrary to Clyde's observation that the slide was slowly moving. A more careful evaluation of the strength parameters was warranted. Due to the relatively thin nature of the clay layer underlying the dense gravelly clay embankment material, it was not possible, either to clearly define the extent or to obtain a large number of high quality samples of the highly plastic slide plane material by the methods of boring and sampling procedures. Accordingly, cone penetration testing technique was adopted. The cone penetrometer used consists of a conical tip with a 60^0 apex angle and a cylindrical friction sleeve; and the CPT was conducted in general accordance with the ASTM specifications (ASTM-D-3441).

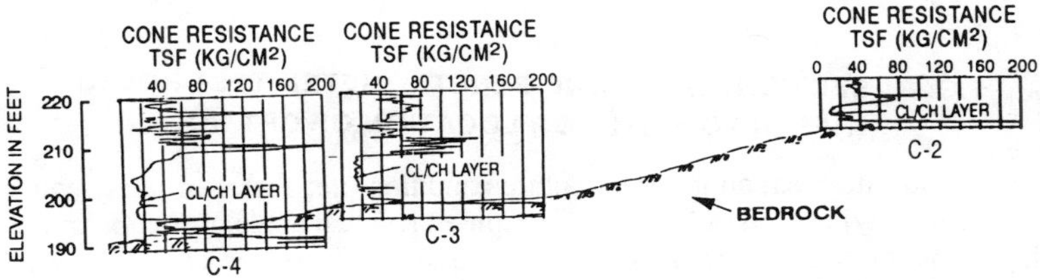

FIG. 7. SECTION THRU CONE-PENETROMETER TEST HOLES (REFER TO FIG. 2 FOR LOCATION OF CPT HOLES)

As can be noted from Fig. 7 and Fig. 8, the CPT data showed a significant drop in cone resistance just above the bedrock and below the embankment fill. The cone resistance for the embankment material shows a typical saw-tooth pattern for granular materials. In general, however, the minimum cone resistance for the embankment was generally between 25ksc to 80 ksc. Within the suspect clay layer, however, the minimum cone resistance was between 10ksc to 20 ksc. The CPT results were of special significance because it was not possible to define the extent and strength of the clay layer as clearly by boring and testing procedures. The friction ratio confirmed the continuity of the medium to highly plastic clay with the layer, thus reinforcing the hypothesis that the original slip surface developed in the lower clay layer. There still remained an unanswered question with regard to the difference in the clay strengths obtained from laboratory tests and interpreted from the CPT results. The CPT data gave a rather low magnitude of shear strength indicative of a residual value. This led to the testing of samples of the clay for residual strength measurements. A number of tests were performed under drained direct shear conditions on the clay soils with reversal of strain directions to induce residual strength; the resulting angle of friction was found to be 16^0.

Using effective stress shear strength parameters for the embankment materials, along with the water table conditions by Clyde in 1930, the stability

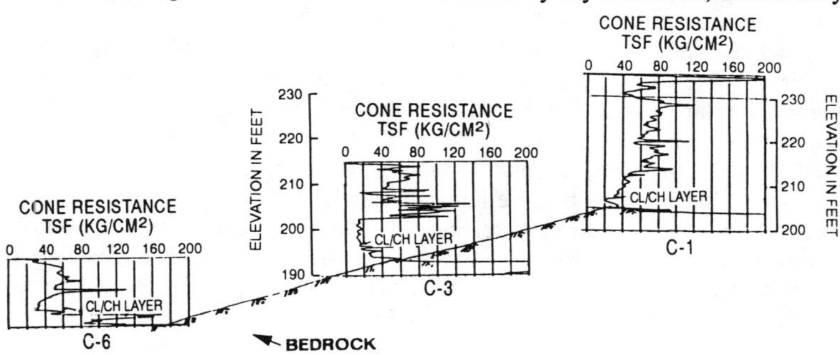

FIG. 8. SECTION THRU CONE-PENETROMETER TEST HOLES (REFER TO FIG. 2 FOR LOCATION OF CPT HOLES)

results indicated that an incipient failure condition (i.e., factor of safety equal to about unity) existed in 1930 for a computed residual angle of friction of the highly plastic slide plane material equal to 16°. This value compares very favorably with the actual laboratory results and the literature correlations (Mitchell, 1976).

The application of the in-situ cone testing technique thus helped not only in defining the failure surface, but in arriving at the proper soil strength parameters. Consequently, it was decided to perform the stability analysis of the embankment using effective strength parameters for the embankment material and a residual angle of friction equal to 16° for the highly plastic clay layer.

Static Stability Analyses of Existing Conditions

As noted earlier, the embankment was repaired during 1922-23 and had been remodeled in 1954, 1956, and 1971. The configuration before the remedial construction in 1985, had essentially remained unchanged since the last remodeling in 1971. The remodeling in each case was of a minor nature. Remodeling in 1954 and 1956 constituted smoothing of the surface; and the remodeling in 1971 involved an addition of a service road in the transverse direction located at mid-height of the slope of the embankment.

The cross-section through the slide area was still considered to be the most critical because of the pre-existing slip surface and the fact that the embankment is thicker there than at any other location. The cross-section and strength parameters used in the analyses are presented in Fig. 9. The stability analyses were performed for three different ground water conditions to evaluate the effect of fluctuations due to either reservoir leakage or surface water infiltration. The results are summarized in Fig. 9 and indicated that in

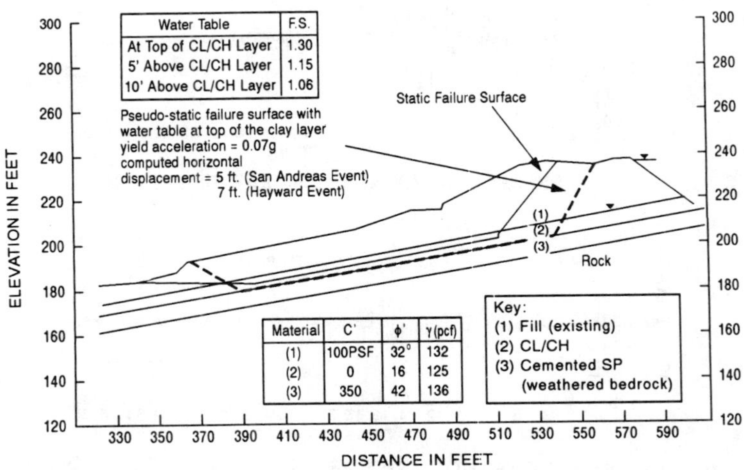

FIG. 9. STATIC STABILITY ANALYSIS - 1981 Cross Section (Effect of Water Level Fluctuations)

the former slide area the existing slope has a rather low margin of safety (FS = 1.3) even for the lowest water table condition assumed. It may be argued that he residual angle of 16° for the clay layer was too conservative as it did not allow for the effects of healing and consolidation of the clay layer after the slide movements ceased. Nevertheless, the influence of a small rise of the ground water within the embankment is found to have a significant impact on reducing the overall factor of safety; a 5-foot increase, decreased the calculated factor of safety to 1.15 and a 10-foot increase to 1.06.

Simplified Dynamic Stability Analyses

In order to evaluate the response of the embankment to design earthquake motions, simplified dynamic stability analyses were performed. The time histories of accelerations of the design earhtquakes used in this study are shown in Fig. 10. These ground motions were reviewed by the late Professor H.B. Seed (1981) who had suggested that the design earthquake motions should be representative of the most severe earthquake likely to occur on either the San Andreas or Hayward Fault. Accordingly, the histories shown in Fig. 10 are respectively for the Hayward and San Andreas Fault. A one dimensional approximation of the critical cross-section was used for two columns of soils; one through the crest at the full height of the embankment and the other through the lower road.

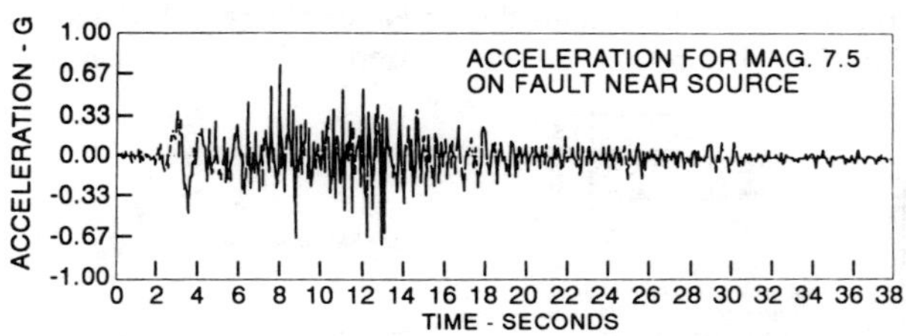

DESIGN EARTHQUAKE FOR NEAR FIELD EVENT

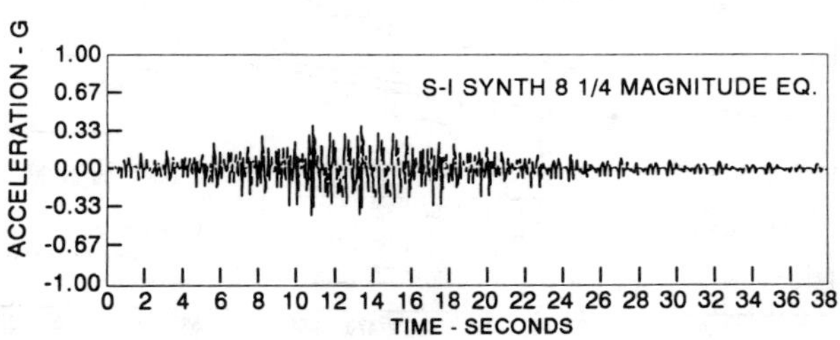

FIG. 10. DESIGN EARTHQUAKE FOR MAJOR FAULT 25 km DISTANT

Using the design earthquake motions, simplified dynamic analyses (Vrymoed et al., 1978) which utilize SHAKE analysis (1972) were performed. A one-dimensional approximation of the critical cross section was developed through the full height of embankment. Corresponding acceleration time histories were computed at the crest and mid-height of the dam.

Displacement Calculations

The response of embankments to earthquake shaking is best expressed in terms of permanent displacements rather than as a factor of safety as is customary for static analyses. Permanent displacement within the embankment are caused when accelerations exceed the minimum acceleration required to cause sliding for some mass of the embankment. This minimum acceleration is called the yield acceleration and is defined as the acceleration level at which the pseudo-static factor of safety is equal to unity. Accordingly, pseudo-static stability analyses were carried out for various potential sliding surfaces within the embankment. Once again, a non-circular failure surface passing through the clay layer was found to be the most critical sliding plane. The yield acceleration for the critical sliding block was then estimated by running several pseudo-static analyses for a range of acceleration values until an acceleration level at which the pseudo-static factor of safety was equal to one was found. This acceleration is the yield acceleration of the sliding block.

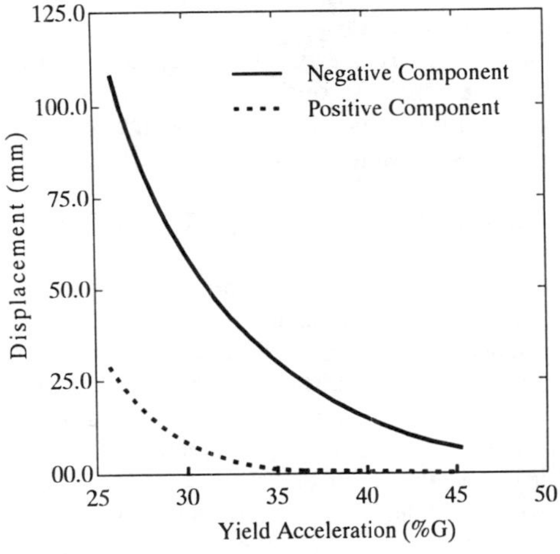

FIG. 11. YIELD ACCELERATION VS DISPLACEMENT

A rather low yield accelertion of 0.07g was computed for the water table at the top of the clay stratum. To compute permanent displacements of the sliding block along the clay layer, the computed yield acceleration level was compared with the design acceleration time histories of the San Andreas and Hayward event. Displacements were computed by performing double integration of the portions of the accelerograms that exceeded yield acceleration values. Separate analyses were performed for both the positive and negative components of the acceleration time histories. The results of these analyses are shown in Fig. 11. As shown, the computed displacements are substantially different for the Hayward Fault depending upon whether the positive (smaller) or negative (larger) acceleration component was used for double integration. Close examination of the acceleration pulses for the design earthquake accelerogram for the Hayward event show that the differences in computed displacements is primarily due to the asymmetry of the fling pulses. Similar double integration calculations for design earthquakes from the San Andreas Fault about 25 kilometers from the dam show only small differences between the positive and negative pulses. The earthquake induced displacements were computed to be about 5 to 7 ft. along the clay layer for the design earthquakes on the San Andreas and Hayward Faults. The Hayward event produced larger displacements. These displacements were considered representative of the 140-foot wide slide area. The consequences of seismically induced displacements of several feet could be serious because of possible encroachment of the failure surface into the northwest corner of the dam. Furthermore, the progressive advancement of the slide could cause cracks to develop which may extend into the reservoir area leading to serious internal erosion that could jeopardize the overall integrity of the embankment. Accordingly, additional studies were made to identify the general nature of possible remedial measures to minimize the hazard associated with seismically induced ground motions.

Alternative Remedial Actions To Improve Stability Of The Dam

It was clear that the most effective means of improving the stability of the embankment would involve removing as much of the weak clay layer as possible and replacing it with a free draining granular material of higher strength. Since the removal of a portion of the embankment by itself creates a temporary stability problem, analyses were performed to evaluate statically safe construction slopes as well as the improvement in earthquake stability due to replacement of the weak clay layer. It was assumed in the analyses that the excavation would be carried down to weathered bedrock and that a minimum thickness of 5 feet of sound, free draining granular material would replace the slide plane material. The material above the 5 ft. thick rock layer was assumed to be existing fill material re-compacted in place or a similar replacement material with at least equal strength characteristics. Alternative schemes were evaluated with the objective of limiting earthquake displacements to less than 2 inches. This limit was established in order to protect against damaging transverse cracking in the embankment.

Alternatives representing progressively greater lengths of clay layer removal were considered. In all of these alternatives, the reconstructed slopes restored the existing embankment shape. Stability analyses were carried out for each of these alternates to determine the static factor of safety for construction slopes and the static factor of safety and the yield acceleration for the reconstructed slopes, and the resulting final displacements during design earthquake conditions. For comparison purposes, displacement computations for each of the remedial alternates were also made using the procedure proposed by Seed (1979). Stability analyses were performed for construction slopes of 2 horizontal to 1 vertical, and 1.5 horizontal to 1 vertical. Assuming that the entire excavation would be completed prior to any fill placement, static construction factors of safety of less than 1.0 resulted. Improvement in this factor of safety could be obtained by staged excavation and backfill procedures.

In the foregoing analyses, it was assumed that the final embankment cross-section would be similar to that of the existing embankment. However, since the computed earthquake displacements for above alternates were still of sufficient magnitude to cause concern (greater than 5 inches), it was necessary to consider a buttressed embankment which involved removal and

Static Factor of safety	Yield Acceleration	San Andreas Event		Hayward Event	
		Peak Design Acceleration	Displacement (Ft.)	Peak Design Acceleration	Displacement (Ft.)
4.3	0.54g	0.41g	0.0* 0.0#	0.77g	0.06* <0.06#

*After seed's method (1979)
#by double integration

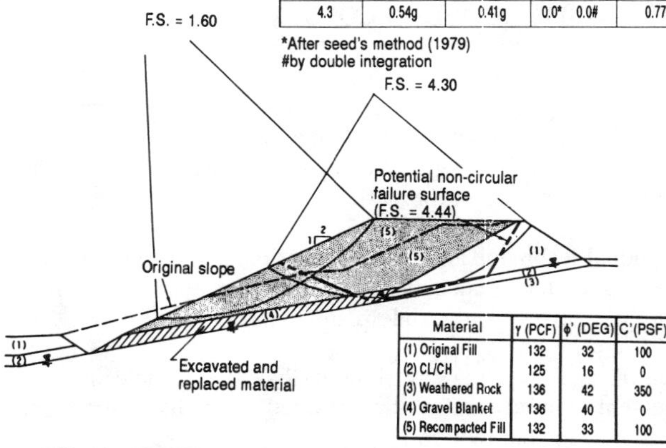

Material	γ (PCF)	ϕ' (DEG)	C'(PSF)
(1) Original Fill	132	32	100
(2) CL/CH	125	16	0
(3) Weathered Rock	136	42	350
(4) Gravel Blanket	136	40	0
(5) Recompacted Fill	132	33	100

FIG. 12. REMEDIAL CONSTRUCTION

replacement of the maximum length of weak clay. The results of this analysis (Fig. 12) clearly showed the effectiveness of this alternative in improving the stability of the embankment as the calculated displacements are less than inches for the controlling Hayward event.

The major concern during construction was the temporary instability resulting from the excavation and clearance of embankment to reach and remove the clay layer. This was handled by emptying the reservoir and by observing the cut slope of 1.75 to 1 for distress. The remedial construction was successfully completed in 1985-86 without any problems associated with the cut slope stability. The photograph (Fig. 13) shows the remedial construction.

Fig. 13. REMEDIALCONSTRUCTION

Conclusions

1. Use of the cone penetrometer test in combination with boring proved to be the most successful means to define the thickness and location of the failure slide which had grown complex because of the unsuccessful remedial methods tried in the past.

2. Observational technique and the results of the insitu cone penetration tests helped to back calculate and verify the appropriate residual strength parameters for the slide plane material.

3. The seismic stability analysis showed that the seismically induced displacements can be significantly impacted by the fling component of near field earthquake motions.

Acknowledgements

The help extended by R. L. Volpe, Geotechnical Consultant, and by Dave Dayton, Brian McCloud and Robert Fiskin of EBMUD during the investigation of this project is gratefully acknowledged. We also appreciate the help of Fred Starr of EBMUD in the preparation of this manuscript. EBMUD is especially thanked for allowing us to publish this paper.

REFERENCES

Bjerrum, L.N., (1966). "Progressive Failures in Slopes of Overconsolidated Plastic Clay and Clay Shales", The Third Terzaghi Lecture presented at the ASCE Conference; Miami, Florida.

Celestino, T.B., and Duncan, J.M., (1980). "NONCIR: A computer Program for Slope Stability Analysis with Automatic Search for Critical Non-Circular Slip Surfaces", Geotechnical Engineering, Dept. of Civil Engineering, University of California, Berkeley.

Clyde, H.S. (1930). "Investigation of Sliding Area Below Clearwell, San Pablo Filter Plant", East Bay Municipal Utility District.

Fishkin, R., and Dayton, D.J., (1972). "San Pablo Filter Plant Clearwell Stability Investigation and ANalysis", East Bay Municipal Utility District.

Marliave, B.H., (1971). "Geologic Review of San Pablo Clearwell", a letter report submitted to Mr. D.G. Larkin, Chief Engineer, East Bay Municipal Utility District.

Mitchell, J.K., (1976). Fundamentals of Soil Behavior, John Wiley & Sons, Inc., New York, pp. 284-285.

Schnabel, P.B., Lysmer, J. and Seed, H.B., (1972). "SHAKE - A computer Program for Earthquake Response Analyses of Horizontally Layered Sites". Report No. 72-12, Earthquake Engineering Research Institute; University of California, Berkeley.

Seed, H.B., (1979). "Considerations in the Earthquake Resistant Design of Earth and Rockfill Dam", Nineteenth Rankine Lecture Geotechnique 29, No. 3.

Vrymoed, J.L. and Calzascra, E.R., (1978). "Simplified Determination of Dynamic Stresses in Earth Dams", Proceedings of ASCE Specialty Conference - Earthquake Engineering and Soil Dynamics", Pasadena, California.

STABILITY EVALUATIONS FOR OLD WATER SUPPLY DAMS IN PENNSYLVANIA

James V. Hamel[1], F.ASCE

ABSTRACT

Numerous small to medium size embankment dams were constructed for water supply in the unglaciated mountainous region of central Pennsylvania over the past two centuries. Most of these older dams were designed by empirical methods and well constructed of locally available materials. Long-term stability of these older dams results from favorable material characteristics and zonation and favorable seepage and drainage patterns. Once this is recognized and demonstrated by subsurface investigations and piezometric observations, evaluation of stability is straightforward. A procedure for evaluating sliding stability of these dams is presented and illustrated with case histories of three dams each more than a century old.

INTRODUCTION

Stability evaluations for existing dams are required in connection with periodic inspections, dam safety criteria, and dam rehabilitations or modifications. In these stability evaluations, heavy emphasis is always placed on visual inspections, observational records, and dam performance histories. For large dams, major projects, and affluent owners, the stability evaluations

---

[1]Consulting Engineer, Hamel Geotechnical Consultants, 1992 Butler Drive, Monroeville, PA 15146 and GTech, Inc., 201 Penn Center Boulevard, Pittsburgh, PA 15235

are often extremely quantitative with extensive field investigations, laboratory testing, and numerical analyses. The quantitative portions of stability evaluations for smaller dams, especially those with less affluent owners, are typically of lesser scope. Regardless of the extent and complexity of the quantitative portion of the stability evaluation for an existing dam, the end product of this evaluation is generally a judgement-based decision regarding dam stability.

Substantial numbers of smaller and older dams exist in the United States and elsewhere. Specifically, in Pennsylvania, numerous small to medium size embankment dams were constructed over the past two centuries to provide water supplies for industry, transportation (i.e., canals and railroads), and municipalities. Many of these old water supply dams are still in service, primarily for municipal water supply, and some of them are being rehabilitated to meet modern criteria.

Most of these older embankment dams were designed by empirical methods before the development of soil mechanics and the evolution of geotechnical engineering. The dams were constructed of locally available soil and rock materials which were excavated, hauled, and placed for the most part by hand labor, animal power, and steam power. Construction progressed slowly by modern standards but this slower pace often provided opportunity for careful selection, zoning, and placement of embankment materials.

With these smaller, older embankment dams in Pennsylvania, as elsewhere, each dam site is unique as a result of its geotechnical framework and each dam is unique as a result of its history of design, construction, and performance. Nevertheless, a certain commonality can be expected for embankment dams located in a certain geological setting where these dams were designed and constructed in a certain technological era. This commonality provides guidance in many areas of dam engineering, including stability evaluation.

This paper was written to illustrate the art of judgement-based stability evaluation for older embankment dams of small to medium size where project budgets are limited and dam information is incomplete. A methodology of stability evaluation developed for older embankment dams in the unglaciated, mountainous region of central Pennsylvania is presented. First the region and its older dams are described briefly. Then the procedure used for evaluation of sliding stability is outlined. Finally, case histories involving three old water supply dams are given.

THE REGION AND ITS OLDER DAMS

Most of the mountainous region of central Pennsylvania lies in the Appalachian Mountain Section of the Valley and Ridge Physiographic Province. Sedimentary rock strata of Paleozoic age have been extensively folded and fractured. Deep weathering has occurred in certain rock units and in certain locations. Solution features exist in some of the carbonate rocks. Only the northern edge of the region was glaciated (Fig. 1).

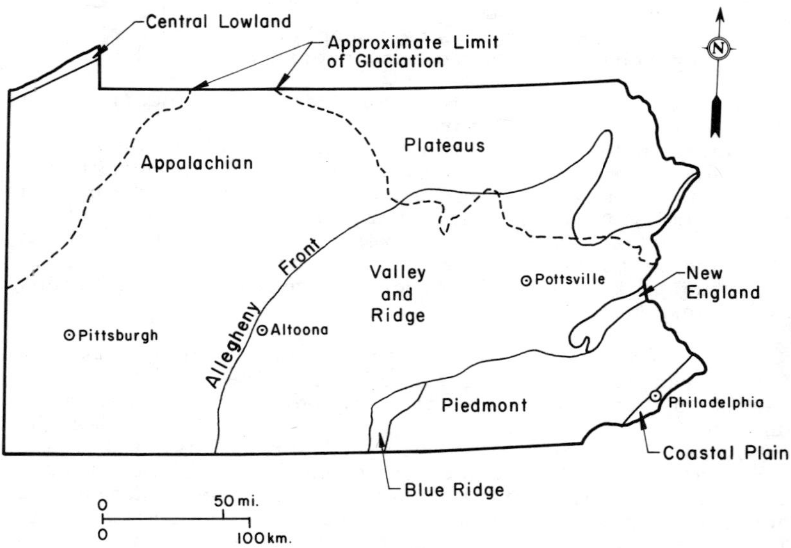

Fig. 1. Map of Pennsylvania with Physiographic Provinces

Most of the valleys are steep and narrow with relatively thin, coarse grained (often bouldery) alluvium deposited during torrential floods on the high gradient streams. Alluvium of this type, along with talus deposits on the valley walls, provided most of the material for dam construction.

Dams constructed in this region during the nineteenth century and early twentieth century were typically zoned embankments with relatively steep slopes. (See, e.g., the cross-section of the original South Fork Dam given by Frank, 1988.) Some of the dams had cores of fine material placed hydraulically between rolled or tamped shells of coarser material. Most of the dams, however, were constructed of rolled or tamped soil materials and dumped or placed rock fill. Hydraulic or semi-hydraulic fill dams are beyond the scope of this paper which is restricted to dams constructed by the latter methods.

In these dams constructed with conventional methods of fill placement, finer materials (generally well-graded granular alluvium with low to negligible clay content) were placed in the central and upstream portions of the embankments. Coarser materials (often free-draining rock fill) were placed in the downstream portions of the embankments. Foundation cutoffs and embankment core walls of puddled clay or concrete were sometimes included. There were typically no abutment cutoffs and little, if any, fractured and weathered rock was removed from the abutments.

The resultant dams tend to have modest heights and crest lengths, steep upstream and downstream slopes, leaky foundations and abutments, and sometimes leaky embankments as well. This leakage is generally benign; otherwise, the dams would not have survived to the present time. Most of the dams have (or, until recent rehabilitation, had) spillways with capacities inadequate by modern standards. Some of the dams have been overtopped by floods of various heights and durations. Despite all these apparent deficiencies, the dams have provided valuable water supplies over periods ranging from approximately 60 to 160 years. During these lengthy service lives, the reservoirs were commonly maintained full or nearly full for long periods of time.

From the standpoint of stability, the dams which survived these long service lives have withstood the test of time. These dams are strong, stable (sometimes leaky) embankments with strong, stable (though typically leaky) foundations and abutments. The long-term stability of these dams results from favorable material characteristics and zonation and from favorable seepage and drainage patterns. Foundation and abutment drainage, along with favorable drainage patterns in the zoned embankments, contribute significantly to overall stability of the dams. Once this is recognized and demonstrated by subsurface investigations and piezometric observations, evaluation of stability is straightforward.

Stability relative to piping (internal erosion) is evaluated empirically on the basis of material characteristics, seepage patterns, and dam performance history. Typically, materials in the embankments, foundations, and abutments of these dams are not susceptible to piping or they are self-healing relative to piping. Otherwise, the dams would have failed by piping (as did some early dams) during or soon after first filling of the reservoirs.

Similar inferences can be drawn relative to sliding stability of older dams which have survived to the present. Unless major changes are made to one of these older dams, there is no reason to expect conditions to change with regard to piping stability or sliding stability, assuming the dam is properly maintained.

EVALUATION OF SLIDING STABILITY

Abundant gravel to small boulder size rock fragments in the embankment materials, particularly the downstream zones, preclude meaningful laboratory strength testing with commonly available apparatus or affordable testing programs. All soil samples from borings are visually inspected and gradation tests are done on selected specimens. Soil exposures at the ground surface and in test pits are also visually inspected. Shear strength parameters are estimated from the literature with due regard for the low levels of effective normal stress prevailing in the relatively low and steep embankment slopes (Hamel, 1975; Holtz and Ellis, 1961; Holtz and Gibbs, 1956; Lambe and Whitman, 1969; Leps, 1970; Sherard, et al., 1963; U.S. Bureau of Reclamation, 1973).

The upstream slopes of these older dams commonly exist under conditions of full or nearly full reservoir with limited potential for rapid drawdown. The downstream slopes typically have high frictional shear strengths and low to negligible pore water pressures associated with gravity drainage. The dam foundations are as strong or stronger than the embankments; so sliding failures along or through the foundations are less likely than sliding failures in the embankment slopes.

A stability chart given by Hoek (1970) has been found adequate for routine evaluation of embankment slope stability under static loading conditions. This chart gives the factor of safety of the most critical circular arc failure surface through the toe of a homogeneous slope of uniform inclination for various conditions of drainage and tension cracks. Extensive use of Hoek's (1970) chart for circular failure surfaces by the writer over the past 18 years has shown this chart to give results in very good agreement with more sophisticated analytical and computer techniques, e.g., Bishop's simplified method of slices (Lambe and Whitman, 1969) for simple homogeneous or quasi-homogeneous slopes.

Central Pennsylvania has low seismicity; i.e., it is in Seismic Zone 1 (U.S. Army Corps of Engineers, 1970). As a result of their gradations and densities, the materials in the foundations and embankments of these older dams generally are not susceptible to liquefaction. Stability analyses for reasonably expected seismic loadings are usually unnecessary if the embankment slopes have adequate stability under static conditions.

LOWER AND UPPER TUMBLING RUN DAMS

Lower and Upper Tumbling Run Dams are located 0.5 mi. (0.8 km) apart on Tumbling Run Creek about 1 mi. (2 km) east of Pottsville, Pennsylvania (Fig. 1). Lower Tumbling Run (LTR) Dam was constructed from 1832-1833 and Upper Tumbling Run (UTR) Dam was constructed from 1835-1836. Both dams were originally constructed to supply water to canals but they are now used for municipal water supply. Available historical information on the dams, along with other pertinent data, is given in Phase I Inspection Reports prepared under the National Dam Inspection Program by Woodward-Clyde Consultants (1978 a,b).

Historical information on UTR Dam is scanty but there are no records of failures or serious problems. LTR Dam failed by overtopping when the spillway was temporarily restricted in 1850 and by internal erosion related to leakage from water supply pipes in the base of its embankment in 1869. LTR Dam was reconstructed after each of these failures.

Both dams had chute spillways excavated in rock along their left abutments (Figs. 2 and 4). These spillways, with capacities of approximately 30% of the Probable Maximum Flood (PMF), were inadequate by current standards. Both spillways were enlarged to pass the PMF in 1987. Stability evaluations were required for both dams during design studies for spillway modifications in 1984.

LTR Dam (Figs. 2 and 3) has a crest length of 430 ft (131 m), a maximum height of 46 ft (14 m), and nominal upstream and downstream slopes of 1.5:1 (horizontal:vertical). UTR Dam (Figs. 4 and 5) has a crest length of 570 ft (174 m), a maximum height of 54 ft (16 m), an upstream slope of 1:1, and a downstream slope of 1.3:1. Both dams are founded on alluvium with a maximum thickness of 20 ft (6 m). This alluvium consists of medium dense to very dense (typically dense) sand and gravel with scattered cobbles and boulders and a basal zone consisting primarily of boulders, cobbles, and gravel. The alluvium is underlain by fractured and steeply dipping sandstone, siltstone, and shale which also form the dam abutments. The basal boulder - cobble alluvium and underlying fractured rock act as a foundation drain beneath each dam.

Both dams have zoned embankments. The central portions of these embankments consist of medium dense to dense, well-graded, semi-pervious mixtures of cobbles, gravel, sand, and silt with minor amounts of clay. The downstream portions of the embankments consist of free-draining rock fill with cobble to boulder size particles. Similar rock fill was inferred to exist in the upstream portion of UTR Dam in view of its steep slope (Fig. 5).

Both dams have water intakes at their upstream toes and water pipes extending beneath their embankments. Fill was placed at a nominal slope of 3:1 from the intakes part way up the upstream slope of each dam. Reservoir levels are controlled by flow over concrete weirs at the tops of the spillway chutes. LTR Reservoir extends downstream from the toe of UTR Dam and the water treatment plant is immediately downstream from LTR Dam (Figs. 4 and 2).

The critical cross-section of each dam (Figs. 3 and 5) was developed from available information including that obtained from subsurface investigations in 1984. The piezometric surfaces on these cross-sections were determined from piezometer readings obtained in all the borings in each dam with full reservoirs during the latter part of 1984. These piezometric surfaces are consistent with surface observations and leakage measurements at both dams. Although the details of embankment zoning were not fully developed and the piezometric data were limited, the cross-sections were judged adequate for purposes of stability evaluation.

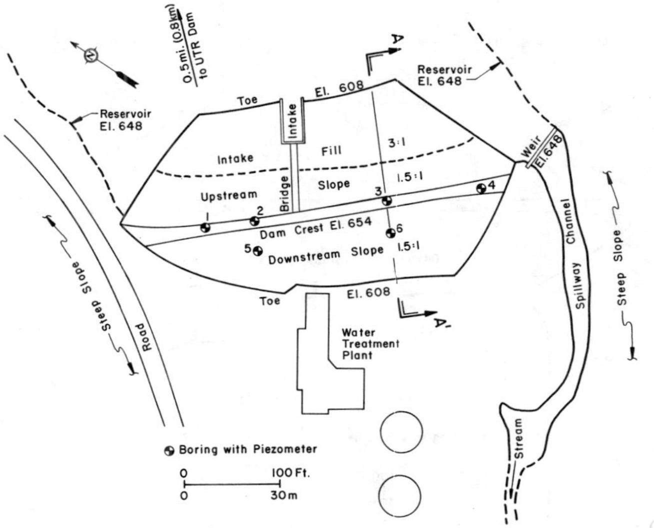

Fig. 2. Lower Tumbling Run Dam - Plan with Borings

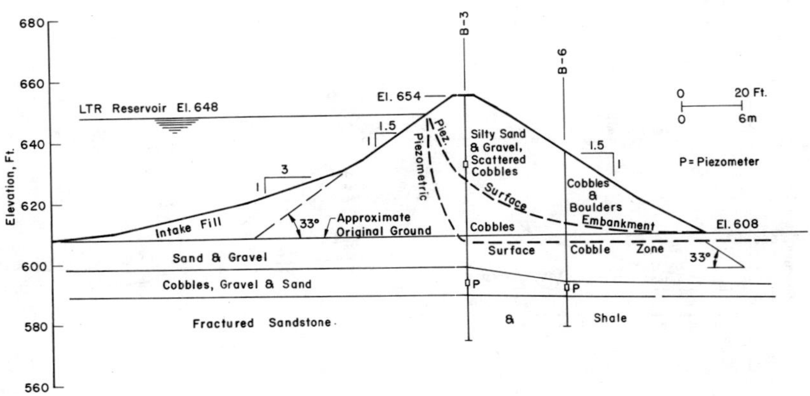

Fig. 3. Lower Tumbling Run Dam - Cross-Section A-A'

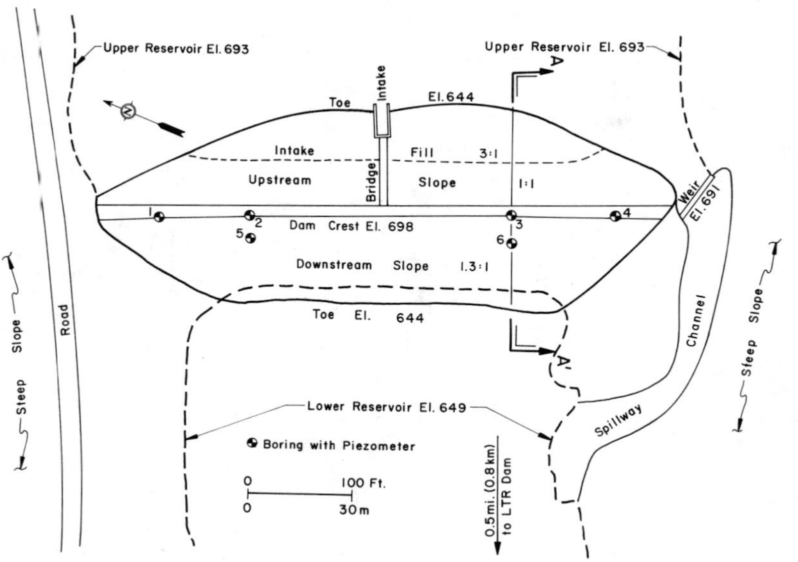

Fig. 4. Upper Tumbling Run Dam - Plan with Borings

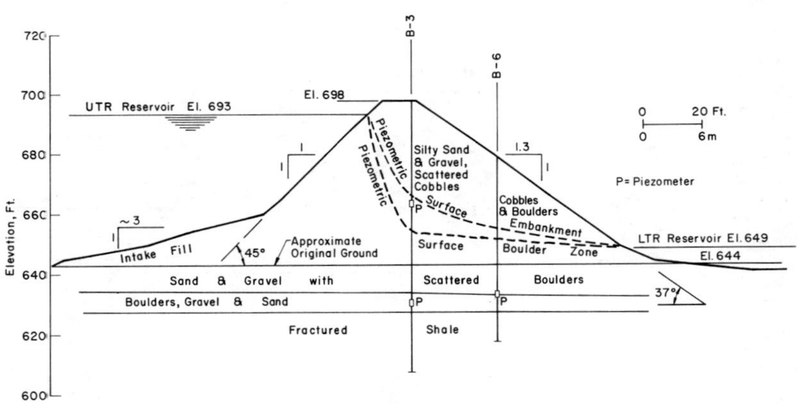

Fig. 5. Upper Tumbling Run Dam - Cross Section A-A'

The upstream slope of UTR Dam has an inclination of 45° (Fig. 5) which implies an effective angle of internal friction ϕ' greater than or equal to 45° with an effective cohesion intercept c' = 0. Similarly, the downstream slope of UTR Dam has an inclination of 37° which implies ϕ' greater than or equal to 37° with c' = 0. LTR Dam has flatter slopes of about 33° which apparently resulted from reconstruction after the 1869 failure.

These slope inclinations, along with visual observations of the materials and data from literature on shear strength of rock fill and other granular materials (Leps, 1970; Holtz and Gibbs, 1956; Sherard, et al., 1963; Lambe and Whitman, 1969; Hamel, 1975) were used to estimate ϕ' values in the range of 40° to 45° with c' = 0 for the embankment materials. The value of $\phi' = 40^{o}$ was considered a conservative lower bound for finer granular materials in the dams and the value of $\phi' = 45^{o}$ was considered a conservative lower bound for rock fill and other coarser granular materials in the dams.

The slopes of LTR and UTR Dams were considered to be fully drained with no tension cracks. The maximum inclinations of the upstream slopes were used in stability calculations. The buttressing effects of the intake fills and of reservoir water (Figs. 2-5) were neglected. Hoek's (1970) chart gave the following factors of safety for static loading conditions.

		Factor of Safety	
Dam	Slope	$\phi' = 40^{o}$	$\phi' = 45^{o}$
LTR	Upstream	1.65	2.0+
	Downstream	1.65	2.0+
UTR	Upstream	1.04	1.28
	Downstream	1.36	1.80

These factors of safety were judged adequate in view of the conservative assumptions employed in the stability evaluations and the favorable performance histories of the dams.

BRUSH MOUNTAIN DAM

Brush Mountain Dam (also called Kettle Dam) is located on Kettle Creek 3 mi. (5 km) east of Altoona, Pennsylvania (Fig. 1). The dam, which was constructed in 1888 to supply water for the Pennsylvania Railroad, is now used for municipal water supply. Available historical information on the dam, along with other pertinent data, is given in the Phase I Inspection Report prepared under the National Dam Inspection Program by D'Appolonia Consulting Engineers (1978). There are no records of failures or serious problems with Brush Mountain Dam though some repairs and modifications have been made over the years.

The dam had a concrete chute spillway along the left abutment (Fig. 6). This spillway, with a capacity of approximately 23% of the PMF, was inadequate by current standards. In 1989, a new concrete chute spillway was constructed on the downstream slope of the dam and a floodwall was installed on the dam crest so that the system could safely pass the PMF. Also in 1989, fractured rock in both abutments was grouted to reduce water losses from the reservoir. A dam stability evaluation was required during design studies for these and other modifications in 1988.

Brush Mountain Dam (Figs. 6 and 7) has a crest length of 800 ft (244 m), a maximum height of 53 ft (16 m), an upstream slope of 2.6:1, and a downstream slope of 1.5:1. The dam is founded on alluvium with a maximum thickness of 36 ft (11 m). This alluvium consists of medium dense to very dense (typically dense) silty sand to sandy silt with abundant gravel, cobbles, and boulders. The alluvium is underlain by fractured and moderately dipping sandstone and shale which also form the dam abutments. A 1914 report in the files of the Pennsylvania Department of Environmental Resources indicates that, in 1912, a concrete cutoff wall was installed in the foundation at the upstream toe of the dam and a concrete facing was placed on the upstream slope of the dam (Fig. 7).

The dam has a zoned embankment. The above-mentioned 1914 report indicates that the upstream half of the dam "was constructed of selected material, placed in thin layers and rolled" while the downstream half was "rough material, stones &c." Borings made in 1986 indicated an upstream earth fill zone and a downstream rock fill zone. The upstream zone consists of well-graded, semi-pervious to impervious mixtures of clayey sandy silt to clayey silty sand with gravel to cobble size rock fragments. The more granular portions of this material are typically medium dense and the more cohesive portions are typically stiff. Overall, the upstream material appeared to have been well-blended and well-compacted. This material, which was virtually ideal for the impervious section of an earth dam, had been blended and placed according to empirical methods pre-dating the development of soil mechanics.

The downstream zone consists of free-draining rock fill with cobble to boulder size particles. The sound of water running through rock fill in the left downstream toe area was noted in the Phase I Inspection Report (D'Appolonia, 1978). The writer heard water flowing through rock fill about 30 ft (10 m) southwest of B-11 (Fig. 6) on March 9, 1988. The sound of running water has also been reported on occasion in rock fill near the right abutment. Most of this water flowing through the downstream rock fill was attributed to abutment leakage and to groundwater discharge from the abutments.

The critical cross-section of Brush Mountain Dam (Fig. 7) was developed from available information including that obtained from subsurface investigations in 1986. The piezometric surface on this cross-section was determined from readings obtained in all piezometers and observation wells in the dam (Fig. 6) with full reservoir during 1986 and 1988. This piezometric surface is consistent with surface observations and leakage measurements at the dam. Although the details of embankment zoning were not fully developed and the piezometric data were somewhat limited, the cross-section (Fig. 7) was judged adequate for purposes of stability evaluation.

Shear strength and density values for the upstream earth fill were estimated from visual inspection of soil samples, triaxial test results on the fine grained portions of similar materials from a nearby dam, and strength data for compacted soils given by the U.S. Bureau of Reclamation (1973). The earth fill material was conservatively estimated to have c' = 200 psf (9.6 kPa) and $\phi' = 30^{\circ}$. A total unit weight of 130 pcf (2080 kg/m^3) was also estimated for the earth fill.

Shear strength values for the downstream rock fill were estimated from visual inspection of the rock fill (hard, angular sandstone fragments in a reasonably dense state) and the rock fill shear strength plot of Leps (1970). Friction angles in the range of 40° to 45° with

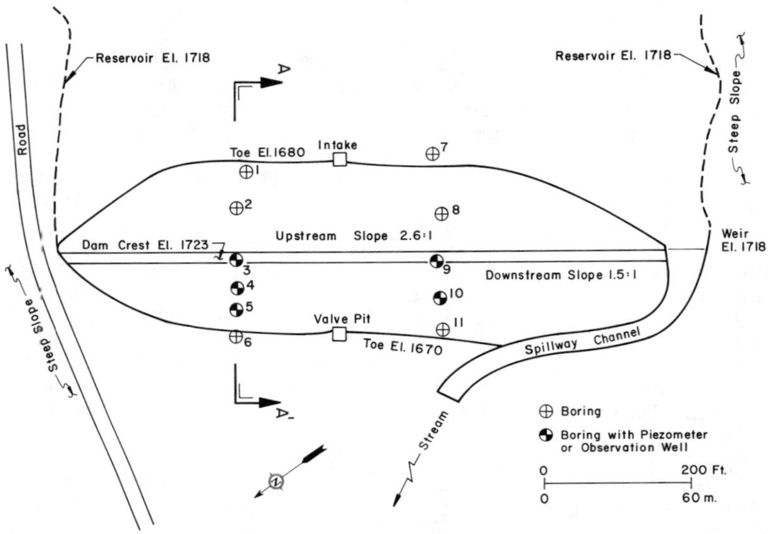

Fig. 6. Brush Mountain Dam - Plan with Borings

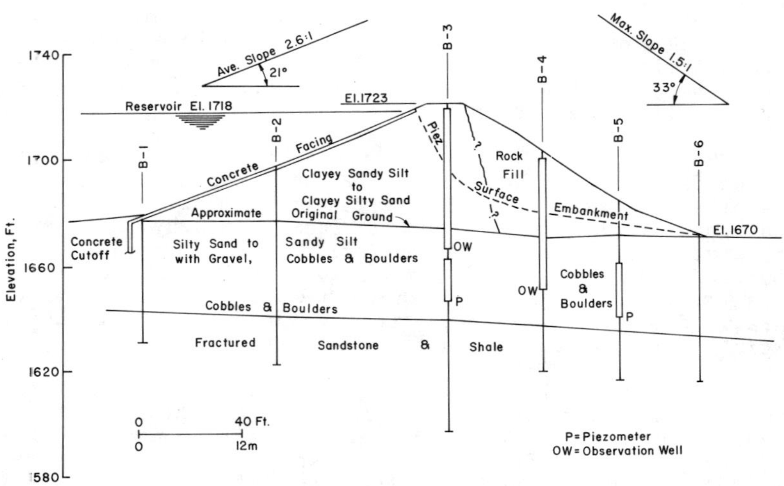

Fig. 7. Brush Mountain Dam - Cross-Section A-A'

c' = 0 were estimated for the rock fill. The 45° value was considered realistic to conservative and the 40° value was considered a very conservative lower bound.

Hoek's (1970) chart was used for the stability analyses. No tension cracks were considered. The upstream slope was analyzed for two bounding conditions of drainage - empty reservoir with fully drained slope and rapid drawdown (conservatively simulated by Hoek's case of horizontal water flow). The stability chart gave a factor of safety exceeding 2.0 for the empty reservoir condition and a factor of safety of 1.35 for rapid drawdown. The downstream slope was considered to be fully drained. The stability chart gave a factor of safety exceeding 2.0 with $\phi' = 45^\circ$ and a factor of safety of 1.60 with $\phi' = 40^\circ$. All of these factors of safety were judged adequate in view of the conservative assumptions employed in the stability evaluation and the favorable performance history of the dam.

SUMMARY

A methodology of stability evaluation for older embankment dams in the unglaciated mountainous region of central Pennsylvania was presented and illustrated with case histories of three water supply dams each more than a century old. The long-term stability of older embankment dams in this region results from favorable material characteristics and zonation and favorable seepage and drainage patterns. Major emphasis is therefore placed on developing the geotechnical framework of each dam site and each dam. Once the geotechnical framework is developed and seepage/drainage patterns are demonstrated by piezometric observations, stability evaluation is straightforward.

Stability relative to piping (internal erosion) is evaluated empirically on the basis of material characteristics, seepage patterns, and dam performance history. Sliding stability of embankment slopes under static loading conditions is routinely evaluated using stability charts and strength values estimated from published data and visual inspection of the embankment materials which typically include abundant gravel to boulder size particles existing under low levels of effective normal stress. The region has low seismicity and materials in the foundations and embankments of these older dams are generally not susceptible to liquefaction. Stability analyses for seismic loading are seldom performed as long as the embankment slopes have adequate stability under static loading.

ACKNOWLEDGEMENTS

The methodology of stability evaluation and the case histories presented herein were developed in collaboration with the late Harry F. Ferguson, M.ASCE, Hon.M.AEG. His contributions to this paper are gratefully acknowledged. Rehabilitation measures for LTR and UTR Dams were investigated and designed by Utility Engineers, Inc., Forty Fort, PA. Rehabilitation measures for Brush Mountain Dam were investigated and designed by Gwin Dobson & Foreman, Inc., Altoona, PA. Mr. Ferguson and the writer served as geotechnical consultants on these projects.

This paper was typed by Karen Lupi and Ruth Konopka of GTech, Inc. Figures were drawn by James Wylie, Technical Illustrator. The manuscript of this paper was reviewed by Joseph J. Ellam, Director, Bureau of Dams and Waterway Management, Pennsylvania Department of Environmental Resources.

REFERENCES

1. D'Appolonia Consulting Engineers (1978). "Brush Mountain Dam," Phase I Inspection Report, National Dam Inspection Program.

2. Frank, W. (1988). "The Cause of the Johnstown Flood," Civil Engineering, Vol. 58, No. 5, pp. 63-66.

3. Hamel, J. V. (1975). "Large Scale Laboratory Direct Shear Tests on Desert Alluvium," Applications of Rock Mechanics, E. R. Hoskins, Jr., ed., ASCE, New York, pp. 385-414.

4. Hoek, E. (1970). "Design Charts for Excavated Slopes," Planning Open Pit Mines, P. W. J. Van Rensburg, ed., Balkema, Capetown, pp. 295-302.

5. Holtz, W. G. and Ellis, W. (1961). "Triaxial Shear Characteristics of Clayey Gravel Soils," Proc. 5th ICSMFE, Paris, Vol. 1, pp. 143-149.

6. Holtz, W. G. and Gibbs, H. J. (1956). "Triaxial Shear Tests on Pervious Gravelly Soils," ASCE, JSMFD, Paper No. 867.

7. Lambe, T. W. and Whitman, R. V. (1969). Soil Mechanics, Wiley, New York, pp. 137-149, 365-366.

8. Leps, T. M. (1970). "Review of Shearing Strength of Rockfill," ASCE, JSMFD, Vol. 96, No. SM4, pp. 1159-1170.

9. Sherard, J. L., et al. (1963). Earth and Earth-Rock Dams, Wiley, New York, pp. 337-338.

10. U.S. Army Corps of Engineers (1970). "Stability of Earth and Rock-Fill Dams," Engineer Manual EM 1110-2-1902, p. 22.

11. U.S. Bureau of Reclamation (1973). Design of Small Dams, 2nd ed., U.S. Government Printing Office, Washington, p. 137.

12. Woodward-Clyde Consultants (1978 a,b). "Lower Tumbling Run Dam," "Upper Tumbling Run Dam," Phase I Inspection Reports, National Dam Inspection Program.

UTEXAS3 EXAMPLE PROBLEMS

Earl V. Edris, Jr.[1], M.ASCE and Dale F. Munger[2], M.ASCE

ABSTRACT

As part of the effort to explain, teach, and guide the Corps of Engineers in the proper use of the slope stability analysis program, UTEXAS3, a series of 13 example problems were developed. These problems illustrate the various program options, analysis procedures, and have been patterned after existing projects. As much as possible, validated parameters were used; however, some variables were chosen for illustrative purposes. The results from limited parameters studies and use of the four analysis procedures available in UTEXAS3 show the sensitivity of various options and procedures on the analysis results. The four analysis procedures available in the program are Spencer's, Simplified Bishop, Corps of Engineers' Modified Swedish, and Lowe and Karafiath's. In addition to discussing the example problems, the presentation will discuss when these example problems could be considered calibrated bench mark problems.

---

[1]Research Civil Engineer; Geotechnical Laboratory, Waterways Experiment Station, U.S. Army Corps of Engineers, ATTN: CEWES-GS-S, 3909 Halls Ferry Rd., Vicksburg, MS 39180-6199

[2]Geotechnical Engineer, U.S. Army Corps of Engineers, North Pacific Division, ATTN: CENPD-PE-GT, P.O. Box 2870, Portland, OR 97208-2870

THE ROLE OF BENCHMARK PROBLEMS IN SLOPE STABILITY COMPUTATIONS

Stephen G. Wright[1], Member ASCE

ABSTRACT

"Benchmark" problems provide a useful means for detecting some of the many different types of errors that may exist in slope stability computations. For this discussion benchmark problems are considered to include any problem that can be used to check the validity of slope stability computations. Benchmark problems can range from artificial problems designed to test a specific numerical algorithm employed in computer software to complex, practical problems that are intended to illustrate proper slope stability computations.

The importance and effectiveness of benchmark problems for detecting errors depends on the type of error. Errors can be grouped into several categories, which are listed below and commented on with respect to the role that benchmark problems may play:

1. Errors in coding computer software or in the basic algorithms implemented in the computer code. Benchmark problems are essential and effective for detecting errors of this type. This is probably the most common use of benchmark problems.

2. Errors due to the method of stability analysis used. Errors likely to be caused by a particular method of slope stability analysis are now well-understood: Experience and extensive research have shown that errors due to the mechanics are

[1]Ashley H. Priddy Centennial Professor in Engineering, The University of Texas, Austin, TX 78712-1076

usually negligible provided that the method of stability computations satisfies complete static equilibrium (e.g. Morgenstern and Price, Spencer, Sarma, etc.). Thus, errors associated with the mechanics can be determined without benchmark problems, although benchmark problems may be useful for illustrating such errors and may serve to alert an uninformed user.

3. Errors inherent in the theory and algorithms employed for the slope stability analyses. All methods of slope stability no matter how correct fundamentally and how correctly they are implemented can produce erroneous results. One example of such erroneous results is the computation of high tensile stresses near the active zone of a slope in cohesive materials. The tensile stresses are correct for the mechanics employed and the assumption of a cohesive material; however, they may not be valid for most soils which cannot be relied upon for tensile strength. In such cases, where unrealistic tensile stresses develop, the problem is usually remedied by introducing into the stability analysis a vertical crack near the upper portion of the shear surface. Another example of erroneous results is the computation of very high compressive or tensile stresses at the toe of the shear surface near the exit point. This condition is generally caused by a shear surface which is excessively steep. Considerations of passive resistance will indicate that a much flatter shear surface should form. The problem is remedied by changing the inclination of the shear surface near the toe of the slope. Both of the conditions of unreasonable stresses described above are inherent in the mechanics of slope stability computations and are caused by selection of an inappropriate geometry for the shear surface, rather than any error in computer software. At least one of the errors of these types can be detected by incorporating appropriate checks into computer software. Benchmark problems can serve to verify that the checks are functioning as intended; however, they can never assure that the user will be alerted of all inherent errors associated with the theory being used.

4. Errors in entering input data into a computer program. These errors can occur regardless of the correctness of computer software. However, good software should be able to check at least some data for validity. For example, negative unit weights, negative shear strength parameters, and excessive artesian

pressures may be detected and an alert issued. For errors in this category benchmarks serve to verify input data error checking in the computer software, but they cannot eliminate errors in input data.

5. Errors in defining the type of problem to be solved. A common error of this type is the use of short-term stability analyses when long-term stability is more critical. Regardless of the correctness of the input values the analysis will not be correct if the wrong problem was solved. Benchmark problems may serve to provide well-illustrated examples of slope stability problems and, thus, might indirectly reduce some errors of this type. There is some promise that in the future expert systems might advise the user in an analysis and benchmarks could also serve the purpose of verifying the expert systems.

6. Errors in selecting and defining input values for the problem. This includes use of erroneous strength values and incorrect soil profile geometry, often caused by unknown conditions in the field. There is again the possibility that in the future expert systems could reduce some errors of this type, and benchmarks could serve to verify the computer code. However, benchmarks will not be adequate to eliminate many problems associated with determination and selection of input values for the field.

Not all of the errors described above can be eliminated by benchmark problems; however, benchmark problems are essential and can eliminate or reduce the occurrence of many of these errors. There are currently a number of sources of good benchmark problems and more will be needed as new features and approaches to slope stability analyses evolve. For example, in recent years the need has developed for benchmark problems to check the validity of stability computations for reinforced slopes and to verify some of the more recently developed schemes for automatically searching for the most critical noncircular shear surface.

COMPARISON OF RIGOROUS SLOPE STABILITY METHODS: STATICAL ASPECTS

Dov Leshchinsky[1]

ABSTRACT

The results of a comparative study of three slope stability methods, which explicitly satisfy all limiting equilibrium equations, are presented. These methods are Morgenstern-Price, Janbu and the variational approach. It is shown that all three methods are equivalent and therefore should render identical safety factor and critical slip surface. However, in the first two methods the results depend on the statical assumptions (i.e., either inclination of interslice forces or location of its resultants). It is also demonstrated that to have a valid comparison (i.e., a valid bench-mark problem) all internal forces should be compared in the first two methods. The third method does not require such an elaborate comparison since it is statically determinate without resorting to interslice statical assumptions. When comparing all three methods the potential mass discritization must be identical. Ten different complex problems reported in the literature are compared to support this study's conclusions. This comparison also indicates the applicability of the variational analysis implying it is, at least, on par with the other two rigorous limiting equilibrium methods. It also shows that the variational method is more efficient than the other two in the sense that no guessing is needed to obtain a statically determinate problem, nor is it necessary to assess the results for statical reasonableness. Hence, bench-mark problems are used to verify and evaluate the applicability of a new slope stability method.

---

[1]Associate Professor of Civil Engineering, University of Delaware, Newark, DE 19716

RELIABILITY AND PROBABILITY IN STABILITY ANALYSIS

John T. Christian[1], F. ASCE, Charles C. Ladd[2], F. ASCE,
and Gregory B. Baecher[3], M. ASCE

ABSTRACT

Although geotechnical engineers have long recognized that they deal with loads and materials that are fraught with uncertainty, the use of formally probabilistic methods for the analysis of slope stability began only about twenty years ago and has had less impact on the practice of geotechnical engineering than would be expected. Many traditionally trained engineers are not familiar with probabilistic concepts, and it has been difficult to incorporate concepts of reliability into regulations and traditions of practice. Also, there has been much confusion over precisely what is meant by reliability and probability of failure. The most effective applications of probabilistic methods are those in which the relative probabilities of failure are evaluated or in which the methodology is used to illuminate the effects of uncertainties in the parameters. Attempts to determine the absolute probability of failure are much less successful. Following a brief review of probabilistic concepts, the paper describes how appropriate probabilistic descriptions of soil parameters can be derived from field and laboratory data and applied in stability analysis. The first-order, second-moment approach is explored, and the techniques are applied to the design of embankment dams. The example illustrates how probabilistic analysis can reveal the relative contributions of the uncertainties about different parameters to the reliability of the embankment. Reliability analysis is especially useful in establishing design values of factors of safety that represent risks that are consistent for different types of failure.

---

[1]Executive Vice President, Stone & Webster Advanced Systems Development Services, P. O. Box 2325, Boston, MA 02107.

[2]Professor of Civil Engineering, Massachusetts Institute of Technology, Cambridge, MA 02139.

[3]President, ConSolve, Inc., 70 Westview St., Lexington, Massachusetts 02173

INTRODUCTION

It is axiomatic among geotechnical engineers that they deal with materials whose properties are difficult to establish and subject to large, unpredictable, and uncertain variations. Even identifying what materials are present can present a significant challenge, and the vagaries of construction processes add to the uncertainties. While many disciplines can deal with such difficulties by working with large factors of safety or conservatism, economic factors close this avenue for many geotechnical engineering projects, certainly for many slopes and embankments.

Geotechnical engineers have dealt with uncertainties by recognizing that risk and uncertainty are inevitable and by applying the observational method (Peck 1969) to maintain control over them. The need for the geotechnical engineer to be involved at all stages of the project is one of the exciting aspects of the field and has been a very important factor encouraging many young engineers to enter the profession. The observational method is applicable only when the design can be changed during construction on the basis of observed behavior. In those cases where the critical behavior cannot be observed until too late to make changes, the designer must rely on a calculated risk. Although prominent engineers have long emphasized the role of the calculated risk in geotechnical engineering (Casagrande 1965), the literature contains relatively few examples of formal evaluation of risk for slopes and embankments, and the profession has been relatively less vigorous in the use of modern probabilistic techniques than have other branches of engineering.

The most effective applications of probabilistic concepts to analysis of the stability of slopes and embankments have used the so-called "first-order, second-moment" methods described by Wu and Kraft (1970), Cornell (1971), Alonso (1976), Tang *et al.* (1976), and Vanmarcke (1977), among others. The techniques have now evolved from purely theoretical investigations into methods that can be applied to practical problems. Nevertheless, their application has not been as widespread as would be warranted by their power or by the original developers' expectations.

At a conference devoted to developments in the understanding of the behavior of slopes and embankments over the last twenty-five years, it is appropriate to ask how probabilistic methods should be used and for what ends. The present paper first reviews some basic, well-known ideas about probabilistically based reliability. It then describes the application of reliability methods to the evaluation of slope stability, primarily by means of a specific example. Finally, it summarizes some of the lessons that have been distilled from such practical applications.

THE SCOPE OF PROBABILISTIC ANALYSIS

Probabilistic analysis is generally understood to deal with the probability of failure of a slope or embankment, given a particular state of knowledge about the loads and the properties of the materials. Probabilistic concepts are also the foundation of several techniques to solve otherwise deterministic problems (*e. g.*, Cherubini and Greco 1987), and the related methods of genetic algorithms and simulated annealing have recently become powerful tools for such analysis. However, they will not be discussed further here; this paper is concerned with attempts to deal with the probabilistic nature of the physical problem itself.

What is meant by the "probability of failure"? The established methods for performing probabilistic analysis of slope stability deal primarily with uncertainty in the analyst's knowledge of the properties of the materials and the loading conditions. They do not address inadequacies in spillway capacity, failure to prevent or detect internal erosion, or errors in construction, and, unless special attention is devoted to the problem, they seldom consider that an inadequate field exploration program might miss a critical geological detail. Nevertheless, these are among the major causes of failure of slopes and embankments. Indeed, Baecher *et al.* (1980) determined that earth dams seem to have probabilities of failure of about 10^{-4} per dam-year for a very wide range of locations and times of construction. Clearly any probability of failure that is to be used in benefit-cost analysis or probabilistic risk assessment must be based on more than the uncertainty in known geotechnical properties. This is one of the principal points raised by Baecher *et al.* (1984) and Christian (1984).

Even if the calculated probability of failure were to include the many imponderables that fall outside of the usual scope of stability analysis, it would still not be clear how to use the resulting numbers. Members of the public, and for that matter most engineers, have great difficulty understanding the implications of events with large consequences and small probabilities of occurrence. Peck *et al.* (1987), as part of a discussion of the societal implications of landslides, give the data of Table 1. Obviously people readily accept the relatively large risks of driving cars or flying to conferences in distant cities. It is well known that acceptance of risk is not purely rational, and simply providing the public with numbers for annual probability of failure of an engineered structure without some evaluation is an invitation to misunderstanding and confusion.

A far more useful approach is to deal with relative probabilities of failure or other measures of reliability. The engineer can then incorporate uncertainty into the decision-making process without trying to establish an absolute value of the probability of failure. This is the approach used here. However, in some cases it may be necessary to provide and explain probabilistic risks to the public, and the engineer involved in such an exercise should recognize that the present methods provide only part of the answer.

Table 1. Annual Probabilities of Death of a Selected Individual from Various Activities (after R. T. Peck *et al.* (1987))

Activity	Probability*
Voluntary individual risks:	
Air travel (crew)	1:1000
Car Travel (1984 British Columbia)**	1:3500
Construction	1:6000
Air Travel (passenger)	1:9000
Involuntary individual risks:	
Fire	1:50 x 10^3
Drowning	1:100 x 10^3
Lightning	1:5000 x 10^3
Structural failure	1:10,000 x 10^3
* Relative to population involved in the activity ** For individual traveling 10,000 mi/yr	

BASIC CONCEPTS

The most effective applications of probability theory to the analysis of slope stability have stated the uncertainties in the form of a reliability index, which is defined as:

$$\beta = \frac{E[F] - 1.0}{\sigma[F]} \quad . \tag{1}$$

In this equation β is the reliability index, $E[.]$ indicates the expected value or best estimate of the variate enclosed in brackets, F is the computed factor of safety, and $\sigma[.]$ indicates the standard deviation of the variate enclosed in brackets. The reliability index describes safety by the number of standard deviations separating the best estimate of F from its defined failure value of 1.0. It can also be considered as a way of normalizing the factor of safety with respect to its standard deviation. The definition in equation [1] does not require that the probability distribution of F be known, but there is an implication that the mean and standard deviation provide a meaningful description of F.

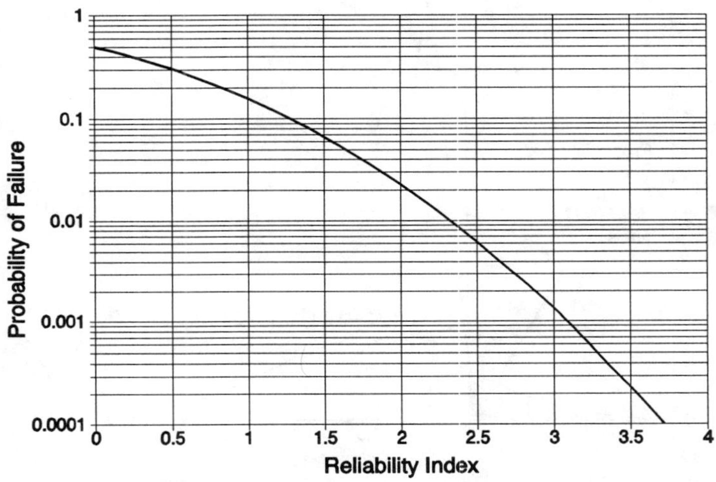

Figure 1. Nominal Probability of Failure for Normally Distributed F as a Function of the Reliability Index.

When the shape of the probability distribution of the factor of safety is known, the reliability index can be related to the probability of failure. Figure 1 shows the relation for the case of a normally distributed factor of safety. Other distributions give similar curves for values of β less than 1.5 to 2.0, but the curves differ substantially for higher values of β. For any consistently made assumption on the probability distribution of F, β is monotonically related to the probability of failure, so that higher values of β imply lower probabilities of failure. Of course, in this context the probability of failure incorporates only those factors that are included in the analysis and excludes the many important factors mentioned in the previous section. Therefore, the probability must be considered a lower bound on the actual probability of failure. Also, all uses of the reliability index tacitly assume that F is distributed in some way that can be described meaningfully by an expected value and a standard deviation.

Whether or not the probability distribution of F is known, the reliability index expresses the uncertainty in the stability of the slope. Figure 2 compares two idealized cases, in both of which F is assumed to be normally distributed about the mean value. It is clear that the case with the mean F of 1.5 and $\sigma[F]$ of 0.5 has a

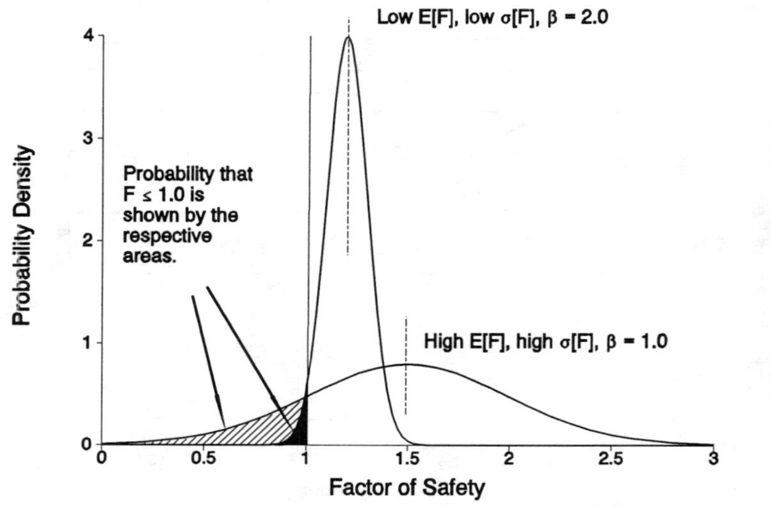

Figure 2. Comparison of Two Situations with Different Means and Distributions of Factor of Safety.

higher overall probability of failure than does the case with a mean F of 1.2 and a smaller $\sigma[F]$ of 0.1. The rational approach is to develop a design to achieve a consistent reliability index rather than relying on some arbitrary value of F alone.

The reliability index can also be expressed in terms of the coefficient of variation of F, defined by:

$$Cov[F] = \frac{\sigma[F]}{E[F]} . \qquad (2)$$

It then follows that

$$\beta = \frac{E[F] - 1.0}{E[F] \cdot Cov[F]} . \qquad (3)$$

Figure 3 shows the relations among β, the expected value of F, and its coefficient of variation. From this figure, or from the underlying equation [3], it is easy to select design values of F that have the same reliability index. For example, to achieve β of 2, when Cov[F] is 0.2, the computed F must be 1.67, but, when Cov[F] is 0.1, the computed F need be only 1.25. Thus, the reliability index expressed in terms of the coefficient of variation of the factor of safety provides an internally consistent

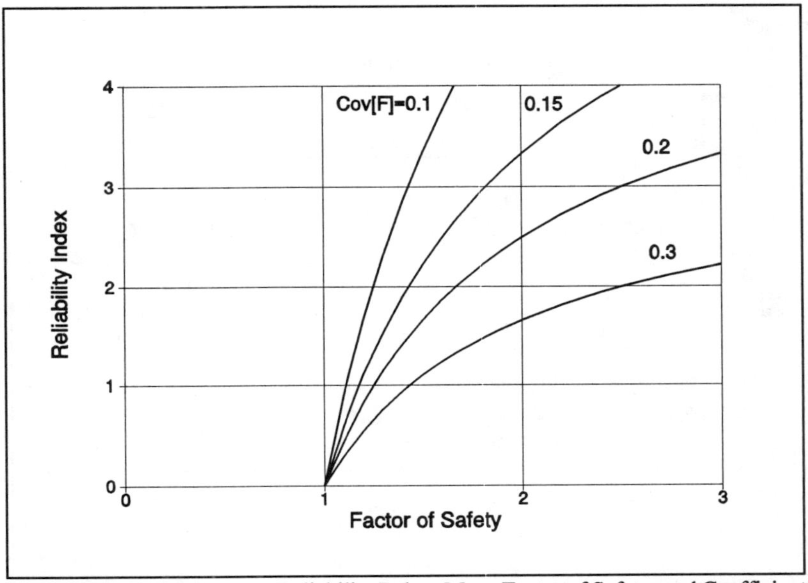

Figure 3. Relation between Reliability Index, Mean Factor of Safety, and Coefficient of Variation of Factor of Safety.

criterion for design. How then is the variation of F to be found?

UNCERTAINTY IN STABILITY ANALYSIS

The uncertainty in the stability of a slope is the result of many factors. Some, such as ignorance of geological details missed in the exploration program, are difficult to treat formally; others, such as the estimates of soil properties are more amenable to statistical analysis. Ladd (1983) describes the application of probabilistic and statistical methods to several practical problems of slope stability. The application of probabilistic and statistical methods to a practical problem will be illustrated by work done by the last two authors as part of activities to recommend techniques for designing embankment dams on soft sensitive clays for the James Bay hydroelectric project in northern Quebec. Ladd *et al.* (1983) and section 3.5 of Ladd (1991) present details of the example design problem used for reliability analysis, the James Bay site B-6.

Figure 4 illustrates how the uncertainties in soil properties can be separated into two categories: those arising from scatter in the data, and those arising from

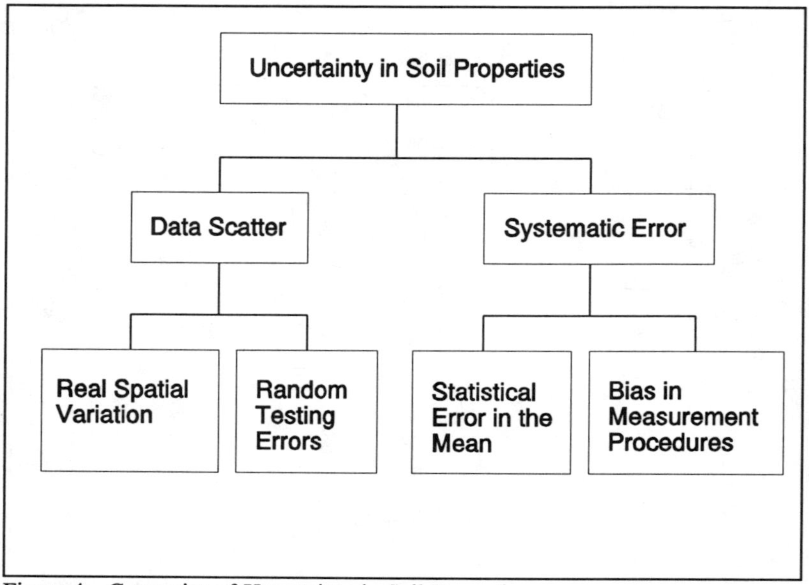

Figure 4. Categories of Uncertainty in Soil Properties.

systematic error in the estimate of the properties. The former consists of inherent spatial variability in the properties and random testing errors in the measurement of the properties. The latter consists of systematic statistical errors due to the sampling process and bias in the measurement process itself. In summary, there are four problems to be faced in selecting design shear strengths:

(1) *Scatter* in the data due to real *spatial* variability within the soil profile,
(2) *Scatter* in the data due to *random* testing errors or noise,
(3) A *systematic error* in the computed mean value of the property due to the limited number of tests performed, leading to a *statistical* uncertainty, and
(4) An *error in the mean* due to *bias* in the measurement, such as those known to occur in field vane testing.

A statistical analysis of these factors not only provides input to the overall probabilistic analysis of stability but also gives insight into their relative importance.

Some definitions and notational conventions need to be established before proceeding further. If the two properties or variates in question are x and y, and they are sampled at n points identified by the subscript i, then the following definitions apply (It is understood that Σ implies summation over all possible values.):

The *mean* or *expected value*, $E[x]$, is

$$E[x] = (1/n) \sum x_i \ . \tag{4}$$

The *variance*, $V[x]$, is

$$V[x] = (1/(n-1)) \sum (x_i - E[x])^2 \ . \tag{5}$$

The *standard deviation*, $\sigma[x]$, is

$$\sigma[x] = (V[x])^{1/2} \ . \tag{6}$$

The *covariance* of two variates, $C[x,y]$, is

$$C[x,y] = (1/(n-1)) \sum (x - E[x]) \cdot (y - E[y]) \ . \tag{7}$$

Their *correlation coefficient*, $\rho_{x,y}$, is

$$\rho_{x,y} = C[x,y] \ / \ (\sigma[x] \cdot \sigma[y]) \ . \tag{8}$$

The *autocovariance* of a single variate over a distance r, $C_r(r)$, is

$$C_x(r) = (1/(n-1)) \sum (x_i - E[x]) \cdot (x_{i+r} - E[x]) \ . \tag{9}$$

The *autocorrelation coefficient*, $R_x(r)$, is

$$R_x(r) = C_x(r) \ / \ V[x] \ . \tag{10}$$

The autocovariance and autocorrelation describe how well the value of a parameter at a given point can predict the value at another point some distance away. Because these are both functions of distance r, it is often convenient to define the *autocorrelation distance*, r_0, as the distance at which $R_x(r)$ due to spatial variability decreases to 37 per cent (the inverse of the base of natural logarithms, e) of the value at $r = 0$. This provides a single number with which to describe the smoothness of spatial variability. A large autocorrelation distance implies a very uniform material that can be described by few, widely separated tests. A small distance implies a material whose properties change rapidly over short distances.

ASSESSING SOIL PROPERTIES AT JAMES BAY

The James Bay hydroelectric project required the construction of approximately fifty kilometers of dikes founded on soft and sensitive clays. Procedures for evaluating the stability of embankments in such circumstances are among the most complicated and controversial topics in geotechnical engineering. Divergent views are particularly prevalent regarding the stability of a strengthened foundation soil during construction, selection of appropriate factors of safety, and evaluation of strength properties. Because of the complexity of the geotechnical problems, an international committee was formed to address and resolve the issues concerning stability of the slopes. One of the first matters to be dealt with was the estimation of the strength of the *in situ* foundation soils.

Data Scatter: Spatial Variability and Noise

Field vane tests conducted with the Nilcon device on the James Bay clays exhibited an unusually large degree of scatter. If μ is Bjerrum's (1972) empirical factor for correcting the bias in field vane data, then a common model of measurement error is

$$\mu \, c_u(FV) = c_u + c_e \tag{11}$$

where c_u is the actual undrained strength, $c_u(FV)$ indicates the value measured by the field vane test, and c_e is a random experimental error with a zero mean. This equation assumes that μ is known precisely. In order to conduct a meaningful stability analysis the engineer needs to know the actual *in situ* shear strength, c_u, including its spatial variability, but the random experimental variations represented by c_e should be eliminated. One way to do this is to use the autocovariance function, provided it does not change with position.

Figure 5 illustrates the ideas behind the analysis. First the expected value, that is the mean or the trend in the data, must be evaluated by some appropriate method such as linear regression. Then the autocovariance is a measure of the statistical description of the variation of the property from the expected value as a function of the distance between the points where the property is measured. Part a of the figure shows the statistical relation for the actual spatial variation in c_u; as would be expected, it decays smoothly with increasing separation between the points. Part b shows the measurement error. It is a spike at a separation of zero because measurement error is purely a local phenomenon. It includes instrument error, operator error, and very small scale variation in soil properties. When the relations of parts a and b are added, the plot of part c results. This is what would be expected from a set of field vane measurements.

Because the autocovariance functions for soil properties can be expected to be different in the vertical and horizontal directions, the vertical autocorrelations should be done using values from the same boring, and the horizontal autocorrelations using data from different borings at the same elevation. Then each pair of data is used in equation [9] to compute the autocovariance as a function of the separation distance r. The results are then averaged or smoothed within each interval of r. The results in the vertical direction for the marine clay at the B-6 site are shown in Figure 5d. The engineer must extrapolate the autocovariance back to the vertical axis to obtain the autocovariance at $r = 0$. The difference between the observed variance and the extrapolated value must be measurement error, noise, or small scale variation. The extrapolation can be done visually, as here, or by fitting an appropriate curve,

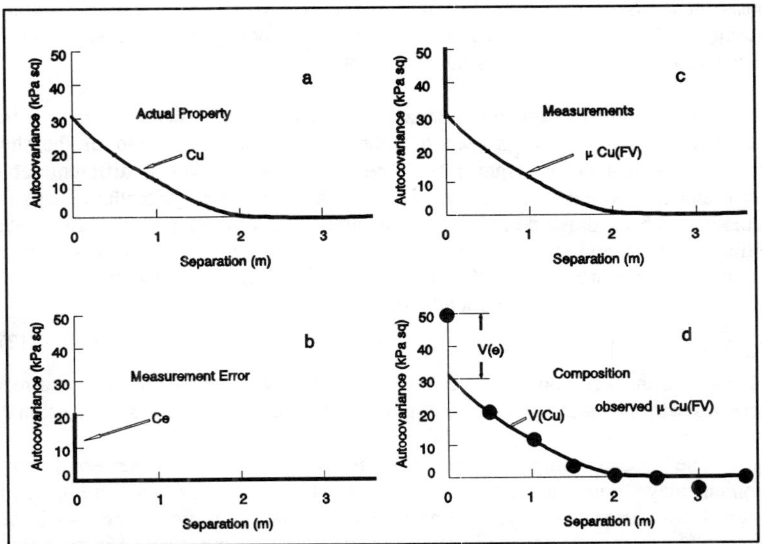

Figure 5. Composition of Autocovariance Function. The actual properties (a) combine with the measurement error (b) to give the total distribution of variances in the measurements (c). This compares with the observed data (d). The figure assumes that $\mu = 1.0$.

although in the latter case the apparent precision of the estimate may be misleading.

It must be emphasized that the purpose of the exercise is to separate the observed variation in measured soil properties into two categories. The first is the actual spatial variance in the field values, which should represent a real effect that occurs in the field and needs to be taken into account in the analysis of slope stability. The second consists of random scatter due to error in measurements and very small scale fluctuations in soil properties. While these certainly do appear in the observed data, they do not affect the behavior of the soil in the field and should not be included in the slope stability analysis.

Systematic Error

In addition to the data scatter and spatial variability discussed above, there are also systematic errors in the estimation of the soil properties. The systematic error contributes to the analytical evaluation of slope stability very differently because the spatial variability averages over the volume of soil but the systematic error is consistent across the volume. Therefore, the systematic error propagates almost unchanged through an analysis and can have a much greater influence on the predictive uncertainty than does the spatial variability.

In the present example two factors contribute to the systematic error. First, a limited number of tests is used to determine the mean or trend of the shear strength. A smaller or larger set of measurements, or even a different set of measurements, would probably yield a different estimate. Standard statistical methods, such as those described by Benjamin and Cornell (1970), can provide confidence limits and variances of estimates obtained by regression analysis. In the present case of a mean value of a parameter x, the variance in the estimate of the mean from n measurements is found from

$$V\{E[x]\} \approx V[x] / n . \tag{12}$$

This leads to the well known result that the standard deviation of the estimated mean decreases with the square root of the number of observations used for the estimate.

The second component of systematic error is bias. The experimental technique may not measure directly the quantity of interest or may measure it in some systematically erroneous way. The need to correct data from field vane tests is well known. The factor μ in equation [11] is Bjerrum's (1972) empirical correction factor as a function of the plasticity index, I_p. However, there is scatter in the data defining this relationship (Ladd *et al.* 1977, Azzouz *et al.* 1983, Ladd 1983). The scatter, which has a coefficient of variation of about 25 per cent, represents the uncertainty in using this correction factor. On the basis of other observations on the marine clay from the present example, μ was taken to be 1.0, but, of course, this would not be true in most cases.

In separating spatial variability and systematic error it is perhaps easiest to think of *spatial* variation as describing the scatter of the soil property *about* the mean trend and of *systematic* error as uncertainty in the *location* of the mean trend itself. The only component of the former is inherent spatial variability after random measurement error has been removed, while the latter consists of measurement bias and statistical estimation error due to a finite number of measurements. The contribution of the spatial variability depends on the size of the region or failure surface over which it is effective in relation to the autocorrelation distance, r_0, and will be discussed in more detail later.

FIRST ORDER PREDICTIVE MODEL TO ESTIMATE UNCERTAINTIES

The so-called first-order, second-moment predictive model starts by considering that a performance variable y can be described by

$$y = g(\underline{x}) + e \tag{13}$$

where y is, for example, the factor of safety, the shear strength, or some other parameter of interest; g is a geotechnical model such as the simplified Bishop method of slices that depends on a vector $\underline{x}$ of geometry and soil properties; and e is the modeling error that describes how the model deviates from the real world. The uncertainty in y can be measured by its variances, and these can be estimated by expanding equation [13] in a Taylor series and truncating to the first terms to yield

$$V[y] \approx \sum_{i=1}^{k}\sum_{j=1}^{k} \frac{\partial g}{\partial x_i}\frac{\partial g}{\partial x_j}\, C[x_i,x_j] + V[e] \tag{14}$$

where $C[x_i,y_i]$ is the covariance of the parameters x_i and x_j contributing to the calculations of the variable y and

$$C[x_i,x_i] = V[x_i] \; . \tag{15}$$

If all the x_i and x_j are uncorrelated for $i \neq j$, then

$$V[y] \approx \sum_{i=1}^{k} \left[\frac{\partial g}{\partial x_i}\right]^2 V[x_i] + V[e] \; . \tag{16}$$

These equations use the variances, or second moments, of the distributions, and they are related by first order approximations. Hence, the approach is called a first-order, second-moment method. Other formulations, using higher moments or higher order approximations, are possible but at a very great increase in analytical complexity and computational effort.

There are two direct ways of evaluating equation [14] or [16]. First, the model, g, can be differentiated formally to give a closed form expression for $V[y]$. However, in many practical cases the computation of the statistics is much too complicated for formal differentiation, and a second approach is needed. This is to calculate divided differences as approximations to the partial derivatives.

The Effect of Uncertainty on SHANSEP c_u Values

Noiray (1982) provides a good example of the first way of dealing with equation [16]. An equation analogous to equation [13] describes the undrained shear strength in the SHANSEP approach (Ladd and Foott 1974):

$$c_u = \sigma'_{vo} \cdot S \cdot \left(\frac{\sigma'_p}{\sigma'_{vo}} \right)^m , \tag{17}$$

If all the variables on the right side of the equation are independent and σ'_{vo} is assumed to be deterministic, then direct substitution and differentiation give the following approximations to the mean and the coefficient of variation of the shear strength:

$$E[c_u] = \sigma'_{vo} \cdot E[S] \cdot \left(\frac{E[\sigma'_p]}{\sigma'_{vo}} \right)^{E[m]} \tag{18}$$

and

$$Cov^2[c_u] = Cov^2[S] + E^2[m] \cdot Cov^2[\sigma'_p] + \ln^2 \left(\frac{E[\sigma'_p]}{\sigma'_{vo}} \right) \cdot V[m] . \tag{19}$$

These equations contain both the spatial and systematic components. Noiray (1982) and Ladd (1983) give an example of applying equation [19] to the foundation stability of a stockpile of limestone for a cement plant. The results include the probability of failure for a very large failure, involving systematic error only, and for a very small failure, involving both uncertainties.

Computing Uncertainty in Factor of Safety

The computation of factor of safety using the method of slices for an actual geometry is an example of a case in which direct differentiation may not be feasible, and the second method of evaluating the variance becomes attractive. The approach is to evaluate y (or F) for the values of the parameters at the condition of interest and then to change each of the parameters in turn by a small amount and re-evaluate y. The differences, divided by the increments in the parameters, provide estimates of the derivatives. The numerical approximation to the derivative is, of course, a central feature of most numerical applications in engineering, and, as in all numerical work, the user must exercise care not to extrapolate a linear analysis too far beyond the region in which actual results exist. The use of the method of divided differences will be illustrated below by a detailed example of the evaluation of the stability of the embankments at site B-6 of the James Bay project.

Other Approaches

Wong (1985) and Wolff(1991) describe a different way to evaluate the propagation of uncertainty in the analysis. This involves interpolating a surface of y through values computed for a discrete set of values of the variables. The usual procedure is to select, for each variable, the mean plus and minus one standard deviation; this will lead to 2^n evaluations if there are n variables. If the variables are independent and are scaled to have zero mean and unit standard deviation, a simple procedure, described by Wong, can be used to compute the statistics of the resulting function y. Of course, the result is an approximation to the actual function and must be used with some caution. Also, as Li (1991) points out, the procedure requires that the variables satisfy the underlying assumptions regarding independence and skewness.

Another procedure that has been used is Monte Carlo simulation. Although it has many advocates and can be quite useful in certain limited applications, it has several major drawbacks. First, it requires many repetitive calculations to obtain adequate results. Second, it is easy to lose track of parametric effects in a mass of output. Third, in some applications the analysis of the uncertainty in the stability of the slope has become confused with a probabilistic solution to the stability problem itself. Even if the analyst prefers to obtain final results by Monte Carlo simulation, it is worth the effort to proceed further with analysis of the contributions of the parameters in order to understand what is important.

The preceding discussion has dealt in some detail with the theory and techniques appropriate to the determination of undrained shear strength profiles from field data such as those obtained in vane shear tests. It should be clear that the same methodology can be applied to other parameters that enter into the calculation of the factor of safety. For example, Bergado and Anderson (1985) have described the probabilistic treatment of pore pressures for slope stability analysis.

APPLICATION TO EMBANKMENT DESIGN

The method of the Taylor series approximation was applied to the design of the single- or multi-stage construction of embankment dams for the James Bay project, and Ladd *et al.* (1983) provide details. Figure 6 shows the cross-section of the dike. The goals of the analysis included developing and understanding of the relative safety of different designs, obtaining insight about the contributions of different parameters to the safety, and establishing consistent design criteria that could be used in subsequent preliminary designs.

The first alternatives considered involved the construction of the embankment in a single stage, either to a height of 6 m or to a height of 12 m with one berm.

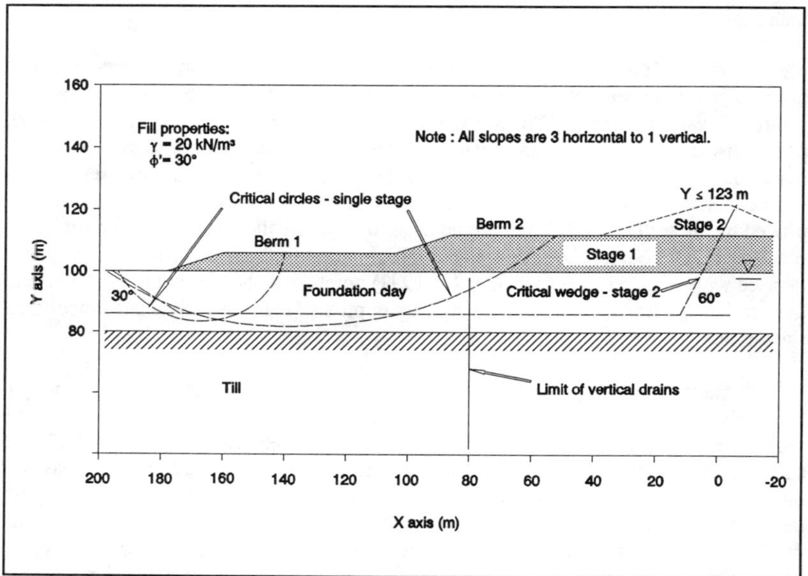

Figure 6. Cross-Section of SEBJ Design Problem (Ladd *et al.* (1983))

Undrained shear strength values would be obtained from field vane tests, and the stability analyses would be done by the simplified Bishop circular arc method of slices. The second alternative called for multi-stage construction. Following the installation of vertical drains, the first stage would consist of the two berms built to a height of 12 m. Then the zone with the drains would be allowed to consolidate to 80 per cent of primary consolidation, and the second stage would be built to a height of 23 m. In this case the undrained shear strength would be obtained from consolidated-undrained shear tests combined with knowledge of the initial and consolidated stress histories as described by Ladd (1991) and later in the text. The stability analysis was done by the Morgenstern-Price method of slices for a wedge-shaped failure pattern.

Figure 7 is a simplified soil profile, which shows that the foundation is composed of 4 m of crust underlain by 8 m of a highly sensitive marine clay underlain in turn by 6.5 m of lacustrine clay, for a total thickness of clay of 18.5 m. It also shows the distribution of vane shear strength and the effective vertical stress history.

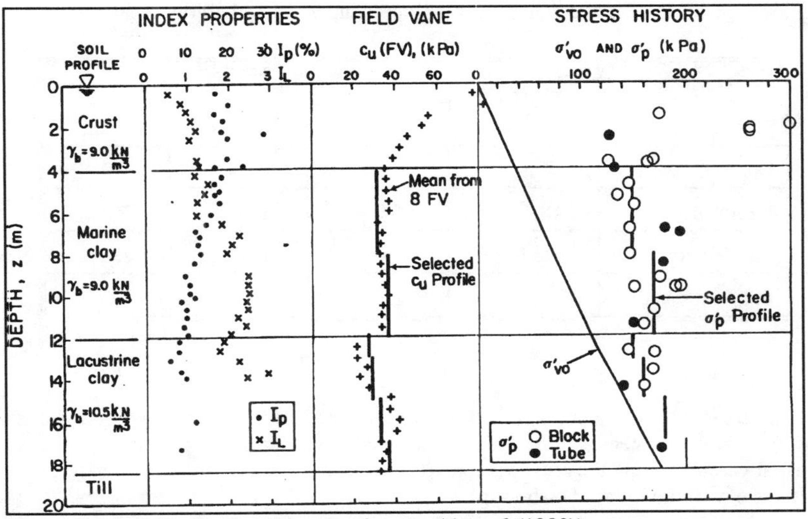

Figure 7. Soil Profile for Cross-Section (Ladd *et al.*(1983))

Uncertainty in Soil Profile and Fill Properties

The single stage cases involved many kilometers of embankment with a modest exploration program for any particular site along the embankment. Thus the analysis had to include uncertainty about the thickness of the crust and the depth to the till as well as the unit weight and friction angle of the fill. The multi-stage case involved a limited length of embankment with an extensive exploration and instrumentation program. Thus the geometry and the soil properties were better known, and the major uncertainty is in the properties of the fill.

The depth of the crust and the depth to the till were both treated as uncertain variables. From the results of field exploration the uncertainties listed in Table 2 were established. The separation of the uncertainty into spatial and systematic components is important because averaging over the failure surface affects the contribution of spatial variability to the results. The depth of the crust affects the undrained strength profile, and its spatial variance was estimated from the field vane data across the entire axis of the dike. The systematic statistical uncertainty results from the finite number of vane soundings and was estimated from equation [12] with $n = 27$.

Uncertainty in the depth to the till affects the depth of the failure surfaces considered in the analysis. Since the critical surfaces are nearly tangent to the top

Table 2. Soil Profile and Fill Property Uncertainties for Reliability Analysis

Variable	Expected Value	Variance		
		Spatial	Systematic	Total
Depth of crust	4 m	0.96	0.04	1.0
Depth to till	18.5 m	0.0	1.0	1.0
Fill density	20 kN/m^3	1.0	1.0	2.0
Fill friction ϕ'	30°	1.0	3.0	4.0

of the till for H = 12 m, this uncertainty does not affect the average across the profile but is a systematic uncertainty even though the estimate of variance reflects spatial variance along the cross-section. The estimate of systematic error reflects the effects of a finite number of borings.

Uncertainty in the density of the fill was taken to be composed equally of spatial and systematic components. This assumption and the estimated magnitudes of the variances were based on judgment. Selection of ϕ' = 30° reflected Canadian practice and its variances were also based on judgment.

Uncertainty in Shear Strength of Foundation Clay

Table 3 presents the uncertainty in c_u used for the single stage analyses, based on the field vane data for the marine and lacustrine clays. The relatively small coefficient of variation selected for the bias in μ for the marine clay (Cov = 0.075) reflected detailed comparison with results from laboratory CK_0U shear tests with different modes of failure, described by Lefebvre *et al.* (1988). This bias was doubled for the lacustrine clay, for which a similarly detailed comparison had not been made. The other components of spatial and systematic uncertainty were developed from the techniques described above. For example, the statistical Cov for the marine clay ($[(0.236)^2/62]^{1/2}$ = 0.030) follows from equation [12]. The reader must bear in mind in reading Table 3 that, as the total variance due to several independent variates is the sum of the individual variances, coefficients of variation and standard deviations are combined by taking the square root of the sum of the squares of the individual components.

For the multi-stage case the shear strengths were established from a combination of undrained strength ratios and the *in situ* stress history as follows. For the overconsolidated intact clay beyond the limits of the vertical drains, no

Table 3. Undrained Shear Strength Uncertainty Based on Field Vane Data

Field Vane Statistics	Clay	
	Marine	Lacustrine
Number of Tests	62	37
Mean, kPa	34.5	31.2
Data Scatter, Cov	0.236	0.272
Spatial Variability, Cov	0.183	0.272
Systematic Error, Cov:		
Statistical	0.030	0.045
Correction factor, μ	0.075	0.15
Sub-total	0.08	0.16
Total Cov	0.20	0.32

consolidation was assumed and c_u/σ_p' ratios were applied to the σ_p' profile shown in Figure 7. For the clay within the limits of the vertical drains, 80 per cent consolidation was assumed under berm number 2, and the c_u/σ_{vc}' ratios were applied to $\sigma_{vc}' = \sigma_{vo}' + 192$ kPa (= 0.80 x 12 m x 20 kN/m^3). Therefore, it is necessary to deal with uncertainty in both c_u/σ_p' or c_u/σ_{vc}' and σ_p' or σ_{vc}'. Morgenstern-Price wedge stability analyses used anisotropic undrained strength profiles for compression, direct simple shear, and extension modes of failure.

The undrained shear strength ratios were obtained from extensive CK_oU triaxial compression and extension and Geonor direct simple shear tests run using the Recompression technique and treated for strain compatibility to account for the effects of progressive failure (Ladd *et al.* 1983, Ladd 1991). Table 4 shows best estimates of these ratios. They were not intended to be conservative. Otherwise the reliability analysis would be inappropriately biased. In this table the subscript C indicates compression failure, D direct simple shear failure on a horizontal plane, and E extension failure.

Table 5 presents the uncertainties in the undrained strength ratios, the vertical preconsolidation and consolidation stresses, and the resulting values of undrained shear strength, τ. Some uncertainties were developed by combining engineering judgment, experience, and experimental data. One of the major strengths of the reliability approach is that, instead of merely assuming conservative values of

Table 4. Undrained Strength Ratios from Laboratory CK_oU Tests

Deposit	State	τ_C/σ'	τ_D/σ'	τ_E/σ'
B-6 Marine	OC	0.26	0.225	0.16
	NC	0.26	0.225	0.16
B-6 Lacustrine	OC	0.225	0.19	0.14
	NC	0.25	0.215	0.12
Note: $\sigma' = \sigma'_p$ for OC clay, and $\sigma' = \sigma'_{vc}$ for NC clay.				

parameters, the engineer can use the best estimates and propagate quantitatively the effects of the uncertainties in those estimates. The estimates are discussed separately in the following paragraphs:

Undrained strength ratios - The coefficient of variation for the intact and consolidated clay were treated collectively for convenience. The spatial coefficient of variation was assumed equal to the average total scatter measured in τ/σ' for each of the three modes of failure. The systematic coefficient of variation had two components: one due to the statistical error calculated from equation [12], and the other the estimated bias. The latter assumed that the triaxial tests underestimated plane strain shear strengths by 10 $\pm$ 5 percent (Cov = 0.05) and that the strain rate used for the tests overestimated the *in situ* strengths by 10 $\pm$ 5 per cent (Cov = 0.05). Thus, these two errors cancel in the mean but cause a bias Cov = $\sqrt{2}$ x 0.05 = 0.07.

Preconsolidation pressures and consolidation stresses - Because very few oedometer tests were run (see Figure 7), the spatial coefficient of variation of σ_p' was assumed equal to that obtained from the field vane data, that is Cov = 0.18 and 0.27 from Table 3 for the marine and lacustrine clays, respectively. The systematic coefficients of variation reflected a combination of statistical error and judgment regarding the likely bias and were higher for the lacustrine clay because of the lack of data within the bottom half of the deposit (see Figure 7). The vertical consolidation stress, σ_{vc}', assumed no uncertainty in σ_{vo}' and therefore reflected uncertainty in the average degree of consolidation, U, which would be estimated from the field piezometer data, and the load imposed by berm 2. The values in Table 5 were all based on judgment and assumed the following: for spatial uncertainty, Cov[U] = 0.075 and Cov[load] = 0.05 to give Cov[σ_{vc}'] = 0.09; for systematic uncertainty, Cov[U] = 0.075 and Cov[load] = 0.0 to give Cov[σ_{vc}'] = 0.075, which was rounded to 0.08.

Table 5. Uncertainty in Undrained Shear Strength Estimates for Multi-Stage Analysis

Parameter	Clay	Coefficient of Variation		
		Spatial	Systematic	Total
Intact Clay:				
τ/σ'_p	Marine	0.067	0.076	0.10
	Lacustrine	0.084	0.092	0.12
σ'_p	Marine	0.18	0.10	0.21
	Lacustrine	0.27	0.15	0.31
τ	Marine	0.19	0.13	0.23
	Lacustrine	0.28	0.18	0.33
Consolidated clay:				
τ/σ'_{vc}	Marine	0.067	0.076	0.10
	Lacustrine	0.084	0.092	0.12
σ'_{vc}	Marine	0.09	0.08	0.12
	Lacustrine	0.09	0.08	0.12
τ	Marine	0.11	0.11	0.16
	Lacustrine	0.12	0.12	0.17

Undrained strength - The coefficient of variation for τ in Table 5 are obtained from

$$Cov^2[\tau] = Cov^2[\tau/\sigma'] + Cov^2[\sigma']. \tag{20}$$

The variance in τ actually used in the analysis then becomes

$$V[\tau] = Cov^2[\tau] \cdot E^2[\tau] \tag{21}$$

where E[τ] is the mean undrained strength within the intact or consolidated layer of clay.

Calculating the Reliability Index

There are three steps to the calculation according to the first-order, second-moment approach:

1) Identifying the uncertainties and estimating the relevant statistical properties as means, variances, and covariances,
2) Determining the partial derivatives of F with respect to the uncertain variables, and
3) Calculating the mean and variance of F and establishing the reliability index β.

Table 6. Variance Composition, Single Stage Dike, H = 12 m

Parameter	$\Delta F / \Delta x_i$	Variance		$(\Delta F/\Delta x_i)^2 \cdot V(x_i)$	
		Systematic	Total	Systematic	Total
ϕ'	0.01	3.0	4.0	0.0003	0.0004
γ_{fill}	0.06	1.0	2.0	0.0036	0.0072
D_{crust}	0.008	0.04	1.0	nil	0.0001
D_{till}	0.056	1.0	1.0	0.0031	0.0031
$c_u(M)$	0.0137	7.6	47.6	0.0014	0.0089
$c_u(L)$	0.0215	24.9	99.7	0.0115	0.0461
			$V[F]$ =	0.0199	0.0658
			Spatial $V[F]$ =	0.0459	
			$V_{ave}[F]$ @ R=0.2 =	0.0199 + 0.0092	
			=	0.0291	
			β =	$(1.453 - 1.0) / \sqrt{0.0291}$	
			=	2.66	

The first of these has been discussed above, and the numbers in Tables 2, 3, and 5 are the result. Since it is not practical to find the partial derivatives of an actual stability calculation analytically, the numerical approximation of divided differences is necessary for the second step. The procedure is approximate, but the linear approximation of the partial derivatives should not introduce significant error. The third step is simply to use the results of the second step directly; details are illustrated in the following pages. First, the effects of uncertainty in the soil profile and soil properties are considered. Then the contributions of model error, representing $V[e]$ in equation [14], are introduced.

It should be noted that the procedure used here minimizes F over the range of failure surfaces and then calculates β. It does not minimize β directly. The minimum F surface and the minimum β surface may not be identical. Oka and Wu (1990) observed that, in a situation with two different but likely modes of failure, reliability calculations based on one mode only can be seriously in error. This is not the situation here.

Single stage embankments - The results for the single stage dike built with one berm to a height of 12 m, without the model error, are displayed in Table 6. The circular arc analyses gave, for the critical failure circle, E[F] = 1.453. The final combination of the spatial and systematic variations accounts for that fact that the spatial variations tend to average out over the failure surface. By integrating the statistically varying shear strength over the failure surface it can be shown that the variance of the average shear strength due to spatial variability is

$$\frac{V[c_{u,ave}]}{V[c_u]} \approx \frac{2r_o}{L} \qquad (22)$$

where r_o is the autocorrelation distance and L is the length of the curve of the intersection of the failure surface with the plane of analysis. For the geometry of this case the reduction ratio becomes 0.2. This is applied to the spatial V[F] of 0.0459.

For the case with a single stage to 6 m with E[F] = 1.500, the critical failure circle is much smaller, and the reduction ratio for the spatial variation in equation [22] becomes 0.7. This leads to significantly larger variation in the computed factor of safety and a much lower value of β. Table 7 shows the uncertainties for this case.

Figure 8 shows graphically how the different parameters contribute to the variance in the factor of safety for the two single stage cases. The bars represent the sum of the systematic uncertainty and the spatial uncertainty *reduced* due to averaging over the failure surface. The importance of the uncertainty in the strength of the lacustrine clay is obvious. The density of the fill and the strength of the marine clay are less important. The 6 m dike has a much smaller value of β than the 12 m dike (1.84 versus 2.66) even though its factor of safety is slightly higher (1.50 versus 1.45). The reduced reliability inde‹ reflects a somewhat larger systematic uncertainty and a much larger spatial uncertainty.

Multi-stage embankment - Table 8 shows the results for the multi-stage dike. The failure surface in this case is large (see Figure 6), but it is mostly horizontal so the appropriate autocorrelation distance for the shear strength is also large. The reduction factor for spatial variation of shear strength becomes 0.07. Because the overall uncertainty is much lower in this case, the reliability index, at 3.69, is the largest of all the cases. As in the single-stage cases, the strength of the lacustrine clay and the density are the most important contributors to the overall uncertainty. Since very little of the failure surface passes through the marine clay, its uncertainty

Table 7. Variance Composition, Single Stage Dike, H = 6 m

Parameter	$\Delta F / \Delta x_i$	Variance		$(\Delta F/\Delta x_i)^2 \cdot V(x_i)$	
		Systematic	Total	Systematic	Total
ϕ'	0.0022	3.0	4.0	nil	nil
γ_{fill}	0.068	1.0	2.0	0.0046	0.0092
D_{crust}	nil	0.036	1.0	nil	nil
D_{till}	0.010	1.0	1.0	0.0001	0.0001
$c_u(M)$	0.020	7.6	47.6	0.0030	0.0190
$c_u(L)$	0.0258	24.9	99.7	0.0166	0.0664
			$V[F]$ =	0.0243	0.0947
			Spatial $V[F]$ =	0.0704	
			$V_{ave}[F]$ @ R=0.7 =	0.0243 + 0.0493	
			=	0.0736	
			β =	$(1.500 - 1.0) / \sqrt{0.0736}$	
			=	1.84	

contributes little. The failure surface passes about equally through the intact and consolidated lacustrine clays, and each contributes about equally to the uncertainty. Although the coefficients of variation are larger for the intact clay, the mean strength is smaller. The contributions of the different parameters are illustrated in Figure 9, which again shows the importance of the lacustrine clay. This figure shows the breakdown of the uncertainty between spatial and systematic contributions, and also shows the important effect of averaging the spatial uncertainty.

Model error - Even if the soil parameters have been estimated correctly, the predictions of an engineering model can be expected to deviate from reality. Estimating the error introduced by this factor can be difficult; the most effective way to deal with the problem is to rely on empirical observations when they exist. An alternative is to use formal analysis in estimating the model error. In the present case three sources of model error were considered: three-dimensional failure compared to two-dimensional analysis (*i. e.*, end effects), failure to find the most critical failure surface in the analysis, and numerical error due to the details of the description of the strength profile and details of the computer programs. Azzouz *et al.* (1983)

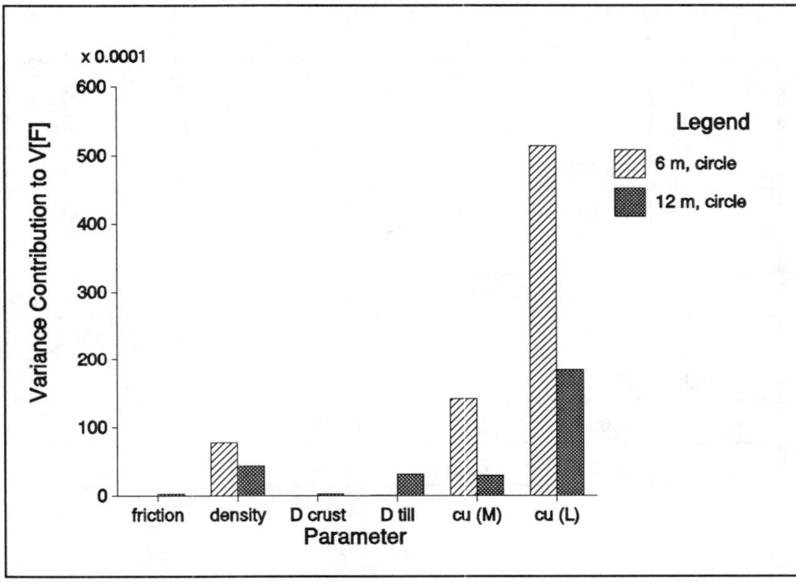

Figure 8. Variance Components for Factor of Safety of Single Stage Dikes.

examined the contributions of end effects in eighteen case histories. They found that the ratio between the three-dimensional and the plane strain F was 1.11 with a standard deviation of 0.06. In the present analysis these numbers were rounded to 1.1 and 0.05. Failure to find the lowest factor of safety is considered to overestimate F by about 5 per cent with a standard deviation of about 5 per cent. The third factor was incorporated by assuming that it contributed 2 per cent to the variability of the results. The combined effect of these three sources is to increase E[F] by 5 per cent and to create an added systematic uncertainty described by

$$Cov = (0.05^2 + 0.05^2 + 0.02^2)^{1/2} = 0.073 \approx 0.07. \tag{23}$$

Summary of results for three cases - The results for all three cases are summarized in Table 9. Although the factor of safety in the multi-stage case is slightly less than the values calculated for the single stage cases, the reliability index is significantly larger. Figure 10 summarizes the overall contributions of systematic and spatial uncertainty. It also shows the effect of the reduction in spatial uncertainty from averaging the soil parameters over the failure surface. The effect is greater the longer the failure surface. Spatial variability becomes almost insignificant for the 23 m high dike constructed in two stages. It should be noted here that, in order to

Table 8. Variance Composition, Multi-Stage Dike, H = 23 m

Parameter	$\Delta F / \Delta x_i$	Variance		$(\Delta F/\Delta x_i)^2 \cdot V(x_i)$	
		Systematic	Total	Systematic	Total
ϕ'	0.0088	3.0	4.0	0.0002	0.0003
γ_{fill}	0.055	1.0	2.0	0.0030	0.0061
Intact Clay:					
$\tau(M)$	0.0018	13.3	41.5	nil	0.0001
$\tau(L)$	0.012	26.3	88.5	0.0038	0.0127
Consolidated Clay:					
$\tau(M)$	0.0021	52.7	111.5	0.0002	0.0005
$\tau(L)$	0.009	62.0	124.4	0.0050	0.0101
			$V[F]$ =	0.0122	0.0298
			Spatial V[F] =	0.0176	
			$V_{ave}[F]$ @ R=0.07 =	0.0122 + 0.0012	
			=	0.0134	
			β =	$(1.427 - 1.0) / \sqrt{0.0134}$	
			=	3.69	

simplify calculations, the values of R used for spatial averaging were based on analyses of the field vane data and therefore should apply only to the undrained strength for the single stage cases with H of 6 and 12 m and part of the multi-stage case involving the intact clay. Different values of R would exist for the fill density, friction angle, and normally consolidated undrained shear strength.

An interesting observation from Table 9 is that introducing model error can have unexpected results. For the single stage dike with H = 6 m, β increases from 1.84 to 1.96. There is little effect for the single stage dike with H = 12 m; β changes from 2.66 to 2.60. For the multi-stage dike β decreases from 3.69 to 3.19. Thus, the computed reliability index can increase or decrease when the effects of model uncertainty are introduced. For a particular configuration and set of properties an increase in $E[F]$ might be offset by an increase in $\sigma[F]$.

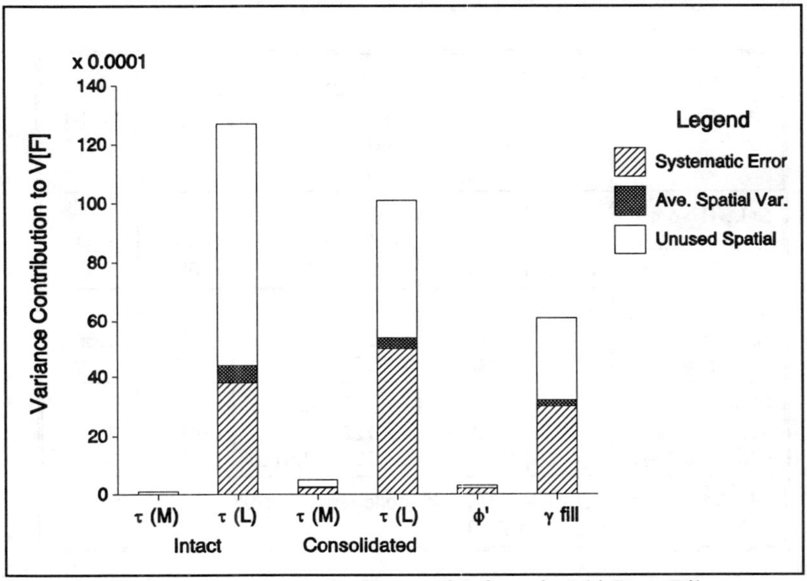

Figure 9. Variance Components for Factor of Safety of Multi-Stage Dike.

Extrapolation to Other Values of Parameters

The above results follow from detailed consideration of a limited set of geometries and values of parameters. In order to develop consistent criteria for design, it is necessary to extrapolate them to other conditions. Two simple possibilities exist. On the one hand, one could assume that for a basic geometry the variance or standard deviation of F is constant and apply equation [1]. On the other hand, one could assume the coefficient of variation is constant and apply equation [3]. The consequences of the two assumptions are shown in Figure 11. Constant variance means that the absolute uncertainty in F does not change with the target value of $E[F]$. Constant coefficient of variation means that the uncertainty in F changes in proportion to $E[F]$. Since the derivatives (or in this case the divided differences $\Delta F/\Delta x_i$) are for the most part nearly constant for different values of F, a projection based on constant variance seems the better choice. This also is the simpler of the two from a computational point of view.

Figure 12 shows the projected nominal probabilities of failure as functions of the computed factors of safety for the example stability problems. The probabilities are computed on the assumption that F is normally distributed, leading to the relation

Table 9. Summary of Reliability Results and Effects of Model Error

Case (No. Stages)	$E[F]$[1]	$V[F]$				$\beta = \frac{E[F]-1}{\sqrt{V[F]}}$
		Reduced Spatial	Prior Syste-matic	Model Error[2]	Total	
Sgl., H=6m						
w/o mod. err.	1.500	0.050	0.024		0.074	1.84
with mod. err.	1.575	0.050	0.024	0.012	0.086	1.96
Sgl., H=12m						
w/o mod. err.	1.453	0.009	0.020		0.029	2.66
with mod. err.	1.526	0.009	0.020	0.011	0.041	2.60
Multi, H=23m						
w/o mod. err.	1.427	0.0012	0.0122		0.0134	3.69
with mod. err.	1.498	0.0012	0.0122	0.0110	0.0244	3.19

[1] $E[F]$ with model error = $E[F]$ w/o model error x 1.05
[2] $V[e] = (0.07 \text{ x increased } E[F])^2$

Table 10. Factors of Safety Corresponding to Probability of Failure of 0.001 and Contributions to Uncertainty (without model error).

Case	Factor of Safety for $P_f = 0.001$	Contribution to Uncertainty	
		Spatial	Systematic
Single stage, H = 6 m	1.84	67 %	33 %
Single stage, H = 12 m	1.53	32 %	68 %
Multi-stage, H = 23 m	1.36	9 %	91 %

plotted in Figure 1. Until now the form of the distribution of *F* has not entered into these calculations. Curves are plotted for results both with and without the effects of model uncertainty included. From this figure the designer can determine the computed factors of safety that correspond to a desired target nominal probability of

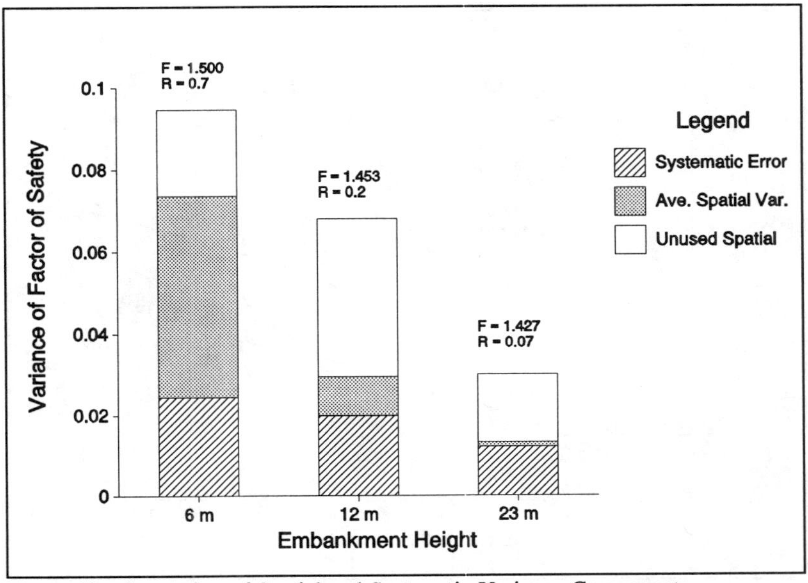

Figure 10. Separation of Spatial and Systematic Variance Components.

failure. For example, Table 10 shows the factors of safety corresponding to a nominal probability of failure of 0.001 during construction.

Effects of types of failure - The uncertainties associated with the three design cases differ from one another because each reflects a different mixture of spatial variability and systematic uncertainty. This has an important implication. A one per cent probability of failure deriving from spatial variability alone implies that, on average, one per cent of the length of a long dike would fail. Conversely, a one per cent probability of failure due to systematic uncertainty alone would imply that, on average, one out of one hundred of a large number of similar dikes would fail along its entire length. In any specific case, such as this one, the probability of failure derives from both spatial variability and systematic uncertainty.

Table 10 shows the contributions of the two components for the three cases. In the single stage case with H = 6 m the distribution of F for different sections along the dike is broad, but the uncertainty on the location of the mean is small. In the multi-stage case the distribution of F along the dike is narrow, but there is proportionally more uncertainty on the location of the mean. In other words, the uncertainty in the case of a single stage dike with H = 6 m deals principally with the

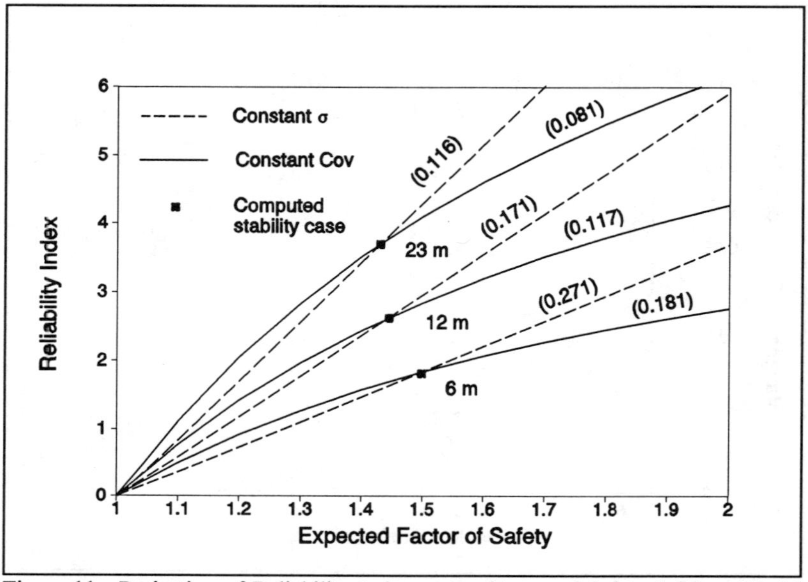

Figure 11. Projection of Reliability Index versus Factor of Safety (without Model Error).

fraction of the dike that will fail, but the uncertainty in the case of a multi-stage dike deals principally with the failure of all or a large portion of the dike.

Target probability of failure - As mentioned above, Baecher *et al.* (1980) determined that the historical rate of failure of dams is about 10^{-4} per dam-year. The details of failures are often imprecise in the catalogues from which this result was computed, but it appears that about one third of the failures were due to overtopping and another one third to internal erosion, piping, or seepage. These numbers are close to what would be expected from the judgment and experience of the profession. The remaining one third of the failures are due to slides and other mechanisms, which include the subject of the present analysis.

The relative contribution of different modes of failure should be kept in mind when a target probability of failure is being selected. It makes little sense to reduce the computed probability of failure due to slope stability problems if other causes of failure are not addressed at the same time. The historical data would suggest that, for a dam with a life of 100 years, the overall probability of failure is 0.01, in the absence of specific attention during design and construction to mitigate the hazards. Since well less than one third of the failures are due to slope instability *per se*, a

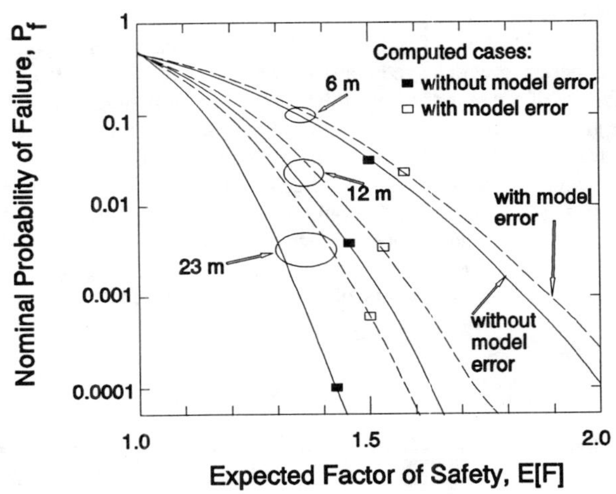

Figure 12. Nominal Probability of Failure versus Computed Factor of Safety for Example Problems

target probability of failure of slope stability failure of about 0.001 seems reasonable for design purposes and consistent with past experience. Figure 12 indicates that to achieve this target probability the stability analysis should recover the factors of safety listed in Table 10. If the effects of model uncertainty were included, the values of desired factor of safety would increase to 1.91, 1.63, and 1.48, respectively. As the computation of the effects of model uncertainty was probably somewhat too conservative and the effects on the desired factors of safety relatively small, this effect was ignored in further results.

Consistent factors of safety - The present analysis was concerned with stability during construction due to failure of the foundation clay. The consequences of such a failure can be expected to consist primarily of the costs of reconstruction. Failure of the embankment after the facility is in operation would involve much larger economic costs as well as other public consequences. This means that different target reliability indices or probabilities of failure are appropriate for different stages in the life of the project. The selection of such values is a difficult task, which presents a great many difficulties to owners, designers, and public officials. Even for the present problem, the consequences of failure are quite different in the different cases. Failure of a small section of a low embankment is much less serious a matter than the failure of a large portion of the 23 m high embankment. A further complication is that the profession has much more experience with rapidly constructed low embankments than with high embankments whose stability relies on the partial consolidation of the foundation clay. The target probabilities of failure and the corresponding consistent factors of safety should be altered accordingly. For example, because experience with staged construction is less fully evaluated and the consequences of failure of a major portion of the completed multi-stage embankment are expected to be greater than those of the failure of the 12 m high embankment, the target probability for the multi-stage case might be reduced to 0.0001, implying a calculated factor of safety of about 1.43. Conversely, the consequences of failure of a portion of the smallest dike might be only one tenth as great. The target probability then becomes 0.01 and the desired factor of safety about 1.63. Based on the revised target probabilities, one obtains the consistent, desired factors of safety shown in Table 11. The range in factors of safety for the different cases is now much smaller than that shown in Table 10 for the same P_f = 0.001. These results

Table 11. Consistent Factors of Safety (without model error).

Case	Target P_f	Consistent Factor of Safety
Single stage, H = 6 m, circular failure surface	0.01	1.63
Single stage, H = 12 m, circular failure surface	0.001	1.53
Multi-stage, H = 23m, wedge failure	0.0001	1.43

led to the recommendation that F = 1.50 be adopted for feasibility studies and preliminary cost estimates for both single and multi-stage dikes.

OTHER PROBABILISTIC ANALYSES

There are many extensions and variations on the basic procedure for employing first-order second-moment techniques to obtain probabilistic insights into the behavior of slopes and embankments. Wong (1985) and Wolff (1991) have proposed the response surface technique as way of describing the parametric variation of the probability of failure or of the reliability index. Instead of a Taylor series combined with a divided difference approximation to the partial derivatives, this technique develops a multi-dimensional approximation to F itself by establishing appropriate interpolation functions. Wolff applied the technique to analyze the reliability of a slope before and after repairs. While the response surface technique seems to be an effective way of computing statistical properties for a complicated function, such as the computed factor of safety, it can be no more accurate than the values of the input parameters, and it is especially important that the effects of averaging on spatial uncertainties be dealt with properly. This is one of Li's (1991) major points.

Castillo and Alonso (1985) also emphasize the importance of treating the spatial variation correctly. They recommend the use of techniques based on random field theory (Vanmarcke 1983) to account for the correlation structure of the variables. They point out that using a single value of a variable, such as the shear strength, instead of different values at different slices has the effect of ignoring the averaging effects and exaggerates the probability of failure. Li (1991) makes the same point. In the present analysis this problem is dealt with by reducing the spatial variability by the factor in equation [22].

Li and Lumb (1987) provide a modified version of the Morgenstern-Price equations suitable for probabilistic analysis. They also recommend the more complicated Hasofer-Lind reliability index, which does not rely on the Taylor series approximation in computing β. Chowdhury *et al.* (1987) have extended the probabilistic analysis to deal with progressive failure. Their approach relied on undrained strengths without consideration of effective stress parameters or stress history. The strength values used in the present paper were adjusted to account for strain compatibility and progressive failure.

Probabilistic methods have shown great promise in the fields of dynamics and earthquake engineering, but it is beyond the scope of the present paper to deal with such developments as the application of random vibration theory to seismic analysis. In most dynamic applications the load is the major probabilistic variable, and the material properties are considered to be deterministic. Yegian *et al.* (1991a, 1991b) describe a very useful application to the calculation of permanent deformations of slopes due to earthquakes.

WHITMAN'S OBSERVATIONS ON THE USE OF PROBABILISTIC METHODS

In his Terzaghi Lecture, Whitman (1984) gave an excellent summary of the state of the art of probabilistic applications in geotechnical engineering. His conclusions remain valid eight years later. He observed:

> The spectrum of slope stability problems is quite broad. Toward one extreme are situations for which there is already extensive practical experience and for which there is considerable agreement as to an appropriate safety factor to use for design. On the other hand, there are problems involving soils for which there is very little direct experience. In this case, there may be considerable doubt as to the proper method for computing safety, and the probability of failure is rather high. Some of the problems of stability in very soft clays . . . fall in this category, as do many slopes in residual soils. In both these situations, reliability theory is of little use in the selection of an appropriate safety factor. Many problems fall into one or the other of these categories, which may explain why there are few, if any, papers in the literature describing the actual application of reliability theory in design decisions concerning a slope.

It is hoped that the present paper provides a partial remedy to the last observation.

Whitman re-examined the stability analysis of the embankment crossing of the Great Salt Lake. He concluded that a probabilistic analysis would justify using very low factors of safety, as was in fact done. He observed that in such a case, where there is almost total control over the project and the consequences of failure during construction are almost entirely economic, the very broad uncertainty in properties might justify, in principle, designing for a factor of safety less than one, but it is unlikely that such a course would be adopted in practice.

Whitman also provided an excellent explanation of the use of event trees for the evaluation of overall risk of geotechnical structures, together with hypothetical examples of well and poorly designed dams. An analysis of this sort must be done if the engineer is to evaluate the overall probability of failure of an engineered structure. Probabilistic risk assessments have been performed for a variety of critical facilities, such as nuclear power plants. However, as Whitman demonstrated, a great deal of judgment goes into assessing the probabilities associated with some events; when a consensus about the appropriate values cannot be reached, the analysis breaks down.

SUMMARY AND CONCLUSIONS

General Conclusions

The review of reliability analysis and its application to the problems of the embankments at the James Bay project leads to several major conclusions. These include:

1) The reliability index β, as defined in equation [1], provides a more meaningful measure of stability than does the factor of safety F. Geotechnical engineers have long recognized that F has little physical meaning and that the choice of a satisfactory value is fraught with difficulty. In contrast, the reliability index describes safety by the number of standard deviations (*i. e.*, the amount of uncertainty in F) separating the best estimate of F from its defined failure value of 1.0. It is a more explicit approach.

2) The uncertainty in the values of the soil properties is a major, in most cases *the* major, contributor to the uncertainty in the stability of slopes and embankments. The uncertainty in soil properties consists of two portions: scatter and systematic error. The first comprises real spatial variability and noise or measurement error. These represent variations about the mean or trend of the parameters. The systematic error includes both a statistical uncertainty in the expected values and the effects of bias. The latter is much the more difficult to evaluate.

3) The structure of the spatial variation can be used to estimate the level of random noise in soil property data and to eliminate it from the calculation of reliability index. The effects of the spatial variability on the computed reliability index are further reduced because the variability is averaged over a region or failure surface and it is only its contribution to the uncertainty in the average that is of interest.

4) Several difficulties arise in practical assessment of scatter and systematic error. There are often not enough data to evaluate the autocorrelation distance statistically. Evaluating the effects of bias introduces many additional concerns. Bias is often very important, and the associated uncertainty can be a significant contributor to the overall uncertainty. Since bias is often ignored in theoretical treatments of analytical procedures and it is difficult to quantify, the engineer must often rely on judgment to establish its contribution. The strength of the reliability analysis is not that one can get a better estimate of each of these uncertainties but that one can deal with them explicitly and coherently.

5) Uncertainties in soil properties yield a lower bound estimate of the probability of failure, not the absolute probability of failure. That requires a more elaborate probabilistic risk analysis involving fault trees or other method of evaluating risk due to all contingencies (*e. g.*, Whitman 1984). However, for most practical applications the calculation of relative probability of failure is sufficient for parametric analysis.

6) A one per cent chance that the whole slope will fail is different from a near certainty that one per cent of a long embankment will fail, and reliability analysis allows such a distinction.

Analysis of the James Bay Embankments

The analysis of the James Bay embankments illustrates a three-step procedure for evaluating the reliability index. First, the uncertainties are identified and the relevant statistical properties are estimated. Second, the partial derivatives of F with respect to the uncertain variables are calculated. Third, the mean and variance of F are calculated and the resulting β evaluated.

Once the uncertainties in the soil properties, geometry, loads, and other contributing factors have been established, it is necessary to propagate their effects through the stability analysis. Equations [14] and [16] provide one way to do so. Equation [19] gives a specific, analytical example for shear strength, but in more complicated cases a numerical evaluation of the derivatives must be used. The first-order, second-moment approach is a very powerful approximate method to deal with the calculations. Several approaches are possible, but the Taylor series combined with a divided difference approximation to the derivatives has proven very effective.

In the analysis the effects of both spatial and systematic uncertainty are considered. The averaging effect tends to reduce greatly the contribution of the spatial uncertainty. In the case of the multi-stage construction spatial uncertainty contributes so little that it could almost be ignored without affecting the results. In the example presented here, the small embankment with a short failure surface has a large uncertainty in F due to spatial variation in undrained strength. This means that several small failures along the length of the embankment are likely. The multi-stage embankment with a long failure surface has almost all the uncertainty in systematic error, which leads to a small probability of failure along a long section of the embankment. When the autocorrelation distance is not known, the analysis can at least give point estimates of the reliability index without providing information on the length of embankment likely to fail.

The example also illustrates the contribution of model error, and the difficulty in determining the magnitude of such bias effects. These effects have seldom been included in reliability analyses. Even though in the present case they are not as important as other contributions, they could become quite significant in other situations.

However the solution is carried out, it is important to deal rationally with the spatial variability of the parameters. In particular, the averaging effect on variations that are due to inherent spatial scatter and are not systematic must be accounted for or the resulting probabilities of failure will be too high. The technique used here is one appropriate method. More formal, detailed methods from random field theory can also be used, but they have found relatively little practical employment because of the large amount of data needed to use them properly.

Recommendations

A major use of reliability methods is to establish consistent targets for the design of an embankment. These are usually expressed in terms of the desired factor of safety. In the present example the analysis was directed at stability during construction and not after impounding water, but the same techniques could be used in that case as well. The desired factor of safety is established by balancing the costs of achieving a degree of safety with the costs associated with failure, multiplied by the probability of failure.

Reliability approaches to the evaluation of the stability of slopes and embankments have evolved to the point that they should be applied more often in practice. Since the analysis requires considerably more effort than does a simple deterministic analysis, the engineer must have a good reason to use the reliability approach. Among the best reasons are: accounting rationally for the propagation of uncertainty in the analysis, not relying exclusively on some arbitrary factor of safety, and dealing coherently with the interaction of several effects and their relative contributions.

ACKNOWLEDGMENTS

The reliability analysis of the embankment dams was supported by the Societé d'Energie de la Baie James (SEBJ), Montreal, Quebec, Canada, as part of activities by a committee of specialists composed of J.-J. Paré (Head), M. Bozozuk, L. S. Brzezinski, O. Dascal, O. Eide, C. C. Ladd, K. T. Law, J. G. Lavallée (Secretary), G. Lefebvre, G. Mesri, P. Rosenberg, and F. Tavenas. Ladd *et al.* (1983) list the members of the Embankment Stability Committee who performed the reliability analysis in cooperation with G. B. Baecher.

APPENDIX I - REFERENCES

Alonso, E. E. (1976) "Risk analysis of slopes and its application to slopes in Canadian sensitive clays," *Geotechnique*, 26 (3), 453-472.

Azzouz, A. S., Baligh, M. M., and Ladd, C. C. (1983) "Corrected field vane strength for embankment design," *Journal of Geotechnical Engineering, ASCE*, 109 (5), 730-734.

Baecher, G. B., Paté, M.-E., and de Neufville, R. (1980) "Risk of dam failure in benefit-cost analysis," *Water Resources Research*, 16 (3), 449-456.

Baecher, G. B., Einstein, H. H., Veneziano, D., and Whitman, R. V. (1984) discussion of "Conventional and probabilistic embankment design," by V. McGuffey, D. Grivas, J. Iori, and Z. Kyfor, *Journal of Geotechnical Engineering, ASCE*, 110 (7), 989-991.

Benjamin, J. R., and Cornell, C. A. (1970) *Probability, Statistics, and Decision for Civil Engineers*, New York, McGraw-Hill.

Bergado, D. T., and Anderson, L. R. (1985) "Stochastic analysis of pore pressure uncertainty for the probabilistic assessment of safety of earth slopes," *Soils and Foundations*, 25 (2), 87-105.

Bjerrum. L. (1972) "Embankments on soft ground: state of the art report," *Proceedings, Specialty Conference on Performance of Earth and Earth-Supported Structures, ASCE,* Lafayette, Indiana, 2, 1-54.

Casagrande, A. (1969) "Role of the 'Calculated Risk' in earthwork and foundation engineering," *Journal of the Soil Mechanics and Foundations Division, ASCE*, 91 (SM4), 1-40.

Castillo, E., and Alonso, E. E. (1985) "Probabilistic versus deterministic modelling: filling a gap," *Proceedings, Eleventh International Conference on Soil Mechanics and Foundation Engineering*, San Francisco, California, 2, 813-818.

Cherubini, C. and Greco, V. R. (1987) "A probabilistic method for locating the critical slip surface in slope stability analysis," *Proceedings, Fifth International Conference on Applications of Probability and Statistics in Soil and Structural Engineering (ICAP5)*, Vancouver, B. C., Canada, published as *Reliability and Risk Analysis in Civil Engineering*, II, 1181-1187.

Chowdhury, R. N., Tang, W. H., and Sidi, I. (1987) "Reliability model of progressive slope failure," *Geotechnique*, 37 (4), 467-481.

Christian, J. T. (1984) discussion of "Conventional and probabilistic embankment design," by V. McGuffey, D. Grivas, J. Iori, and Z. Kyfor, *Journal of Geotechnical Engineering, ASCE,* 110 (7), 991-993.

Cornell, C. A. (1971) "First-order uncertainty analysis of soils deformation and stability," *Proceedings, First International Conference on Applications of Probability and Statistics in Soil and Structural Engineering (ICAP1)*, Hong Kong, 129-144.

Ladd, C. C. (1983) "Geotechnical exploration in clay deposits with emphasis on recent advances in laboratory and in situ testing and analysis of data scatter," *Journal of Civil and Hydraulic Engineering*, Taiwan, 10 (3), 3-35.

Ladd, C. C. (1991) "Stability evaluation during staged construction," *Journal of Geotechnical Engineering, ASCE,* 117 (4), 540-615.

Ladd, C. C., Dascal, O., Law, K. T., Lefebvre, G., Lessard, G., Mesri, G., and Tavenas, F. (1983) "Report of the embankment stability subcommittee," *Committee of Specialists on Sensitive Clays on the NBR Complex*, SEBJ, Montreal, Canada, Annexe II.

Ladd, C. C., and Foott, R. (1974) "New design procedure for stability of soft clays," *Journal of the Geotechnical Engineering Division, ASCE,* 100 (GT 7), 763-786.

Lefebvre, G., Ladd, C. C., and Pare, J.-J. (1988) "Comparison of field vane and laboratory undrained shear strengths in soft sensitive clays," *Vane Shear Strength Testing in Soils: Field and Laboratory Studies*, ASTM STP 1014, 233-246.

Li, K. S. (1991) discussion of "Embankment reliability versus factor of safety: before and after slide repair," by T. F. Wolff, *International Journal for Numerical and Analytical Methods in Geomechanics*, 15 (12), 893-896.

Li, K. S., and Lumb, P. (1987) "Probabilistic design of slopes," *Canadian Geotechnical Journal*, 24 (4), 520-535.

Noiray, L. (1982) "Predicted and measured performance of a soft clay foundation under stage loading," S. M. thesis in Civil Engineering, Massachusetts Institute of Technology.

Oka, Y., and Wu, T. H. (1990) "System reliability of slope stability," *Journal of Geotechnical Engineering, ASCE,* 116 (8), 1185-1189.

Peck, R. B. (1969) "Advantages and limitations of the observational method in applied soil mechanics: Ninth Rankine Lecture," *Geotechnique*, 19 (2), 171-187.

Peck, R. T., Morgan, G. C., and Anderson, L. R. (1987) "Philosophy of landslide risk evaluation and acceptance," *Proceedings, Fifth International Conference on Applications of Probability and Statistics in Soil and Structural Engineering (ICAP5)*, Vancouver, B. C., Canada, published as *Reliability and Risk Analysis in Civil Engineering*, II, 946-952.

Tang, W. H., Yucemen, M. S., and Ang, A. H.-S. (1976) "Probability-based short term design of slopes," *Canadian Geotechnical Journal*, 13 (3), 201-215.

Vanmarcke, E. H. (1977) "Reliability of earth slopes," *Journal of the Geotechnical Engineering Division, ASCE*, 103 (GT11), 1227-1246.

Vanmarcke, E. H. (1983) *Random Fields: Analysis and Synthesis*, Cambridge, Mass., M. I. T. Press.

Whitman, R. V. (1984) "Evaluating the calculated risk in geotechnical engineering," *Journal of Geotechnical Engineering, ASCE*, 110 (2), 145-188.

Wu, T. H., and Kraft, L. M. (1970) "Safety analysis of slopes," *Journal of the Soil Mechanics and Foundations Division, ASCE*, 96 (SM2), 609-630.

Wolff, T. F. (1991) "Embankment reliability versus factor of safety: before and after slide repair," *International Journal for Numerical and Analytical Methods in Geomechanics*, 15 (1), 41-50.

Wong, F. S. (1985) "Slope reliability and response surface method," *Journal of Geotechnical Engineering, ASCE*, 111 (1), 32-53.

Yegian, M. K., Marciano, E. A., and Ghahraman, V. G. (1991a) "Seismic risk analysis for earth dams," *Journal of Geotechnical Engineering, ASCE*, 117 (1), 18-34.

Yegian, M. K., Marciano, E. A., and Ghahraman, V. G. (1991b) "Earthquake-induced permanent deformations: probabilistic approach," *Journal of Geotechnical Engineering, ASCE*, 117 (1), 35-50.

APPENDIX II - NOTATION

The following symbols are used in this paper:

$C[.,.]$ = covariance of two variates enclosed in brackets
$C_x(r)$ = autocovariance over a distance r
$Cov[.]$ = coefficient of variation of variate enclosed in brackets
c_e = random experimental error
c_u = undrained shear strength
$c_u(FV)$ = undrained shear strength from field vane test
D_{crust} = thickness of crust
D_{till} = depth to till
$E[.]$ = expected value of variate enclosed in brackets
e = random error
F = factor of safety
$g(x)$ = arbitrary function of x
H = height of embankment
L = length over which spatial variation is averaged
(L) = indicates value for lacustrine clay
(M) = indicates value for marine clay
m = exponent in SHANSEP equation
n = Number of data points or tests
P_f = probability of failure
R = reduction factor to account for averaging of spatial variation
$R_x(r)$ = autocorrelation coefficient over distance r
r = distance for correlation
r_o = autocorrelation distance
S = coefficient in SHANSEP equation
$V[.]$ = variance of variate enclosed in brackets
x = a variate or variable
x_i = value of x at a particular point
y = a variate or variable
β = reliability index
γ = unit weight
μ = correction coefficient for field vane data
$\rho_{x,y}$ = correlation coefficient between x and y
$\sigma[.]$ = standard deviation of variate in brackets
σ_p' = effective preconsolidation pressure
σ_{vc}' = effective vertical consolidation stress
σ_{vo}' = effective initial vertical stress
τ = shear strength; subscripts C, D, and E indicate triaxial compression, direct simple shear, and triaxial extension, respectively
ϕ' = effective friction angle

FS=1.5: IS IT APPROPRIATE FOR EMBANKMENT DESIGN?

by Scott A. Ashford,[1] AM, Lawrence H. Roth,[2] F,
Sandra L. Madsen,[3] AM, and Donald G. Anderson,[4] M. ASCE

ABSTRACT

This case history discusses an issue that practicing engineers face on all embankment design projects: What is the appropriate minimum factor of safety (FS)? An historical perspective shows that over a period of 55 years, FS=1.5 is a paradigm, evolving from a target value, to a codified value, and, finally, to a mandatory minimum expected by review agencies and regulators. However, the literature indicates that FS should vary with the quality and quantity of data, the technical approach, and judgment. The case history illustrates these points, concluding that FS=1.5 is a *good place to start*, but codification limits the designers' ability to exercise geotechnical engineering judgment.

INTRODUCTION

This paper presents a case history of embankment design illustrating an issue that practicing engineers face on all projects involving slope stability: what is the appropriate minimum factor of safety (FS)? Or more importantly, is there a single, acceptable minimum value? FS=1.5 has been practically "codified" as the generally accepted minimum allowable FS for embankment design; on public works projects, proving FS $\geq$ 1.5 is almost mandatory, regardless of the sophistication or thoroughness of the technical approach. Selecting this value as a minimum acceptable, or "target," FS at the outset

---

[1]Project Engineer, CH2M HILL, Emeryville, CA

[2]Director of Geosciences, CH2M HILL, Emeryville, CA.

[3]Project Engineer, CH2M HILL, Emeryville, CA.

[4]Senior Geotechnical Engineer, CH2M HILL, Seattle, WA.

may make an otherwise stable slope appear "unsafe," depending on the quantity and quality of the data, and the analytical procedures followed.

Unfortunately, a codified value for FS cannot adequately account for the unknowns in the analyses (e.g., variations in shear strength and stratigraphy, pore-water pressures, type and accuracy of the analytical technique). It should be the responsibility of the design engineer to select a minimum acceptable FS based on site conditions, adequacy of the data, and analytical procedures chosen for embankment design. Review agencies, and in some cases regulators, must be afforded the opportunity to review and comment on the target FS in light of how they judge the appropriateness and completeness of the approach. A codified FS, however, reduces or eliminates the role of judgment for the designer, reviewer, and regulator.

The first part of this paper presents an historical perspective on the development of 1.5 as the generally accepted minimum allowable FS for embankment design in the U.S. Following this, a case history illustrates the difficulties and frustrations of trying to meet a codified minimum FS. In this case, a rigorous approach was used to design a large highway embankment on a relatively soft, compressible foundation, and to enable comparison of a calculated FS with a target FS established by a review agency. A less rigorous and less costly approach, however, could have also been used to meet the same target FS, but with greater uncertainty. Following the case history, conclusions are made regarding the question, FS=1.5: Is it appropriate for embankment design?

HISTORICAL PERSPECTIVE

Evolution of FS=1.5. There are surprisingly few citings of a minimum acceptable FS in the literature. Yet, over the years, FS=1.5 has evolved into a paradigm as illustrated in this historical perspective.

At the First International Conference on Soil Mechanics and Foundation Engineering, Terzaghi (1936) discussed the factor of safety for embankment stability, but his brief paper makes no mention of a minimum acceptable FS. Similarly, no other authors at the conference found it necessary or appropriate to define a minimum acceptable value. Seven years later, Terzaghi (1943) stated in *Theoretical Soil Mechanics*, that for stability computations for the $\phi=0$ condition, "The slope angle of the sides of the cut should be so selected that the factor of safety with respect to sliding is equal to 1.5." This represents the first significant citing of a minimum acceptable value.

Five years later, Taylor (1948), in his text, *Fundamentals of Soil Mechanics*, does not define a minimum acceptable FS for embankment stability. In an example embankment stability calculation, the calculated FS is 1.37. Taylor writes that though this value appears low, "It is a typical value, however, and many embankments, which according to engineering experience are safe, have safety factors smaller than this value." Taylor also states "there is no such thing as *the* factor of safety and that when *a* factor of safety is used its meaning should be clearly defined."

In the same year, another text, *Soil Mechanics in Engineering Practice* (Terzaghi and Peck, 1948), expands on Terzaghi's earlier writings and proposes a minimum acceptable FS for embankment stability: "Under normal circumstances, the foundation conditions are not considered satisfactory unless the factor of safety with respect to base failure [of an embankment] during or immediately after construction is at least equal to 1.5." For earth dams, the text states, "The theoretical factor of safety with respect to slope failures should never be less than 1.3 and should preferably be 1.5." The authors go on to say, "Since our knowledge of the conditions for the stability of clay embankments is still incomplete, ..., a theoretical factor of safety of 1.5 should be regarded as the minimum requirement."

Fifteen years later Sherard et al (1963), writing in *Earth and Earth-Rock Dams*, postulate a minimum acceptable FS. The authors write, "At the present state of our ability ..., it is not possible or reasonable to establish any hard and fast criteria for a minimum tolerable safety factor." They go on to discuss the difficulties in doing so, including the various methods available for determining shear strength and the different types of analytical procedures used by practicing engineers. Following this, the authors state, "In spite of these differences, a safety factor of 1.5 has been adopted as a minimum tolerable value for the full reservoir condition on the great majority of projects. The probable end result of this evolutionary process, assuming that no failures occur to alter present thinking, will be that a safety factor of 1.5 will be considered satisfactory when computed with a method of calculation in which side forces are considered ... and with pore-pressures estimated from a steady-state flow net."

Perhaps this evolutionary process was nearing completion when Duncan and Buchignani (1975) gave minimum acceptable values of FS varying from 1.25 to 2.0 as "guidance." They suggest using FS=1.5 when "cost of repair [is] much greater than cost of construction, or [there is] danger to human life or other valuable property if the slope fails" and when the "uncertainty of strength measurements" is small.

Codification. Nearly 40 years passed from the First International Conference to Duncan and Buchignani's manual, and it seems 1.5 evolved into a *good place to start* for the minimum acceptable FS for embankment design. Of importance, all authors referenced above stressed that judgment needed to be used in conjunction with their recommendations. One could assume for a given project, selecting 1.5 or, for that matter, any value as the minimum acceptable FS would be dependent on the quality and amount of subsurface information, the understanding of the groundwater and strength conditions, and the rigorousness of the analytical procedure.

What helps move 1.5 from being *a good place to start* to being a codified value is its adoption by many governmental agencies and regulatory bodies as *the minimum acceptable FS*. For example, in the *Materials Manual* by the California Department of Transportation (Caltrans, 1973), 1.5 is given as the minimum acceptable FS. In the U.S. Navy's *DM-7.01* (USN, 1986), the value of 1.5 is a *mandatory* minimum acceptable FS. California Integrated Waste Management Board's *Closure/Postclosure Regulations* (1990), state that "The [foundation] report must indicate a factor of safety for the critical [landfill] slope of at least 1.5 under dynamic conditions." Absent from these "codes" are instructions for the appropriate procedures to analyze slope stability and determine FS. With many options available for analysis, a specified target minimum FS should also be accompanied by detailed descriptions on the methods to calculate it, as noted by Taylor (1948).

It is likely that these *codified* values were intended to be somewhat flexible, but the fact that they are in print truly limits flexibility for all parties in the design process including the geotechnical engineer, the reviewing agency, and the regulator. Recognize too that little guidance is provided regarding how and when engineering judgment may be used to select a more appropriate minimum FS.

Factors Affecting FS. At the first Stability and Performance of Slopes and Embankments conference, held at Berkeley in 1966, Lowe (1966) presented a state-of-the-art paper entitled "Stability Analysis of Embankments." In it, Lowe notes three aspects that influence uncertainty in stability analyses: analysis of forces, selection of shear strengths, and application of the first two to specific loading conditions. Lowe states that the purpose of any minimum target FS is to account for uncertainty in the analyses, shear strength being the primary source of uncertainty.

The ordinary method of slices, which satisfies only overall moment equilibrium, is a common procedure that both Lowe (1966), and Duncan and Buchignani (1975) say is adequate in most circumstances. However, this method generally gives a lower FS than other more rigorous methods which satisfy all conditions of equilibrium (Wright et al, 1973), such as the method proposed by Janbu (1957). It seems logical, then, that the procedure selected for analysis of forces should therefore influence the value selected for the target minimum FS.

Lowe uses the unconsolidated-undrained (UU) strength for the end-of-construction case for embankment design. However, Ladd, in his Terzaghi Lecture (1991), states that the consolidated-undrained (CU) strength is a better representation of field conditions in this case. Furthermore, Mayne (1988) and Ladd (1991) indicate that there is much more variability in the results of UU triaxial than there is in CU triaxial tests. This suggests that the strengths used in the analysis should also influence the value selected for the target minimum FS.

Lowe (1966) states that, as a "cardinal rule," the tests for shear strength should duplicate field conditions. In the case of a foundation failure of an embankment, Ladd (1991) suggests the direct simple shear (DSS) strength best represents the average field conditions. However, UU and isotropically consolidated undrained (CIU) triaxial compression tests are routinely performed to evaluate shear strength for slope stability because these tests are less costly and are easier to run. These tests often give higher values of shear strength than DSS (Mayne, 1988). It appears that test methods should also influence the value selected for the target minimum FS.

There are many other factors that should be taken into account in selecting a minimum acceptable FS. For example, Duncan and Buchignani (1975) indicate that selection of FS should at least partially depend on the following factors: the uncertainty in slope geometry, the cost of modifying the slope, the cost of the consequences of failure, and whether the slope is temporary or permanent. Clearly these factors cannot be accounted for in a single codified value.

Alternatives. One alternative to a deterministic approach with a codified FS is a limit state approach; for example, the Det norske Veritas' code for design of offshore structures (DnV, 1981) and the *Canadian Foundation Engineering Manual* (1985). In the DnV approach, safety is still verified by determining that the design load will not exceed the design resistance. However, the design load is calculated by multiplying a characteristic load by a load coefficient found in the code. The design resistance is calculated by dividing characteristic strength by material coefficients, also found in the code.

There are two notable conditions in the DnV approach. First, conservative values must be selected for characteristic strength; the code states that greater conservativism must be used when there is much scatter in the raw data, or when limited data are available. Second, if effective stress analyses are performed, the code stipulates that soil strength must be determined using laboratory shear strengths with pore-pressure measurements. On the other hand, if total stress analyses are made, the code requires using soil strengths based on shear strength tests that match field stress conditions as closely as possible. Additionally, the code recommends that test results be interpreted using stress paths. This code provides guidance on the analytical procedures required by the reviewing agency, yet considerable room is left for the use of engineering judgment.

Another alternative is a probabilistic approach, also recommended by DnV, which states that target probability levels should be selected on a case-by-case basis (1981). Reliability and risk analysis in geotechnical engineering was a frequent topic of discussion in journals throughout the 1970s and early 1980s. Whitman's 1981 Terzaghi Lecture, "Evaluating Calculated Risk in Geotechnical Engineering," gives a comprehensive look at the role of reliability in our field (Whitman, 1984). Whitman describes reliability analysis as a way to balance uncertainty with safety while allowing the engineer to exercise judgment more clearly. Whitman explains that reliability analysis is particularly well suited to slope stability and discusses FS=1.5. Because various methods are used by different engineers for stability analyses, "... one slope with a reported factor of safety of 1.5 may actually have little margin of safety, while another with the same reported factor of safety may be virtually proof against failure." Whitman also concludes that "(the) use of FS=1.5 for all slope stability problems implies wide differences in reliability."

Whitman (1984) explains that assignment of numerical value for risk is essential to reliability analysis, and goes on to say: "...there is a real danger that criteria for allowable risk might become fixed and inflexible, thus demanding a precise evaluation of risk beyond what can realistically be achieved. The ... danger is real, and when it happens, important discussions about the objectives and priorities of society are reduced to unprofessional squabbles over the details of an analysis." Similarly, it appears codifying FS establishes a fixed and inflexible criterion which limits the designer's ability to exercise engineering judgment as illustrated in the following case history.

CASE HISTORY

Background. The Great America Parkway Interchange and SR-237 Realignment Project is a large highway project requiring considerable geotechnical design. The project includes elevating 2 miles of existing highway on a 1-million-cubic-yard embankment fill up to 40 feet high on the southern margin of San Francisco Bay, just north of San Jose, California. Several bridge structures will cross existing city streets, a railroad, and creeks, necessitating 10 bridge abutments in the embankment fill. Proposed embankment side slopes are 2H:1V (horizontal to vertical) and abutment slopes are 1.5H:1V. Consolidation settlement is a concern for the bridge structures, and most abutments will require preloading and surcharge prior to bridge construction, resulting in slopes 45 feet high during construction. The fill is unprecedented in the area, and a rigorous geotechnical exploration and laboratory testing program was undertaken.

Exploration and Testing. Nearly 6,000 linear feet of drilling and sampling and 5,000 feet of cone penetration testing were performed to help define subsurface conditions. In addition, the geological history of the Bay margin was reviewed, and the performance of other, smaller embankments in the area were evaluated to better characterize expectations of behavior. Several types of strength tests were performed as well as index and consolidation testing of recovered soil samples. Strength tests included field vane; UU and CIU triaxial compression; and direct simple shear (DSS).

Soil Conditions. Based on the exploration, the site consists primarily of layers of lean clay with varying degrees of overconsolidation caused by alternating sea level stands, desiccation, and groundwater overdraft. There is a relatively thin surface layer of desiccated San Francisco Bay mud, underlain by several layers of Holocene- and Pleistocene-age alluvial clay, and Pleistocene-age Bay clay. A soil profile for the upper 50 feet of the site is presented in Figure 1. Average test results for the soil units shown on the profile are presented in Table 1. Since the embankment fill was to be imported, the designers could specify desirable material properties. Reasonable values for these properties were assumed for the analysis as shown in Table 1.

Stability Analyses. Although the project was funded by a local sales tax levy, Caltrans will be the owner of the constructed highway. This put Caltrans in the role of the review agency, and it stipulated that the factor of safety for the embankment fills meet two requirements. First, during construction, the minimum acceptable calculated FS was to be 1.25 for slopes where "failure would not be catastrophic with respect to impedance of traffic or dangerous to human life or contiguous structures" (Caltrans, 1973). Second, when the freeway was open to traffic, the minimum acceptable FS was to be 1.5, for slopes where "failure could endanger human life, present a serious traffic hazard, or damage costly contiguous structures" (Caltrans, 1973).

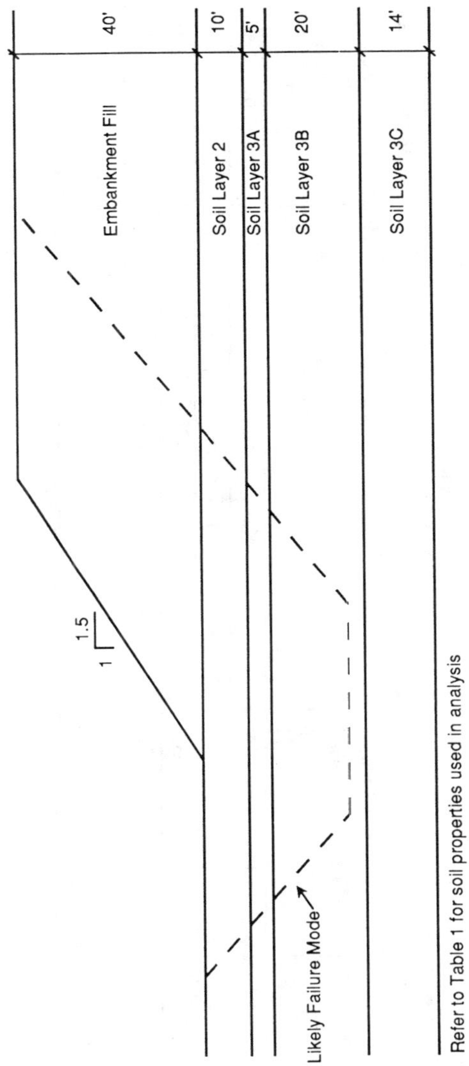

FIGURE 1
TYPICAL SOIL PROFILE AT ABUTMENT

Table 1
Soil Properties

Soil Layer	Saturated Unit Weight (pcf)	Natural Water Content (%)	LL[1]	PI[1]	Design Shear Strength (psf)				
					Before Consolidation			After Consolidation	
					DSS	CIU	UU	DSS	CIU
2	118	34	49	26	920	1,530	1,450	1,500	2,560
3A	120	33	34	21	1,150	1,490	1,880	1,590	2,065
3B	120	32	42	19	920	1,220	740	1,110	2,265
3C	130	22	42	23	1,800	2,900	2,480	2,450	3,435

Embankment fill has saturated unit weight = 125 pcf, and friction angle = 35 degrees.
[1]LL=Liquid Limit, PI=Plasticity Index

The stability of embankment slopes was determined using undrained strengths and analysis methods similar to those suggested by Ladd (1991). DSS strengths were selected to model the foundation soil since the likely mode of failure was through a weak foundation layer. Soil conditions were relatively uniform across the site, and the results of the laboratory testing indicated that the soil exhibited normalized properties. The SHANSEP procedure was used to define shear strength of the foundation soils (Ladd and Foott, 1974). DSS correlations with CIU tests were developed in the laboratory program and verified using published correlations (Mayne, 1988) because extensive DSS testing was economically prohibitive.

The computer program STABL (Siegel, 1975) was used to calculate the FS for abutment slopes using Janbu's method of slices (Janbu, 1957). The FS was slightly less than 1.5 for some abutment slopes when SHANSEP and DSS strengths were used. Table 2 shows the results of stability analyses for a typical abutment using DSS strengths.

Table 2
Stability Analysis Results[a]

Strength Basis	Factor of Safety: Before Consolidation	Factor of Safety: After Consolidation
DSS	1.2	1.4
CIU	1.6	2.1
UU	1.3	--

[a]Based on Janbu

The designers were satisfied with FS less than 1.5 based on the quantity and quality of soil strength data and on the ability of the analyses to model anticipated field conditions. However, the designers faced a dilemma trying to meet a codified value for FS using procedures that were not contemplated at the time the codified value was established. In their judgment, the abutments were safe and, using a poker analogy, trying to achieve FS=1.5 was like continuing to draw while holding four aces.

The designers could have used analytical procedures involving greater uncertainty, or that were less representative of field conditions to obtain FS=1.5. For example, Table 2 also shows the results of stability analyses for the same abutment using both CIU and UU strengths. Even though the FS resulting from CIU strengths exceeded 1.5, the slope would not have been any "safer."

To achieve FS = 1.5, the designers adjusted surcharge heights and varied preload schedules to gain strength from consolidation. In some cases, strength gains from consolidation were insufficient to meet FS = 1.5. In these cases, the designers slightly increased the DSS strengths based on observed trends in CIU and UU strength data.

DISCUSSION

Two elements of the analyses merit discussion: shear strength and factor of safety. In Table 1, there is considerable variation between the shear strength for each layer depending on which type of laboratory test is selected. Note that there is no clear correlation between the UU-based strengths and the others. In addition, there is a much wider data spread for the UU tests than for the CIU and DSS tests. This agrees with conclusions made by Mayne (1988) and Ladd (1991) that there is greater uncertainty in the results of UU tests. Based on this alone, it seems inappropriate to require the same minimum acceptable FS for analyses based on UU tests versus those based on CIU or DSS tests.

The analyses needed to demonstrate FS=1.5 when the highway would be open for traffic. To achieve this, the simplest procedure was to first check FS without accounting for strength gains due to consolidation. If adequate (i.e., FS≥1.5), then no further analyses were necessary. If not, then more detailed analyses accounting for strength gains due to consolidation were required. Using CIU strengths from Table 1 as the basis for analysis, the embankment was "safe" after the first check. In contrast, the initial DSS-based analyses did not meet FS=1.5 even accounting for strength gains from complete consolidation. The results for the UU-based analysis fell in between. In the end, FS=1.5 was met by accounting for strength gains because of consolidation under surcharge, and by modifying design shear strengths based on trends observed in all strength data.

Although analyses were made for the same soil, the calculated factors of safety are not equivalent. Each includes various amounts of uncertainty, which should be assessed when a target FS is selected. Obviously, all three analyses should not be required to meet the same FS; yet, each must if the minimum acceptable FS is codified. In this case, a DSS-based analyses with FS=1.4 was, in the designers' judgment, acceptable with regard to public safety. Unfortunately, additional analyses were necessary to satisfy the codified value of FS=1.5.

CONCLUSIONS

The historical perspective provided in the literature illustrates the evolutionary process that has resulted in 1.5 as the generally accepted minimum allowable FS. However, a codified value fails to account for factors requiring consideration before selecting a minimum acceptable FS for a project. Furthermore, using a "minimum factor of safety" can create an uncomfortable dilemma for designers, and the case history shows again what we have known before: no single factor of safety exists.

The factor of safety has often been called the "factor of ignorance" because our ignorance of the true behavior of an embankment and its foundation demands a factor of safety in the first place. Codifying a minimum acceptable FS does not allow for varying degrees of ignorance, or conversely, for using increased reliance on judgment based on greater knowledge. The limit state and probabilistic approaches are able to account for varying degrees of ignorance. Unfortunately, these approaches suffer the same potential downfall if their requirements also become fixed in code. Codifying FS does not reduce ignorance. In fact, it may unwittingly contribute to ignorance by reducing or removing the ability to exercise engineering judgment in slope stability analyses for embankment design.

The question remains, "FS=1.5: Is it appropriate for embankment design?" The answer appears to be yes... and no. Many embankments have been designed and constructed successfully using FS=1.5. Because of this, engineering judgment suggests that this is a *good place to start*. However, as Whitman (1984) points out, some slopes with FS=1.5 are likely far safer than others. And, as presented in this paper, the factor of safety is highly dependent on the method of testing, selection of strength, and analysis of forces. Factors of safety of less than 1.5 should be acceptable on projects where uncertainty is low, and values greater than 1.5 should be required where uncertainty is high. Designers, in concert with review agencies and regulators, should exercise their engineering judgment to select a target FS compatible with the acknowledged limitations of the data and the analytical models. Future investigators should help characterize a target FS by providing additional insight on the sources of uncertainty, and on how to account for uncertainty in selecting a target FS. Judgment says that it should be occasionally greater, occasionally smaller; therefore, FS=1.5 is not always the appropriate minimum acceptable factor of safety for embankment design.

REFERENCES

California Integrated Waste Management Board (1990). *Closure/Postclosure Regulations*, Title 14, California Code of Regulations, Division 7, Chapter 3, Article 7.8, 19-21.

California Department of Transportation (1973). *Materials Manual Volume VI: Foundation Exploration, Testing and Analysis Procedure.*

Canadian Geotechnical Society (1985). *Canadian Foundation Engineering Manual*, Bitech Publishers, Vancouver, B.C.

Det norske Veritas (1981). *Rules for the Construction and Inspection of Offshore Structures*, 8-60.

Duncan, J. M., and Buchignani, A. L. (1975). *An Engineering Manual for Slope Stability Studies*, Geotechnical Engineering Report, Department of Civil Engineering, University of California at Berkeley.

Janbu, N. (1957). "Earth Pressure and Bearing Capacity Calculations by Generalized Procedure of Slices," *Proceedings of the Fourth International Conference on Soil Mechanics and Foundation Engineering*, Vol. 2, London, England, 207-212.

Ladd, C. C. (1991). "Stability Evaluation During Staged Construction," *Journal of Geotechnical Engineering*, ASCE, 17(4), 540-615.

Ladd, C. C. and Foott, R. (1974). "New Design Procedure for the Stability of Soft Clays," *Journal of the Geotechnical Engineering Division*, ASCE, 100(7), 763-784.

Lowe, J., "Stability Analysis of Embankments" (1966). *Proceedings of the Stability and Performance of Slopes and Embankments Conference*, Soil Mechanics and Foundations Division, ASCE, 1-33.

Mayne, P. W. (1988). "Determining OCR in Clays from Laboratory Strength," *Journal of Geotechnical Engineering*, ASCE, 114(1), 76-92.

Sherard, J. L., Woodward, R. J., Gizienský, S. F., and Clevenger, W. A. (1963). *Earth and Earth-Rock Dams*, John Wiley and Sons, New York, N.Y.

Siegel, R. A. (1975). *Computer Analysis of General Slope Stability Problems*, Technical Report No. 75-8, Joint Highway Research Project, Purdue University, West Lafayette, Indiana.

Taylor, D. W. (1948). *Fundamentals of Soil Mechanics*, John Wiley and Sons, New York, N.Y.

Terzaghi, K. (1936). "Critical Height and Factor of Safety of Slopes Against Sliding," *Proceedings of the First International Conference on Soil Mechanics*, Cambridge, Mass., Vol. 1, 156-161.

Terzaghi, K. (1943). *Theoretical Soil Mechanics*, John Wiley and Sons, New York.

Terzaghi, K. and Peck, R. B. (1948). *Soil Mechanics in Engineering Practice*, John Wiley and Sons, New York, N.Y.

United States Navy (1986). *Design Manual 7.01; Soil Mechanics*, Naval Facilities Engineering Command, Alexandria, Virginia.

Whitman, R. V. (1984) "Evaluating Calculated Risk in Geotechnical Engineering," *Journal of the Geotechnical Engineering Division*, ASCE, 110(2), 145-216.

Wright, S. G., Kulhawy, F. H., and Duncan, J. M. (1973). "Accuracy of Equilibrium Slope Analysis," *Journal of the Soil Mechanics and Foundations Division*, ASCE, 99(SM10), 783-791.

Use of Reliability Methods for the Sequential Analysis of a Small Dam[1]

Eric C. Drumm[2], Richard M. Bennett[2]
and
William E. Manrod III[3], Members ASCE

Abstract

A case study analysis of a small dam is described. A limiting equilibrium analysis indicated that the dam was safe, but the factor of safety for steady state conditions at maximum pool was lower than the desired limits. The critical failure surface did not result in overtopping, or intercept the phreatic surface. Thus, even if a failure occurred along the identified critical surface, the dam would be stable. A sequence of failures would have to occur to cause overtopping or a piping failure.

Based on the assumed location of the first critical failure surface, an adjusted cross-section was approximated. A second stability analysis of the altered cross-section yielded an additional factor of safety. Since the overall safety level of the dam will always be greater than any of

---

[1]"The submitted manuscript has been authored by a contractor of the U.S. Government under contract DE-AC05-84OR21400. Accordingly, the U.S. Government retains a nonexclusive, royalty-free license to publish or reproduce the published form of this contribution, or allow others to do so, for U.S. Government purposes."

[2]Institute for Geotechnology, Department of Civil Engineering, The University of Tennessee, Knoxville, TN 37996-2010.

[3]Engineering Division, Oak Ridge site,* Martin Marietta Energy Systems, Oak Ridge, TN 37831-7233. *Managed by Martin Marietta Energy Systems, Inc. for the U.S. Department of Energy.

the individual safety levels, the individual factors of safety must be combined to obtain an overall measure of dam safety. Probabilistic methods were utilized to evaluate the safety of this sequential failure in terms of a reliability index. This provided a rational means of combining the results from the sequential analysis into a single measure of safety. The analysis indicated that the dam was indeed safe, avoiding a costly and environmentally difficult retrofit.

Introduction

The evaluation of earth dams for slope stability includes not only consideration of the computed factor of safety, but also the anticipated mode of failure. Limiting equilibrium methods (Hirschfeld et al., 1973) are used to evaluate potential failure surfaces and determine the factor of safety (FS) for given material properties, water table and surface loading conditions. Although the equilibrium equations are only valid for conditions at failure (FS = 1), the surfaces with the lowest FS's are identified. To assess the dam safety, the resulting FS's are compared with requirements specified by various regulatory agencies, (U.S. Army Corps of Engineers, 1970; Dodson, 1974; National Research Council, 1983). In practice, failure surfaces are often identified with FS's below the desired value, yet still greater than one. These analyses suggest that although the dam is safe, a costly retrofit or modification to the dam cross section is required to provide the desired margin of safety. However, if the mass of soil/rock in the critical failure surface is small and does not significantly alter the dam cross section, the dam may still be stable even if the identified failure occurs. An additional slope failure through the altered cross section must occur for the dam to fail. Although not possible with traditional limiting equilibrium methods of analysis, a rational expression of the dam safety for such a sequential failure process is needed.

For the case of a sequence of failures, the overall safety level for the dam is greater than the safety level corresponding to any individual failure surface. However, it is difficult to state an overall factor of safety for the slope, since the sequence has resulted in several values for FS. Reliability techniques can be used to obtain an overall reliability index representing the combined safety level of the dam with respect to such sequential failures. Such an approach is briefly described here, and discussed in terms of a case study of a small earth dam.

Slope Failure Modes

There are several possible modes of downstream slope failure in an earth dam, but only two cases significantly threaten the overall dam (safety) stability. The first case, illustrated in Figure 1a, is a failure surface that reduces the height of the dam and results in overtopping. Figure 1b depicts the second case, which occurs when the failure surface intersects the phreatic surface. This failure mode can alter the flow regime and may result in excessive hydraulic gradient and piping. Either of these modes is considered dam failure.

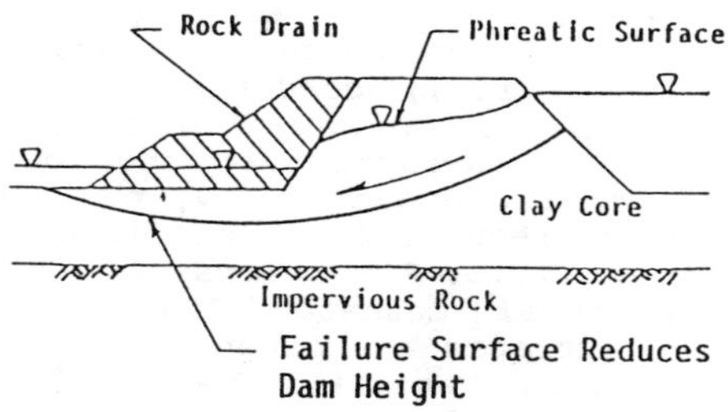

(a)

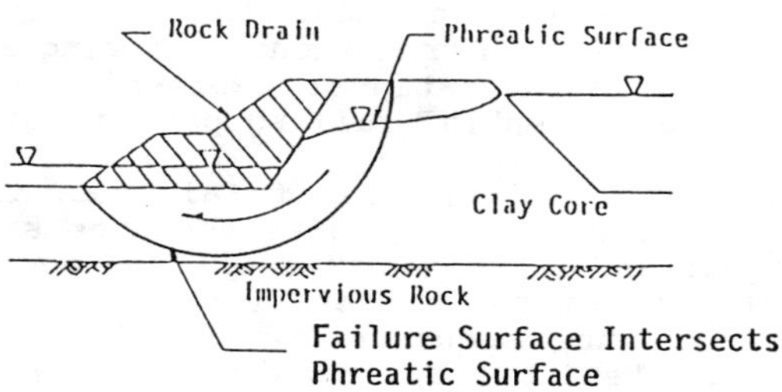

(b)

Figure 1 Slope Failure Modes that Threaten Dam Safety

A localized failure within a protective downstream stone berm, such as shown in Figure 2a, may have little effect on the overall stability of the dam. The phreatic surface has not been affected and the dam cross section has changed very little. Thus, these instabilities are usually disregarded if the volume of material within the failure is small.

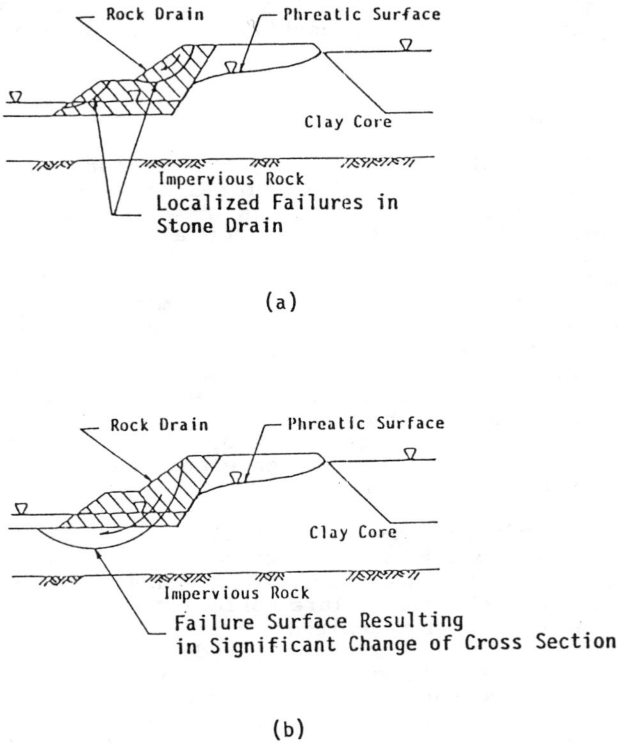

Figure 2 Slope Failure Modes that do not Threaten Dam Safety

A failure surface such as that illustrated in Figure 2b must be carefully evaluated to determine the effect on the overall stability of the dam. Although the height of the dam has not been reduced and the phreatic surface has not been interrupted, movement along the failure surface significantly changes the cross section of the dam.

Additional analysis of the affected cross section may be required, since the factor of safety obtained for the initial failure does not represent the overall safety of the dam.

This staged or sequential failure process may occur over time, allowing detection during a regular dam inspection program. Under some conditions, however, the failure may progress from one stage to the next with little warning. It is important, therefore, to know the overall safety level for this sequential type of failure.

Limiting Equilibrium Analysis of a Small Dam

The stability of a small dam impounding hazardous sediments was evaluated. A typical cross section of the dam and downstream rock berm is shown in Figure 3. The computer code PCSTABL4 (Lovell et al., 1983; Carpenter, 1985) was used for the analysis. PCSTABL4 permits the evaluation of general non-circular failure surfaces by the Janbu method (1954), which may be more appropriate for nonhomogeneous embankments than methods based on circular failure surfaces. The analysis assumes the existence of plane strain conditions in the embankment. It is also assumed that the effects of seepage forces can be neglected. The impact of this assumption is limited in the cross section investigated here, since the rock berm eliminates the surface of seepage and the gradient of the phreatic surface is small.

The limiting equilibrium analysis indicated that for steady state conditions at maximum pool, the FS was 1.35. The critical failure surface is shown in Figure 3. Although the critical failure surface resulted in a FS less than the value of 1.5 desired for long term stability, the height of the dam has was not reduced and the phreatic surface was not interrupted. Thus, it was concluded that the actual safety of the dam was greater than that indicated by the analysis, since the integrity of the dam was not compromised by the identified failure mode.

For the dam to actually fail, an additional slope instability would have to occur. To evaluate the stabilty of the dam under such a sequence of failures, several assumptions were made. First, it was assumed that the initial failure occurred along the surface corresponding the initial critical failure surface, and resulted in an adjusted cross-section as shown in Figure 4. An additional limiting equlibrium analysis on the adjusted cross-section was then conducted, to estimate the factor of safety that would exist after the first "failure." The results suggested localized failures within the stone berm with

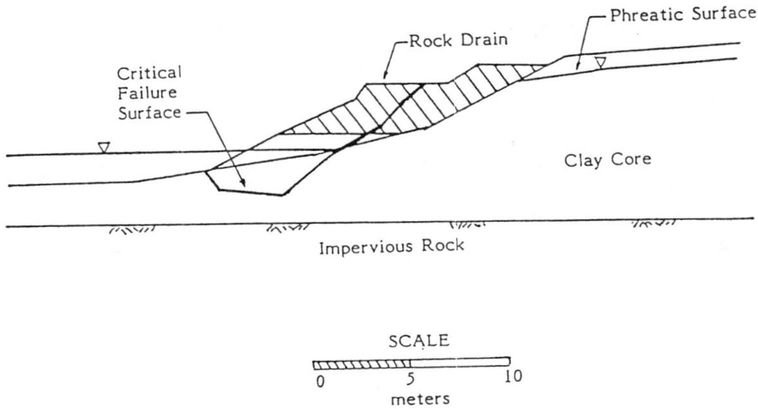

Figure 3 Cross Section of Dam and Initial Failure Surface

factors of safety slightly greater than 1.0. A deep-seated failure surface within the silty clay core that intercepted the phreatic surface resulted in a factor of safety of 1.55.

Although limiting equlibrium methods are not well suited for defining the location of the critical failure surface, this is a means to approximate the factors of safety that may result during a sequence of failures. Limiting equlibrium methods only evaluate the stability of a selected surface, and the equlibrium equations are valid only for conditions at failure (FS = 1). A better estimate of the location of the governing failure surfaces can be obtained from methods that consider the deformation characteristics of the soil, but these methods are not widely used in practice. Consequently it was assumed that the critical failure surface could be obtained from the limiting equlibrium analysis. The uncertainty associated with this assumption was then included in the probabilistic analysis, which is described below.

This approach permits the evaluation of a sequence of slope failures, and results in a FS for each surface. However, it is difficult to state an overall factor of safety for the dam, since the sequential failures resulted

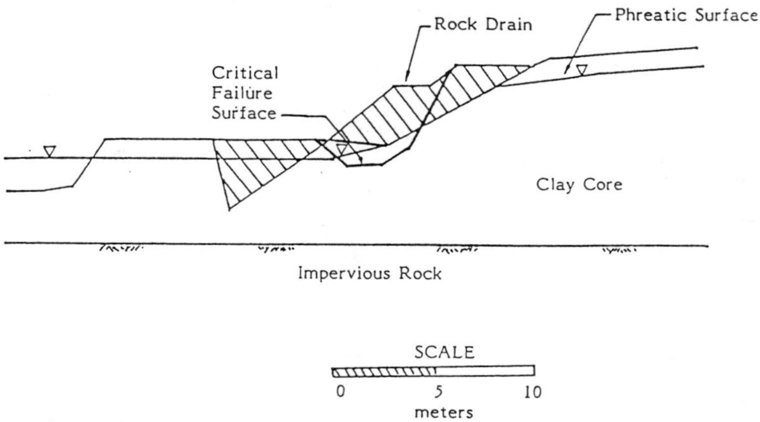

Figure 4 Assumed Cross Section after Initial Failure

in multiple values of FS. Reliability techniques can be used to obtain an overall reliability index, representing the combined safety level of the dam.

Probabilistic Analysis

The friction angles of the two materials were assumed to be the random variables in the probabilistic analysis. The mean values of the friction angles were assumed to be the nominal values used in the stability analysis. Based on values reported in the literature, a coefficient of variation of 0.25 and a coefficient of skewness of 0.41 were assumed (Athanasiou-Grivas, 1985; Biernatowski and Pula, 1988). A beta distribution (McGuffey et al., 1981) with a lower bound of 0.0 was used to model the random variables. These parameters are summarized in Table 1. The friction angle in the soil core and in the rock fill were assumed to be statistically independent. Due to the steady state drained loading condition, the cohesion of both materials was taken to be 0.0.

Table 1. Statistical Parameters for Beta Distribution of Shear Strength

Variable	Mean	Coefficient of Variation	Coefficient of Skewness
Drained friction angle	30.0	0.25	0.41
Rockfill friction angle	35.0	0.25	0.41

The probabilistic analysis of each slope failure was conducted in a manner similar to the method developed by Alfaro and Harr (1981), Chowdhury and Athanasiou-Grivas (1982), and Tang et al. (1985). The governing failure surface obtained from the deterministic analysis was divided into a number of vertical slices, corresponding to a linear portion of the failure surface as obtained from the limiting equilibrium analysis. The driving force of each slice is due to the weight of the slice, while the resisting force is the frictional force along the failure surface. The transfer of shear forces between the slices along vertical boundaries was neglected, resulting in factors of safety slightly greater than those obtained using Janbu's method (Athanasiou-Grivas and Asaoka, 1982). This approach resulted in factors of safety of 1.46 and 1.98 for the original and modified dam cross sections, respectively.

In addition to the uncertainty in the soil parameters, there is also uncertainty in the model used to predict slope failure. This uncertainty arises because: limiting equilibrium analyses are strictly valid only for a FS of 1.0, other possible failure surfaces may exist, other shapes for the adjusted cross section are possible, and plane strain conditions may not be valid for the second failure. Therefore, an additional random variable was introduced to account for this model error. The random variable was assumed to have a mean value of 1.0, a coefficient of variation of 0.10 for the first failure and 0.20 for the second failure, and to follow a normal distribution.

Second-order reliability methods (Madsen, 1985) were used to determine the reliability index for the failure of each slope in the sequential failure. The results yield a reliability index of 1.47 for the first failure and 1.94 for the second deep-seated failure in the modified slope (Figure 3).

The overall probability of failure of the dam is the probability of the intersection of the individual failure events. Both slopes have to fail for the dam to reach a critical condition. It was assumed that the model errors were statistically independent and the friction angles for different failure surfaces were perfectly correlated. This yielded an overall reliability index of 2.26. The assumption that the friction angles for different failure surfaces are perfectly correlated results in a lower bound to the reliability index. It is thus apparent that the overall safety level of the dam is significantly greater than that indicated by either failure surface alone.

Summary and Conclusions

The computed safety of earth dams with respect to slope stability is governed not only by the factor of safety, but also by the type of failure. When the critical surface significantly alters the cross section of the dam, but does not reduce the dam height or intercept the phreatic surface, a second, or possibly third, failure must occur for the dam to reach an unstable state. The overall probability of failure of the dam is then the probability of the intersection of the individual failure events. All sequences or slopes must fail for the dam to reach the critical condition. This sequential failure process results in a parallel system which increases the dam safety. The overall safety level for the sequential failure will be higher than the safety level of any individual failure.

In the case study examined here, a comparison of the reliability index of the individual failures with that of the overall analysis indicates that the safety level of the dam is significantly greater than that indicated by either failure surface alone. The safety levels of the sequential failures were rationally combined using established probabilistic methods. The resulting safety level was deemed adequate. The sequential and probabilistic analysis allowed a rational decision to made on the safety of the dam.

Appendix I. References

Athanasiou-Grivas, D. and Asaoka, A. (1982). "Slope Safety Prediction under Static and Seismic Loads." *Journal of the Geotechnical Engineering Division*. ASCE, 108(GT5), 713-729.

Athanasiou-Grivas, D. (1985). "Seismic Safety Analysis of Rigid Retaining Walls" Structural Safety and Reliability. Proceedings ICOSSAR '85, International Association for Structural Safety and Reliability, New York, NY, III:311-III:319.

Alfaro, L. and Harr, M.E. (1981). "Reliability of Soil Slopes," *Frost Action and Risk Assessment in Soil Mechanics, Transportation Research Record 809*, pp. 78-82.

Biernatowski, K., and Pula, W. (1988). "Probabilistic Analysis of the Stability of Massive Bridge Abutments Using Simulation Methods." *Structural Safety*. 5(1), 1-15.

Carpenter, J.R. (1985). "PCSTABL4 User Manual." *Report to U.S. Department of Transportation*. Federal Highway Administration, Office of Research Development and Technology, Washington, D.C.

Chowdhury, R.N., and Athanasiou-Grivas, D. (1982). "Probabilistic Model of Progressive Failure of Slopes." *Journal of the Geotechnical Engineering Division*. ASCE, 108(GT6), 803-819.

Dodson, E.L. (1974). "National Dam Safety Program, Inspection Guidelines - Stability Aspects," *Proceedings*, Safety of Small Dams, New England College, Henniker, New Hampshire, August 4-9, pp. 50-64.

Hirschfeld, R.C., Poulos, S.J., and Bertram, G.E. (1973). *Embankment Dam Engineering; Casagrande Volume*, Wiley, New York, 454 p.

Lovell, C.W., Sharmay, S., and Carpenter, J.R. (1983). "Introduction to Slope Stability Analysis with STABL4" Report to Federal Highway Administration, Revised 1985.

Janbu, N. (1954). "Application of Composite Slip Surfaces for Stability Analysis," *Proc. European Conf. on Stability of Earth Slopes*, Sweden, Vol. 3, pp. 43-49.

Madsen, H.O. (1985). "First Order vs. Second Order Reliability Analysis of Series Structures." *Structural Safety*. 2(3), 207-214.

McGuffey, V., Iori, J., Kyfor, Z. and Athanasiou-Grivas, D. (1981). "Statistical Geotechnical Properties of Lockport Clays," *Frost Action and Risk Assessment in Soil Mechanics*, Transportation Research Record 809, pp. 54-60.

National Research Council (1983). Safety of Existing Dams: Evaluation and Improvement, Committee on Safety of Existing Dams, Water Science and Technology Board, Commission on Engineering and Technical Systems.

Tang, W.H., Chowdhury, R., and Sidi, I. (1985). "Progressive Failure Probability of Soil Slopes." Structural Safety and Reliability. Proceedings of ICOSSAR '85, International Association for Structural Safety and Reliability, New York, NY, III:363-III:372.

U.S. Army Corps of Engineers (1970). Engineering and Design Stability of Earth and Rock-Fill Dams, EM 1110-2-1902, Department of the Army.

Evaluation of Failure Potential in Mudstone Slopes Using Fuzzy Sets

Der-Her Lee[1] and C. Hsein Juang[2], M., ASCE

Abstract

This paper presents a qualitative evaluation scheme for assessing slope failure potential. The main thrust of the evaluation scheme is an evaluation-tree consisting of two levels of factors that are known by experts to affect stability of slopes. Selection of these factors and weights among them are based on results of a survey of expert opinions on slope failures in the southwestern Taiwan *mudstone* area. A slope failure potential map is prepared and shown for the mudstone area studied.

Introduction

Rapid economic and population growth in Taiwan, Republic of China in recent years has changed the paradigm on land-use planning in the island--the *mudstone-dominant* sloping land in southwestern Taiwan, considered a useless and hazardous area in the past, is becoming a precious land for massive housing development projects. The new paradigm creates an urgent need for new regulations on land use which must address, among others, the issues of damage potential caused by slope failures and preventive and remedial measures for mitigating such failures.

Proper land-use planning, which is generally recognized as an effective means to mitigate the potential hazards caused by slope failures, demands a slope failure potential (SFP) map. An SFP map can be used to avoid sites with high failure potential. In case of a *prescribed* site, one can easily decide if detailed slope stability analysis is warranted. Note that an area SFP map is never intended as a substitute of a detailed geotechnical analysis of a specific slope. On the other hand, it is unwarranted to obtain expensive, detailed soil parameters so as to perform conventional stability analyses for purposes of preparing an SFP map of a large area.

---

[1]Department of Civil Engineering, Cheng Kung University, Tainan, Taiwan, ROC.
[2]Department of Civil Engineering, Clemson University, Clemson, SC 29634-0911.

A low-cost, rapid procedure for assessing and mapping slope failure potential, taking into account of the factors that are known to affect the stability of slopes, would be desirous in this case. One such procedure has been developed by Juang et al (1992).

The method by Juang, et al. (1992) uses a qualitative evaluation scheme with a fuzzy set analysis for assessing and mapping slope failure potential. In their evaluation scheme, an evaluation tree consisting of two levels of factors that are known by experts to affect stability of natural slopes was established and used. The overall failure potential of a slope is a *weighted average* of the failure potential ratings of the slope according to each of the factors adopted. The rating criteria are qualitative in nature, and failure potential is rated and recorded in linguistic terms such as *very high*, *high*, *medium*, *low*, and *very low*. These linguistic terms or *soft data* are represented by fuzzy numbers to account for inherent uncertainties of these data. These fuzzy data are then processed to yield a final fuzzy set that represents the overall failure potential of the slope. A slope failure potential index (SFPI) was defined based on the final fuzzy set. The SFPI ranging from 0 to 1 gives an indication of slope failure potential. A contour plot of the SFPI values obtained for slopes in a large area provides a convenient way for preparing an SFP map.

The factors adopted in Juang et al's (1992) study were established mainly based on a comprehensive review of expert opinions and knowledge expressed in the literature (Koshiba, 1980; Sato and Terada, 1985; Takeshita, 1986; Watari, 1978; Yamada, et al, 1983, among many others) and their experience on slope stability problems in the Mt. So-San area, an area dominated by alluvial deposits. As such, they suggested that "caution should be exercised when adopting these factors and weights for evaluation and mapping of slope failure potential in other areas or regions not having similar geology, environment, and climate conditions.

In the slope damage mitigation project on which this paper is based, it was decided to conduct a survey of opinions of geotechnical engineers knowledgeable in the area studied, a *mudstone-dominant* sloping land in the southwestern Taiwan. A questionnaire was prepared based on the evaluation tree developed by Juang et al. (1992) with the intent to account for local conditions and variations, particularly those pertinent to the mudstone area studied. Results of this survey are used to tailor an evaluation tree and procedure suitable for assessing and mapping slope failure potential of the southwestern Taiwan mudstone area. This paper documents results of this research project in the sections that follow.

Survey and Analysis of Expert Opinions

The purposes of the questionnaire are: 1) to identify factors that affect the stability of slopes in the southwestern Taiwan mudstone area, and 2) to determine the relative importance or weight for each of the factors adopted. The experts were asked to review the list of factors compiled by Juang et al. (1992) for its applicability in the mudstone area studied. The experts were encouraged to add or delete whatever factors they deemed appropriate. They were also asked to assess the weights among the factors adopted. In addition, questions regarding the criteria for rating the stability of a slope according to each factor were asked.

The weights among the adopted factors and the criteria for rating the stability of a slope were assessed in linguistic terms in the survey. The opinions expressed in linguistic terms are referred to herein as soft data. These soft data are translated into fuzzy numbers. A simple average of a set of opinions represented by these fuzzy numbers (Schmucker, 1984; Dong and Wong, 1987; Juang, et al., 1991a) yields a combined opinion. The resulting combined opinion (for the weight assigned to a particular factor or for a rating grade of slope failure potential) is a fuzzy number. This fuzzy number is then translated back to a proper linguistic term using the concept of *Euclidean distance* defined below:

$$d_i = \left(\int [\mu_c(x) - \mu_i(x)]^2 \, dx \right)^{1/2} \tag{1}$$

where

d_i = Euclidean distance between the fuzzy set that represents the combined opinion and the fuzzy set that represents the prescribed linguistic term i,

$\mu_c(x)$ = membership function defining the fuzzy set that represents the combined opinion,

$\mu_i(x)$ = membership function defining the fuzzy set that represents the prescribed linguistic term i, and

i = 1, n where

n = number of prescribed linguistic terms (in this study, n = 5; for example, for rating of the failure potential, the terms prescribed are *very high*, *high*, *medium*, *low*, and *very low*).

The best translation is the linguistic term that yields the least Euclidean distance. This procedure is used to analyze all the gathered opinions. The results obtained from analyses of the gathered opinions provide a basis for establishing an empirical method for an *indication* of failure potential of the slopes in the mudstone area studied. These results, including factors, weights and rating criteria, are summarized in Fig. 1 and Tables 1 through 5.

Figure 1 shows the evaluation tree adopted for assessing and mapping slope failure potential in the mudstone area studied. The factors adopted were grouped into two levels: primary and secondary levels. In the primary level, there are four categories of factors, and in the secondary level, there are 5 to 7 factors in each of the four categories. These categories and factors are itemized in Table 1, along with weights assigned to them. The rating criteria for factors in each category are shown in Tables 2, 3, 4 and 5, respectively.

It is noted that these factors are identified from the survey of opinions of experienced geotechnical engineers in the mudstone area studied. The rating criteria and weights among the adopted factors are also determined based on expert opinions. While the factors, weights, and rating criteria are deemed appropriate for the area studied, more studies are needed to further examine these variables. In particular, data from field measurements and observations should be collected and used to calibrate expert experience. At present, caution should be exercised when the results presented here are to be applied to other areas without similar conditions.

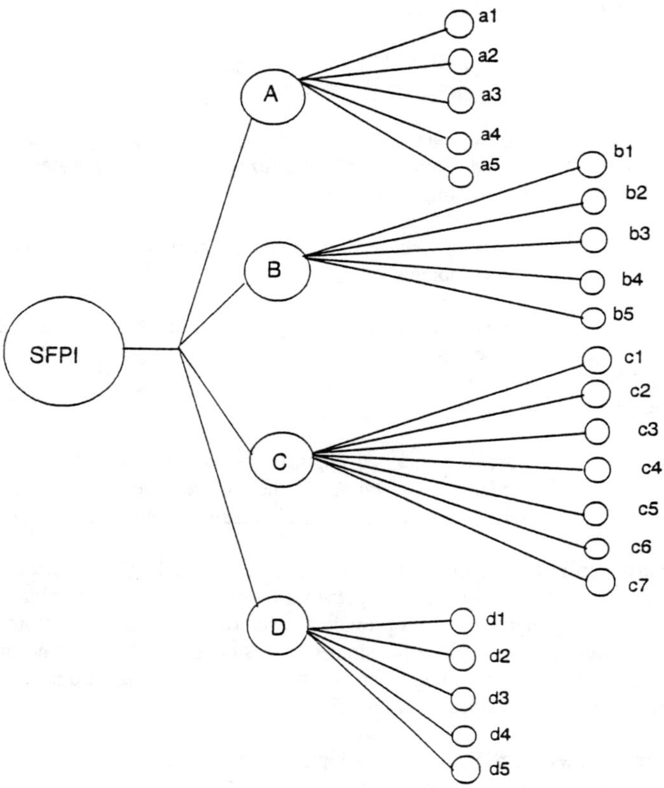

Figure 1 The SFP Evaluation Tree Used

Table 1 Factors Adopted for SFP Evaluation in the Mudstone Area Studied

Primary-Level Factor[1]	Weight[2]	Secondary-Level Factor[1]	Weight[3]
A: Topography	EI[4]	a1: slope gradient	EI[4]
		a2: slope direction	I
		a3: shape of vertical cross section	I
		a4: location of slope	VI
		a5: height of slope	EI
B: Geology	VI	b1: geological origin	I
		b2: bed rock strength	VI
		b3: spacing of discontinuities	EI
		b4: condition of fault	VI
		b5: type of overburden	VI
C: Environment	EI	c1: thrust of formation	I
		c2: type of vegetation	I
		c3: density of vegetation	VI
		c4: type of land use	VI
		c5: type of protection	VI
		c6: surrounding failure	I
		c7: drainage facility	EI
D: Meteorology	EI	d1: discharge from slope	VI
		d2: permeability of top soils	I
		d3: maximum daily precipitation	VI
		d4: maximum hourly precipitation	VI
		d5: presence of GWT	EI

[1]Each factor is proceeded by a symbol which is used to identify and locate that factor in Figure 1.
[2]Weight among the primary-level factors.
[3]Weight among the secondary-level factors.
[4]Grades for weight: EI - extremely important, VI - very important, I - important, NVI - not very important, UI - unimportant.

Table 2 Criteria for Assessing SFP based on Topography Factors

Secondary-level Factor under Topography	Criteria for Evaluating Slopes*				
	E	D	C	B	A
Gradient (deg)	<20	20-30	>70	30-40; 60-70	40-60
Direction (toward)	north	northeast; northwest	east; west	southeast	south
Shape of vertical cross section	___	___	straight; concave	convex	complex
Location	___	___	central; toe	upper	shoulder
Height (m)	0-2	2-5	___	5-20	>20

*Rating Grade (for failure potential): A- very high, B- high, C-medium, D- low, and E- very low.

Table 3 Criteria for Assessing SFP based on Meteorology Factors

Secondary-level Factor under Meteorology	Criteria for Evaluating Slopes*				
	E	D	C	B	A
Discharge from slope	none	___	small	medium	large
Permeability of top soils	___	___	high (sand)	medium (silt)	low (clay)
Maximum daily precipitation (mm)	0-20	20-50	50-100	100-150	>150
Maximum hourly precipitation (mm)	0-5	5-10	10-20	20-30	>30
Presence of ground water table	___	no GWT	___	when raining	stable GWT

Table 4 Criteria for Assessing SFP based on Environment Factors

Secondary-level Factor under Environment	Criteria for Evaluating Slopes*				
	E	D	C	B	A
Type of formation of slope (formed by)	___	___	landslide; previous slope failures	erosion; man-made (cut slope)	man-made (fill slope)
Type of vegetation	___	___	grass; trees	bamboo	no cover
Density of vegetation (% of land covered)	very high (>75%)	high (50-75)	medium (25-50)	low (10-25)	very low (<10)
Type of land use	___	forest	pasture; orchard; recreation zone	residential area; school	roadways; slope protection facilities
Type of protection facility	___	RC retaining wall; concrete crib wall	reinforced earth method; natural slope (no protection)	rock buttress; spreading of slurry	man-made slopes (no protection)
Surrounding slope failure (cases)	none	___	some (1-2)	several (3-5)	many (>5)
Drainage facilities	___	two-way (vertical and horizontal)	one-way (vertical)	one-way (horizontal)	no facility

*Rating Grade (for failure potential): A- very high, B- high, C-medium, D- low, and E- very low.

Procedure for Assessing and Mapping Slope Failure Potential

The area (or region) to be mapped for its slope failure potential needs to be divided into a large number of small sub-areas or cells. For each cell, the failure potential is assessed using the criteria established in Tables 2, 3, 4 and 5. The ratings obtained according to these criteria are then combined to yield an overall failure potential. This is done by repeatly applying the following equation at each node of the evaluation tree shown in Fig. 1:

$$R = \left[\prod_{j=1}^{n} (r_j)^{w_j} \right]^{\beta} \tag{2}$$

where

R = "combined" rating of slope failure potential based on all factors on a given branch of the evaluation tree,

Table 5 Criteria for Assessing SFP based on Geology Factors

Secondary-level Factor under Geology	Criteria for Evaluating Slopes*				
	E	D	C	B	A
Geological origin	____	igneous hard rock	metamorphic hard rock	sedimentary hard rock; igneous soft rock	sedimentary soft rock
Bed rock strength (q_u, kgf/cm^2)	extremely strong (>1000)	strong (500-1000)	medium (250-500)	weak (100-250)	extremely weak (<100)
Spacing of discontinuity (cm)	> 200	60-200	20-60	6-20	< 6
Condition of fault (distance from the area studied, m)	none (>1000)	inactive (50-1000)	inactive (<50)	active (50-1000)	active (<50)
Type of overburden	coarse gravel	gravel	sand	____	silt, clay no overburden

*Rating Grade (for failure potential): A- very high, B- high, C-medium, D- low, and E- very low.

r_j = rating of failure potential according to the factor i,
w_j = weight of the factor i as compared with other factors on the same branch of the evaluation tree,
n = number of factors in the given branch of the evaluation tree, and

$$\beta = \frac{1}{\sum_{j=1}^{n} w_j} \qquad (3)$$

For the evaluation tree shown in Fig. 1, the entire evaluation process involves repeated use of Eq. 3--the equation is used to combine ratings four times in the secondary-level and once in the primary level. Note that in this study, effects of these factors on the failure potential are assumed to be independent.

Since ratings and weights are assessed and recorded in linguistic terms (see Tables 1 through 5), these terms need to be translated into fuzzy numbers before further processing. Fuzzy numbers are *normal* and *convex* fuzzy sets, and are characterized by membership functions (Zadeh, 1975). Table 6 defines membership functions for the ratings and weights fuzzy numbers used in this study. They are of triangular shape. While a smooth *π-curve* membership function was suggested by Juang et al. (1992), no significant difference is found in the end results by using triangular functions (instead of *π-curve* functions) for the problem studied. Triangular functions offer greater speed, and are deemed appropriate for the problem at hand. Interested readers are referred to Juang et al. (1991b) for further discussion of membership functions.

Table 6 Membership Functions for the Rating and Weight Grades Used

Rating Grade (see Table 2)	Weight Grade (see Table 1)	Membership Function, f(y) (Defined over a real interval, [0,1])
E	UI	$f(y) = 4(y - 0.75),\quad 0.75 \le y \le 1.00$
D	NVI	$f(y) = 4(y - 0.50),\quad 0.50 \le y \le 0.75$ $= 4(1 - y),\quad 0.75 \le y \le 1.00$
C	I	$f(y) = 4(y - 0.25),\quad 0.25 \le y \le 0.50$ $= 4(0.75 - y),\quad 0.50 \le y \le 0.75$
B	VI	$f(y) = 4(y),\quad 0.00 \le y \le 0.25$ $= 4(0.50 - y),\quad 0.25 \le y \le 0.50$
A	EI	$f(y) = 4(0.25 - y),\quad 0.00 \le y \le 0.25$

With the input data being fuzzy numbers, the evaluation of Eq. 2 involves *fuzzy computations*. While direct implementation of fuzzy computations using the *extension principle* (Zadeh, 1975) may be possible, it was shown by Dong and Wong (1987) better results can be obtained by conventional arithmetic. In this study, the method based on Monte Carlo simulation technique, outlined by Juang et al. (1991a; 1992) was used. However, it was found, through extensive comparisons, results obtained by using Dong and Wong's algorithm (Dong and Wong, 1987), known as the vertex method, are in good agreement with those obtained from Juang et al's method (1991a).

The final output of Eq. 2 is a fuzzy number that represents the overall failure potential. Although a fuzzy output is expected (since the input is fuzzy), it is necessary to *scale* the resulting fuzzy numbers so that the failure potential is easy to interpret and the SFP map is easy to prepare. A slope failure potential index (SFPI) defined by Juang et al. (1992) is used. The SFPI is defined as:

$$\text{SFPI} = (A_L - A_R + 1) / 2 \tag{2}$$

where

A_L = area enclosed by the universe and to the left of the membership function of the final fuzzy number obtained, and

A_R = area enclosed by the universe and to the right of the membership function of the final fuzzy number obtained.

The values of SFPI fall between 0 and 1. If each and every one of the cells covering the entire area studied is evaluated and the SFPI value obtained, then contours of the SFPI values can be plotted and an SFP map can be prepared.

Mapping of Slope Failure Potential in the Southwestern Taiwan Mudstone Area

The area selected for a demonstration on the use of the developed evaluation and mapping procedure is in the southwestern Taiwan mudstone area. It is a strip-shape area along the south route of the Second Freeway in Kaohsiung County, with a total length of about 2.6 km and width of about 0.3 km. This area provides a good representation of the entire southwestern mudstone area.

The information used in the evaluation are obtained from 1) collection and review of relevant data such as air photos, topography maps, geology maps, and the literature of the area, and 2) field investigation. The data compiled are those that are required in using the evaluation tree shown in Fig. 1 and the criteria listed in Tables 2 to 5. The source data obtained for this study is documented in a National Science Council (NSC) report by Lee (1991). A flow chart depicting the entire evaluation process is shown in Fig. 2.

One major variable in preparing an SFP map is the size of the cell selected. In general, in an area where change in topography is large, the size of the cells must

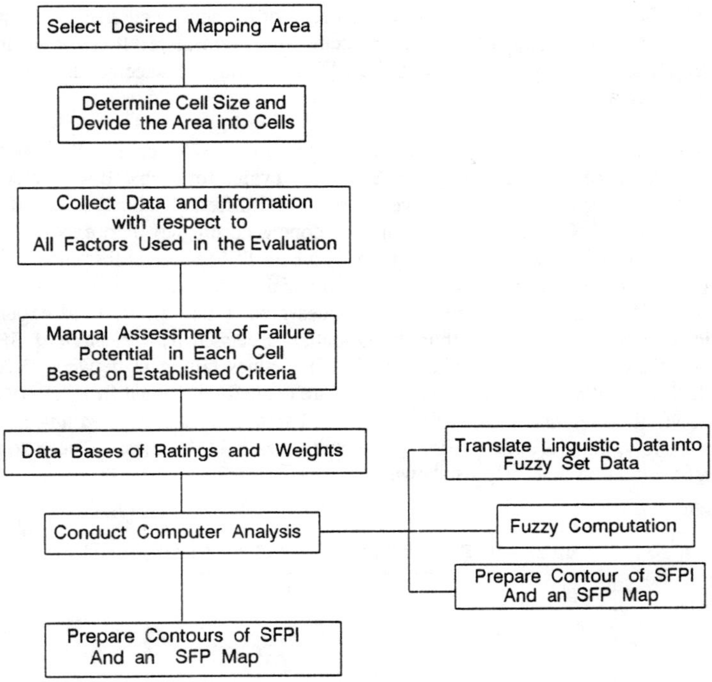

Figure 2 Flow Chart of the Procedure for Evaluating and Mapping SFP

be small in order to reflect such variation; in an area where the topography is rather uniform, the size can be larger to save time and efforts. While it is generally true that accuracy of an SFP map depends on size of the cells, economical and practical reasons preclude use of very small sizes. In this research, it was decided to conduct a sensitivity study of the effect of cell size in a test area, 200 m by 200 m in size. The cell sizes used are: 50m x 50m, 40m x 40m, 30m x 30m, 25m x 25m, 20m x 20m, 15m x 15m, and 10m x 10m. Contours of the calculated SFPI values are plotted for each of the seven sizes used. The results (not shown here, see Lee, 1991) show that the shape of the contours changes as the size decreases, and that when the size reaches 25m x 25m, the shape of the contours becomes steady--no significant change is observed when the size is further reduced. To further show this size effect, four cross sections were selected from the test area. Along each cross section, failure potential is evaluated using each of the seven cell sizes. The results are all the same in four different cross sections. A typical result is shown in Fig. 3. Based on this sensitivity study, a size of 25m x 25m was selected and used in the entire study area.

The entire study area was divided into about 1250 cells. Following the evaluation procedure depicted in Fig. 2, the SFPI value for each cell was calculated. The contours of the SFPI values were plotted and different failure potential zones were mapped. Note that the zonation was done according to field experience. In particular, three rules are used: 1) in zones where SFPI value is less than 0.34, no failure was observed or known, thus the potential is classed as *very low* if SFPI is less than 0.34, 2) in zones with SFPI values greater than 0.70, field experience indicates a frequent failure, thus the potential is classed as *very high* if SFPI is greater than 0.70, and 3) for areas with SFPI values falling in between 0.70 and 0.34, three classes, *high*, *medium*, and *low* are created with equal SFPI interval. In addition, the *medium* class is further divided into two classes, *medium-high* and *medium-low*, since a large portion of the study area is in this class. These zonation rules result in an SFP map as shown in Fig. 4.

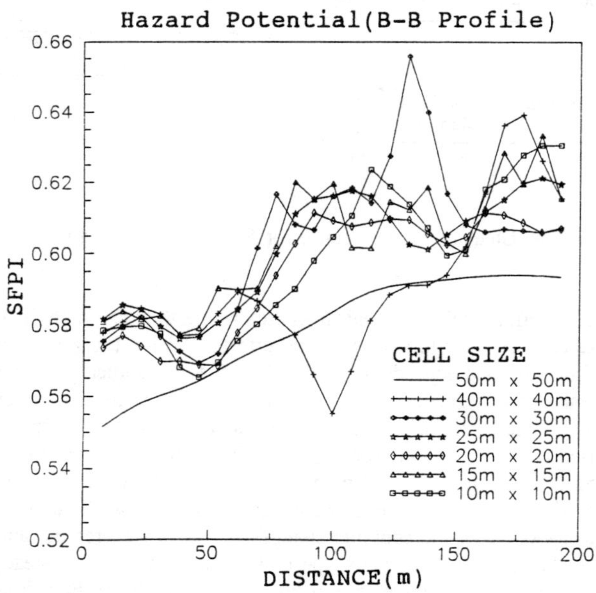

Figure 3 Effects of Cell Size on the Failure Potential

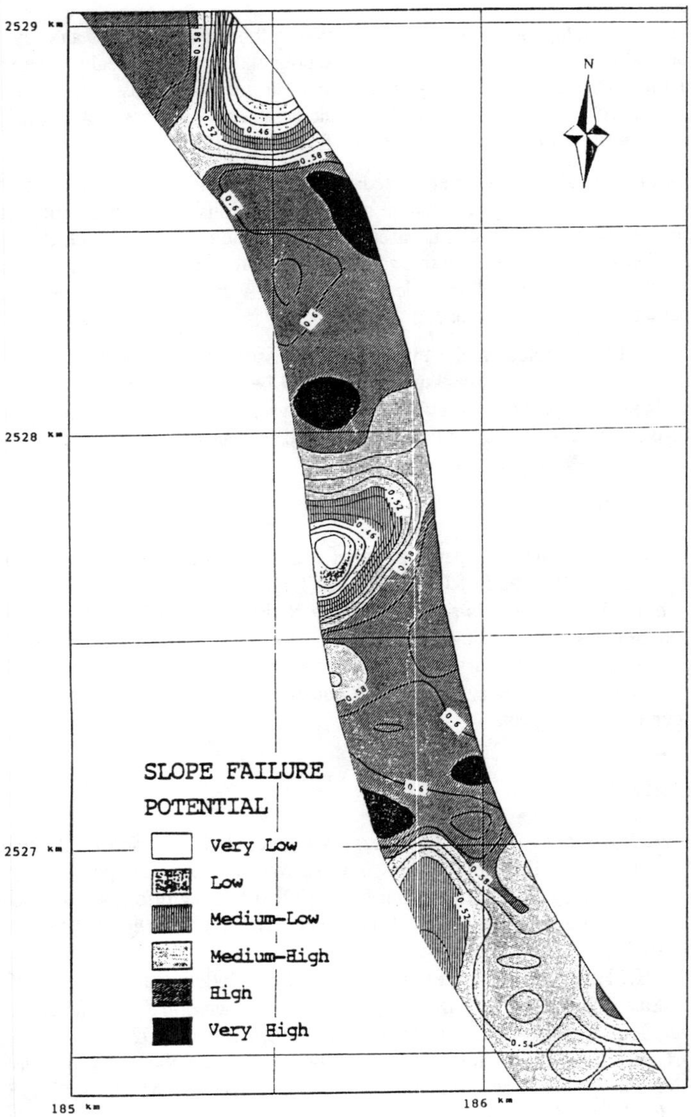

Figure 4 The Final SFP Map Obtained for the Study Area

Conclusions

1. The methodology and procedure presented in this paper provides an easy means for evaluating and mapping slope failure potential on an area scale. The validity of the presented procedure is supported by limited field observations and past experience in the area studied. This conclusion confirms that of an earlier study by the writers in another area with different conditions.
2. Factors, weights, and criteria used in the evaluation are quite subjective, and many of them are quite dependent on local conditions. While the procedure and the methodology are proven, based on limited studies, to be useful, the factors, weights and criteria used in this study should be re-examined for its applicability in other areas. Some modification to these variables may be necessary to account for local conditions.
3. The factors adopted in the evaluation are assumed to be independent in this study. The effect of this assumption should be evaluated, when more data are available. In particular, multi-variate analyses may be used to determine the correlations among the adopted factors with respect to failure potential.

Acknowledgment

The study on which this paper is based was partially supported by the National Science Council, Republic of China through Grant NSC-79-0414-P006-10B to the first author. The research is a joint effort by National Cheng Kung University and Clemson University. Assistance provided by Chyi Sheu, Chen-Ping Lin, and Yung-Sheng Chan during the course of this study is appreciated. Thanks must also be extended to the ASCE reviewers for their constructive comments which help clarify the paper in many aspects.

References

Dong, W. and Wong, F.S. (1987) "Fuzzy Weighted Averages and Implementation of the Extension Principle" *Fuzzy Sets and Systems*, Vol. 21, pp. 183-199.

Juang, C.H., Huang, X.L., and Elton, D.J. (1991a) "Fuzzy Information Processing by the Monte Carlo Simulation Technique" *Civil Engineering Systems*, Vol. 8, pp. 19-25.

Juang, C.H., Huang, X.H. and Elton, D.J. (1991b) "Modeling and Analysis of Non-Random Uncertainties - Fuzzy Set Approach" *Internation J. for Numerical and Analytical Methods in Geomechanics*, Vol. 16, March, 1992.

Juang, C.H., Lee, D.H., and Sheu, C. (1992) "Mapping Slope Failure Potential Using Fuzzy Sets" *Journal of Geotechnical Engineering*, ASCE, Vol. 118, No. 3, March, pp. 475-494.

Kobashi, S. (1980) *Slope Stability*, Kajima Publishing Co., Japan (in Japanese).

Lee, D.H. (1991) "The Damage Potential of Slope Land for Southwestern Mudstone Area in Taiwan" *Final Report to the National Science Council, No. 79-0414-P006-10B*, Taiwan, Republic of China (in Chinese).

Sato, T. and Terada, M. (1985) "Data Processing and Prediction of Slope Failure Based on Fuzzy Set Theory" *Annual Report of Natural Hazard Mitigation Disaster Prevention Research Institute*, Kyoto University, Japan, No. 28, B-2, PP. 63-86 (in Japanese).

Schmucker, K.J. (1984) *Fuzzy Set, Natural Language Computations, and Risk Analysis*, Computer Science Press, Rockville, MA.

Takeshita, K. (1986) *Prediction and Prevention of Slope Failure Hazards*, Hakua Shobou Book Co., Japan (in Japanese).

Watari, M. (1978) "The Causes and Remedies of Landslide and Slope Failure" *Doboku Sekou*, Japan, Vol. 19, No. 2, pp. 43-51 (in Japanese).

Yamada, G., Watari, M. and Kobashi, S. (1983) *The Practical Conditions and Remedies of Landslide and Slope Failure*, 7th Edition, Sankaidou Publishing Co., Japan (in Japanese).

Zadeh, L.A. (1975) "The Concept of a Linguistic Variable and its Application to Approximate Reasoning" *Information Science*, Vol.8, pp. 199-249.

STABILITY EVALUATION OF WASTE LANDFILLS

by Richard A. Mitchell, Associate Member, ASCE[1] and
James K. Mitchell, Fellow, ASCE[2]

INTRODUCTION

Modern waste landfills serve a variety of functions, including storage of the maximum amount of wastes in a minimum amount of space, isolation of the waste materials from the surrounding environment, and the provision of safe and useable land areas after closure. Until recently, attention in the design, construction, filling, and post-closure monitoring and maintenance of new landfills has focused mainly on the prevention of unacceptable levels of contamination in the surrounding groundwater and atmosphere, and the closure plans for old landfills have centered on similar concerns. However, there have been recent stability failures within waste piles, through landfill foundations, and along liner systems. In addition, the filling and closure plans for landfills on very soft foundations may be seriously impacted by the potential for stability failure. Consequently, stability of landfills is now a major concern in the design, construction, filling, and closure of waste landfills.

In this paper we examine the types of stability problems associated with the siting, design, construction, filling, and closure of waste landfills. The scope includes potential failure mechanisms, evaluation of material properties, stability analysis methods, safety factors, and stability failure risk mitigation. Our concern is with both hazardous/toxic solid waste landfills and municipal waste

---

[1]Engineering Geologist, CH2M HILL, 6425 Christie Avenue, Suite 500, Emeryville, CA, 94608
[2]Edward G. Cahill and John R. Cahill Professor of Civil Engineering, University of California, Berkeley, California 94720

landfills, as similar regulatory criteria and design considerations may apply to each.

TYPES OF STABILITY FAILURES AND THEIR CONSEQUENCES

Two typical types of solid waste landfills are shown schematically in Figure 1. It should be noted that Figure 1 does not show the details of the liner systems, leachate and gas collection systems, and monitoring systems that form essential parts of new waste landfills. Analysis of these systems requires an understanding of the stress-deformation properties of the components and component interfaces for these systems. In Figure 1(a) a waste pile is built over the existing ground surface. Ultimately, it reaches some maximum design height and a cover is placed as a part of final closure. Figure 1(b) shows the general configuration when waste is placed into a deep excavation or existing valley or depression. For waste repository construction in which the waste is to be placed in an excavation, it is likely that the owner will want to use the steepest possible side slopes in order to maximize the available waste storage volume for a given site area. In each case usual practice is to insure that the ground water level is maintained below the bottom of the waste.

Several failure modes and mechanisms are possible for the landfills in Figure 1 during excavation, during filling, and after closure. Important in the identification and assessment of the potential failure modes is that both covers and liners for modern landfills are typically multilayer composites composed of both soil and geosynthetic materials. A schematic diagram of a typical cover design is shown in Figure 2 and a composite double liner system used for an existing hazardous waste storage facility is shown in Figure 3. Each of the systems shown contains several interfaces whose resistance against in-plane shear may be low and thus act as potential planes for failure.

Potential failure modes are summarized by the schematic diagrams shown in Figure 4, and can be described as follows:

1. *Sidewall slope and base failures, Figure 4(a).* This type of failure can occur during the excavation of a repository and during the placement of liner systems prior to the commencement of filling operations. Although only a rotational failure mode is shown in Figure 4(a) for a base failure, the possibility for wedge and block failures owing to the presence of weak layers or seams should not be overlooked. The potential for and safe designs against these types of failures can be assessed using the usual methods of geotechnical investigation and slope stability analysis, and are not discussed further here.
2. *Pullout of liner system components from anchor trenches and downslope movement, Figure 4(b).* Downslope movement of the liner system components is resisted by interface friction between

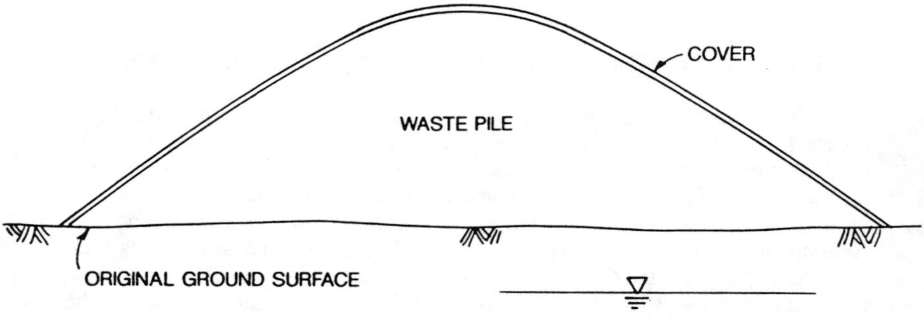

a) Waste Pile Over Existing Groundsurface

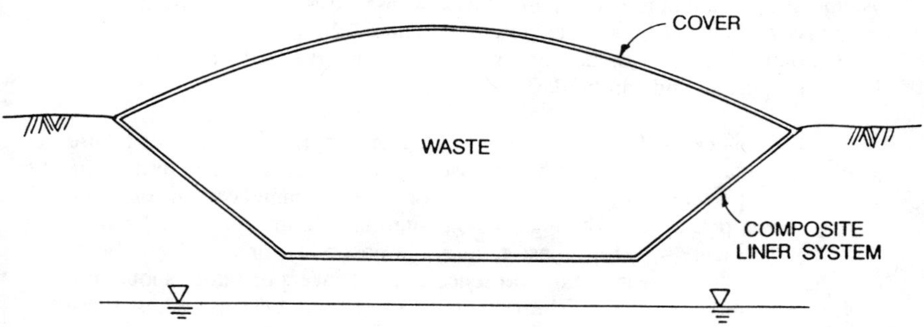

b) Landfill Within Excavated Area or Existing Valley

FIGURE 1
TWO GENERAL TYPES OF SOLID WASTE LANDFILL

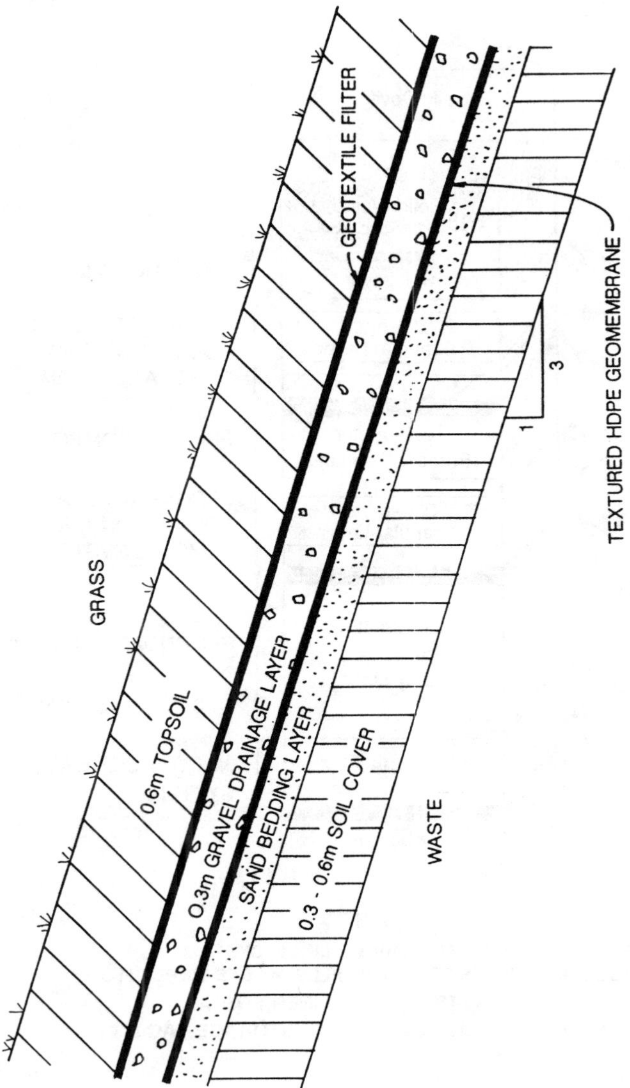

FIGURE 2
TYPICAL SOLID WASTE LANDFILL COVER DESIGN

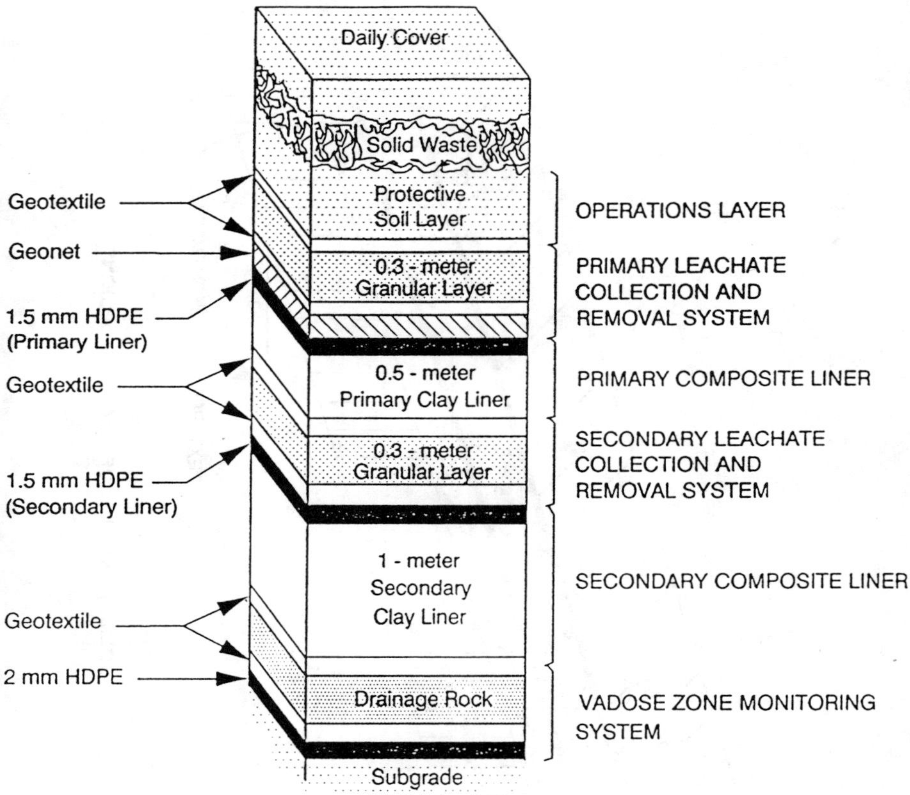

FIGURE 3

COMPOSITE DOUBLE LINER SYSTEM USED AT THE KETTLEMAN HILLS WASTE REPOSITORY - AN EXAMPLE OF A LINER FOR A MODERN HAZARDOUS WASTE STORAGE FACILITY

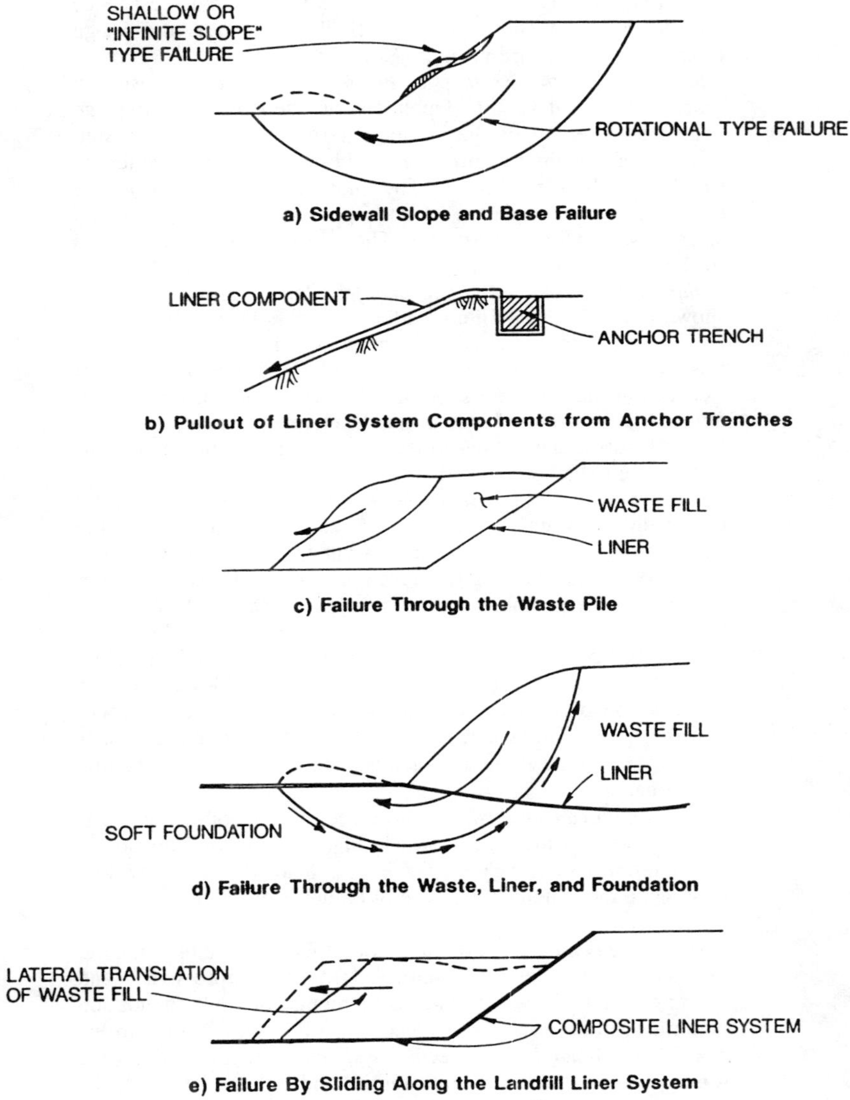

a) Sidewall Slope and Base Failure

b) Pullout of Liner System Components from Anchor Trenches

c) Failure Through the Waste Pile

d) Failure Through the Waste, Liner, and Foundation

e) Failure By Sliding Along the Landfill Liner System

FIGURE 4
SCHEMATIC DIAGRAMS OF POTENTIAL FAILURE MODES IN LANDFILLS

the liner components themselves, between the lowermost liner layer and the excavation side wall, and by pullout resistance afforded by the anchor trench. Methods for analysis and design of anchor trenches are given by Koerner (1990).

3. *Failure through the waste pile, Figure 4(c).* After a waste fill reaches a critical height, dependent on the unsupported slope angle and strength of the waste material, a failure will result. Analysis of stability for this case can be made using conventional stability analysis methods. The most difficult aspect is the selection or determination of the appropriate unit weight and strength values for the waste fill. These properties are considered further later in this paper.
4. *Failure through the waste pile and foundation, Figure 4(d).* As shown in the figure, a failure surface can pass through the waste, the liner, and the underlying foundation soil. The potential for this mode of failure will be greatest when the foundation soils are weak, as in the case of a soft clay. The possibility of this type of failure is a controlling factor in the development of a closure plan for a landfill that is described in a separate paper to this conference (Mitchell et al., 1992).
5. *Failure along the liner system, Figure 4(e).* The waste mass slides essentially as a unit along low strength interfaces within the composite liner system. The stability against failure of this type is controlled by the shearing resistances of the interfaces in the liner system and the geometry and unit weight of the waste fill. The Kettleman Hills landfill stability failure described by Mitchell et al. (1990), Seed et al. (1990), and Byrne et al. (1992) resulted from this mode of sliding. Stability failures of this type can generally be prevented by avoiding large elevation differentials and the use of flatter side slopes. However, flatter side slopes reduce the volume available for waste storage for a given landfill plan area.
6. *Failure within caps and covers.* When composite soil-geosynthetic cover systems of the type shown in Figure 2 are used on slopes, low interface shearing resistance can lead to instability and downslope movements. This type of failure is examined further in a later section.
7. *Excessive settlement.* Although not strictly a stability failure, excessive settlements resulting from compression and decomposition of waste and settlements of the foundation materials can be disruptive to leachate collection and monitoring systems and landfill gas collection systems. Settlement within the waste pile can also exert large tensile and pullout forces on the side slope liner and may cause tearing. In addition, non-uniform

settlements can cause distortions that crack and rupture covers and linings. If water gains access to the landfill the stability may be adversely affected.

All of the failure modes may be initiated by seismic as well as static loading. In addition, seismic loading of filled waste repositories is of concern because the dynamic loading creates relative movements between the waste and the foundation. These displacements could result in tearing of the liner and loss of its sealing function. Of all the modes of failure, those that breach the liner system are of the greatest concern because of the possibility that leachate will then escape into the underlying soil and groundwater.

MATERIAL PROPERTIES AND THEIR DETERMINATION

The evaluation of the stability of landfills is done in very much the same way as the analysis of any other type of geotechnical stability problem. For situations that can be reasonably represented in two dimensions, reliable methods of analysis and computer programs for their application are readily available. For situations requiring consideration of three-dimensional effects, analysis methods are currently less well developed. However, in both cases the most critical aspect of the evaluation is the certainty with which the relevant properties are known. Thus, while uncertainties about analysis may lead to errors perhaps of the order of 10 to 20 percent, uncertainties in strengths may easily result in errors that are twice as great. Consequently, the selection of proper values for the strengths of the waste and foundation material, and of the shearing resistances along the interfaces within the liner and cover systems is the most critical part of any stability study. The greatest difficulties and uncertainties are associated with evaluation of the strength and stress-deformation properties of the liner system materials and interfaces and of the waste fill.

Properties of Hazardous and Municipal Waste Fills

Hazardous wastes are usually contained within steel containers, boxes, etc., and stacked within a repository. Soil cover is used to fill the spaces in and around the waste containers and as daily cover. Characterization of the resulting mass for analysis can be done only approximately. In at least one case, the average unit weight was determined by excavation of several truckloads and was found to be 18.8 kN/m^3. The shearing strength of such a composite fill can probably be conservatively estimated as equal to the strength of the backfill soil at a low relative compaction.

Relatively little is known about the geotechnical engineering properties of municipal solid waste. Part of the paucity of data may be explained by the fact that such data have not been in demand until recently. Of more significance, however, are the difficulties in credible sampling and testing of refuse. This difficulty is further compounded by the fact that refuse composition and properties are likely to vary erratically within a landfill and are also likely to change with time due to decomposition.

Density and Moisture Content. The unit weight of refuse varies depending on a number of factors including initial composition, compactive effort, decomposition, settlement, and moisture content. Although the density of the refuse will typically increase with depth, considerable variability should be expected within relatively short distances. Landva and Clark (1987) report that typical unit weights for refuse generally range from about 6.3 kN/m^3 to 11.8 kN/m^3.

The moisture content of municipal solid waste is highly dependent on a number of interrelated factors. Some of these factors include the initial composition of the waste, local climatic conditions, landfill operating procedures (e.g. use of daily cover), the effectiveness of any leachate collection and removal system, the amount of moisture generated by biological processes within the landfill, and the amount of moisture removed with landfill gas. The water content of municipal solid waste in the United States reportedly ranges from about 10 to 50 percent (Huitric et al., 1979). In regions where evapotranspiration exceeds precipitation, a typical moisture content is on the order of 25 percent.

Shear Strength. Three general approaches have been used to estimate the shear strength of the waste in landfills: (1) laboratory testing; (2) back-calculation from field tests and operational records; and (3) in-situ testing. Laboratory testing has been performed on reconstituted samples from landfills, samples in which various substitutions were made (e.g., sand for stones, crushed glass for glass containers, lint for cloth, etc.), and on samples collected from drive samplers. In one test, Fang et al. (1977) tested compacted bales of refuse. Direct shear has been used for most tests although the Earth Technology Corporation (1988) performed a multistage triaxial test on a sample of refuse obtained with a drive sampler. A summary of typical results from laboratory tests is shown in Figure 5(a).

A number of back-calculated shear strengths used for design are based on a field load test made in the Los Angeles area by Converse et al. (1975). The shear strengths that were back-calculated from this test assumed failure of the slope (i.e., a factor of safety equal to 1). However, since the slope did not fail, the back-calculated strengths were conservative by an unknown amount. Different combinations of cohesion and friction angle back-calculated by various authors based on the Converse test results are shown in Figure 5(b).

Limited attempts have been made to estimate the shear strength of refuse by in-situ testing. Earth Technology Corporation (1988) reported the results (Figure 5[c]) of vane shear and standard penetration tests that were performed at a landfill near Los Angeles. However, because the vane shear device is small (6.35 cm) compared with the inclusions (e.g. tires, wood, carpet, etc.) that make up refuse, the shear strength data may not be representative of

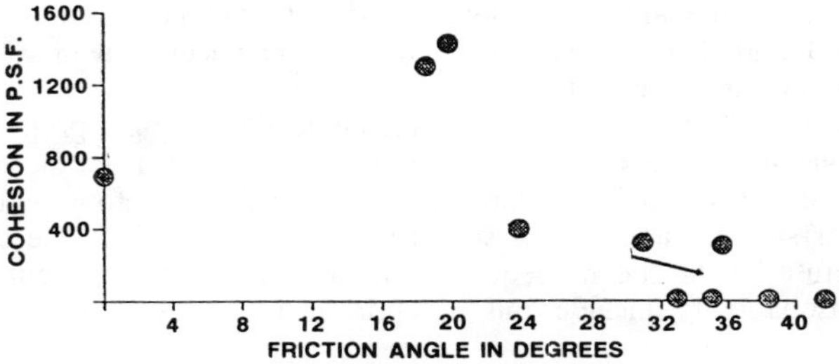

(a) Strength Parameters from Laboratory Shear Tests

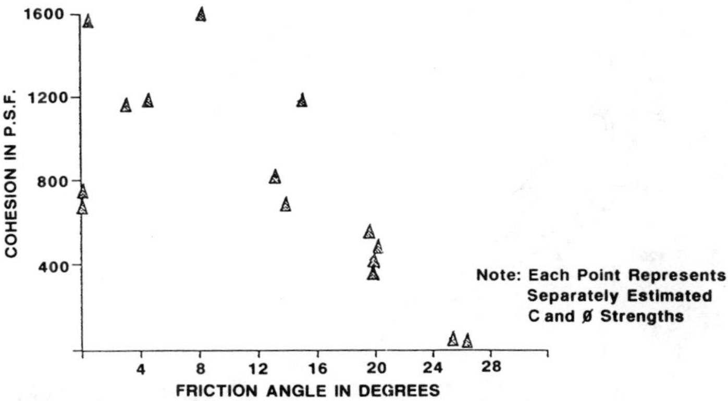

(b) Strength Parameters Estimated by Back Calculation from a Field Load Test and from Performance Records

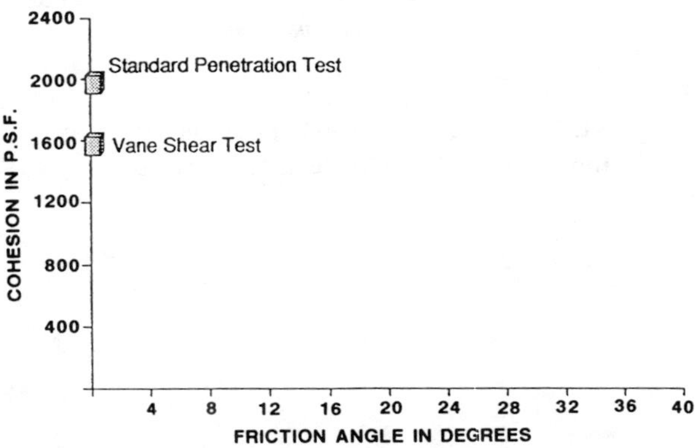

(c) Strength Parameters from In-Situ Testing

SOURCE: SINGH AND MURPHY, 1990

FIGURE 5
SHEAR STRENGTH OF REFUSE BASED ON DIFFERENT ANALYSIS METHODS

actual conditions. Similarly, since there are no published correlations between refuse strength and blow counts, there is considerable uncertainty in assigning a strength based on this information.

Singh and Murphy (1990) present recommended strength properties for municipal solid waste based on a summary plot of the results from the different tests (Figure 6). There is large scatter in the data, and therefore, assigning a specific shear strength to refuse is somewhat subjective. While the cohesive nature of refuse has not been adequately characterized at the present time, it appears reasonable to include this component in an evaluation of shear

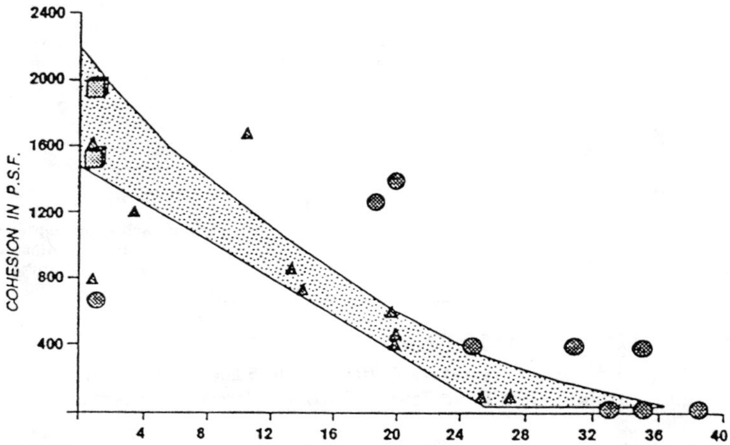

SOURCE:
SINGH AND MURPHY,
1990

FIGURE 6
SUMMARY PLOT FOR ALL TEST WITH RECOMMENDED PARAMETERS SHADED

strength. Although true cohesion (or bonding) between particles is unlikely, a significant cohesion intercept may result from interlocking or overlapping of waste constituents. This interpretation is supported by some laboratory test results and the common observation that vertical cuts in refuse can stand unsupported to considerable heights.

Properties of Composite Liner and Cover Interfaces

Figures 2 and 3 show that there are several interfaces along which slip may occur in a landfill liner or cover. These include geotextile-geonet, geotextile-geomembrane, geonet-geomembrane, and compacted clay-geomembrane. Until rather recently, little information was available in the open literature concerning typical values of interface friction angle. Among the published values were those obtained by Martin et al. (1984). The measured values for three sands, four types of geomembrane, and four geotextiles show that: (1) the frictional resistances between cohesionless soils and geomembrane materials, 17° to 27°, are less than the angle of internal friction of the cohesionless soils themselves, which are usually in the range of 30° to 50°; (2) the interface friction angles between geomembranes and geotextiles are quite variable, 6° to 28°, depending on type of materials; and (3) sand-to-geotextile friction angles, 23° to 30°, are less variable and generally higher than the other combinations tested.

With the failure in March 1988 of Class I hazardous waste Landfill Unit B-19 at Kettleman Hills in California, which was clearly indicated as a slip along and/or in the composite double liner system, it became necessary to determine the interface shearing resistances of specific combinations of geomembranes and geotextiles, geomembranes and geonets, and geomembranes and compacted clay liner material in order to properly evaluate the conditions at the time of failure[3]. The results of both direct shear tests and pullout tests done for this purpose are presented by Mitchell et al. (1990a).

Since the Kettleman Hills failure there has been intense interest in the development of new apparatus and testing procedures for the evaluation of interface strength for different composites, and a more extensive data base is now available. In addition, textured geomembranes have been developed for use against soils and other geosynthetics when greater strengths are required.

Geosynthetic-geosynthetic interfaces. The results of the geosynthetic interface shear tests reported by Mitchell et al. (1990) in connection with the initial Kettleman Hills failure investigation are particularly instructive examples of the values and variabilities of interface shear resistances. Values of residual interface friction angle for different geosynthetic combinations are summarized in Table 1. Peak values of interface friction angle were ½° to 2° greater than

---

[3]It was not known until after removal of the waste pile from above the liner, more than two years after the failure, that the actual failure was primarily along the secondary compacted clay-flexible membrane liner interface.

the residual friction angles. As the peak strengths occurred at very small interface shear displacements (typically less than 3 mm) that are likely to be exceeded by deformations occurring during construction and fill placement operations, the residual interface strength is recommended as that most representative for stability analysis and design purposes. The direct shear and pullout tests used to determine the values in Table 1 were done under normal stresses on the failure surface ranging from 158 to 480 kPa, which corresponded to waste fill heights of 8.4 to 25.4 m above the liner. The measured values were essentially independent of the value of normal stress used.

Among the significant features of the results in Table 1 are that: (1) the values of interface friction vary over ranges of about 2° to 3° for most geosynthetic combinations[4]; (2) "polishing" of interfaces as a result of prior slip leads to a reduction of about 1° in interface friction angle; (3) dry and submerged values of interface friction angle may differ slightly; (4) the orientation of a geonet relative to its direction of sliding along a HDPE geomembrane influences the interface friction angle by up to about 2°[5]; and (5) overall, the geosynthetic combinations listed in Table 1 have sufficiently low interface friction angles; i.e., 7° to 10° that the danger of sliding along the liner system may be the limiting factor in the design and filling plan for a waste landfill. These low values are the more significant when composite liners are used on repository side slopes which are at inclinations of 1 on 3 horizontal to vertical (18.4°) or 1 on 2 (26.6°).

Geomembrane-compacted clay interfaces. The government-mandated requirements for landfill liner systems in the U.S.A. have undergone significant changes over the past ten years. The key elements of the current composite double liner system are the two geomembrane-compacted clay combinations. The geomembrane is intended as the primary barrier for leachate containment. The compacted clay immediately beneath the geomembrane provides a backup to retard leachate flow in the event of leakage through the geomembrane. To fulfill this function it is necessary that the geomembrane and compacted clay be in close contact, without intermediate layers or materials that could diffuse leachate widely over the surface of the clay. Although a composite liner

---

[4]The significantly greater variation in residual friction angles measured in the different tests for HDPE liner against HDPE liner was surprising, and the precise reason has not yet been determined, although similar variations have been noted in other studies. Furthermore, variations of comparable magnitude were found in tests of polyethylene sheet against HDPE membrane as a result of differences in pre-cleaning procedures, interface polishing, temperature, and time under stress prior to shear.

[5]In Table 1 "aligned shear" means that the geonet ribs were oriented within 15° of the shear direction; "transverse shear" means that the geonet rib orientation was oblique to the shear direction by more than 15°.

provides excellent protection against escape of leachate, it introduces new interfaces that can be the source of stability problems.

Table 1
Summary of Interface-Shear-Strength Tests: Kettleman Hills Repository

Interface Components	Conditions	Direct Shear Tests: Residual Friction Angle: ϕ_r	Pullout-Box Tests: Residual Friction Angle: ϕ_r
HDPE liner/geotextile	Dry, unpolished	9.5° to 12.5°	9.5° (1 Test)
	Dry, partly polished	9.0° to 11.0°	---
	Dry, polished	8.5° to 10.5°	8.0° (1 Test)
	Wet, unpolished	8.0° to 10.0°	7.0° to 10.5°
	Wet, polished	7.0 to 9.5°	6.5° to 9.0°
HDPE liner/geonet (transverse shear)	Dry	7.0° to 8.0° +	--
	Submerged	7.0° to 10.0°	8.0° to 9.0°
HDPE liner/geonet (aligned shear)	Submerged	5.0° to 8.0°	6.0° to 8.0°
Geotextile/geonet	Dry	>20°	--
	Submerged	10° to 14° (+)	--
HDPE liner/HDPE liner	Dry	6.0° to 13.0°	--
	Submerged	6.0° to 11.0°	7.0° to 13.5°

The focus for many years has been on the preparation and placement of the compacted clay so that it will be at a state that insures minimum hydraulic conductivity. The general relationships between compaction conditions and hydraulic conductivity have been well-established by a number of studies; e.g., Mitchell et al. (1965), Daniel and Benson (1990). The lowest values of hydraulic conductivity are obtained for compaction to a high degree of saturation and a high density; i. e., wet of optimum using a large compactive effort. Specifications for clay liners commonly require compaction above some minimum density and water content to insure that the hydraulic conductivity will be below some limiting value, typically 1×10^{-9} m/s. Little attention has been given until recently to the other mechanical properties associated with compaction to these specified conditions.

The failure of the Kettleman Hills hazardous waste landfill in 1988 served to draw attention to the low strengths along geomembrane interfaces, however. As a part of the same testing program that produced the data summarized in Table 1, tests on compacted clay-HDPE geomembrane interfaces gave undrained shearing strengths of the order of 40 kPa (Table 2) for compaction to conditions thought at the time of testing to best represent the in-situ state at the time of failure; namely, an average relative compaction of 94 percent based on Standard Proctor and a water content about five percent wet of optimum, followed by soaking under a light surcharge and then rapid direct shear after application of surcharge pressures comparable to those in the field. Subsequent tests by Byrne et al. (1992) gave more probable values of 30 to 35 kPa at the time of failure.

Table 2
Friction Angles or Shear Strengths on Critical Interfaces in the Kettleman Hills Landfill Liner System

Interface	Residual Friction Angle (ϕ_r) or Residual Undrained Shear Strength (τ_r) Along Saturated Base	Residual Friction Angle (ϕ_r) Along Dry Slopes
HDPE Liner/Geotextile	8° ± 1°	9° ± 1°
HDPE Liner/Geonet	8.5 ± 1°	8.5 ± 1°
HDPE Liner/Saturated Clay	45 ± 12 kPa	N/A

Compressive testing of smooth (nontextured) HDPE-compacted clay interfaces has shown that the interface strength, expressed as an equivalent friction angle, varies greatly over the range of possible as-compacted conditions (Seed and Boulanger, 1991). Some of Seed and Boulanger's results are shown in Figure 7 for smooth HDPE against compacted clayey till containing 5 percent bentonite. Two sets of results are shown: the upper diagram is for samples sheared as compacted, and the lower diagram is for tests on samples soaked under a small normal stress (12 kPa) prior to application of surcharge and undrained shear.

Seed and Boulanger (1991) divided the measured friction angles into three zones within the as-compacted water content-dry density space. Zone I, which includes the range of water contents and dry densities that would most probably be specified to insure low hydraulic conductivity, includes the lowest strengths, Zone II produces strengths of intermediate value, and Zone III gives the highest as-compacted strength values. The differences in strength among the zones are reduced as a result of pre-soaking, but this is primarily a result of significant reductions in the strengths of samples in Zones II and III, whereas, the strengths of samples compacted in Zone I are not appreciably reduced.

It is clear, therefore, that those compaction conditions that produce the lowest values of hydraulic conductivity also produce the lowest values of undrained strength, a situation that is quite consistent with the known relationships between soil properties and soil structure. Although it may be possible to increase the shearing resistance directly along the compacted clay-HDPE geomembrane interface through the use of a textured geomembrane, it remains to be seen whether or not this will really improve the situation, because slip might occur, for clays compacted to high density and high water content, along a surface entirely through the clay parallel to the interface. To evaluate this possibility undrained direct shear tests on clay samples compacted over the appropriate range of water contents could be used.

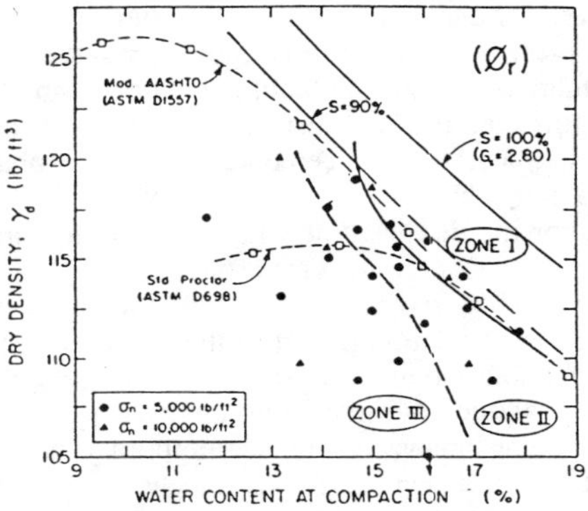

ZONE I Ø = 5°-8°

ZONE II Ø = 11°-18°

ZONE III Ø = 17°-24°

(a) Samples Sheared As Compacted

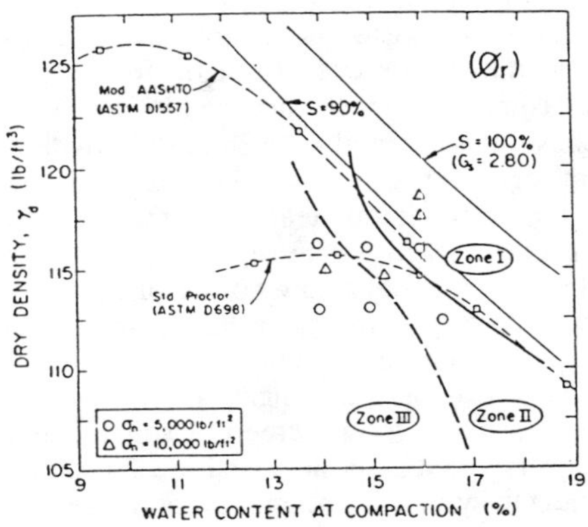

ZONE I Ø = 8°-9°

ZONE II Ø = 11°-13°

ZONE III Ø = 12°-14°

(b) Samples Sheared After Initial Presoaking

SOURCE: SEED AND BOULANGER, 1991

FIGURE 7
INTERFACE SHEAR STRENGTHS FOR SMOOTH HDPE-COMPACTED CLAYEY TILL (with 5% Bentonite) INTERFACES AS FUNCTION OF COMPACTION AND SOAKING CONDITIONS:

Selection of Properties for Analysis and Design

Given the variabilities of materials and the variations in strengths for any given material or interface combination, there is always concern over the appropriate values to use for analysis and design. In our opinion, the approach taken should depend on the particular problem being studied. If an analysis is being made of the stability of an existing landfill, or a back analysis of a failure is required, then strengths that are the most representative of the actual in-situ values should be used. Worst case scenarios and the probability of their occurrence should also be evaluated.

On the other hand, for the design of new facilities and for the development of filling plans, conservative estimates of properties should be used. The actual values in any case must be chosen with regard to: (1) the factor of safety, which may either be mandated by regulation or left to the designer; (2) the variability in properties; (3) the possibility that properties may change with time; (4) the type of failure being analyzed and the consequences if it occurs; and (5) whether the condition being analyzed will be temporary or permanent. In connection with (3), it is noted that for failure mechanisms that pass through municipal waste, the waste material properties may change with time as a result of consolidation under increased heights of fill and due to biological decomposition of the waste.

Thus far, deformations appear not to have been given extensive consideration in the analysis of waste landfill stability, except as they may be caused by earthquake loadings. However, strengths for design analyses should be chosen that are compatible with acceptable levels of deformation. Unfortunately, what constitutes acceptable levels of deformation is not generally known. The amount of shear distortion that can be tolerated within a composite double liner system without tearing of the liner components is in need of determination, especially for waste repositories in seismic areas.

Slope and foundation deformations may not be of particular concern from an overall stability standpoint in many cases; however, if they are large, the functioning of leachate collection and monitoring systems could be impaired. Large post-closure deformations, as might occur under conditions of decomposing waste or consolidation and creep of soft foundation soils, could both rupture landfill covers and disrupt new land use functions.

The hydraulic conductivity-shear strength paradox discussed in the last section must be given careful consideration in the selection of strength properties and preparation of compaction specifications. A procedure has been proposed recently by Daniel and Benson (1990) for the development of a compaction specification that is based mainly on hydraulic conductivity, but which allows for consideration of strength in the specification of acceptable compaction ranges.

STABILITY ANALYSES

Methods

The usual methods for slope stability analysis are generally applicable for the analysis of waste landfill stability. The characteristics and limitations of these methods are reviewed by Duncan (1992) in a separate paper to this conference. Limit equilibrium methods are usually used; however, finite element analyses have also been used for the study of deformations and stress distributions. The analyses are generally made by assuming two-dimensional plane cross sections that are anticipated to give the lowest factor of safety in the most probable direction of sliding. Wedge equilibrium methods are used when a probable failure mode includes plane surfaces of sliding, which is often the case when low strength liner interfaces are present. Rotational failure analyses assuming circular sliding surfaces are commonly done using computer programs based on Bishop's modified method, Janbu's simplified method, or Spencer's method. The latter can be used for non-circular surfaces as well.

Regardless of the method used, it is important to bear in mind the following factors listed by Duncan (1992):

1. The accuracy can only be as good as the accuracy with which slope geometry, unit weights, shear strengths, and pore pressures are known.

2. Methods that satisfy all conditions of equilibrium are quite accurate for any condition, except when numerical problems are encountered, and give results that differ "by no more than about 6 percent from what can be considered the correct answer."

3. Force equilibrium methods give factors of safety that are sensitive to the assumed inclinations of side force between slices.

In the case of waste landfills, the strength and unit weight properties of the waste may only be known within broad limits. As the analysis of the stability of waste landfills against sliding along the liner system interfaces often are done using force equilibrium methods, the third point above may have a significant influence on the calculated factors of safety.

In most geotechnical stability problems, the properties of the material on both sides of the failure surface are usually about the same. On the other hand, in the case of waste landfills, especially those containing municipal waste, the unit weight and strength of the waste may be much less, and the compressibility may be much greater than the corresponding values for the foundation soil. Thus for analysis of failure along liner interfaces, the properties on opposite sides of the potential failure surface can be widely different. The potential role of differential straining in the initiation of failure was considered by Byrne et al. (1992) in their analysis of the Kettleman Hills landfill failure. The extent to which large differences between material properties on opposite sides of a potential failure surface may influence the results has not yet been thoroughly investigated; however, finite element studies are in progress to further evaluate their possible significance (Kramer, 1992).

The incompatibility between strains that cause failure in soils and in refuse has been considered by Singh and Murphy (1990) for analysis of potential failure surfaces that pass through both the waste and a soft foundation material. For example, they propose that for a refuse fill more than about 70 m high with relatively steep side slopes; e.g., 1.5H-2.0H to 1V, placed over soft or weak soils having an undrained shear strength less than 100 kPa, the stability along a potential failure surface such as shown in Figure 8 should be computed neglecting the shearing resistance contributed by the fill. Although this approach may be over-conservative for estimating the factor of safety against complete failure, it may be suitable for limiting excessive deformations which could disrupt drainage, leachate, and gas collection systems, and could cause excessive deformation of the liner.

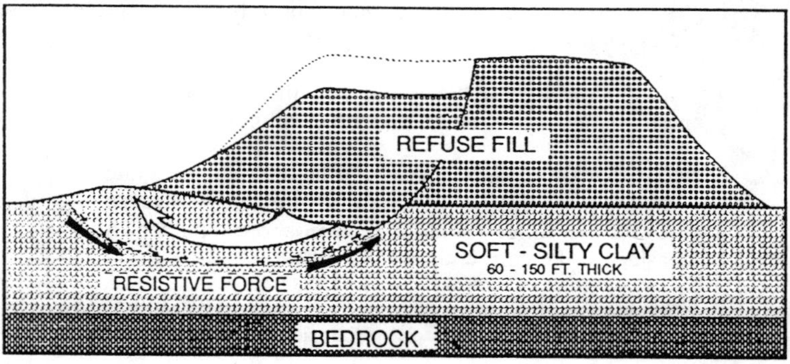

FIGURE 8
POSSIBLE FAILURE MODE FOR A HIGH WASTE FILL ON A SOFT FOUNDATION
(Shear strength of refuse is neglected in computation of stability)

Three-Dimensional Considerations

Waste landfills are three-dimensional structures. The results of analyses of the stability failure of the Kettleman Hills hazardous waste landfill by sliding of the waste pile along the repository liner system by Mitchell et al. (1990) and Seed et al. (1990) suggested that three-dimensional effects could have been a contributing factor, because approximate three-dimensional analyses gave factors of safety that were about 10 to 15 percent less than obtained from two-dimensional wedge analyses. This was a rather surprising result because it had generally been thought that three-dimensional stability analyses would always give higher factors of safety than two-dimensional analyses. In fact, Duncan (1992) in a separate paper to this conference, made a careful review of studies of three-dimensional slope stability that have been done over the past 25 years and concluded that "the factor of safety calculated using 3D analyses will always be greater than, or equal to, the factor of safety calculated using 2D analyses."

The physical reasoning that led Mitchell et al. (1990) and Seed et al. (1990) to conclude that the three-dimensional stability was more critical than would be computed using a two-dimensional stability analysis method at Kettleman Hills can be explained by reference to Figure 9. The driving forces for sliding are caused by the active force from block 1 plus the components of the active forces from blocks 2 that act in the sliding direction. The latter forces may be significant for a system in which the interface friction angle is low compared to the slope angle, and their magnitudes, for a given set of properties, depend on the side slope divergence angle θ. Whether or not the factor of safety is larger or smaller for divergent side slopes ($\theta>90°$) than for parallel side slopes ($\theta=90°$) depends on whether the added driving force from blocks 2 are greater than the added sliding resistance provided by the increased size and surface area of the passive block (block 3) as the divergence angle increases.

In a subsequent analysis of the Kettleman Hills failure using the Janbu method; i.e., neglecting the vertical component of interblock forces and satisfying overall vertical and horizontal force equilibrium, as well as more accurate values of properties than were available at the time of the analyses by Seed et al. (1990), Byrne et al. (1992) found slightly lower values for the two-dimensional factor of safety than for the three-dimensional factor of safety. On the other hand, using different assumptions for the interblock force directions, Chang (1992) found, for a force equilibrium method, that lower factors of safety were obtained for the three-dimensional analysis than for what might appear to be the most critical planar cross section.

The reasons for these differences are probably attributable to the differences in assumptions concerning the directions of interblock forces and the fact that a two-dimensional solution is not a correct representation of the conditions in Figure 9 for any value of the divergence angle. For example, as noted above, changes in the inclinations of interblock forces can give significant differences in the computed factor of safety when methods that satisfy only force equilibrium are used (Duncan, 1992). Fortunately, all the analyses gave results that can account for the observed failure.

The probable maximum differences between factors of safety computed by two-dimensional and three-dimensional methods, of the order of 15 to 20 percent, are sufficiently small that the much simpler two-dimensional methods should be suitable for use in practice, provided conservative safety factors are used when the particular geometry is indicative of three dimensional contributions to the driving force. More complete analyses and explanations of these issues are planned for subsequent publications.

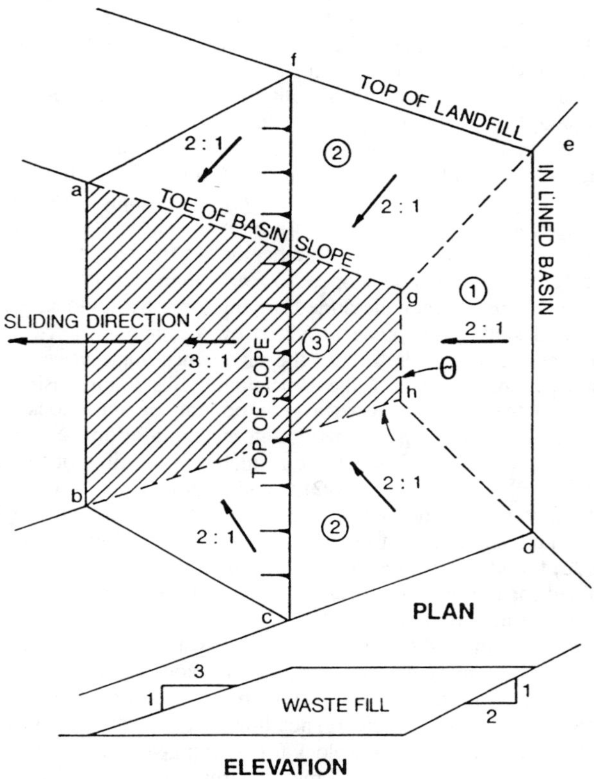

FIGURE 9
SCHEMATIC OF LINED
WASTE REPOSITORY

Stability of Landfill Covers

Regulations require that at final closure, landfills be covered by a system that will prevent infiltration of water from above or escape of gas from below. A typical composite system containing grass on the ground surface, topsoil separated from a gravel drainage layer by a geotextile, a textured HDPE geomembrane underlain by a sand bedding layer, and a soil cover was shown in Figure 2. As the top of most waste piles will be sloping and the slopes may be relatively steep, the stability of the cover must be considered. As is the case for composite liner systems, there are several interfaces, so both the soil and interface strengths are important. Infinite slope analyses can be used for evaluation of safety factors.

Provision for drainage of infiltration by rain water is frequently provided by a gravel drainage layer above the geomembrane, and drains spaced along the slope are also occasionally used. If the gravel drainage blanket does not have a sufficiently high hydraulic conductivity to prevent the development of hydraulic gradients within the gravel layer, seepage forces[6] are then generated that can have a significant destabilizing effect. Similarly, if the drainage system capacity is insufficient to prevent the development of hydraulic heads within the gravel blanket, there can be loss of stability. These factors can be especially important on steep slopes where the factor of safety will probably be low to begin with.

SEISMIC CONSIDERATIONS

The behavior of waste landfills and their seismic stability are issues of major importance in California and other earthquake-prone regions of the United States. Unfortunately, too little is known at present to be able to specify generally applicable methods of analysis and design. Furthermore, the levels of seismicity that should be considered during the operational and post-closure phases of a landfill have not been established. As current practice in the seismic analysis and design of the landfills is described by Seed and Bonaparte (1992) in a separate paper to this conference, only brief summary and comment is given here.

Among the issues for which insufficient information is available are:

1. The seismic response of landfills.
2. The dynamic strengths and stress-deformation behavior of liner and cover system interfaces.
3. The dynamic strengths and stress-deformation properties of refuse fill.
4. Deformations that liner and cover systems can withstand without tearing of the liner components.

---

[6]The seepage force per unit volume is equal to the hydraulic gradient times the unit weight of water.

5. Appropriate methods of analysis for seismic stability and deformations during earthquakes.

To date most seismic evaluations of stability have been done using pseudostatic methods of analysis. Deformational analyses have been made using the yield acceleration method of Newmark (1965) as applied by Makdisi and Seed (1978). The results are limited, however, by lack of information about yield accelerations and shear modulus properties of the wastes. Some data on the latter have been reported by Sharma et al. (1990) and Singh and Murphy (1990), which indicate shear wave velocities ranging from 30 to 275 m/s for some landfills in California. It would appear from this broad range, however, that site specific measurements or correlations will be needed that take into account the type of waste and its density. By way of comparison the shear wave velocity in soft San Francisco Bay mud is about 125 m/s.

The behavior of 10 solid waste landfills during the October 1989, magnitude 7.1 Loma Prieta earthquake, which was centered in the Santa Cruz Mountains of California, was summarized by Orr and Finch (1990). These landfills were located both near the epicenter and on soft San Francisco Bay mud at sites up to 100 km away. The estimated Modified Mercalli Intensities ranged from V to VII, and the estimated peak accelerations were from 0.10g to 0.45g. Only minor damage was reported, with the most common type being minor cracking of landfill slope surfaces. These cracks were little different than those caused by consolidation and decomposition of the wastes. There were no slope failures. However, many landfill gas recovery systems were affected by temporary loss of power, and several above-ground pipes broke due to the earthquake stress. All leachate control and gas recovery systems were back in operation within 24 hours. The duration of the Loma Prieta earthquake was unusually short, and this may have limited the landfill damage.

FACTORS OF SAFETY

General agreement on values of factors of safety against different types of landfill failure has not yet been reached. Typically, earth slopes and embankments are designed for factors of safety of 1.3 to 1.5 under static conditions. Although comparable values may be appropriate for potential failure surfaces through repository side-walls or the foundation, they may not be reasonable for failure through slopes composed of waste or for failure along liner interfaces, owing to uncertainties about the influences of such factors as creep deformations and waste consolidation and creep. For a factor of safety of 1.5, the average stress level would be 67 percent; however, substantially greater stress levels can exist locally because of non-uniform stress distribution along a potential failure surface. If these higher stresses can cause excess deformations, then higher factors of safety should be used. Finite element analyses that reasonably model the waste properties could be helpful in identifying such conditions.

California regulations require that solid waste landfills with side slopes steeper than 3H:1V have a factor of safety of at least 1.5 under dynamic loading based on a pseudostatic analysis that assumes the peak acceleration resulting from the Maximum Probable Earthquake[7]. If a pseudostatic factor of safety of 1.5 cannot be attained, a dynamic deformation analysis is required and it must be demonstrated that the design can withstand the estimated deformations. However, there are no currently defined regulatory limits or generally accepted criteria regarding what constitutes tolerable deformation, and owing to lack of information on the dynamic properties of waste, the results of such analyses are likely to be of limited usefulness and reliability.

RECENT EXPERIENCES

Three recent cases in California focus attention on the dominant influences that slope stability may have on the design, construction, operation, and closure of waste landfills. The Keller Canyon Landfill, now under construction in Contra Costa County east of Oakland, California, illustrates the need for thorough evaluation of stability in the design of a modern municipal waste landfill in a seismic area of high ground relief. The Kettleman Hills hazardous waste landfill stability failure shows the critical role of low interface strengths in a liner system. Finally, for a landfill located adjacent to San Francisco Bay, low strength soft mud and peat underlying the landfill are impacting the design of a suitable cover for closure.

As the Kettleman Hills and the landfill closure cases have been described in considerable detail elsewhere, and in separate contributions to this conference, they are included herein by reference, and only a few salient conclusions are presented in this paper. For details of the Kettleman Hills failure and failure analyses, see Mitchell et al. (1990a,b), Seed et al. (1990), and Byrne et al. (1992). The closure problem and possible solutions for the landfill founded on soft mud and peat is described by Mitchell et al. (1992).

Stability Considerations in the Design of Keller Canyon Landfill

This new municipal waste landfill in Northern California will be constructed in a south-to-north trending canyon. A toe berm will be constructed across the lower end of the canyon and fill placement will generally proceed by phases in an up-canyon direction. Most of the landfill will be underlain by a composite liner system consisting of an underdrain, compacted clay with a permeability of

---

[7]The Maximum Probable Earthquake is the maximum earthquake that a site is likely to experience in a 100 year period, but no lower than the maximum that has occurred during historic time.

10^{-8} m/sec or less, a 1.5 mm textured HDPE liner, and a granular leachate collection and removal system (Figure 10).

Stability of the landfill and the toe berm presented concerns for design because the landfill is located in a seismically active area, portions of the refuse fill will be constructed on the side slopes of the canyon, and the landfill will be underlain by a composite liner with multiple interfaces of varying strength. Static and dynamic stability analyses were therefore performed on representative critical cross sections through the toe berm and through the landfill. Typical cross sections and conditions used for these analyses are shown in Figures 11(a) and 11(b), respectively.

Seismic Design Criteria. Current regulations specify that Class II landfills and appurtenant facilities be designed to withstand the maximum credible earthquake (MCE) without damage that would render the landfill environmental control structures or systems inoperable. However, as widely accepted criteria for allowable earthquake-induced deformation of landfills do not yet exist, and because a design to completely prevent permanent displacement is impracticable, seismic design was based on limiting displacements to levels that were considered tolerable for the landfill. Seismic deformation less than or equal to one-half the thickness of the clay layer component of the landfill liner was considered tolerable because:

1. The composite liner (0.61 m of compacted clay overlain by a 1.5 mm textured HDPE FML) should provide containment following an MCE event resulting in displacements of one-half the liner thickness or less because at least 0.3 m of low permeability clay would remain below the leachate collection system and refuse.
2. The leachate collection and removal layer above the composite liner is a redundant system that included collection pipes in a 0.46 m thick granular drainage blanket that discharge at the toe berm of the landfill. This system should be functional after an MCE event because the drainage blanket alone would have enough capacity to minimize leachate head on the composite liner if the collection pipes were ruptured.
3. In the event the liner and leachate collection system were compromised, the underdrain system would provide a means of detecting and collecting leachate release.

Strength Properties Used for Analysis. The stability analyses were based on strengths determined from a number of different laboratory tests performed at various stages in the field investigation and design process. The strengths of the compacted clay toe berm and the compacted clay portion of the liner were evaluated based on the results of unconsolidated-undrained triaxial tests performed on samples compacted to 94, 97, and 100 percent relative compaction (standard Proctor) with moisture contents varying from 3 percent wet of optimum to 1 percent dry of optimum. Figure 12 shows test results from

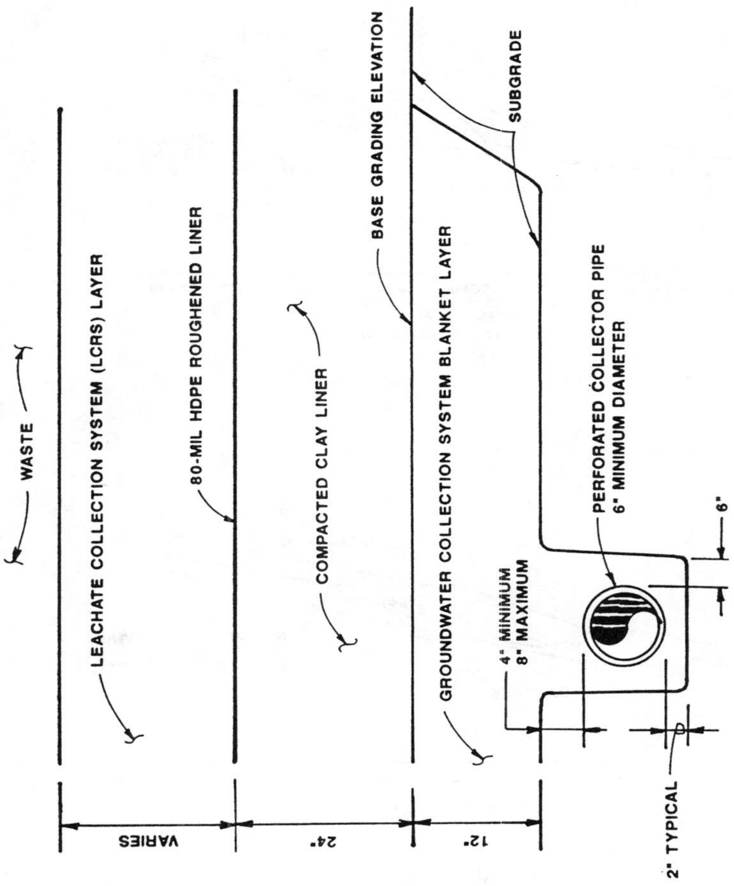

FIGURE 10
GROUNDWATER COLLECTION SYSTEM AND LINER DETAIL
KELLER CANYON LANDFILL

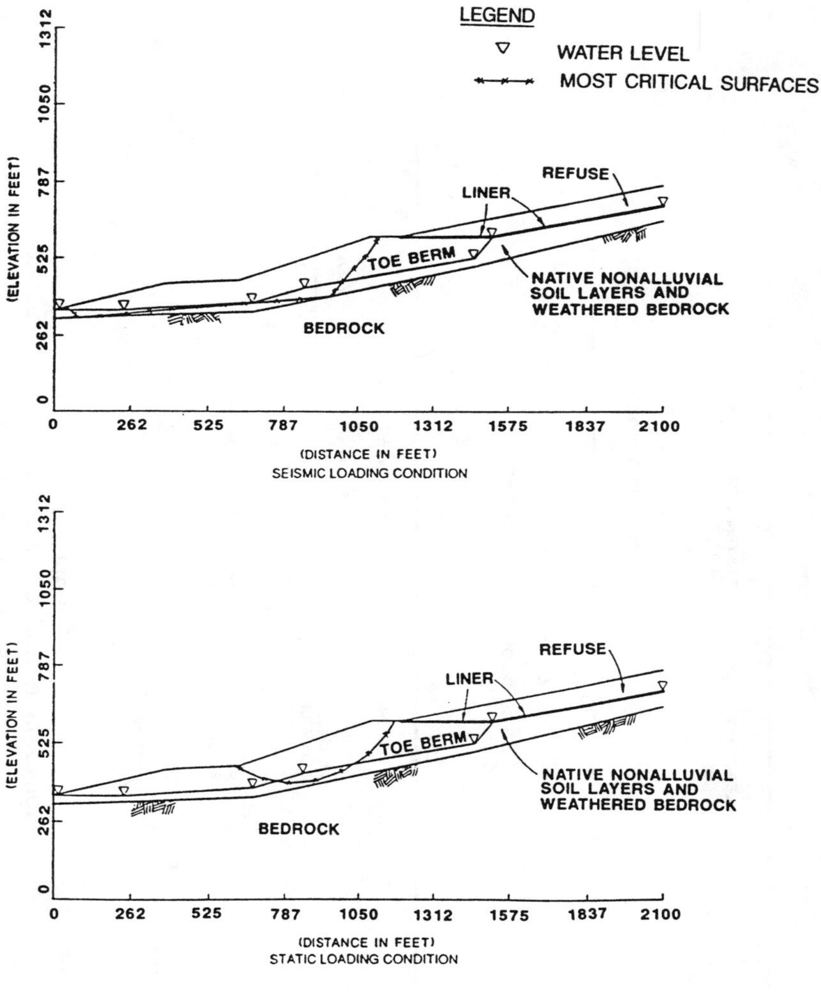

FIGURE 11 (a)
TOE BERM STABILITY ANALYSIS
CROSS-SECTION B1-B1'
KELLER CANYON LANDFILL

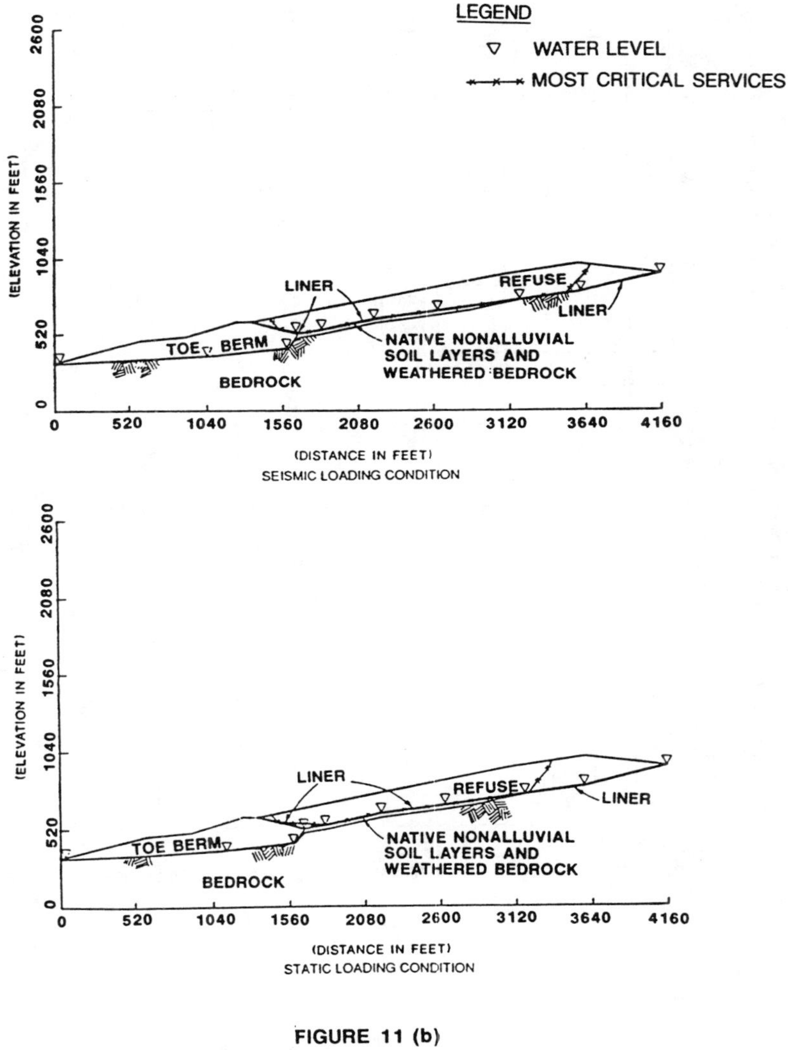

FIGURE 11 (b)
REFUSE STABILITY ANALYSIS
CROSS-SECTION F1-F1'
KELLER CANYON LANDFILL

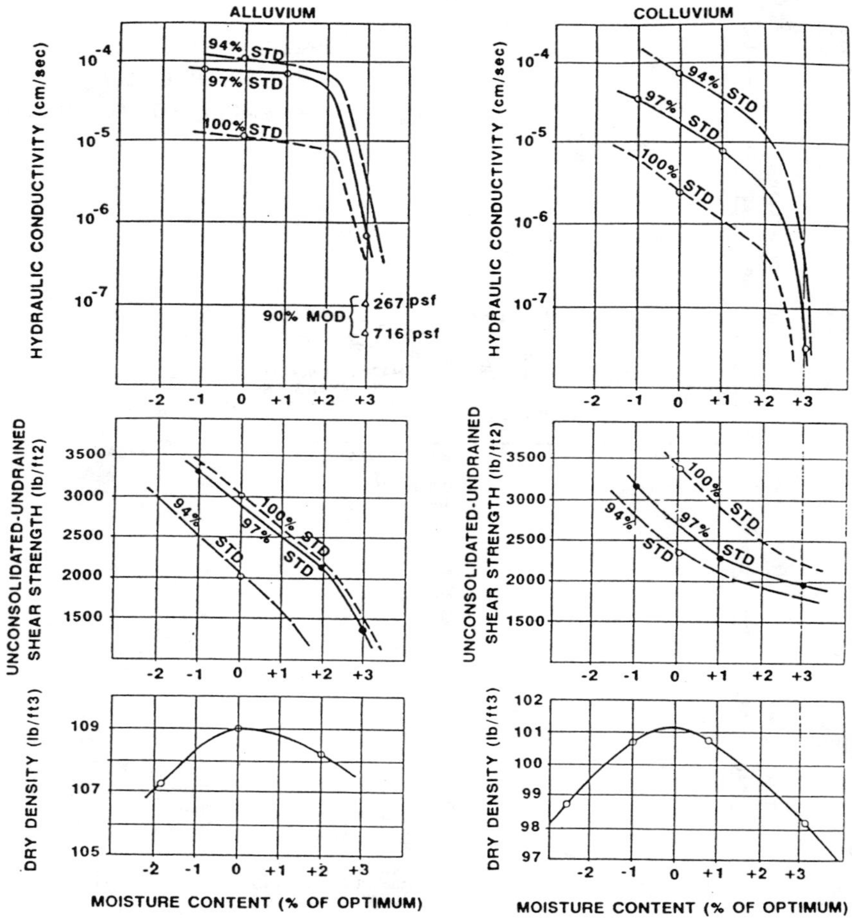

FIGURE 12

TYPICAL COMPACTION, SHEAR STRENGTH , AND HYDRAULIC CONDUCTIVITY TEST RESULTS (Test method ASTM D698) KELLER CANYON LANDFILL

two site soils (alluvium and colluvium) that will be used for liner construction. This figure clearly shows the decreased strength values associated with decreased permeability and higher moisture contents.

The strength of the compacted clay-textured HDPE interface was evaluated based on the results of strain-controlled consolidated drained direct shear tests (Table 3). Consistent with typical clay soil behavior, there was a significant decrease in strength with increasing saturation of the material. Initial stability analyses were based on test results assuming the compacted clay liner was placed approximately 3 percent wet of optimum. However, subsequent analyses were performed to account for saturation at or near the compacted clay-textured HDPE interface. The strengths of other liner interfaces were evaluated based on available data in the literature. Based on this review, the textured clay-HDPE interface was considered to represent the most critical surface for stability analyses.

Table 3
Liner Interface Friction Values Used for Analysis of Keller Canyon Landfill Stability

Interface	Condition	Source	Friction Angle (Degrees)	Cohesion Intercept (Pa)
Compacted Clay[a]-Textured HDPE	Optimum Moisture Content (Unsaturated)	Direct Shear Test	31.8	6.5
Compacted Clay[a]-Textured HDPE	Optimum Moisture Content +3% (Unsaturated)	Direct Shear Test	28.4	16.9
Compacted Clay[a]-Textured HDPE	Optimum Moisture Content (Saturated)	Direct Shear Test	9.5	8.8
Compacted Clay[a]-Textured HDPE	Optimum Moisture Content +3% (Saturated)	Direct Shear Test	8.5	14.6

[a]Clay compacted to 91 percent relative compaction in accordance with ASTM D1557 (modified Proctor)

Analysis Results. Two representative cross sections were analyzed to evaluate static safety factors, pseudostatic yield acceleration, and estimated dynamic deformation of the toe berm at its final condition and the results of these analyses are shown in Table 4. The static factor of safety for the toe berm is greater than 2, which exceeds the generally recommended value of 1.5 for earth fills and embankments. The pseudostatic yield acceleration to decrease the factor of safety to 1 ranged from about 0.18 to 0.20. For both sections analyzed, the greatest seismic deformation resulted from the maximum credible earthquake on a fault located about 1.6 km from the site. As indicated in Table 4, the maximum estimated seismic deformation ranged from less than 75 mm to less than 150 mm. Because these estimates were based on a conservative approach, actual displacements that might occur during a large earthquake on one of the faults near to the site would probably be less.

Table 4
Summary of Static and Pseudostatic Slope Stability Analyses and Deformation Analyses for the Final Configuration of the Toe Berm

Section	Max. Fill Height Above Most Critical Surface (m)	Long-Term Final/ Buildout Static Safety Factor	Seismic Yield Acceleration	Est. Max. Seismic Deformation (mm)
B1-B1′	41	2.0	0.22g	<150
B3-B3′	73	2.3	0.21g	<75

Results of static analysis of the refuse fill (Table 5) indicate adequate factors of safety against failure along the landfill liner for both clay moisture conditions that were analyzed. Although adequate factors of safety were calculated, Table 5 highlights the decrease in stability that occurs with increasing saturation of the compacted clay.

Table 5
Summary of Static Safety Factors for Final Refuse Configuration

Section	Max. Fill Height Above Most Critical Surface (m)	Static Safety Factor Assuming Compacted Clay at Optimum Moisture Content Plus 3%	Static Safety Factor Assuming Compacted Clay Saturation
B3-B3′	73	2.3	1.7
F1-F1′	53	2.2	1.7
F3-F3′	26	2.2	1.8

Stability Failure of the Kettleman Hills Waste Landfill

The unexpected Kettleman Hills Waste Landfill failure in March 1988 focused attention on several important issues relating to the safe design and filling of waste landfills:

1. The liner systems that are presently used for the containment and removal of landfill leachate may contain interfaces having very low strengths; i.e., friction angles of 8 degrees or less.
2. The interface between a HDPE geomembrane and compacted clay may have a very low shearing resistance, as discussed earlier in this paper, and emphasized by Byrne et al. (1992) in their description and analysis of the failure.
3. Under certain conditions, the geometry and properties of landfills and liner systems may make the stability in three dimensions more critical than would be apparent from the results of two dimensional analyses. However, the difference in safety factors for force equilibrium analyses that use the same assumptions for two and three-dimensional analyses

should be less than 15 to 20 percent. For cases where three-dimensional stability may be critical, designs to assure stability during filling can be made by increasing the safety factors used for simple two-dimensional analyses.

4. Landfill filling plans should be developed in such a way that an adequate factor of safety can be maintained for all fill heights and geometries.

Closure of an Existing Landfill Founded on Soft Mud and Peat

In May 1989 a California regulatory agency issued closure requirements for a landfill located adjacent to San Francisco Bay and founded on soft clay (Bay mud and peat). These requirements prescribed one of two final covers be constructed, and that the cover be designed for a static factor of safety of 1.5. However, the results of stability analyses showed it would not be possible to attain the required safety factors under short-term static loading conditions. The analyses further suggested possible failure of the landfill and foundation soil if either of the two covers were constructed in accordance with the agency-mandated schedule.

Since no feasible stabilization alternatives to increase the factors of safety to the prescribed values were identified, an alternative stability criteria and closure schedule was proposed to the regulatory agency. This alternative provided landfill dewatering prior to construction of the cover to increase factors of safety. Although landfill dewatering alone would not result in a factor of safety of 1.5 or greater, the after-construction safety factors would be about 10 to 15 percent greater than those calculated for landfill slopes under current conditions.

The stability analyses performed for this closure design showed how site specific conditions may limit the feasibility of meeting certain mandated requirements and how strict adherence to prescribed requirements without an adequate understanding of the site conditions, site history, material properties, and possible failure mechanisms can have significant consequences. Thus, it may be necessary to propose alternatives designs, construction schedules, and/or stability criteria to allow construction of closure slopes in a safe and environmentally sound manner.

CONCLUSIONS

Stability problems associated with the siting, design, construction, filling, and closure of waste landfills are of equal importance to those associated with the design and construction of liner and cover systems to prevent leakage and contamination of the surrounding soil, groundwater, and atmosphere. In this paper potential failure mechanisms, evaluation of material properties, stability analysis methods, safety factors, and stability failure risk mitigation have been examined. Three case histories illustrate many of the important issues.

Among the important specific conclusions that can be drawn concerning the stability of waste landfills are the following:

1. Several types of stability failure are possible, including: (1) sidewall slope and base failures, (2) pullout and downslope movement of liner systems, (3) failure through the waste pile, (4) failure through the waste pile and foundation, (5) failure along the liner system, (6) failure within caps and covers, and (7) failures caused by excessive settlement.
2. The selection of proper values for the strengths of the waste and foundation material, and of the shearing resistances along the interfaces within the liner and cover systems is the most critical part of any stability study.
3. Relatively little is known about the geotechnical engineering properties of municipal solid waste. Typical unit weights for refuse generally range from about 6 to 12 kN/m^3. Shear strengths have been determined by a variety of methods, and a summary of values is given in Figure 6 in terms of recommended ranges of cohesion and friction angle.
4. Interface friction angles between geosynthetics such as HDPE geomembrane and geonet and between geotextile and HDPE geomembrane may be as low as 8 degrees.
5. The interface shear resistance between compacted clay liner material and HDPE geomembrane used in a composite liner system may be very low for the wet of optimum compaction conditions usually used to obtain low hydraulic conductivity in the compacted clay. This low strength may be the controlling factor in the overall stability of a partly filled waste repository or waste pile and must be taken into account in the development of filling plans and final designs.
6. Acceptable deformations in waste landfills are largely unknown, and more research is needed to determine values above which liner systems are ruptured.
7. Usual geotechnical methods for limit stability analysis are generally applicable for analysis of waste landfill stability; however, consideration of incompatibility between failure strains in soils and in refuse may be warranted in some cases.
8. Under certain conditions, the three-dimensional stability during the filling of waste repositories with divergent side slopes may be less than that for a similar repository with parallel side slopes. However, this difference is sufficiently small that two-dimensional analyses may be used provided conservatism is incorporated into the interpretation of the results of the analysis.
9. Much remains to be learned about the seismic response of waste landfills, the dynamic strengths of waste landfill components, acceptable levels of deformation, and the appropriate methods of analysis for seismic stability.

REFERENCES

Byrne, J.R., Kendall, J. and S. Brown (1992), "Cause and Mechanism of Failure, Landfill B-19, Phase 1-A, Kettleman Hills Facility, Kettleman City, California," Proceedings ASCE Specialty Conference on Stability and Performance of Slopes and Embankments - II, University of California, Berkeley, June 29 - July 1.

Bjerrum, L. (1972), "Embankments on Soft Ground," ASCE Specialty Conference on Performance of Earth and Earth-Supported Structures, Proceedings, Vol. 2, Purdue University.

Chang, M-H. (1991), "Three-Dimensional Slope Stability Analysis of Lined Waste Landfills," Ph.D Dissertation in Civil Engineering, University of California, Berkeley.

Converse, Davis and Associates (1975), "Slope Stability Investigation, Proposed Final Slope Adjacent to the Pomona Freeway," Report Prepared for Operating Industries Inc, Monterey Park, California.

Daniel, D.E. and Benson, C. H. (1990), "Water Content-Density Criteria for Compacted Soil Liners," Journal of Geotechnical Engineering, A.S.C.E., Vol. 116, No. 12, pp. 1811-1830.

Duncan, J.M. (1992) "State-of-the-Art: Static Stability and Deformation Analysis," Proceedings ASCE Specialty Conference on Stability and Performance of Slopes and Embankments - II, University of California, Berkeley, June 29 - July 1.

Earth Technology Corporation (1988), "Instability of Landfill Slope, Puente Hills Landfill, Los Angeles County, California," Report Submitted to Los Angeles County Sanitation District.

Fang, et al. (1977), "Strength Testing of Bales of Sanitary Landfill," Fritz Engineering Laboratory, Lehigh University, Department of Civil Engineering.

Huitric, Raskit, and Haug (1979), "Moisture Retention of Landfilled Solid Wastes," Second National Conference on Hazardous Materials Management, San Diego, California.

Koerner, R. M. (1990), Designing With Geosynthetics, 2nd Edition, Prentice-Hall, Englewood Cliffs, NJ.

Kramer, S. (1992), "Research Progress Report on Three-Dimensional Stability of Waste Impoundments," Department of Civil Engineering, University of Washington, Seattle.

Landva, A.O. and J.I. Clark (1987), "Geotechnical Testing of Waste Fill," Report Submitted to the Institution of Civil Engineers, London.

Makdisi, F. I. and Seed, H.B. (1978), "Simplified Procedure for Estimating Dam and Embankment Earthquake-Induced Deformations," Journal of the Geotechnical Engineering Division, A.S.C.E., Vol. 104, No. GT7, pp.849-867.

Martin, R. B., Koerner, R. M., and Whitty, J. E. (1984), "Experimental Friction Evaluation of Slippage Between Geomembranes, Geotextiles and Soils," Proceedings of the International Conference on Geomembranes, Denver, Colorado, June 20-23, 1984, pp. 191-196.

Mitchell, J. K., Hooper, D. R., and Campanella, R. G. (1965), "Permeability of Compacted Clay," Journal of the Soil Mechanics and Foundations Division, A.S.C.E., Vol. 91, No. 4, pp.41-65.

Mitchell, J. K., Seed, R. B., and Seed, H.B. (1990a), "Kettleman Hills Waste Landfill Slope Failure. I: Liner System Properties," Journal of Geotechnical Engineering, A.S.C.E., Vol. 116, No. 4, pp. 647-668.

Mitchell, J. K., Seed, R. B., and Seed, H. B. (1990b), "Stability Considerations in the Design and Construction of Lined Waste Repositories," ASTM STP 1070, Geotechnics of Waste Landfills - Theory and Practice, A. Landva and G. D. Knowles, ed., American Society for Testing and Materials, pp. 209-224.

Mitchell, R. A., Hatch, S. E. and Siegel, R. (1992), "Closure Design for a Landfill Founded on Soft Mud and Peat," Proceedings ASCE Specialty Conference on Stability and Performance of Slopes and Embankments - II, University of California, Berkeley, June 29 - July 1.

Newmark, N. M. (1965), "Effects of Earthquakes on Dams and Embankments," Geotechnique, Vol. 15, No. 2, pp.139-160.

Orr, W. R. and Finch, M. O. (1990), "Solid Waste Landfill Performance During the Loma Prieta Earthquake," ASTM STP 1070, Geotechnics of Waste Fills - Theory and Practice, A. Landva and G. D. Knowles, editors, American Society for Testing and Materials, pp. 22-30.

Seed, R. B. and Bonaparte, R. (1992), "Seismic Analysis and Design of Waste Fills: Current Practice," Proceedings ASCE Specialty Conference on Stability and Performance of Slopes and Embankments - II, University of California, Berkeley, June 29 - July 1.

Seed, R. B. and Boulanger, R. W. (1991), "Smooth HDPE-Clay Liner Interface Shear Strengths: Compaction Effects," Journal of Geotechnical Engineering, A.S.C.E., Vol. 117, No. 4, pp. 686-693.

Seed, R. B., Mitchell, J. K., and Seed, H. B. (1990), "Kettleman Hills Waste Landfill Slope Failure. II: Stability Analyses," Journal of Geotechnical Engineering, Vol. 116, No. 4, pp. 669-690.

Seed, H.B. and Makdisi, F.I. (1977), "A Simplified Procedure for Estimating Earthquake-Induced Deformations in Dams and Embankments," Earthquake Engineering Research Center, Report No. UCB/EERC 77/19.

Sharma, H. D., Dukes, M. T., and Olsen, D. M. (1990), "Field Measurements of Dynamic Moduli and Poisson's Ratios of Refuse and Underlying Soils at a Landfill Site," ASTM STP 1070, Geotechnics of Waste Fills - Theory and Practice, A. Landva and G.D. Knowles, editors, American Society for Testing and Materials, pp. 57-70.

Siegel, R. A., Robertson, R. J., and Anderson, D. G. (1990), "Slope Stability Investigations at a Landfill in Southern California," ASTM STP 1070, Geotechnics of Waste Fills -Theory and Practice, A. Landva and G. D. Knowles, editors, American Society for Testing and Materials, pp. 259-284.

Siegel, R.A. (1975), "Computer Analysis of General Slope Stability Problems," Technical Report 75-8, Joint Highway Research Project, Purdue University.

Singh, S. and Murphy, B. (1990), "Evaluation of the Stability of Sanitary Landfills," Geotechnics of Waste Fills - Theory and Practice, ASTM STP 1070, A. Landva and G. D. Knowles, editors, American Society for Testing and Materials, pp. 240-258.

Spencer, E. (1967), "A Method of Analysis of the Stability of Embankments Assuming Parallel Inter-Slice Forces," <u>Geotechnique</u>, Vol. XVII, No. 1, pp. 11-26.

CAUSE AND MECHANISM OF FAILURE
KETTLEMAN HILLS LANDFILL B-19, PHASE IA

R. John Byrne[1], J. Kendall[2] and S. Brown[3]

BACKGROUND

On March 19, 1988, there was a failure in Landfill B-19, Phase IA at the Kettleman Hills Class 1 hazardous waste treatment, storage and disposal facility. The facility is located near Kettleman City, California and is operated by Chemical Waste Management, Inc. (CWMI). The construction of Phase IA was completed in February 1987 and approximately 580,000 cubic yards of waste and other material had been placed at the time of the failure. The surface topography of the waste as of March 15, 1988, just four days before the failure, is shown on Figure 1a. At this stage, the waste had reached a maximum height above the landfill base of about 90 feet. As indicated on Figure 1, the waste was located in the northern part of the landfill on a relatively flat base and enclosed by 2H:1V excavated slopes to the northwest and southwest, and a 3H:1V excavated slope to the northeast. The landfill cell was open to the southeast, with a waste slope of slightly less than 3H:1V dipping in that direction.

The entire waste mass slid a horizontal distance of about 35 feet towards the southeast, and vertical slumps of up to 14 feet along the sideslopes of the landfill were observed after the failure. The nature of the movements and the pattern of surface cracking observed suggested that the main body of the waste had moved as a relatively monolithic unit, consistent with the kinematic constraints associated with slip on a piecewise linear surface, and that failure

---

1 Principal, Golder Associates Inc., 4104-148th Avenue NE, Redmond, Washington 98052

2 Project Manager, Chemical Waste Management, Inc., 35251 Old Skyline Road, Kettleman City, California 93239

3 Project Manager, Chemical Waste Management, Inc., 35251 Old Skyline Road, Kettleman City, California 93239

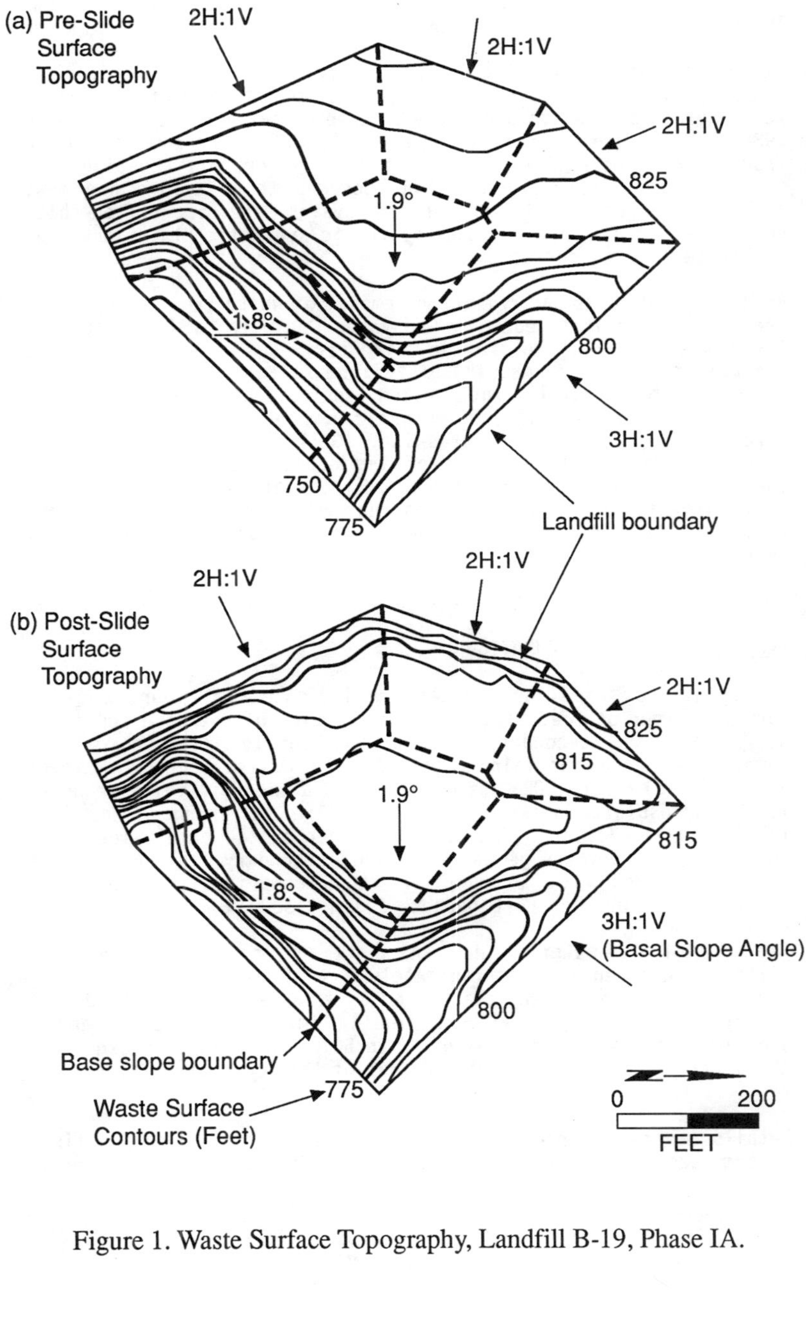

Figure 1. Waste Surface Topography, Landfill B-19, Phase IA.

had most likely occurred primarily within the landfill liner system. The post-slide topography of the waste is shown on Figure 1b.

Immediately following the failure, Professors R.B. Seed, J.K. Mitchell and H.B. Seed of the University of California, Berkeley, were engaged by CWMI to undertake a cause of failure study. In July 1988, they issued a report (Seed et al., 1988) which summarized the results of their field observations, laboratory testing and engineering analyses. That study deduced that failure had most likely occurred within the liner system and identified both geomembrane/clay and various synthetic/synthetic interfaces as candidate sliding surfaces.

By March 1990, CWMI had commenced removal of waste from the Phase IA area and had contracted with GeoSyntec Consultants (formerly GSI Environmental, Inc.) to provide third party CQA observation of the removal of the waste, mapping of failure features within the waste and within the landfill liner system, and collection and testing of samples of both the soil and synthetic components of the liner. In October 1990, Golder Associates Inc. (GAI) was engaged by CWMI to prepare a definitive analysis of the cause and mechanism of the Phase IA failure, based on the updated information obtained from the field investigations/observations during waste removal, together with the additional data developed from the associated laboratory testing.

MECHANISM OF FAILURE

LANDFILL LINER CONFIGURATION

The details of the liner system are indicated on Figure 2. The landfill liner on the relatively flat portion of the base consisted of primary and secondary clay/60 mil Gundle HDPE geomembrane composites, each overlain by a Leachate Collection and Removal System (LCRS). Each LCRS was comprised of a one-foot-thick granular layer encapsulated between geotextile (Trevira Spunbond No. 1145 fabric) filter fabric. The landfill liner on the basal sideslopes consisted of a primary geomembrane and a secondary clay/geomembrane composite, each with an overlying LCRS comprised of geotextile filter fabric over an HDPE geonet (Polynet 3000).

The clay liner material was an admix of onsite claystone and siltstone amended with approximately 2 to 5 percent by weight of bentonite. The admix was classified as a CH material (Unified Soil Classification System) with a liquid limit of 60 to 70 and a plasticity index on the order of 40 to 50. The nominal thickness of the primary clay was 1.5 feet and that of the secondary clay was 3.5 feet.

Details of the interface between the base and the sideslope liner systems are also shown on Figure 2c. This figure shows the lateral

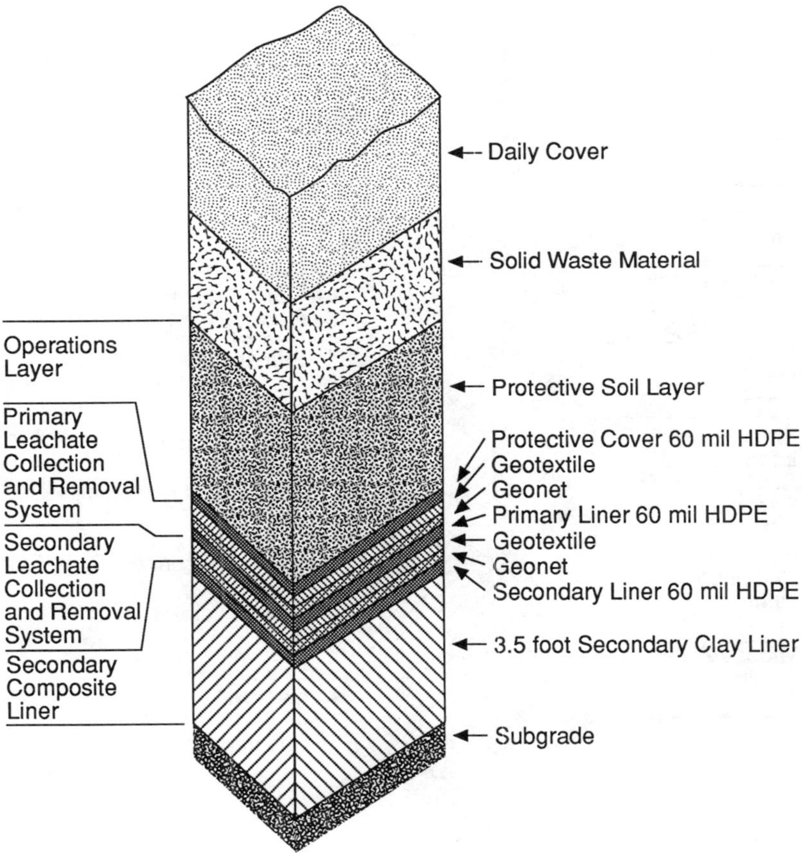

Figure 2a. Schematic Illustration of Sideslope Liner.

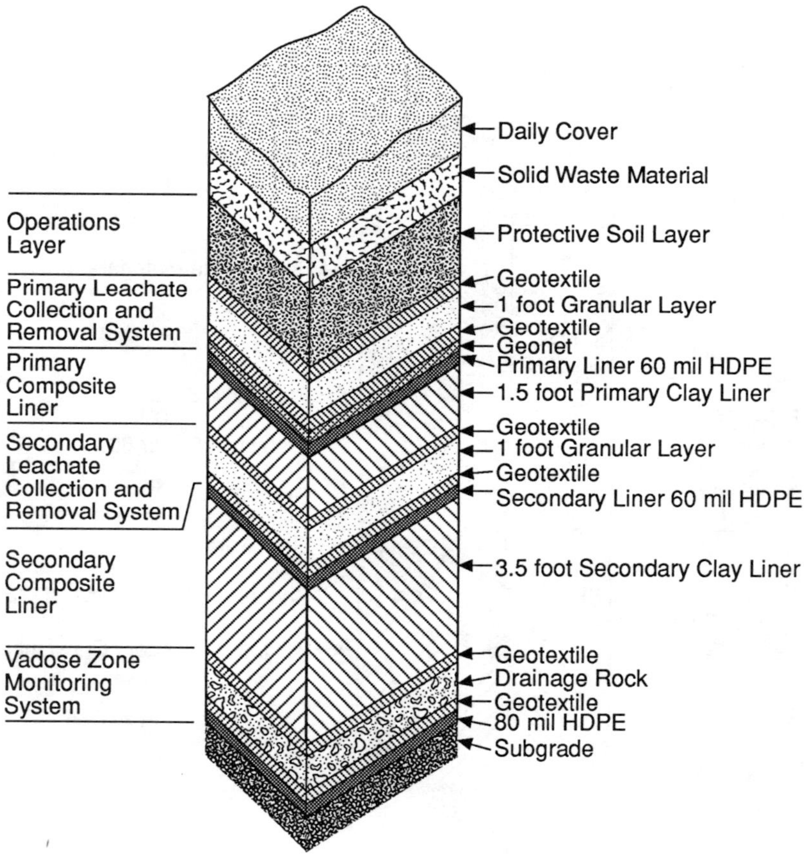

Figure 2b. Schematic Illustration of Base Liner.

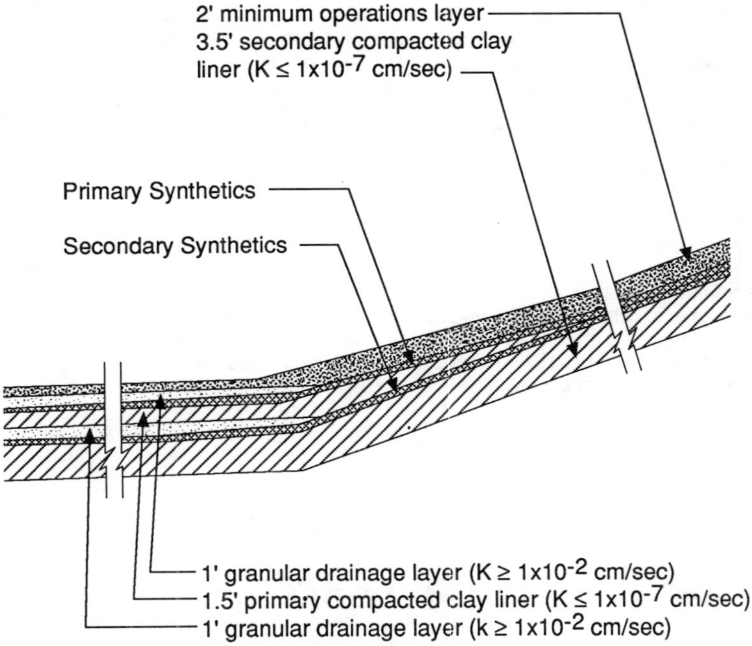

Figure 2c. Sideslope Toe Region Detail.

extent of the basal primary and secondary granular drainage layers, the tapering of the primary clay base liner in the toe region of the sideslopes, and the transitioning to an all-synthetic sideslope liner system (except for the secondary clay liner component).

HISTORY OF MOVEMENTS

At about 6:30 am on Saturday, March 19, 1988, eyewitness reports indicated an approximate 1/2-inch wide crack across the truck ramp on the north corner of the landfill. At that time, no evidence of cracking was noted in the waste above either the 2H:1V or 3H:1V slopes. By about 9:30 am, a 3 to 4-inch wide crack with a vertical offset of 6 to 8 inches was noted in the waste running along the top edge of the 2H:1V slopes. By about 1:30 pm, the waste had translated and slumped into its final configuration.

Pre- and post-failure surveys indicated that the main body of the waste located above the relatively flat base had moved approximately 35 feet horizontally in a direction 130 (±5) degrees E of N, essentially parallel to the line of intersection between the landfill base and the 3H:1V sideslope on the northeast side of Phase IA. The waste located on the 2H:1V basal sideslopes to the west/northwest slumped an average of about 12 to 14 feet. This magnitude of slump is compatible with the apparent dip of these sideslopes in the direction of sliding. The waste located above the 2H:1V basal sideslope to the southwest slumped an average of 6 to 8 feet. This magnitude of slump is also compatible with the apparent dip of this sideslope in the direction of sliding.

The Phase IA vertical sump riser located at the northeast end of the vadose zone monitoring trench (Figure 3) also appeared to have translated some 35 feet in the direction of the failure, and have remained intact. During subsequent excavation of the waste and exposure of the sump riser, it was confirmed that the riser and its concrete base pad had moved en masse and remained intact and essentially vertical, indicative of minimal distortions within the waste at this location.

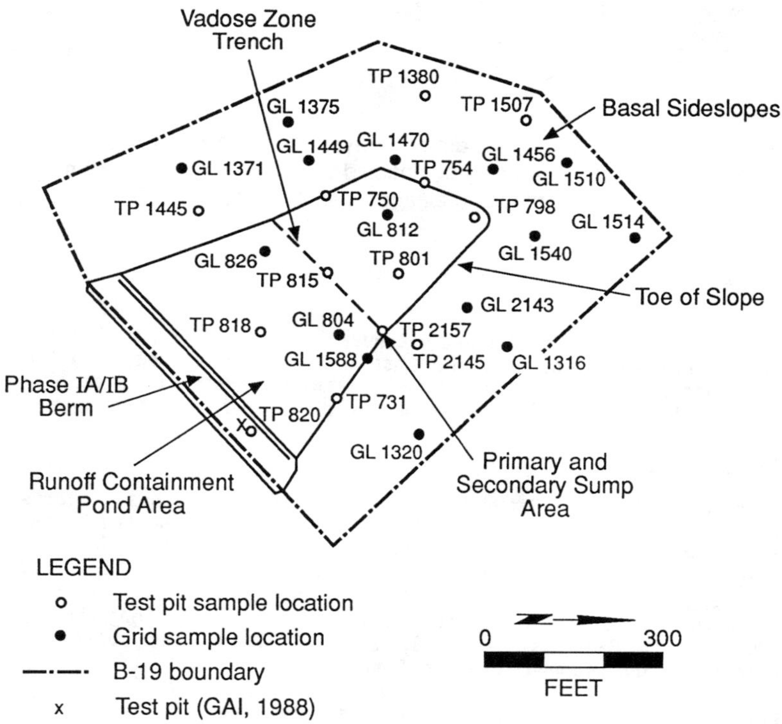

Figure 3. Landfill B-19, Phase IA, Test Pit and Grid Locations Map.

FIELD INVESTIGATIONS

In October 1989, GeoSyntec Consultants began providing CWMI with QA/QC monitoring services in connection with the transfer of waste from the northern to the southern area of Landfill B-19. Those services included documenting the condition of the waste and the liner in the Phase IA area, as part of the overall investigation into the cause and mechanism of the failure. The actual field activities undertaken and methodologies employed are summarized in a final report prepared by GeoSyntec Consultants for CWMI (GeoSyntec, 1991).

Field activities performed during removal of the Phase IA waste and exposure of the liner system included mapping of the waste for evidence of cracking/shearing, evaluation of the waste bulk density, characterization of the condition of the liner at the specific locations identified on Figure 3, sampling of the liner materials, mapping of liner failure features, and the development of a photographic and video log of all significant field activities.

POST-FAILURE CONDITION OF THE LINER SYSTEM

Landfill Base

Detailed inspection of the flat basal liner system (sample locations 801, 815, 818, 804, 812 and 826; Figure 3) indicated that essentially all components of the liner above the secondary clay moved en masse, with little evidence of relative movements within or between any of the liner components. Relative movements appeared to be generally confined to the secondary geomembrane/secondary clay interface, with striations on the secondary clay surface showing an orientation of approximately N130E, parallel to the direction of sliding indicated from the surface monitoring. There was no evidence of sliding within the secondary clay itself.

Northeast Sideslope and Sump Area

Sideslope mapping indicated that all of the synthetic components on this slope had remained fixed in the anchor trench at the top of the slope, and had translated in the direction of sliding to the southeast, at the base of the slope. No evidence of sliding within the secondary clay was indicated except where the sump riser trench penetrated the clay. The movements of the secondary geomembrane were best demonstrated by the distorted pattern of the panel seams which indicated that the secondary geomembrane at the bottom of the 3H:1V slope moved along strike about 30 to 40 feet relative to top of slope. Hence, the relative movements within the liner system on this slope appeared again to be concentrated primarily at the secondary clay/secondary geomembrane interface.

The synthetics located above the secondary geomembrane generally appeared to have moved in a similar manner to the secondary geomembrane. All synthetic layers showed a pattern of wrinkles/waves oriented obliquely to the slope, consistent with the nature of the relative translation experienced within the synthetics between the top and the bottom of the slope.

Northwest Sideslopes

In the upper part of these slopes, the secondary liner system (geotextile, geonet and geomembrane) remained fixed in the anchor trench, whereas the primary liner system pulled out of the anchor trench and moved approximately 25 feet downslope with the waste. At about 30 to 50 feet downslope of the B-19 perimeter boundary, corresponding generally to a noticeable grade change in the secondary clay liner, the secondary synthetics had ruptured and sliding again appeared to have occurred primarily at the secondary clay/secondary geomembrane interface over the majority of these slopes (Figure 4). The width of this tear in the secondary geomembrane, in the general direction of sliding, was on the order of 25 to 30 feet. The tear also extended across the full width of the northwest sideslopes and into the southwest sideslope.

Figure 4. Secondary Geotextile Tear on Northwest Sideslopes, Landfill B-19, Phase IA

Northwest Sideslopes - Toe Area

Two Test Pit Sample locations (754 and 798 - Figure 3) were positioned at the northwest end of the landfill, near the intersection of the northwest sideslopes with the flat base. In this area, the failure mechanism appeared to be somewhat more complex than simple sliding at the secondary geomembrane/secondary clay interface, due to the kinematic constraints posed by the relatively sharp change in the orientation of the liner. In the location of test pit 754, the primary clay wedge extending up the sideslope appeared to have moved approximately 20 to 25 feet in the direction of sliding. This movement is somewhat less than the overall movement experienced by the landfill mass, indicating that not all of the movement within this area occurred within the underlying secondary liner system. Further evidence that the secondary liner system behaved as something of a "hard spot" at this location is also provided by the pattern of distortions and failures within the secondary synthetics. A pattern of relatively severe compression wrinkles, corresponding to a shortening of up to 10 feet, was observed immediately upslope of the toe area. A corresponding pattern of tension tears (typical width of 5 feet) was mapped some 20 feet downslope of the toe area.

The failure mechanism therefore appeared to consist of secondary geomembrane/secondary clay interface sliding both upslope and downslope of the toe zone, with a more complex failure mechanism in the toe area (involving sliding on additional surfaces) to achieve a kinematically acceptable failure mode. Mapping in this toe area generally also showed striations at the primary geomembrane/primary clay interface, oriented in the approximate direction of the mass movement and extending almost 50 feet south of the toe.

At Test Pit Sample location 798, horizontal displacements of the various components of the liner system in the slope toe area were determined by comparing the post-failure locations of identifiable interfaces (e.g., northern boundaries of the primary and secondary drainage gravel layers, and of the primary clay) with the as-built information. This comparison showed that both the primary gravel and primary clay layers in the area of test pit 798 had moved about or slightly in excess of 30 feet, with the secondary gravel displacing on the order of 20 to 25 feet. This determination again suggests that not all relative movements in this general area occurred at the secondary clay/secondary geomembrane interface, and that the failure surface passed through some of the overlying liner materials to satisfy kinematic constraints.

Southwest Sideslope

The condition of the liner system on this sideslope was similar in some respects to that on the northwest sideslopes, and in other respects to that on the northeast sideslope. The primary synthetic components of the liner had again pulled out of the anchor trench along the northern-most 300 feet of the slope, and moved approximately 15 feet downslope. The secondary synthetics had generally remained fixed within the anchor trench at the top of the slope, and had translated approximately 30 feet in the direction of sliding at the bottom of the slope. Hence, sliding again appeared to be concentrated primarily at the interface of the secondary clay and the secondary geomembrane, except for the upper portions of the slope where sliding occurred at the primary membrane/secondary geotextile interface.

Because the general direction of the mass movement was not coincident with the strike of the southwest sideslope (as for the northeast sideslope), the liner synthetics also experienced a down-dip or stretching component of relative movement. This movement resulted in along-strike oriented tension failures within the secondary geomembrane, in addition to the shear distortions noted above.

At the toe of the slope, mapping of test pit 750 showed wrinkles in the primary synthetics and folds in the primary clay (oriented generally along the strike of the slope), deformation and tearing of the secondary synthetics, and localized mounding of the secondary clays. These conditions are again indicative of a more complex failure mechanism in the vicinity of the toe of slope, where there is a down-dip component of movement, in order to satisfy kinematic requirements.

PHASE IA FAILURE MECHANISM

The surface surveys conducted pre- and post-failure, together with the detailed mapping of the waste and liner system during Phase IA waste removal, established quite clearly the kinematics of the Phase IA failure. Sliding occurred at the interface between the secondary clay and the secondary geomembrane, over the majority of the base of the landfill. Relative movements also occurred above this interface at limited specific locations, such as in the area of the intersection between the flat base and the 2H:1V basal sideslopes to the northwest and southwest. Kinematic constraints at these discontinuities in the slope of the predominant failure surface clearly resulted in a somewhat more complex failure mode in these areas, with sliding on multiple surfaces and distortions within limited zones of material. Failure also occurred by sliding between the primary geomembrane and secondary geotextile in the upper portions of both the northwest and southwest sideslopes.

The main body of the waste moved en masse to the southeast (direction of sliding N130E). This direction of movement is indicated by both the surface displacements and the orientation of the striations on the secondary clay surface. The direction of sliding was essentially parallel to the strike of the northeast 3H:1V basal sideslope. Therefore, the waste located above both the relatively flat base section and the northeast sideslope moved in an essentially horizontal direction, and behaved as passive or stabilizing influences for the failure mode. The waste located above the remaining basal sideslopes (i.e., 2H:1V sideslopes to the northwest and southwest) also moved in the direction of sliding, with vertically downward components of motion roughly proportional in magnitude to the apparent dips of these slopes in the direction of the mass movement. This waste provided the active or destabilizing force for the failure mode (Figure 5). The movements apparently ceased when the active blocks had slumped sufficiently to reduce the driving force and re-establish equilibrium.

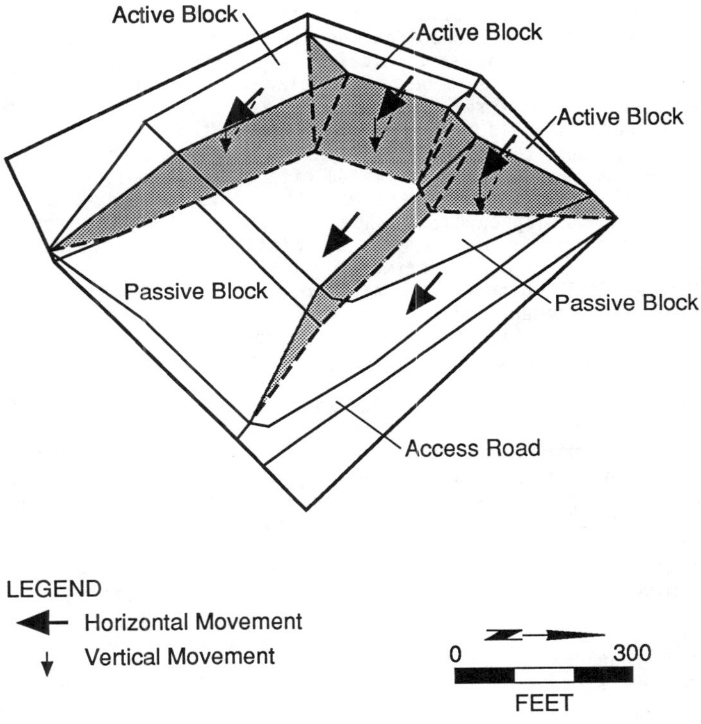

Figure 5. Displacement Vectors for Waste Blocks During Failure of Landfill B-19, Phase IA.

POST-FAILURE CONDITION OF SECONDARY CLAY

As failure was observed to have occurred predominantly along the secondary clay/secondary geomembrane interface, the condition of the secondary clay was of interest, particularly within the relatively near vicinity of the interface with the geomembrane. Quality Assurance records for the placement of the secondary clay (Golder

Associates, 1987) show that the placed moisture content of the clay of the secondary liner ranged from about 27 to 33 percent, with a median value of almost 30 percent. Based on data from the post-failure investigation, average moisture contents near the top of the secondary clay layer but below any desiccated zone ranged from something on the order of 29 to 30 percent in the area of the upper sideslopes (based on a very limited data set), to between 26 and 29 percent in the flat basal portion of the liner (GeoSyntec, 1991). Simple one-dimensional consolidation analysis for the secondary clay liner indicates that for a vertical pressure of 10,000 psf, applied over a period of one year and left in place for a further two years, a reduction in moisture content near the top of the clay of less than two percent would be anticipated. Hence, the measured moisture contents appear to be generally consistent with the as-placed moisture contents and the estimated effects of consolidation. As would be expected, the more heavily loaded clays located at the base of the landfill appeared to show, on average, somewhat greater consolidation and associated loss of moisture than did the clays of the upper sideslopes. The primary clay liner materials in the base of the landfill had an average moisture content somewhat lower than that of the secondary clays, presumably as a result of the lower placed moisture content (mean of 28.6 percent) and the higher degree of consolidation associated with the thinner primary clay liner.

Detailed moisture content profiles were developed for the secondary clays at many of the mapping/sampling locations, with moisture contents determined at vertical intervals of 0.5 inches through the liner. On the sideslopes, the clays in the 1 to 2 inches immediately adjacent to the geomembrane were generally significantly drier (typically 21 to 25 percent moisture content) than the underlying material. In addition, many of these locations demonstrated desiccation cracking within the clays (typically 1/8-inch wide and up to 2 to 3 inches deep). The field mapping program also indicated that the potential for thermal drying and clay desiccation on the sideslopes appeared to be related to the nature of the contact between the clay and the overlying geomembrane. Where the synthetics had lifted from the clay as a result of wrinkling etc. associated with the failure, drying and desiccation appeared most severe. The clay surface appeared to be generally least affected where the clay and geomembrane were still in close contact. It should also be noted that not only were the sideslope liners exposed to the elements for considerable periods prior to and during filling of Phase IA, but most of the sample locations were again left exposed (i.e., no soil cover above the synthetics) for a substantial time between waste removal and moisture content sampling. The sideslope secondary clay moisture contents measured in the near-vicinity of the secondary geomembrane are not, therefore, considered typical of conditions existing at the time of failure.

Over the flat portion of the landfill base, the secondary clay moisture content profiles in the near vicinity of the geomembrane showed a somewhat varied pattern. The protective working layer in this area was left in place at each sampling location until the mapping and sampling activities occurred, thereby minimizing the potential for thermally-induced moisture content changes in the clays in the period between excavation and sampling. Approximately half of the sampling locations showed a lowered level of moisture in the first one to several inches adjacent to the secondary membrane, with the moisture contents in this drier interface region being as much as 4 to 5 percent lower than those in the underlying clay. At the remaining sample locations, the moisture contents tended to decrease slowly with distance from the clay/geomembrane interface, with average moisture contents in the near vicinity of the geomembrane averaging between 30 and 31 percent. There appeared to be no particular pattern to the different moisture content profiles observed, and the precise details of construction (e.g., time of exposure of clay before placement of the membrane, nature of the conditioning of the placed-clay prior to membrane installation, time of exposure of the secondary membrane before placement of the secondary gravel layer, placed moisture content of the clay, etc.) at each location may have been the primary determinant of the observed conditions. The secondary clay on the flat base of the Phase IA liner was generally in good condition, with evidence of some minor surficial desiccation cracking at a couple of sampling locations.

Based on the measured moisture contents both during placement of the clay and following its excavation, and on the determination that the potential for dissipation of excess pore pressures in the vicinity of the interface would have been minimal in the period prior to the failure, it was considered appropriate to evaluate the undrained shear strength of the secondary clay/secondary geomembrane interface for clay material prepared over the range of as-placed moisture contents.

LABORATORY STUDIES

A program of interface strength testing was performed as part of the original investigations into the cause of failure (Seed et al., 1988). All synthetic/synthetic interfaces present within the liner system were tested, in addition to strength testing of the clay materials and the clay/geomembrane interface. The initial post-failure studies reported by Seed et al., (1988) were performed under a limited time frame, however, and thus were unable to fully investigate the influence of varying clay compaction and moisture conditions on the shear strength characteristics of the clay/geomembrane interface. As the investigations described above established that the Phase IA failure occurred predominantly at the interface between the secondary clay and the secondary geomembrane, the program of laboratory testing performed during this study focussed primarily on establishing the strength characteristics of the clay/geomembrane interface of the secondary liner system. In addition, a limited program of testing of other components of the liner was undertaken to confirm the original Seed et al., (1988) data. Details of the test procedures, together with the results, are given in GeoSyntec (1991). The results are summarized below.

Geomembrane/geonet interface:

The geomembrane/geonet interface direct shear tests, conducted over an approximate normal stress range of 3,000 to 9,000 psf, were performed on samples of geomembrane and geonet taken from the primary liner at the base of the landfill. Residual shear strengths corresponded to a friction angle of about 8 degrees, which compared very well with the corresponding data of Seed et al., (1988). Peak shear strengths corresponded to a friction angle of about 11 degrees, compared to the corresponding 9 degrees from the Seed et al., (1988) testing. Field observations showed that under the sustained loading of the waste, the geonet tended to imprint itself into the underlying geomembrane. The peak shear strengths obtained from the current testing program apparently reflect this behavior somewhat. In fact, the undisturbed and imprinted geomembrane/geonet interface in the field might be expected to be significantly stronger than indicated by the GeoSyntec laboratory data. Hence, it appears that the peak shear strength of the geomembrane/geonet interface may actually increase with time due to the progressive embedment of the geonet into the geomembrane. The above program of testing indicated even higher peak and residual friction angles at the highest normal stress levels (i.e., 9,000 psf) used.

Geomembrane/geotextile interface:

The geomembrane/geotextile interface direct shear tests, conducted over a normal stress range of 3,000 to 9,000 psf, were performed on samples of geomembrane and geotextile taken from the secondary liner at the base of the landfill. Peak shear strengths corresponded to a friction angle of about 14 degrees, which is slightly higher than the peak shear strengths of 12.5 degrees obtained by Seed et al., (1988) for unpolished dry interfaces. The Seed et al., (1988) data also indicated that shear strengths were reduced by typically 1 to 2 degrees for a wetted surface condition, and further reduced when a polished geomembrane sample was used. Residual shear strengths of about 8 degrees were obtained for the testing reported by GeoSyntec Consultants, which compares well with the corresponding large-displacement residual shear strengths measured by Seed et al. (1988). The load-displacement relations observed during testing of this interface showed a very rapid rise to the peak strength followed by reduction to residual conditions within typically less than one inch of shear displacement (Figure 6a). Hence the geomembrane/geotextile interface behaved in a quite brittle manner with the development of residual conditions for relatively small interface displacements.

Geomembrane/clay interface:

Quick (undrained) direct shear tests of the secondary clay-secondary geomembrane interface were performed, with the clay prepared at a

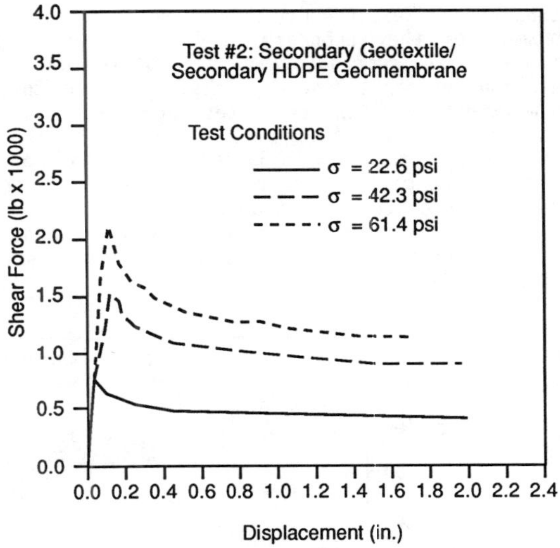

(a) Typical Geomembrane/Textile Interface Test

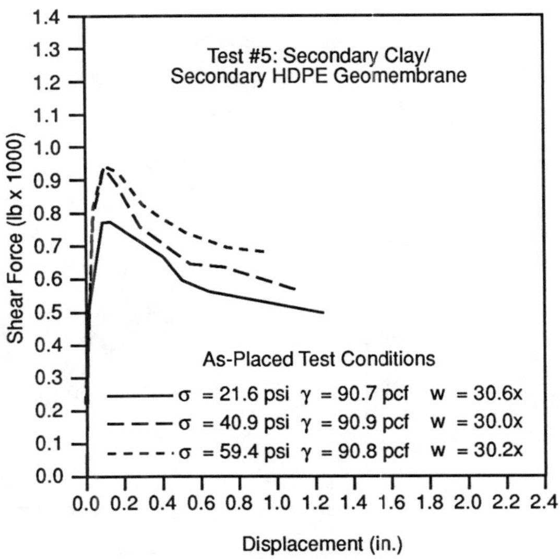

(b) Typical Geomembrane/Clay Interface Test

Figure 6. Typical Direct Shear Test Data.

range of dry densities and moisture contents representative of the as-placed condition. Measured dry densities and moisture contents during placement of the secondary clay are shown on Figure 7, together with a linear regression analysis of the data indicating the mean relation between placed dry density and moisture content. In order to maintain the quantity of testing at reasonable levels, the clay samples were prepared over a representative range of moisture contents from about 28 to 33 percent, and at corresponding dry densities in each case so as to lie approximately on the mean dry density-moisture content relation referenced above. The exact preparation condition of each of the test samples is also indicated on Figure 7. The samples were tested over a normal stress range from about 3,000 to 9,000 psf.

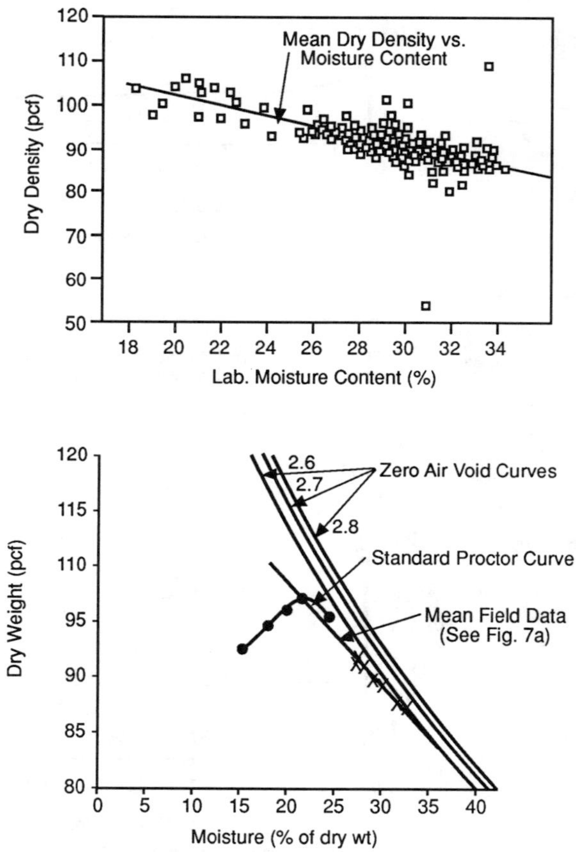

(a) Dry Density vs. Moisture Content Landfill B-19, Phase IA Secondary Clay Field Tests

(b) Dry Density vs. Moisture Content Secondary Clay/Geomembrane Interface Laboratory Tests

Figure 7. Secondary Clay Condition Field and Laboratory.

Measured peak and residual strengths of the clay/geomembrane interface are summarized on Figure 8. Peak shear strength values from the testing program showed no significant dependence on the normal stress level, and appeared to decrease significantly with increased moisture content. A best fit linear relation between peak undrained interface shear strength and clay moisture content is shown on Figure 8, and suggests a mean undrained shear strength of about 930 psf corresponding to the mean placed moisture content of 29.7 percent. Since the shear strength-moisture content relation appears to be reasonably linear, the mean shear strength can be considered to be the shear strength corresponding to the mean moisture content.

Residual shear strength values for the interface showed generally similar behavior, decreasing fairly rapidly with increased moisture content of the clay (Figure 8). The data set again indicated a generally linear relation between the undrained strength and moisture content, but in this case suggested a relatively weak relation between the measured residual strength and the level of normal stress. For the mean placed moisture content of 29.7 percent, a best fit residual undrained shear strength relation is shown on Figure 9 and indicates a shear strength of about 700 psf at overburden pressures corresponding to the maximum depth of waste. A typical load-displacement relation for the clay/geomembrane interface test is shown on Figure 6b, and again indicates relatively rapid development of peak strength and subsequent reduction to residual conditions within typically one inch of interface slip. This behavior suggests that it is unlikely that peak shear strengths can be simultaneously mobilized over significant portions of the landfill liner.

STABILITY ANALYSES

SLIDE GEOMETRY

Figure 1a shows the pre-slide topography of the B-19, Phase IA landfill, four days before the failure. Also shown on Figure 1 is a representation of the top of the secondary clay, corresponding to the surface of sliding. The base of the landfill consisted of a relatively flat bottom section (dipping at approximately 3 percent to the east, north of the seepage collection trench, and at 3 percent to the north, south of the seepage collection trench), with approximately 2H:1V sideslopes on the southwest and northwest sides of the landfill, and 3H:1V sideslopes on the northeast side. The post-slide topography, immediately following the failure, is shown on Figure 1b.

MATERIAL PROPERTIES

All stability analyses were performed assuming an average unit weight of the waste fill of 110 pcf, as determined from the bulk density determinations made during excavation of the waste. The same average unit weight was taken for the basal operations layer,

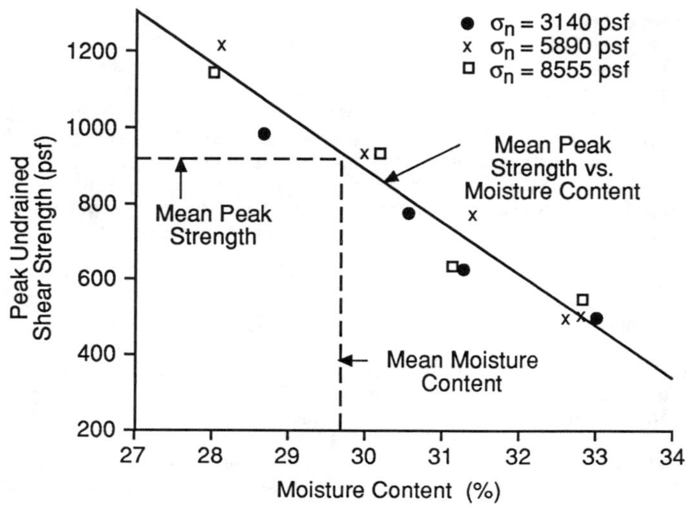

(a) Peak Shear Strength

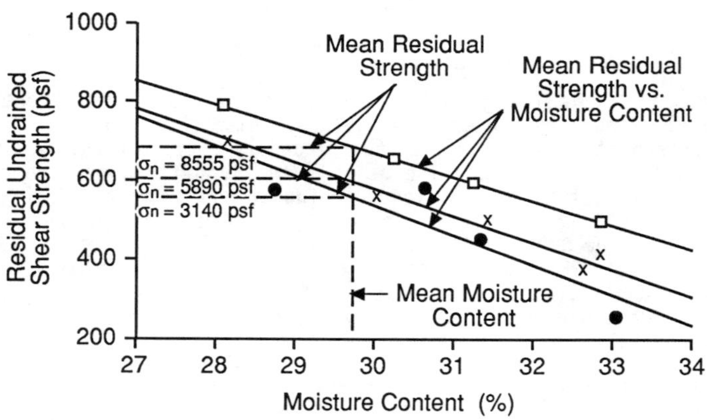

(b) Residual Shear Strength

Figure 8. Clay/Geomembrane Interface Strength Data.

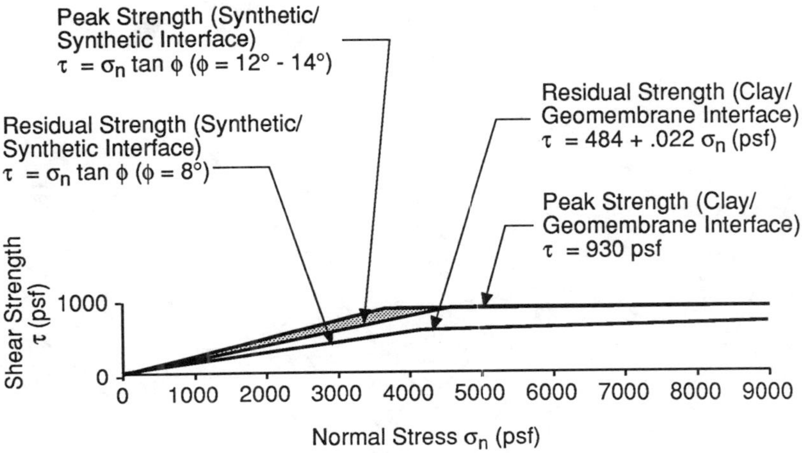

Figure 9. Liner Strength Relations.

the primary and secondary gravel drainage layers, and the primary clay overlying the sliding surface.

The shear strength parameters for the sliding surface were derived from the previously reported laboratory test data and are summarized on Figure 9.

STABILITY ANALYSES

The geometrical configuration of the waste at the time of the failure was such that there existed some uncertainty as to the location of a representative two-dimensional cross-section for stability analysis. Of some significance in terms of the difficulty of a two-dimensional representation of the Phase IA failure, was the presence of the haul road along the northeast side of the landfill. It was therefore decided that the Phase IA failure analysis should utilize a slope stability methodology that incorporated three-dimensional effects. A conventional limiting equilibrium technique, based on the Janbu approach of neglecting the vertical component of interslice shear forces, and satisfying overall vertical and horizontal equilibrium of the sliding mass at the limit condition, was employed. The slope stability computer program employed (program LF, Golder Associates, 1989) also incorporated a pre-processing package that enabled calculation of the volumes of material above each of the basal planes, directly from the survey information.

The three-dimensional representation of the base and basal sideslopes used in the stability analysis, is shown on Figure 10. The top of the secondary clay was characterized by six planes, four

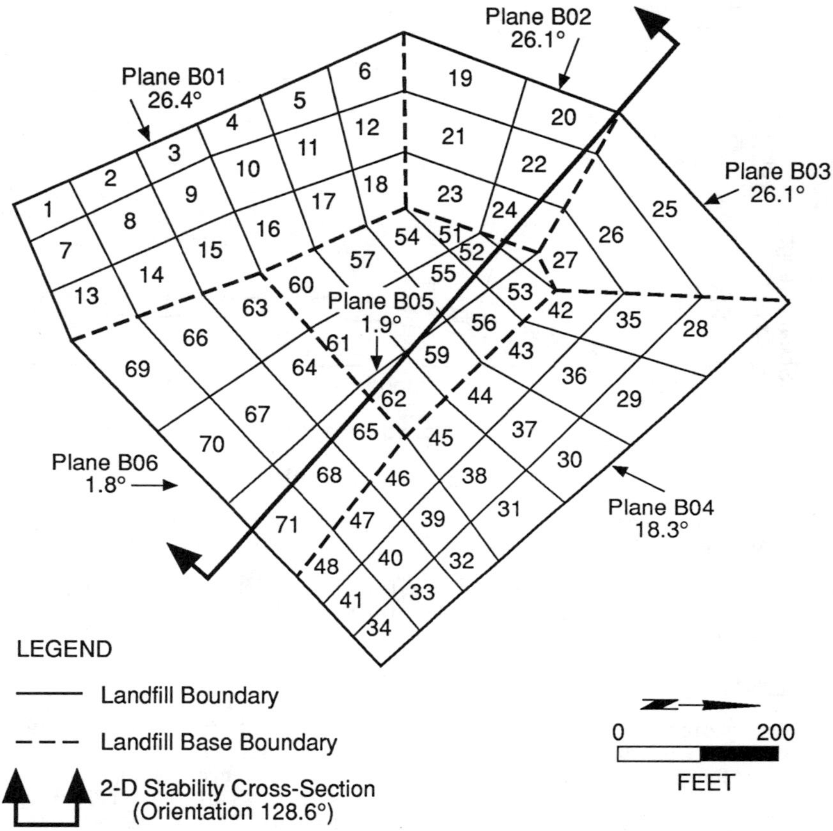

Figure 10. Base Elements for 3-D Stability Analysis.

representing the basal sideslopes and two representing the essentially flat base of the landfill. The planar representation of the landfill base was derived from the as-built construction drawings for the top of the secondary clay. Each of the basal planes was then subdivided into a number of smaller elements, in order to enable a better representation of the stress distributions over the basal planes and permit the use of a generalized non-linear shear strength relation. The individual elements of each of the basal planes are also indicated on Figure 10. In addition, Figure 10 shows the location of a cross-section through the landfill, used to perform comparative two-dimensional stability analyses. The cross-section is oriented approximately in the observed direction of failure and passes through the central region of the landfill waste.

RESULTS OF ANALYSIS

The results of the stability analyses are summarized in Tables 1 (3-D analyses) and 2 (2-D analyses). Assuming the development of residual strength conditions over the entire sliding surface, the predicted factor of safety (3-D analysis) of the pre-failure geometry is 0.85. The corresponding calculated factor of safety from the 2-D analyses (Table 2) is 0.81. These relatively low factors of safety based on realistic residual strength values would appear to be consistent with the large displacements suffered by the landfill waste, once movement was initiated and residual strengths mobilized over the entire sliding surface. Applying the liner residual strengths to the post-failure geometry indicated a factor of safety in the range of 1.08 for the 3-D analysis (Table 1), and 1.02 for the 2-D analysis (Table 2). These values provide reasonable support for the residual strengths determined by the laboratory testing. The predicted extent of the zone of sliding on the clay/geomembrane interface (as opposed to the synthetic/synthetic interface), is shown on Figure 11. The predicted extent of the clay/geomembrane failure surface appears to be generally consistent with results of the field observations.

Assuming the simultaneous mobilization of peak strengths over the entire sliding surface, the predicted factor of safety from the 2-D and 3-D analyses lies within the range of about 1.2 to 1.3. However, the shapes of the load- displacement curves indicated on Figure 6 imply that the simultaneous mobilization of peak strengths over the entire sliding surface would not be anticipated. In practice, differential straining over any potential sliding surface would result in peak strengths being developed over only a portion of that surface, at any particular point in time. Hence, factors of safety based on peak strength would not provide a useful index of stability.

Limiting equilibrium models are unable to accurately account for strain softening material behavior of the type determined for the critical interfaces of the Landfill B-19, Phase IA liner system. However, the mechanism by which differential straining within the liner can result in the initiation of failure, can be demonstrated as follows. It is reasonable to assume, for example, that essentially residual strength conditions were developed within the liner over Planes B01, B02 and B03 (Figure 10), during placement of the waste. Relatively small differential movements are required to develop residual strengths, and these could have occurred in association with compaction and settlement of the waste during placement, for example. If it is also conservatively assumed that peak shear strengths are simultaneously mobilized over the remainder of the sliding surface (Planes B04, B05 and B06 of Figure 10), then the calculated factor of safety lies within the range of 1.05 to 1.12, as shown in Tables 1 and 2. Hence, taking account of the shapes of the shear load-shear displacement relations for the various liner interfaces, and recognizing the potential for

TABLE 1. Results of 3-D Stability Analyses
Pre-Failure Waste Geometry

Clay/Geomembrane Interface Strength	Sliding Surface Details	Synthetic/Synthetic Interface Strength	Factor of Safety
Su = 930 psf (Peak)	All Planes	ϕ = 12° (Peak)	1.25
Su = 930 psf (Peak)	All Planes	ϕ = 14° (Peak)	1.32
Su = 930 psf (Peak)	(Planes B04, B05, B06 - Fig 10)	ϕ = 12° (Peak)	1.08
Su = 484 + .022σ_n (Resid.)	(Planes B01, B02, B03 - Fig 10)	ϕ = 8° (Resid.)	
Su = 930 psf (Peak)	(Planes B04, B05, B06 - Fig 10)	ϕ = 14° (Peak)	1.12
Su = 484 + .022σ_n (Resid.)	(Planes B01, B02, B03 - Fig 10)	ϕ = 8° (Resid.)	
Su = 484 + .022σ_n (Resid.)	All Planes	ϕ = 8° (Resid.)	0.85

Post-Failure Waste Geometry

Clay/Geomembrane Interface Strength	Sliding Surface Details	Synthetic/Synthetic Interface Strength	Factor of Safety
Su = 484 + .022σ_n (Resid.)	All Planes	ϕ = 8° (Resid.)	1.08

TABLE 2. Results of 2-D Stability Analyses
Pre-Failure Waste Geometry

Clay/Geomembrane Interface Strength	Sliding Surface Details	Synthetic/Synthetic Interface Strength	Factor of Safety
Su = 930 psf (Peak)	All Planes	ϕ = 12° (Peak)	1.17
Su = 930 psf (Peak)	All Planes	ϕ = 14° (Peak)	1.20
Su = 930 psf (Peak)	(Planes B05, B06 - Fig 10)	ϕ = 12° (Peak)	1.05
Su = 484 + .022σ_n (Resid.)	(Plane B02 - Fig 10)	ϕ = 8° (Resid.)	
Su = 930 psf (Peak)	(Planes B05, B06 - Fig 10)	ϕ = 14° (Peak)	1.06
Su = 484 + .022σ_n (Resid.)	(Plane B02 - Fig 10)	ϕ = 8° (Resid.)	
Su = 484 + .022σ_n (Resid.)	All Planes	ϕ = 8° (Resid.)	0.81

Post-Failure Waste Geometry

Clay/Geomembrane Interface Strength	Sliding Surface Details	Synthetic/Synthetic Interface Strength	Factor of Safety
Su = 484 + .022σ_n (Resid.)	All Planes	ϕ = 8° (Resid.)	1.02

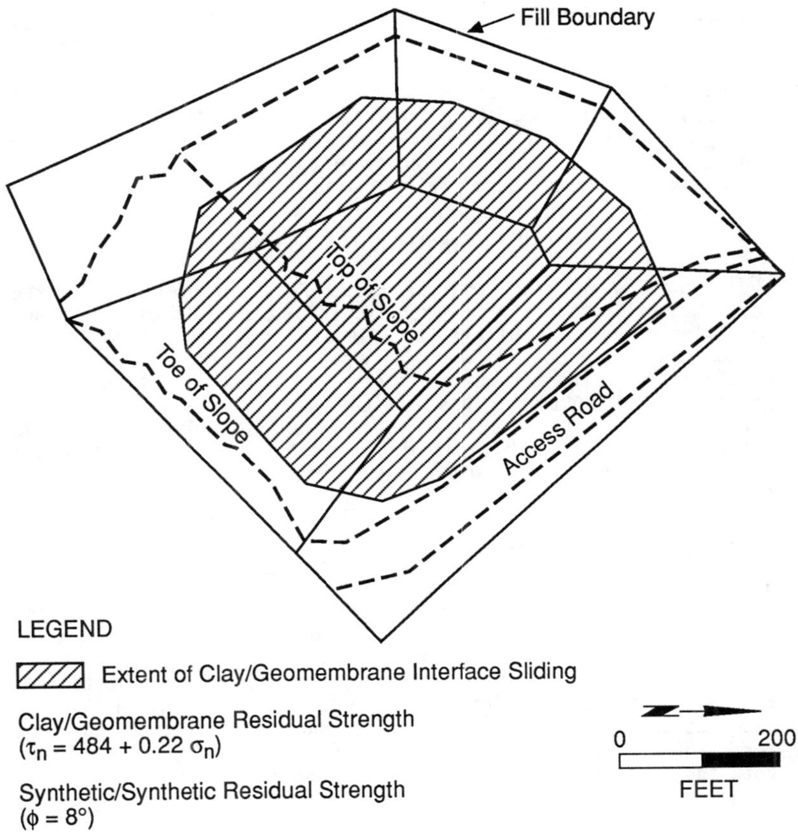

Figure 11. Sliding Surfaces Indicated From 3-D Stability Analysis.

differential shear displacements over the extent of the sliding surface, it is considered that the measured liner interface strengths are very consistent with the initiation of a failure in Phase IA of Landfill B-19. Once failure was initiated and residual strength conditions developed over the entire sliding surface, there existed a significant imbalance between driving and resisting forces. Hence, significant displacements were required before driving forces were sufficiently diminished to re-establish equilibrium.

The direction of failure predicted from the 3-D stability analyses was in the range of 116 to 122 degrees. The strike of the northeast 3H:1V basal sideslope was 128 degrees, which corresponded approximately to the average observed direction of failure. The analyses therefore indicate a critical failure orientation of about 10 degrees different from observations. However, the analyses also indicated that the calculated factor of safety was very insensitive to the assumed direction of failure within this range.

CAUSE OF FAILURE/CONCLUSIONS

The mechanism of the Landfill B-19, Phase IA failure at the Kettleman Hills facility on March 19, 1988, consisted of slip along multiple interfaces within the landfill liner system. The predominant surface of sliding appeared to be the secondary clay/secondary geomembrane interface. In the upper portions of the northwest and southwest basal sideslopes, the sliding surface corresponded to the primary geomembrane/secondary geotextile interface. In the toe regions of the northwest and southwest basal sideslopes, where the waste mass moved either directly or transversely away from those slopes, the failure mechanism was more complex and involved slip on multiple interfaces together with zones of distortion as a result of kinematic constraints to sliding on a continuous liner surface in these areas. In all likelihood, slip also occurred at interfaces other than the secondary clay/secondary geomembrane interface, over other limited areas of the base of the landfill. Such slip surfaces would be anticipated in areas of relatively low waste overburden, such as the southeast toe of the waste slope. The driving or destabilizing forces for the failure mechanism observed came from the waste located above the 2H:1V basal sideslopes located to the northwest and the southwest, and the resisting or stabilizing forces came from the waste located above both the essentially flat base of the landfill and the 3H:1V basal sideslope located to the northeast (Figure 5).

Low liner interface strengths relative to the constructed geometry of the landfill were clearly the underlying cause of the failure. Synthetic/synthetic interface strengths measured in the laboratory indicated residual friction angles as low as 8 degrees. Peak shear strengths of such synthetic/synthetic interfaces tend to be mobilized at quite small displacements, with a subsequent relatively rapid drop to residual conditions.

The predominant surface of sliding during the failure appeared to be the geomembrane/clay interface of the secondary liner system. The first important issue with respect to the strength of this interface is that it behaved in an essentially undrained mode during the approximately one year of waste loading prior to the failure. This undrained behavior was a result of the relatively low permeability and associated low coefficient of consolidation of the secondary clay liner, together with the significant thickness of the secondary clay. Hence excess pore pressures developed during loading remained largely undissipated in the near-vicinity of the secondary geomembrane, and shear strength gains associated with increased normal effective stress levels could not be realized. The strength of the clay/geomembrane interface of the primary liner has been interpreted to be generally higher than that of the corresponding secondary liner clay/geomembrane interface, because the thinner primary clay enabled some dissipation of excess pore pressure during the period before the failure. This difference was undoubtedly at least part of the reason that failure occurred in the secondary rather than in the primary liner system. The primary clay also appears to have been placed at a slightly lower moisture content and a higher dry density than the secondary clay, and these placement conditions would also tend to increase the strength of the primary liner system.

Undrained shear strength testing of the clay/geomembrane interface indicated that the shear strength is sensitive to the placed condition (moisture content and dry density) of the clay. Mean peak shear strengths were indicated as being essentially independent of the applied level of normal stress over the range tested, and as having a value of about 930 psf. Mean residual shear strengths were generally significantly lower, and indicated as being weakly dependent on the applied normal stress level, as shown on Figure 9. As for the synthetic/synthetic interfaces, the clay/geomembrane interface demonstrated mobilization of peak strengths at relatively small displacements, with a quite rapid subsequent decrease to residual conditions.

For the configuration of the landfill just prior to the failure, three-dimensional stability analyses based on peak shear strengths indicate calculated factors of safety on the order of 1.25 to 1.30. Considering the peak-residual nature of the shear load-shear displacement relations for the various liner interfaces, together with the potential for differential shear displacements over the sliding surface, it is clear that stability assessments on the basis of peak strengths are inappropriate. Furthermore, the laboratory-determined shear strengths are considered to be generally consistent with the initiation of a failure within the Phase IA, B-19 liner system. The calculated factor of safety (3-D model) based on the measured residual strengths is 0.85. This relatively low safety factor is consistent with the observed occurrence of large displacements, following failure initiation and the attainment of residual strength conditions over the entire slip surface.

The Landfill B-19 failure therefore appears to be explained by application of relatively conventional analysis and testing methods. Laboratory testing must, however, be conducted under conditions representative of the actual field conditions. Three-dimensional stability analyses are convenient for evaluating the stability of relatively complex geometric configurations such as exhibited by the Phase IA waste prior to the failure. Their use eliminates uncertainties associated with identifying a representative cross-section (location and orientation) for two-dimensional stability analysis. It should not be inferred, however, that three-dimensional stability analyses are necessary for the design of such systems. The likely benefits of a three-dimensional analysis can usually be determined by inspection.

The failure demonstrates that specifications for the placement of liner clay must focus not only on achieving specific permeability requirements, but also on developing liner shear strengths that are adequate to support both the interim and final geometric configurations of the landfill. Since strength and permeability aspects of liner performance tend to require conflicting or opposing placement conditions for the liner clay, specifications might be relatively stringent. This applies to both the as-placed condition of the clay and the procedures for addressing post-placement disturbance to the clay (e.g., rainfall events). Laboratory strength testing should be conducted over the full range of anticipated field conditions (moisture content, dry density, degree of saturation), and these field conditions must be subsequently verified by an appropriate field quality control program.

REFERENCES

GeoSyntec Consultants, (1991). "Draft Final Report Landfill B-19, Phase IA Investigation, Kettleman Hills Facility, Kettleman City, California." Prepared for Chemical Waste Management, Inc., January.

Golder Associates, (1987). "Secondary Clay Liner Construction Report, Landfill B-19, Phase IA, Kettleman Hills Facility, Kettleman City, California." Prepared for Chemical Waste Management, Inc., February.

Golder Associates, (1989). "Landfill - 3-D Landfill Stability Program." in-house computer program.

Seed, R.B., Mitchell, J.K., and Seed, H.B., (1988). "Slope Stability Failure Investigation: Landfill Unit B-19, Phase IA, Kettleman Hills, California." Report No. UCB/GT/88-01; Department of Civil Engineering, University of California, Berkeley, July.

ACKNOWLEDGEMENTS

The work reported herein was based on the efforts of several organizations and individuals. Golder Associates prepared the "Cause and Mechanism of Failure Report" which forms the basis for this paper. Chemical Waste Management, Inc., facilitated performance of the work by coordinating the activities of the various parties involved in the failure investigation, and responding promptly to requests for information. GeoSyntec Consultants performed the field and laboratory studies associated with the failure investigation, developed well-organized summaries of the very substantial volumes of data generated, and provided constructive input and support to the overall failure investigation effort. Professors R. Seed and J. Mitchell of the University of California, Berkeley, participated in various review meetings that established the overall direction of the failure study, and provided review of this failure analysis report. Professor G. Leonards of Purdue University, served as Golder Associates' senior review consultant for this study.

STABILITY ANALYSIS OF A MUNICIPAL SOLID WASTE LANDFILL

Jonathan D. Howland, A.M. ASCE[1] and Arvid O. Landva[2]

ABSTRACT

Stability analyses of municipal solid waste (MSW) landfills pose some unique challenges in terms of material characterization. These include waste strength evaluation, scale effects between laboratory testing and field behavior, and strain compatibility between the waste and the underlying soil. This paper presents a case study of analyses performed to evaluate the stability of a proposed vertical expansion of an existing municipal solid waste landfill in Upstate New York.

Published field and laboratory data regarding the strength of municipal solid waste are reviewed and analyzed to develop conventional Mohr-Coulomb strength parameters for limiting equilibrium stability analyses. An alternative method is used to analyze the results of the published waste strength case histories. Rather than statistically analyzing sets of back-calculated cohesions and internal angles of friction from the case histories, the reported data are evaluated in terms of shear stresses and normal stresses. Good consistency was found between the diverse data sets.

The overall stability of the proposed vertical landfill expansion was analyzed for short-term (during construction) and long-term (after completion) conditions. The assumption of undrained loading of the soil during landfilling (i.e., instantaneous placement of waste) was found to be overly conservative for this site as it led to calculated factors of safety less than one. It was therefore necessary to account for the increase in strength of the soils beneath the landfill due to dissipation of excess pore pressures during waste placement. Taylor's approximate method for evaluating consolidation settlements for loads applied uniformly over time was generalized to estimate excess pore pressures as a function of time for soils subjected to an arbitrary load time-history. An effective stress analysis using the estimated excess pore pressures for the end-of-landfilling condition indicated that the stability of the landfill would be adequate for the proposed final grades and loading rates.

[1] Associate Geotechnical Engineer, Dunn Corporation, 12 Metro Park Rd., Albany, New York 12205 U.S.A.

[2] Professor of Civil Engineering, Department of Civil Engineering, University of New Brunswick, Fredericton, New Brunswick E3B5A3 Canada

INTRODUCTION

Vertical expansions of municipal solid waste (MSW) landfills have become common for three reasons: 1) the continued need for landfilling as part of solid waste management; 2) the closure of the multitude of small, local "dumps" because of their inadequate protection of the environment; and, 3) the great difficulty (lengthy process, high cost, adverse public opinion and uncertain outcome) in siting and permitting new "state-of-the-art" solid waste management facilities. Vertical expansions of existing, permitted facilities avoid a number of the problems associated with the development of a new site. However, vertical expansions also pose some significant technical challenges in design. One of those challenges, addressed in this paper, is the analysis of the overall stability of landfills designed for steeper slopes and greater heights.

The permitting, design, construction, operation and closure of municipal solid waste landfills in the State of New York are governed by NYCRR Part 360. These regulations were extensively revised and expanded in 1988. Regarding stability, Section 360-2.12(c)(4) on Landfill Siting simply states that:

> "(i) A landfill must not be located in unstable areas where inadequate support for the structural components of the landfill exist or where changes in the substrate below or adjacent to the landfill may result in failure of the facility. An application for expansion of an existing landfill must demonstrate that adequate support exists or can be engineered to support any additional loads that may be generated by continued operation of the facility.
>
> (ii) Factors to be considered when determining unstable areas include: soil conditions that may result in differential settling and subsequent failure of dikes, berms, or containment structures: geologic or man-made features or events that may result in sudden or gradual failure of dikes, berms or containment structures: the anticipated loading due to height and density of waste deposition; and the stability characteristics of the waste to be deposited."

Regarding design, Part 360 allows final grades of up to 3(Horizontal) : 1(Vertical) or 33 percent. So, given a proposed landfill expansion designed to the allowable grades, the geotechnical engineer is charged with analyzing the stability of the proposed design.

Figure 1 shows a schematic profile of the landfill under consideration. On this figure can be seen: waste originally placed at the site over a period of approximately 10 years; new landfill cells with a leachate collection system (LCS) and low permeability clay liner; a piggy back liner of compacted clay placed over the old waste; and, the proposed final grades for which the stability analyses presented herein were performed. At the time of analysis, the landfill was approximately 60 feet high.

Also shown on Figure 1 is the generalized stratigraphy beneath the landfill. The soil layer thicknesses shown on this figure and used for analysis are approximately the maximum values observed in the borings drilled at the site. From the ground surface downward are: an upper glaciolacustrine clay and silt layer; an upper glacial till layer; a lower glaciolacustrine clay and silt layer; a lower glacial till layer; and, the underlying bedrock.

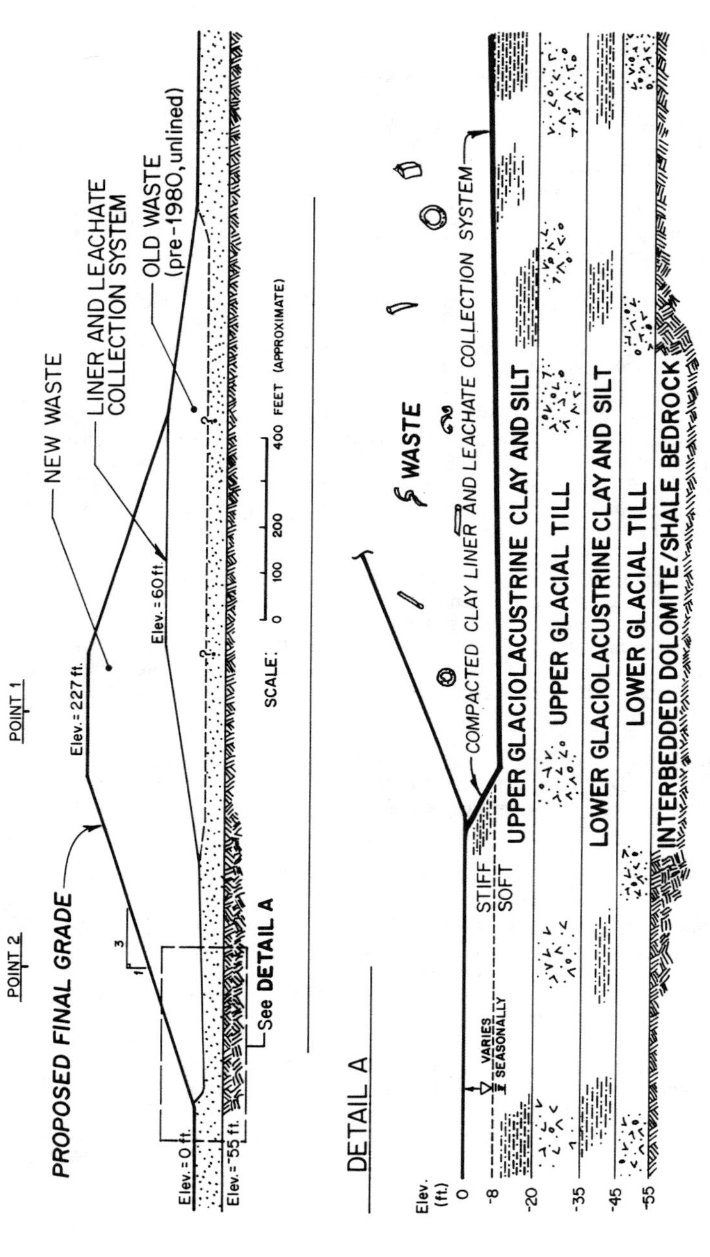

Figure 1. GENERALIZED LANDFILL and SUBSURFACE PROFILE

The lacustrine clay layers are characterized as moderately plastic, medium stiff to soft and lightly over consolidated (OCR ~ 2), except for the upper 8 feet of the upper layer which is very stiff and moderately over consolidated (OCR ~ 5), probably due to desiccation. The clay is varved with alternating silty and clayey seams. Also, there are frequent discontinuous partings and seams of silt or fine sand. The till layers are stiff/dense, poorly sorted, with the major component being slightly to moderately plastic fines.

The bedrock consists of interbedded dolomite and shale with low RQD values in the upper 10 feet. The depth to bedrock varies from 40 to 55 feet. The groundwater table varies seasonally within about 10 feet of the ground surface.

The liner and leachate collection system for this landfill were designed by others in the late 1970's as two feet of compacted soil with a permeability of less than 1×10^{-5} centimeters per second (cm/sec). The liner was constructed throughout the 1980's as two to three feet of compacted clay with a maximum permeability of 1×10^{-7} cm/sec (4×10^{-8} cm/sec average).

The landfill operating permit requires the leachate level in the collection wells to be maintained at 7 feet or more below the average surrounding grade. This operating procedure results in an inward groundwater gradient through the liner for much of the landfill. The influence of this seepage gradient was considered to be negligible for the stability analyses presented herein.

Material Properties

This section presents the development of strength parameters for use in the stability analyses. The two materials considered, municipal solid waste and natural soils, are thought to have significantly different strength mobilization rates, that is, strain compatibility is a concern. While this is not the ideal case for limiting equilibrium stability analyses, the assumption is made that the strength parameters developed for the MSW are peak values. Therefore, residual (large strain) strength parameters were used for the natural soils.

Native Soils

Both total and effective stress parameters were developed for the natural soils at the site. Total stress parameters were based on unconsolidated, undrained triaxial tests, a total stress interpretation of consolidated, undrained triaxial tests and the correlation of Standard Penetration Test N-values to unconfined compressive strength (Peck *et al.,* 1974). Effective stress parameters were based on consolidated, undrained triaxial tests with pore pressure measurements and consolidated, drained direct shear tests. On a total stress basis the soils were, in general, not observed to strain soften. However, on an effective stress basis (q versus p'), some decrease in the principal stress ratio with strain was noted. That is, the deviator stress at large strains (15 percent) was maintained by decreasing excess pore pressures.

The peak drained friction angle of the lacustrine clay soils was observed to be approximately 26 degrees, which is typical of these soils in northern New York State. A large strain, drained friction angle of 21 degrees was used for these soils in this analysis.

Compacted Clay Liner

The landfill liner system was constructed during the 1980's with locally obtained lacustrine clays. Quality control/quality assurance testing consisted of field moisture content and density measurements and laboratory permeability tests. No strength or consolidation testing was performed on liner samples.

Mitchell *et al.* (1990), Richardson and Reynolds (1991) and Koerner (1990) underscore the importance of evaluating the stability of landfill liner systems, particularly when geosynthetics have been incorporated in the design. Since no geosynthetics, other than geotextile filters around the gravel leachate collection lines, were incorporated in the liner design, a separate analysis of the liner stability was not performed.

For the stability analyses presented herein, the compacted clay liner was assumed to have a strength at least equal to that of the underlying soft lacustrine clay. This assumption was based on the observation that the liner was compacted at a moisture content of about 25 percent to a dry density of approximately 100 pounds per cubic foot (pcf). The soft lacustrine clay has an *in-situ* moisture content and dry density of about 30 percent and 90 pcf, respectively.

Municipal Solid Waste

While the measurement and interpretation of soil strength parameters for stability analyses is a routine part of geotechnical engineering practice, the development of strength parameters for municipal solid waste is not. On one hand, the strength of MSW appears to increase with increasing normal (confining pressures), like a soil. On the other hand, its high organic and "fiber" contents when placed make it behave more like a Radforth (fibrous) peat (Landva and La Rochelle, 1983) than a mineral soil. The fiber content of MSW will remain much the same, but the putresible organic content will decrease with time. Therefore, some changes in geotechnical properties must be expected. Also, MSW can be extremely heterogeneous, so that a sample taken at one location may not be indicative of the "average" properties.

The era of waste placement may also be a factor. Consider, for example, the increased use of plastics for packaging in the 1960's. Landva and Clark (1990) note that the friction angle between plastic bags is approximately 9 degrees. While plastics in the waste stream does not appear to reduce the average bulk strength of MSW to 9 degrees, Landva and Clark (1990) measured a friction angle of 24 degrees for a shredded refuse with a high plastic content.

Municipal solid waste placed in the 1990's may again behave differently than MSW of previous eras. Permitted landfill air space has become increasingly valuable and, therefore, increased compactive effort is seen as a way to get more waste into the same space. Also, stricter enforcement of daily soil cover requirements will tend to increase the average density of the landfill mass. It is unknown whether the increased density of the layered waste and daily cover will be offset by a corresponding increase in strength. Increased recycling will also have an unknown effect on waste composition and strength.

Method of Waste Strength Analysis

Given the concerns and uncertainties identified above, the challenge is to develop reasonable strength parameters for use in the required stability analysis. In the present case, site specific sampling and testing was ruled out as there are no generally accepted waste testing standards or equipment. Instead, waste strength data reported in the literature (both direct laboratory and field testing and back-calculations of failures and load tests) were evaluated.

The general approach used here to compare the various reported MSW strength data was to plot mobilized shear strength against average normal stress, rather than averaging reported values of Mohr-Coulomb strength parameters, cohesion (C) and angle of internal friction (Phi). While interpreting data in terms of shear strength versus normal stress is conventional practice for direct measurements, i.e. laboratory tests, strength back-calculations from failures or load tests are often reported as pairs of C and Phi that satisfy equilibrium, which is the result of having one known (factor of safety = 1.0) and two unknowns (C and Phi). Based on this type of analysis, Singh and Murphy (1990) concluded that a Mohr-Coulomb characterization of waste strength may be inappropriate and that waste strength should be conservatively ignored for stability analyses.

For an individual waste strength case study, back-calculated pairs of C and Phi that satisfy equilibrium were plotted as shear stress versus normal stress (Figure 2). The value of C calculated for Phi assumed to be equal to zero is the average shear strength mobilized along the failure surface (Point A, Figure 2). For a material with a linear strength envelope, the crossing point of the C - Phi pairs satisfying equilibrium (Point B, Figure 2) indicates the average normal stress along the failure surface (Point C, Figure 2).

The crossing point (Point B, Figure 2) is considered to be the one consistent datum derived from an individual case study. The crossing point from each back-analysis case study was transferred to a summary plot of shear stress versus normal stress in order to develop a strength envelope for municipal solid waste. The average normal stress based on the crossing point was checked by estimating the location of the failure surface and calculating the average normal stress based on the reported unit weight of the waste and any applied loads.

Municipal solid waste strength data reported in the literature are described below and summarized on Figure 3. Good agreement between the two estimates of average normal stress was generally found except for one case noted below. Other data were evaluated as part of this study, but are not reported herein, because they were deemed inapplicable for various reasons, such as excessive processing of the material to accommodate conventional geotechnical testing equipment.

Case 1 - New Jersey Landfill Failure

Oweis (1985) and Dvirnoff and Munion (1986) report the failure of the Global Landfill in New Jersey. The landfill had a maximum height of approximately 115 feet and was constructed on a very soft tidal marsh deposit. The failure occurred after approximately 50 feet of new waste was placed over 15-year old waste in a 4 to 5 month period. The failure apparently initiated in the tidal marsh deposit and propagated upward through the waste. A significant amount of rain was reported over a three-day period prior to the failure.

The failure was back-analyzed using an unspecified method of slices and the results were reported as pairs of C and Phi corresponding to a factor of safety equal to one. The average normal stress indicated by the crossing point of the C - Phi pairs is

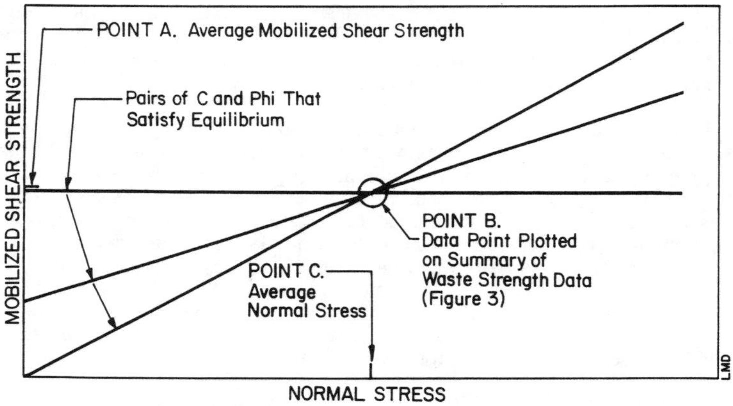

Figure 2. METHOD OF ANALYZING AN INDIVIDUAL WASTE STRENGTH CASE HISTORY

approximately 20 percent less than the average normal stress along the failure surface in the waste estimated with the reported waste unit weight of 60 pcf. Based on this discrepancy, the results of this case study are plotted on Figure 3 as a triangular area that covers the calculated range of average normal stress.

Case 2 - California Landfill Load Test

Oweis (1985) presents a back-calculation of the shear strength mobilized by constructing a soil test fill on top of a MSW landfill located in southern California. The response of the landfill to the applied load was monitored with slope inclinometers: a maximum lateral movement of about 17 inches was measured. At the end of the load test, the movements of the landfill markedly decreased and the landfill was apparently stable.

Pairs of mobilized C and Phi, corresponding to a factor of safety equal to one, were calculated with an unspecified method of slices for a circular failure surface. This point is plotted on Figure 3 with an upward arrow to indicate an unknown, higher ultimate waste strength value and average normal stress at failure. The unit weight used for the calculations was reported to be 45 pcf. Singh and Murphy (1990) review how a number of engineers have used the results of this load test.

Note that the slope inclinometer data indicate movement throughout the waste beneath the test fill. The maximum shear strains, however, appear to be in the upper part of the waste. A sliding block or wedge analysis may be more appropriate for this load test.

Case 3 - Laboratory Direct Shear Tests

Landva and Clark (1990) present the results of laboratory strength tests performed on a variety of wastes. The results of direct shear tests on two primarily municipal

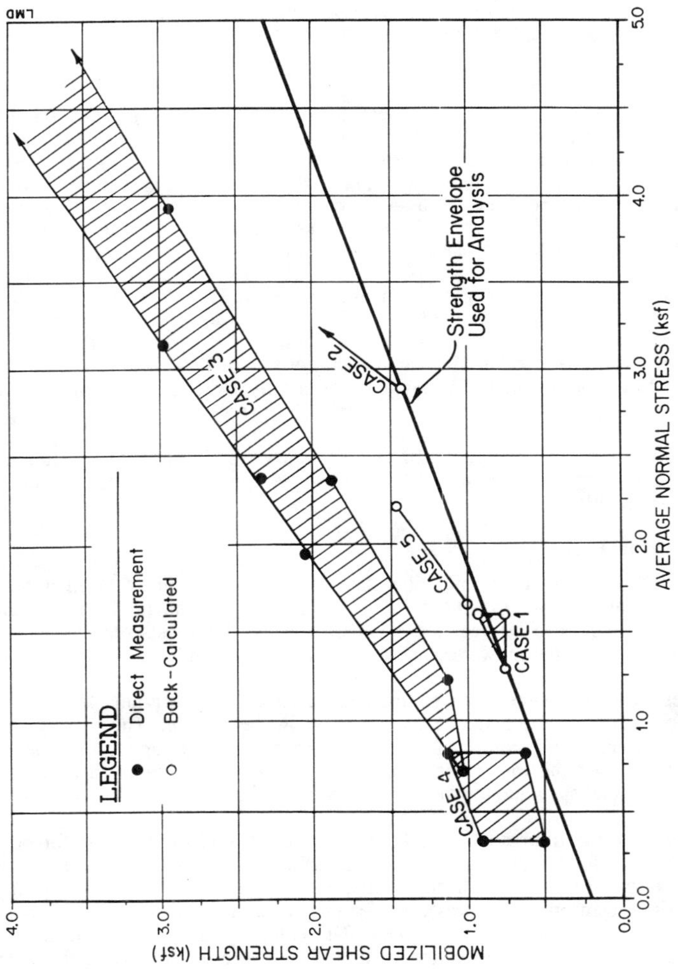

Figure 3. SUMMARY of MUNICIPAL SOLID WASTE STRENGTH DATA (REFER TO TEXT FOR AN EXPLANATION OF EACH CASE)

wastes are presented as a range on Figure 3. The arrows indicate testing at normal stresses greater than those shown on the figure. The wastes were 10 to 15 years old and sampled with 6 and 11-inch solid stem augers.

The tests were performed in a large direct shear box measuring approximately 17 by 11 inches along and across the shear direction, respectively. Normal loads of up to 12 kips per square foot (ksf) were applied and samples were sheared to a maximum displacement of about 4 inches. Pore pressures were monitored during shear and the shearing rate was adjusted, if necessary, to maintain drained conditions.

Case 4 - Field Direct Shear Tests

Richardson and Reynolds (1991) recently reported (after this analysis was originally performed) on the failure of a MSW landfill located in central Maine. The landfill was constructed over a period of about 15 years on a thick deposit of soft, sensitive, marine clay and silt, and had attained a height of about 60 feet at the time of failure. The failure was rapid and preceded by a period of heavy rainfall. The failure was attributed to: a higher-than-expected waste unit weight; and, excavation near the toe of the original landfill for construction of a new cell.

No back-calculated waste strengths are presented, possibly due to the uncertainty introduced by the strain softening behavior of the underlying clay. However, large scale direct shear tests were performed after the failure. Six multi-point direct shear tests were performed on the landfill with a five foot square shear box. Normal load was provided by dead weights. The range of the results is plotted on Figure 3. The age of the waste tested is unknown, but presumed to be recent.

A dozen test pits of approximately 10 cubic yards were also excavated after the failure. The average total density was found to be approximately 95 pcf, much higher than originally anticipated.

Case 5 - Plate Load Tests

Eliasson (1942) reports load-deflection data from two typical plate load tests performed on sanitary landfills near New York City. A total of 44 tests were conducted on 7 month to 10 year old wastes. The two load tests reported in the paper and summarized herein were conducted on 7 and 30 month old wastes.

The load tests were conducted by applying dead loads up to 8000 pounds per square foot (psf) to a 1 foot square plate located 4 feet below the surface. The shape of the load-deflection curves was similar for both tests, becoming non-linear at a load of about 5000 psf. At the maximum load of 8000 psf, the younger waste settled approximately 5 inches, while the older waste settled about 3 inches.

For this analysis, the ultimate failure load was extrapolated to 9000 psf for both tests. Since the estimated ultimate load was the same for both load tests, these tests are treated as one case for the back-calculation. The extended bearing capacity formula (Brinch Hansen, 1970) was used for back-calculating the mobilized waste strength to account for the relatively deep plate embedment and the influence of strength mobilized above the elevation of the plate. Note that the strength of the backfill above a footing is typically ignored in most bearing capacity predictions.

The results of the back-calculation are plotted as two points on Figure 3. The higher point corresponds to no strength mobilized in the waste above the plate, while the lower point represents the waste strength calculated assuming full mobilization of the waste above the plate.

Summary - Municipal Solid Waste Properties

Based on the data summarized on Figure 3, two conclusions were made. First, the strength of municipal solid waste is primarily frictional in nature. Second, the results of back-analyses indicate a lower waste strength than the results of direct measurements. The reason for this difference is unknown, but could be the greater influence of peat-like fibrosity on smaller scale direct measurements relative to back-calculations from large scale failures or load tests. Alternatively, waste strengths back-calculated from failures of landfills on soft foundations may be inaccurate because of uncertainties about the soil behavior (Case 1).

The strength envelope shown on Figure 3 was selected for use in the stability analyses presented in the following section. This strength envelope, defined by Mohr-Coulomb parameters of C equal to 200 psf and Phi equal to 23 degrees, is a lower bound to the data. The envelope reflects the back-calculated strength data and is therefore consistent with the stability analyses to be performed.

Note that, in general, the various factors hypothesized at the beginning of this section have not been systematically proven or quantified. Therefore, it is not recommended that the strength parameters selected above be used by reference for other analyses. It is recommended that the general methodology of comparing available directly measured and back-calculated data on a mobilized shear strength versus normal stress basis be used. Great care should be exercised in selecting waste strength parameters for use in the stability analysis of high landfills, particularly when the anticipated stresses exceed the stress ranges for which data are available.

Excess pore pressures in the waste were assumed to be zero for all analyses because of the presence of the leachate collection system at the base of the landfill and reported high waste permeabilities (Landva and Clark, 1990). Note that research in progress by the second author of this paper indicates that waste permeability may decrease significantly at high confining pressures.

A waste unit weight of 45 pcf was used for the stability analysis. This value was provided by the owner of the landfill based on measurements of waste and daily cover weights and surveyed airspace volumes. The selected unit weight appeared to be consistent with values summarized by Bromwell (1978), but toward the lower end of data reported by Landva and Clark (1990) based on measurements from large test pits (approximately 10 cubic yards). The influence of the waste unit weight on stability is considered in the following section as well as some recently developed data.

STABILITY ANALYSIS

The general approach used for the stability analysis of the landfill for the proposed final grades was to evaluate the most critical conditions first and, if adequate stability was obtained, to conclude the analysis. Analyses were performed using the computer program PC-STABL, version 5M developed at Purdue University

(Achilleos, 1988) over the past 15 years. This program employs the Modified Bishop method of slices to analyze circular failure surfaces, the Simplified Janbu method to analyze block-type failure surfaces and Spencer's full-equilibrium method for either type of failure surface. The program models soils with either total stress or effective stress parameters and allows pore pressures to be simulated in a number of ways, including individual piezometric surfaces for each soil type.

Hundreds of potential failure surfaces were evaluated by specifying surface initiation and termination ranges and minimum allowable surface elevations. The surface generation parameters were adjusted until a zone of critical surfaces was found with the lowest factors of safety.

UNDRAINED ANALYSIS

A stability analysis of the landfill was first performed using undrained soil strength parameters that conservatively ignore dissipation of excess pore pressures in the subsoils during waste placement. The excess pore pressures in the waste were assumed to be zero. The minimum calculated factor of safety was 0.8 for a sliding block surface through the upper lacustrine clay layer. The minimum factor of safety for circular surfaces was 1.2 for a surface passing through the lower lacustrine clay layer.

DRAINED ANALYSIS

Based on the results of the undrained analysis above, it was concluded that dissipation of excess pore pressures from the subsoils during waste placement must be considered. Long-term conditions were modelled using effective stress strength parameters for the soils and hydrostatic pore pressures corresponding to the groundwater table. The minimum calculated factor of safety was 1.6 for a circular surface passing through the upper lacustrine clay layer. Sliding block surfaces had factors of safety greater than 1.8.

PARTIALLY DRAINED ANALYSIS

Based on the positive results of the long-term, drained analysis, it was concluded that the landfill could be constructed to the proposed final grades, but the stability would depend on the waste placement rate and dissipation of excess pore pressures from the subsoils during loading. An effective stress analysis of the stability during construction was performed by estimating the excess pore pressures in the foundation soil layers as a function of waste placement rate and time.

To verify this method of analysis, undrained conditions were reanalyzed using effective stress strength parameters and excess pore pressures equal to the applied vertical load. The minimum factors of safety were about 10 percent less than those calculated by total stress analysis. Based on these results, it was concluded that the method of analysis was reasonable and conservative.

Pore Pressure Dissipation

The excess pore pressures during land filling were estimated using Taylor's approximate method (Lambe and Whitman, 1969) for calculating consolidation settlements versus time for loads applied linearly over a finite construction period, rather than the instantaneous loading assumed for conventional consolidation theory (simulated by the undrained stability analysis reported above).

Taylor's method gives an empirical, closed-form solution for settlement versus time. The extension of this method to estimating pore pressure dissipation is based on the observations that:

1) the settlement ratio (settlement at a given time divided by ultimate settlement) is proportional to the average consolidation ratio;
2) the average excess pore pressure ratio is equal to one minus the average consolidation ratio;
3) pore pressures within the consolidating layer are related to the average excess pore pressure by conventional consolidation theory; and,
4) the effect of varying waste placement rates could be calculated by summing the excess pore pressures from the different loads through superposition in time.

The average excess pore pressure (u_e) in a consolidating layer can be calculated as:

$$u_e(t) = u_o \cdot t/t_1 \cdot [\, 1 - U(\, T(\, t/2\,)\,)\,] \quad \text{for } t < t_1$$

$$u_e(t) = u_o \cdot [\, 1 - U(\, T(\, t\text{-}t_1/2\,)\,)\,] \quad \text{for } t \geq t_1$$

where t - time from the beginning of load application
t_1 - length of time to apply the total load
u_o - initial excess pore pressure for instantaneous loading
T, U - time factor, average consolidation ratio for instantaneous loading

The extended Taylor's method was used in this analysis by subdividing the waste loading into one year periods and summing the estimated pore pressure from each load component. The following assumptions, which tend to over-estimate the excess pore pressures, were made:

1) the initial excess pore pressure is equal to the applied load above a point. In reality, the soils beneath the landfill have a pore pressure parameter A approaching one only for very low strains (about 1 percent) in undrained triaxial tests. For larger strains the A parameter approaches zero. Therefore, the excess pore pressures will be over-estimated beneath the outer portion of the landfill; and,
2) only two-way vertical drainage occurs upward to the LCS and to the underlying very permeable bedrock. In reality, lateral drainage will occur because the horizontal permeability of the lacustrine clay layers (but not the compacted clay liner) is approximately two orders of magnitude greater than the vertical permeability. Lambe and Whitman (1969) present a chart solution for axisymmetrically loaded areas of limited extent. After dividing the horizontal dimension of the landfill by the square root of the horizontal-to-vertical permeability ratio, the axisymmetric solution indicates that lateral drainage would allow significant additional dissipation of the excess pore pressures.

The consolidation coefficient (c_v) for the upper lacustrine clay, upper till and lower lacustrine clay are approximately 0.002 cm^2/sec, while the lower till has a c_v value four times greater. For this analysis the three upper layers (35 feet including the compacted clay liner) were assumed to consolidate at the same rate. The lower till was assumed to be drained.

The estimated maximum and average excess pore pressures for two points beneath the landfill are shown as a function of time on Figure 4. Point 1 (Figure 1) corresponds to the centerline of the landfill where the existing waste was placed over a long period of time. Point 2 (Figure 1) corresponds to a recently developed cell where the existing waste was placed relatively rapidly. The excess pore pressure profiles across the landfill are shown on Figure 5 at two selected times: about the time of the analysis (1991); and, the projected end of waste placement (1994).

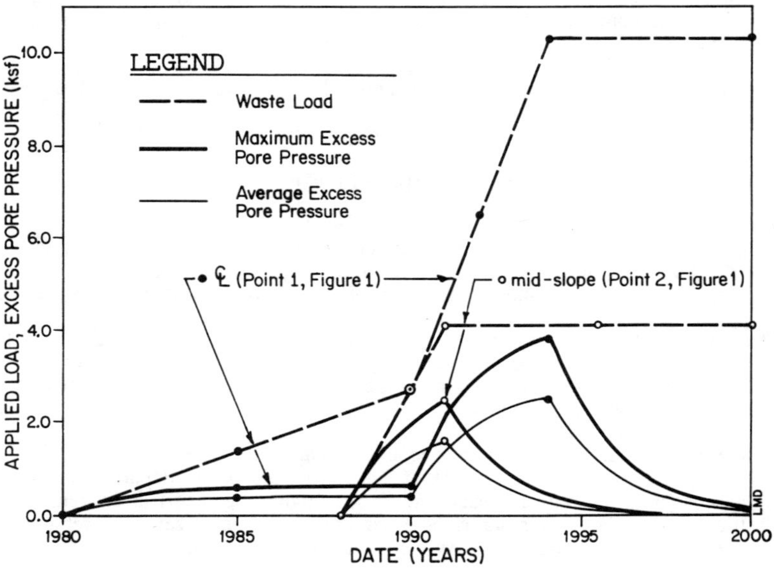

Figure 4. LOAD and EXCESS PORE PRESSURE vs. TIME

Stability Analysis

The stability of the landfill was analyzed with the effective stress soil parameters and the generalized excess pore pressure profile shown on Figure 5. The average estimated excess pore pressure was assigned to the upper and lower lacustrine clay layers. The maximum estimated excess pore pressure was assigned to the upper till layer. The lower till layer was assigned zero excess pore pressure because it had a consolidation coefficient approximately four times greater than the other soils.

The lowest factor of safety calculated for the partially-drained, during-construction condition was 1.5 for a toe circle passing through the upper lacustrine layer. The lowest calculated factor of safety for sliding block-type surfaces was also 1.5 for a surface passing through the upper lacustrine clay.

WASTE UNIT WEIGHT

Waste unit weight data reported by Richardson and Reynolds (1991) show that MSW can be much heavier than previously thought. While unit weight will have little influence on the drained stability of primarily frictional materials, it can affect the results of undrained and partially-drained analyses. The stability calculations for the subject landfill were redone with a waste density of 95 pcf. The factor of safety decreased from 1.5 to 1.4 for partially drained conditions and from 1.6 to 1.5 for long-term (drained) conditions. Based on a recent survey of the landfill, the owners calculated a waste unit weight of approximately 75 pcf.

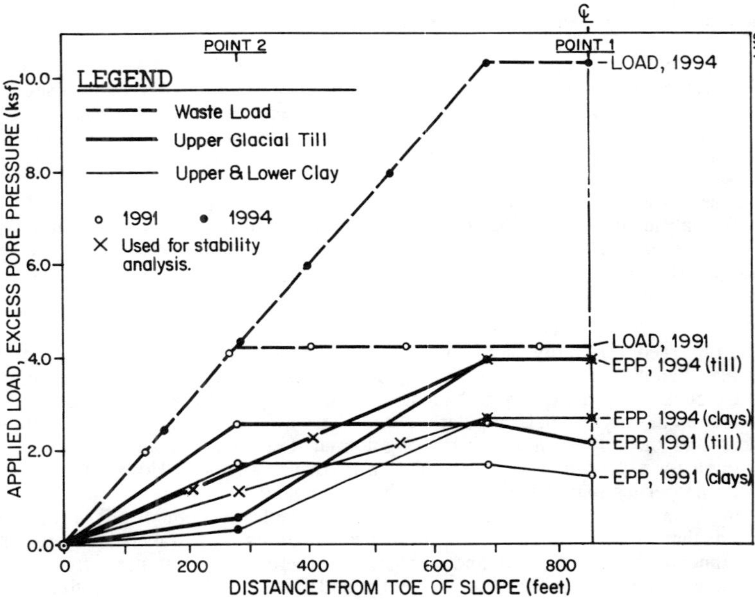

Figure 5. ESTIMATED EXCESS PORE PRESSURE (EPP) in SOILS BENEATH LANDFILL

SUMMARY - STABILITY ANALYSIS

Based on the analyses discussed above, it was concluded that the stability of the landfill would be adequate for the proposed final grades and the projected waste placement rate. If unsatisfactory results had been obtained from the approximate, partially drained analysis presented above, more sophisticated, numerical analyses, such as coupled stress distribution and consolidation, could have been performed. Given the conservative and acceptable results from the approximate analysis, the additional effort required for the numerical analyses was not deemed necessary for this investigation.

Note that the average normal stress in the waste for the critical failure surfaces was approximately 3 ksf. This corresponds to the upper limit of the average normal stresses for the back-calculated data reviewed herein. Great care should be used analyzing the stability of landfills where the stresses exceed those from which the waste strength parameters were developed.

CONCLUSIONS

This paper has presented a case study of the stability analyses performed for the vertical expansion of a municipal solid waste landfill. The focus of the paper was the development of strength parameters for the waste and the estimation of excess pore pressure dissipation in the subsoils during waste placement.

For this study, the compacted clay liner was considered to be at least as strong as the underlying soft lacustrine clay. Therefore, the liner was not analyzed as a separate mode of failure. This mode of failure must, in general, be considered, especially for liner designs that incorporate geosynthetics (Koerner, 1990; Richardson and Reynolds, 1991; Mitchell *et al.*, 1990).

The selection of waste strength parameters was based primarily on the results of back-analyses from waste failures and full scale load tests reported in the literature. These strength values were lower than the strength values reported for direct field and laboratory measurements. Also, the back-calculation derived parameters were considered to be consistent with the stability analyses performed.

The observed variability of the waste strength data reviewed was found to be no greater than that of naturally occurring, mineral-based soils. However, conventional soil mechanics can explain and predict variations in soil strength based on such factors as density, particle angularity, etc. For waste, the observed linear trend of increasing shear strength with increasing normal stress does explain the variation of strength for a particular waste tested in a particular manner. However, no waste index properties have been identified that systematically account for the different strength envelopes observed for different wastes and test methods. It is therefore urged that waste strength data from back-analyses and direct measurements continue to be investigated and published. There is a great need for reliable methods of describing MSW (e.g., Landva and Clark, 1990) and evaluating MSW strength on a site specific basis. This will become particularly important as landfills are built to heights that produce stresses that exceed the data currently available.

ACKNOWLEDGMENTS

The authors would like to thank Messrs. David A. Myers, P.E. and Kevin S. Dufek of the Dunn Corporation for review of and assistance with the analyses presented herein. Professor F. Peter Tolcser, P.E. of Hudson Valley Community College assisted with the interpretation of the waste strength data. The support of Dunn Corporation is gratefully acknowledged, particularly the careful assistance of Ms. Dianna Gilbert, Ms. Lisa Bertrand and Ms. Laura Dineen with the preparation of this manuscript.

REFERENCES

Achilleos, E. (1988). "User Guide for PC STABL 5M," Report JHRP-88/19, School of Civil Engineering, Purdue University, Lafayette, IN.

Brinch Hansen, J. (1970). "A Revised and Extended Formula for Bearing Capacity," Bulletin No. 28, Danish Geotechnical Institute, Copenhagen, Denmark.

Bromwell, L.G. (1978). "Properties, Behavior and Treatment of Waste Fills," Seminar, Methods of Soil Improvement, Metropolitan Section, ASCE, New York, NY.

Dvirnoff, A.H. and Munion, D.N. (1986). "Stability Failure of a Sanitary Landfill," International Symposium on Environmental Geotechnology, H. Fang, Ed.

Eliasson, R. (1941). "Load Bearing Characteristics of Landfills," Engineering News Record, Vol. 369, pp. 103-105.

Koerner, R.M. (1990). Designing with Geosynthetics, Prentice Hall, Englewood Cliffs, NJ.

Lambe, T.W. and Whitman, R.V. (1969). Soil Mechanics, Wiley, New York, NY.

Landva, A.O. and Clark, J.I. (1990). "Geotechnics of Waste Fills - Theory and Practice," Geotechnics of Waste Fills - Theory and Practice, STP 1070, A.O. Landva and G. D. Knowles, Eds., ASTM, Philadelphia, PA.

Landva, A.O. and La Rochelle, P. (1983). "Compressibility and Shear Characteristics of Radforth Peats," Testing of Peats and Organic Soils, STP 820, P.M. Jarrett, Ed., ASTM, Philadelphia, PA.

Mitchell, J.K., Seed, R.B. and Seed, H.B. (1990). "Stability Considerations in the Design and Construction of Lined Waste Repositories," Geotechnics of Waste Fills - Theory and Practice, STP 1070, A.O. Landva and G. D. Knowles, Eds., ASTM, Philadelphia, PA.

Oweis, I.S. (1985). "Stability of Sanitary Landfills," Seminar, Geotechnical Aspects of Waste Management, Metropolitan Section, ASCE, New York, NY.

Peck, R.B., Hanson, W.E. and Thornburn, T.H. (1974). Foundation Engineering, Wiley, New York, NY.

Richardson, G. and Reynolds, D. (1991). "Geosynthetics Considerations in a Landfill on Compressible Clays," Proceedings, Geosynthetics '91 Conference, Industrial Fabrics Association, Minneapolis, Minnesota.

Singh, S. and Murphy, B.J. (1990). "Evaluation of the Stability of Sanitary Landfills," Geotechnics of Waste Fills - Theory and Practice, STP 1070, A.O. Landva and G. D. Knowles, Eds., ASTM, Philadelphia, PA.

APPENDIX A - UNITS

1 foot = 12 inches = 0.33 yards = 0.305 meters = 30.5 centimeters
1 pound/ft^3 = 0.157 kN/m^3; 1 kip/ft^2 = 1000 pounds/ft^2 = 47.9 kPa

The Design Of Landfill Slopes

Ibraheem Alshunnar[1]

Abstract

The design of landfills involves the evaluation of the stability of slopes with different geologies and geometries. The stability of cut slopes, existing natural slopes, embankments, and waste slopes during operation or after closure may need to be evaluated. Also, the stability of the liner system and the landfill cover has to be evaluated. The geometry and/or the geology of these slopes may change during the construction and/or the operation of the landfill. The stability of a slope should be evaluated for all possible geometries and geologies throughout the life of a landfill. The failure to evaluate a slope configuration or to account for a design parameter may cause a slope failure. While every site and design has its own characteristics some of the most common cases and design consideration are covered in this paper.

The following design considerations for the design of landfill slopes are discussed in details: 1) Ground water condition before and after the construction of the landfill. 2) Subsurface soil condition. 3) Construction sequence. 4) Adjacent sites conditions and history. 5) Site topography. 6) External loads. 7) Liner geometry and composition. 8) Waste filling sequence. The effect of these considerations on the stability of slopes are discussed. Solutions for possible instability problems are also discussed. Taking such considerations into account should increase the reliability of the results, decrease the cost of the analyses, and decrease the time required to perform the analyses.

Introduction

The stability of slopes is very crucial in the

----

[1]Senoir Engineer, NTH Consultants, 38955 Hills Tech Drive, Farmington Hills, Michigan 48331-3432

design of landfills. The capacity of a landfill is dependent on the geometry of the soil side slopes and waste final slopes. A slope failure may cause the loss of lives and cost millions of dollars in loss of dumping space and construction time, and remediation measures. Many landfill slope failures have been reported around the country resulting in losses of millions of dollars.

The reliability of the results of a slope stability analyses is dependent on the reliability of the models used to present design slopes. The purpose of this paper is to discuss some considerations in the design of landfill slopes. Such considerations should increase the reliability of the results, decrease the cost of the analyses, and decrease the analyses time.

A slope can be characterized by geology, geometry, and activity. Landfill design involves the design of slopes with different geologies, geometries, and activities. The stability of cut slopes, existing natural slopes, embankments, and waste slopes during operation and after closure need to be evaluated. Also, the stability of the liner system and the landfill cover need to be evaluated. In general, any alteration of a slope geology, geometry, or activity should be evaluated. In the case of a landfill there are many construction and/or operation activities that may cause such alteration. Some considerations to account for these changes is discussed in this paper.

Ground Water Condition:

In the design of landfill slopes the ground water condition before and after construction should be evaluated. The excavation of a cut may cause a change in flow direction and an increase in seepage forces.

In areas where artisan condition occurs the excavation of a cut slope may reverse the flow direction or increase upward flow. For example a downward flow occurs in the site shown in Figure (1). The excavation of a landfill with cell floor below the piezometric head of the confined aquifer will cause an upward flow below the cell floor. Seepage water, if not properly collected and removed, may cause soil softening and/or reduce soil/liner interface friction triggering a liner/waste slope failure during the operation of the landfill. A seepage collection system could be utilized to collect and dispose seeping water.

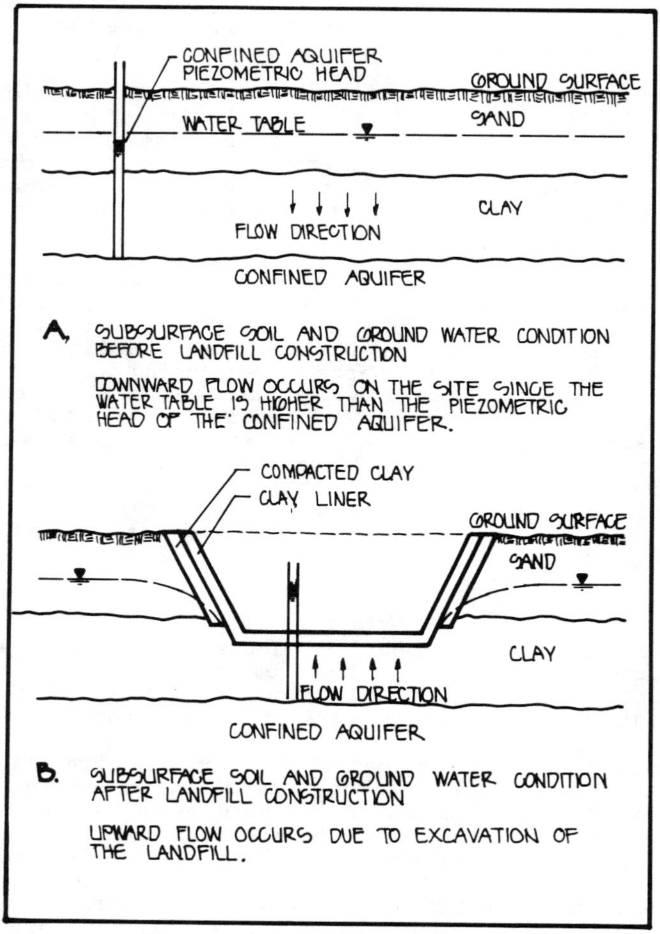

Figure (1) Example On Ground Water Condition Before and After Landfill Construction

In areas where the water table is higher than the proposed cell floor the excavation will increase the seepage force. The value of this force is dependent on the height and the inclination of side slopes. A cut-off trench could be utilized to reduce these forces.

Subsurface Soil Condition:

The subsurface soil condition is the most important factor in the design of slopes. Some consideration related to the stability of landfill slopes will be discussed here.

In landfills, it is common to excavate cut slopes in extensive clay deposits. In this case the occurrence of water bearing sand seams, pockets, or layers should be investigated. Such occurrence may cause slope instability during construction. This condition is more common in sites where artisan condition occurs.

In some cases the landfill design involves the construction of a berm around the landfill. If the berm is constructed using cohesive material of high undrained shear strength and the foundation soil is soft; a reduced fill shear strength should be used in the slope stability analyses. The explanation is that shear deformation in the soft foundation may cause tensile stresses in horizontal direction in the more rigid fill resulting in tension cracks that may trigger slope instability. This phenomena is called stress-strain incompatibility. An example of the stress strain curve is shown on Figure (2). Interested readers are referred to Chirapuntu and Duncan, (1976).

Also in the case of the placement of waste on soft subsoils the stability of waste/soil slopes should be investigated. In this case the filling of a landfill is equivalent to the construction of an embankment above soft foundation and a waste/soil slope failure is possible, Figure (3). Erdogan et el.(1986) investigated a case where a 1 vertical to 7 horizontal waste slope was required to provide adequate stability.

A landfill occupies a large area where different soil conditions may occur. localized locations with low shear strength parameters should be investigated and generalized conditions may not be applicable to the whole site.

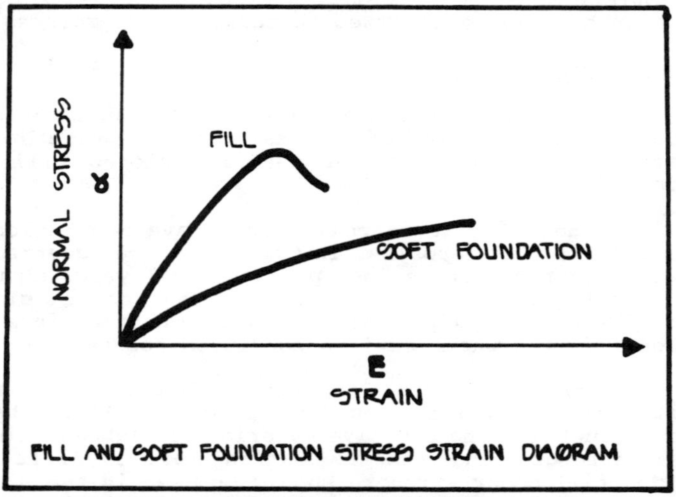

Figure (2) Example Of Stress Strain Incompatability.

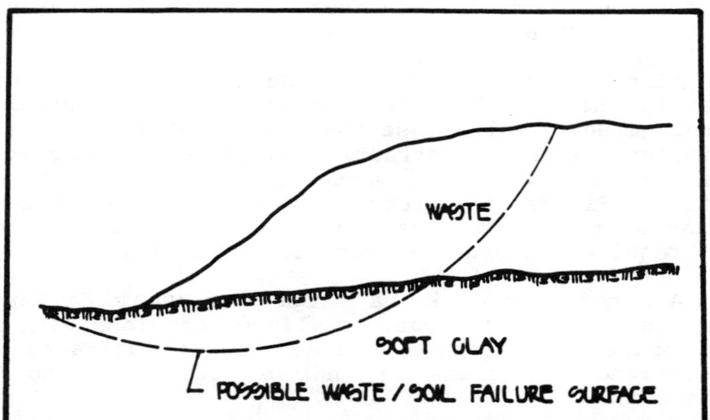

Figure (3) Example Of Slope Stability Problems For Landfills Constructed On Soft Subsoils.

Construction Sequence:

The construction activities and sequence should be studied carefully. An intermediate loading condition or geometry could be the most critical case. Three examples are discussed here.

A landfill is composed of two large cells. Due to the use of air structure to cover the landfill it will be constructed in two phases. In the first phase, as shown in Figure (4), cell I will be excavated and filled. In the second phase cell II will be excavated and filled. It can be noticed that an intermediate condition could be critical for cell I/cell II slope after the excavation of cell II. This is due to the change in slope geometry and geology.

Landfill construction activities may include digging excavations at the cell floor to install collection pipes and/or sumps (i.e. leak detection sumps). The effect of these excavations on the stability of the side slopes should be carefully investigated. Unsupported excavations may cause high local strains that may cause reduction in soil shear strength a distance from the excavation. Such reduction may cause slope instability.

Many landfills utilize composite double liner systems. In this case a landfill is first excavated to the secondary liner elevation and then a primary liner system is constructed. The secondary and the primary liner slope configuration should be included in the analyses.

During the construction of a landfill it is very common to encounter unsuitable material that has to be excavated and replaced for regulatory purposes. Examples of unsuitable material are sands, concrete and bricks. During the removal and replacement of these materials a critical slope configurations may occur. It is very important that the design engineer follow the construction sequence of the landfill and account for all intermediate conditions in his design.

Adjacent Sites Conditions and History:

The conditions of sites surrounding the landfill should be evaluated. In many cases a proposed landfill is located next to an old or an active landfill or could be an extension for an existing landfill. The

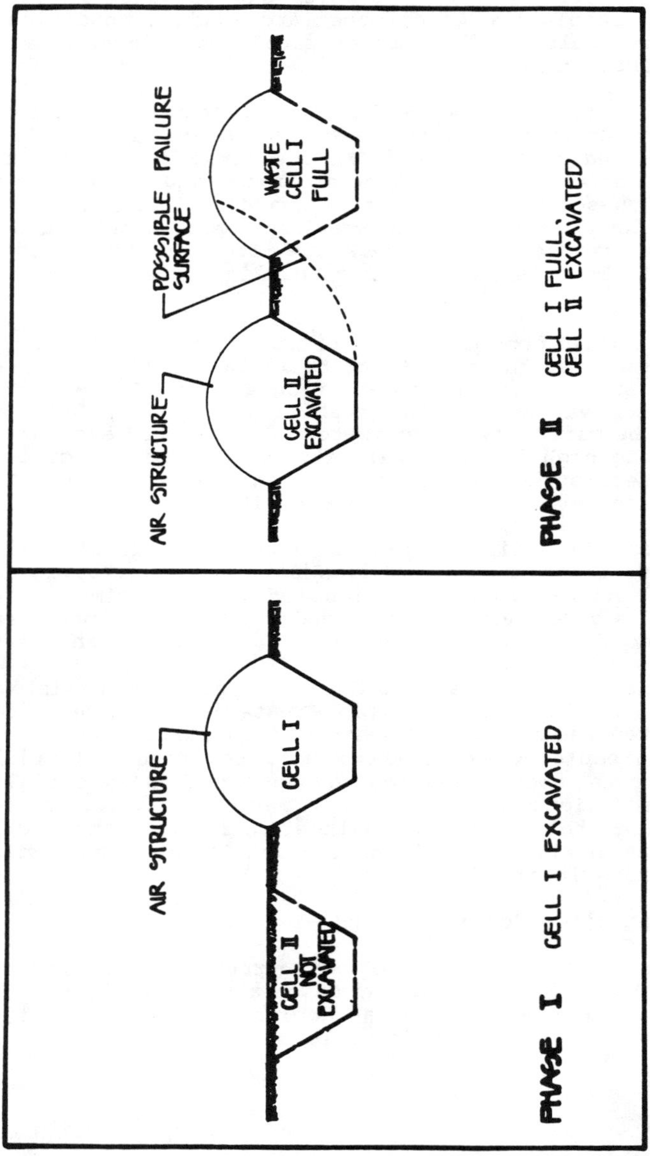

Figure (4) Example On Slope Stability Cases For Landfills Constrcuted In Two Stages.

slopes of the proposed landfill should be designed accounting for additional driving forces that could be introduced by these landfills. The proper waste elevation of an active landfill should be used in the analyses. An old landfill may not have a leachate collection system. This may cause high pressures against the proposed slopes.

The history of the adjacent sites could be very important for the stability analyses. Information about the performance of nearby slopes or structure may provide valuable information for the design engineer. Information about slope movements in adjacent sites should be evaluated very carefully.

General Topography:

The topography of the site should be studied carefully. The stability of existing natural slopes after excavation and after filling the landfill should be investigated. As shown on Figure(5.a) the occurrence of soft layer could cause the failure of a natural slope after the excavation of the landfill. Also as shown in Figure (5.b) filling of the landfill may increase the driving forces causing slope instability.

External Loads:

External loads increase the driving forces reducing the safety factor against sliding. Different external loads may need to be included in the analyses of landfill slopes. Examples are equipment loading and stock pile of excavated soils that may be utilized later in the construction of the landfill. The effect of external loads is more significant for shallow slopes and long term stability.

Liner Geometry and Configuration:

The stability of the liner system plays an important rule in the design of landfill slopes. The inclination and height of a side slope may be determined by the stability of the liner system. In general a liner system is composed of different layers of soils and geo-synthetics. The interface between these layers is a possible sliding surface,Figure(6). The stability of the liner system should be evaluated using representative friction parameters. These parameters are dependent on the expected overburden

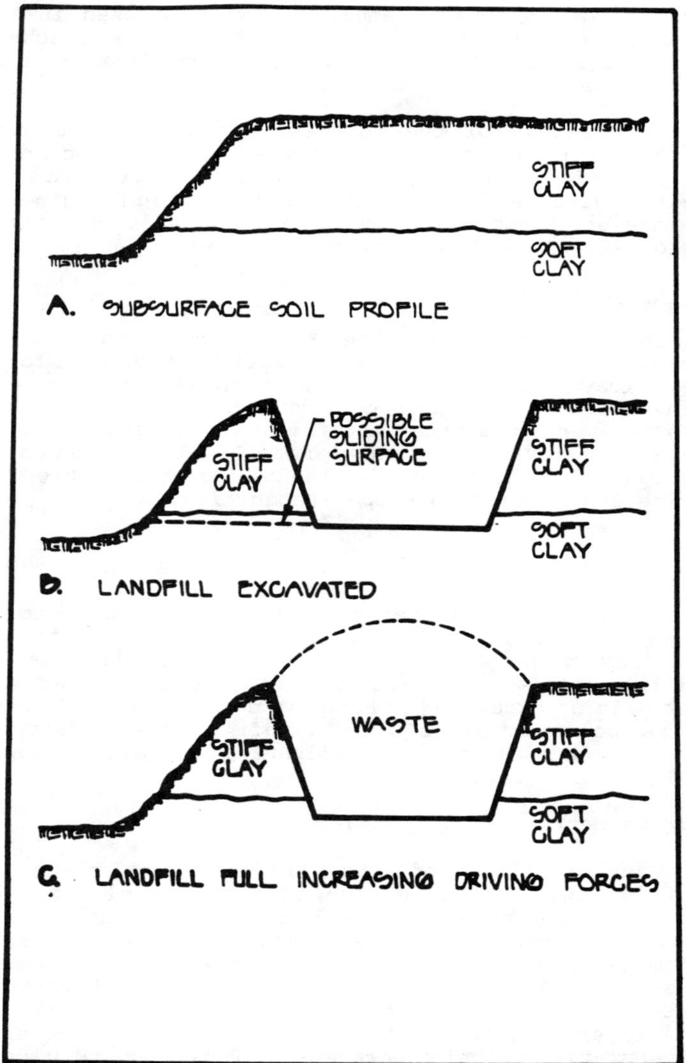

Figure (5) Examples On The Stability Of Natural Slopes Adjacent To Landfills.

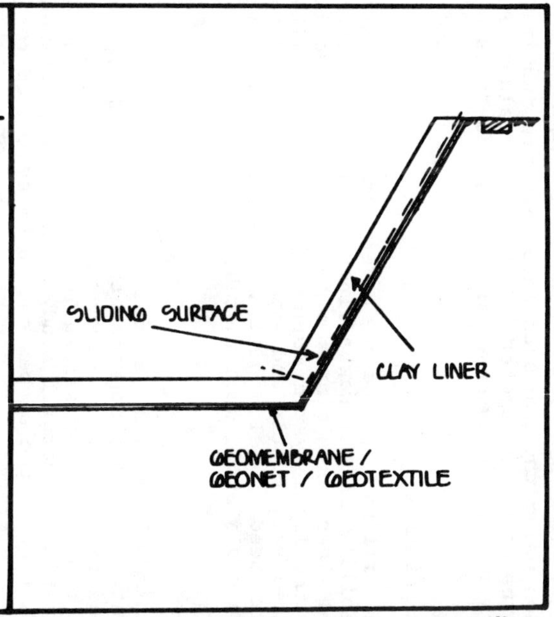

Figure (6) Example On Liner Instability.

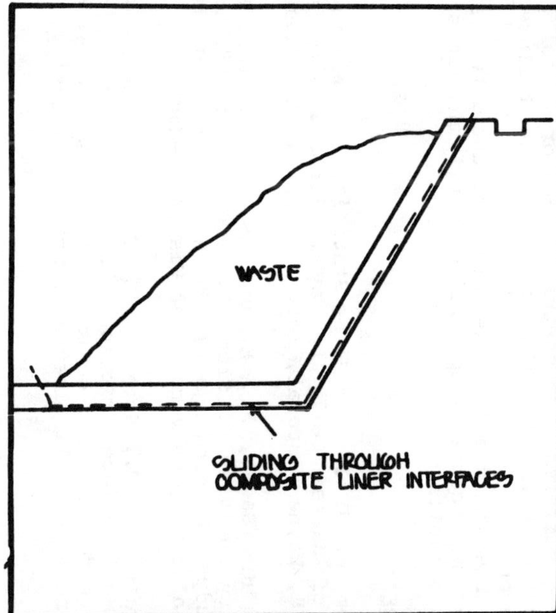

Figure (7) Example On Instability Problems Due To Filling Sequence.

pressure, type of interface, type of materials used for the construction of the liner system, surface roughness, foundation soil loading condition and other factors. Laboratory testings that model the proposed liner condition using the proposed materials are required to obtain the design parameters. For typical values, refer to the work done by Seed et el. (1988) and Koutsourais et el. (1990).

Filling Sequence:

The filling sequence has a very important effect on the stability of the waste/liner slopes. Constructing high and steep waste slopes may cause slope instability thought the liner system,Figure(7). Such instability is considered very costly and expensive. For more details refer to Seed et el. (1988). Well planned filling sequence could avoid such instability.

Conclusion:

The design of landfill slopes is a very complex and important task. Some considerations in the design were discussed in this paper. However, every site and design has its own characteristics. The geotechnical engineer should perform a careful study of all design elements. The stability of the slopes should be evaluated for all possible different geometries and/or geologies throughout the life of the landfill. The failure to evaluate one
case or the failure to take into account one design parameter may cause slope failure and the loss of millions of dollars.

REFERENCES:

Bergstrom, W., R. 1990. " Slope Stability Investigation, Cell C of Master Cell VI, Wayne Disposal Site # 2 Landfill", Wayne Disposal, Inc., Ypsilanti, Michigan.

Erdogan, H., Sadat, M., M. and Hsieh, N., N. 1986. "Stability Analyses Of Slope Failures In landfills: A Case Study", Proceedings of the Ninth Annual Madison Waste Conference, University Of Wisconsin-Madison.

Koutsourais, M., M., Sprague, C., J., and Pucetas, R., C. 1990. "Interface Friction Study Of Cap and Liner Components for Landfill Design", Proceedings Of The 4th GRI Seminar: Landfill Closures", Drexel University,

Philadelphia, PA 19104.

Seed, R.,B., Mitchell, J., K. and Seed , H., B. 1988. "Slope Stability Failure Investigation: Landfill Unit B-19,Phase I-a, Kettleman Hills,California" Report NO. UCB/GT/88-01, University Of California-Berkeley.

Suphon, C.and Duncan, J. , M. "Role Of Fill Strength In the Stability Of Embankments On soft Clay Foundations",Office Of Research Services, University Of California-Berkeley.

Deformation of Fill Slopes Caused by Wetting

Iraj Noorany[1], Member, ASCE, Joel A. Sweet[2],
and Ian M. Smith[3], Member, ASCE

Abstract

Post-construction wetting of compacted fills causes ground heave or settlement, and lateral deformation of slopes. This paper presents a method for estimating the amount of fill deformation as a function of soil properties, fill geometry and the depth of wetting. Swell/compression soil properties are measured on anisotropically loaded triaxial samples compacted to the original fill moisture-density conditions and subjected to wetting. The swell/compression strains thus determined are then used as input in a finite-element analysis to compute the effect of wetting on the fill geometry. Laboratory test results and analysis are presented for a fill slope in a residential development in San Diego, California. The analytical model presented can be used for confined fills as well as fill slopes.

Introduction

This study was made to investigate the effects of post-construction wetting on compacted fill slopes, and was prompted by growing concern regarding heave/settlement and lateral deformation of compacted fill slopes in many residential and commercial developments in California. Lateral slope deformation is initiated by deep wetting caused by a combination of precipitation and excessive irrigation, and is evidenced by the opening and gradual widening of joints of patio slabs and walkways, separation of slabs from foundations, development of cracks in floor slabs and foundations - with crack alignments being parallel to the top of the slope, - and sometimes tilt of building and patio slabs. The consequences of this type of fill deformation can vary from minor distress to major structural damage, depending on the soil type and compaction conditions, fill depth and geometry, slope height and inclination, clear distance of building and appurtenant structures from the slope, and the extent of wetting.

[1]Prof. of Civ. Engrg., San Diego State Univ., San Diego CA 92182, and Visiting Prof., Univ. of California, San Diego

[2]Consulting Engineer, P.O. Box 1117, Del Mar CA 92014

[3]Prof. of Geotech. Engrg., Univ. of Manchester, Manchester M13 9PL UK

Deformation Mechanism and Method of Analysis

Structural fills are typically placed in 6- to 8-inch layers and compacted with heavy equipment. The immediate product, a well-compacted fill, is strong and compresses very little as additional layers of fill are built up. This during-construction deformation is inconsequential because the fill is completed to the final elevation designated on the construction plans. The subsequent post-construction wetting of the fill causes swelling in the near-surface zone, and compression in the deeper zone of the fill; the cumulative effect of swell or compression of different layers is manifested as ground surface heave or settlement (Justo et al., 1984; Nwabuokei and Lovell, 1986; Lawton et al., 1989; Brandon, et al., 1990; Noorany and Stanley, 1990.)

In an infinitely large fill of uniform thickness subject to uniform wetting, the ground surface heave or compression will be uniform. However, in a non-uniform fill or slope, the stress conditions are not geostatic, and the swell/compression strains vary widely within the fill; the cumulative effect of swell or compression of different elements in this case is manifested as ground heave or settlement as well as lateral deformation of the ground in the direction of the slope.

In a typical slope, the amount of volume change (swell or compression) caused by wetting of different soil elements at different locations depends on the following factors:

a. The initial as-compacted moisture content and dry density (or relative compaction, RC); these parameters can be determined from data obtained during placement of the fill;

b. The anisotropic stress condition acting on the element immediately prior to wetting; this is difficult to determine accurately because of the locked-in residual prestress caused by the foot pressure of the compaction equipment (Nwabuokei and Lovell, 1986; Duncan and Seed, 1986; Duncan et al., 1991). Because of this compaction prestress, compacted soil behaves like an overconsolidated material; horizontal and shear stresses at any depth are not accurately known, whereas vertical stress can be computed from soil unit weight and additional stresses imposed by any structure on the ground surface. The degree of prestress is high at the top of the slope and decreases with depth as vertical stress increases; thus, the ratio of lateral stress to vertical stress, K, also decreases with depth.

The combination of factors described in (a) and (b) above, as well as restraints offered by adjacent elements, control the amount of vertical and lateral strains for each element upon wetting. Deformation of the ground surface and slope can be calculated from a finite-element analysis.

Effective stress analysis of this problem requires knowledge of both the as-compacted pore-water and pore-air stresses in an unsaturated soil, and the changes in these stresses caused by wetting (Maswoswe, 1985; Schreiner, 1988). Solutions in terms of effective stresses have been suggested by Fredlund (1985); Lloret et al. (1986); and de Miranda (1988).

Because of the difficulties in measuring or predicting pore-air and pore-water stress in a compacted fill, a total stress model has been used in this study. For this model, a laboratory test program was carried out to measure vertical and lateral strains caused by wetting of anisotropically loaded compacted soil samples in triaxial tests. The swell/compression strains applicable to various fill depths were then input in a finite-element analysis in order to compute the post-wetting geometry of the slope. The analysis was applied to a fill slope that had undergone significant lateral deformation as a result of post-construction wetting.

Soil Properties

Laboratory tests were performed on two soils from the site of a residential development (Fairway Vistas) in the Rancho Bernardo area in San Diego, California. The original mass grading for this project was done in 1980, with some additional grading in 1984. Forensic explorations were made in 1990, after distress features indicative of possible lateral deformation of slopes were noticed. Three test borings and several test pits were excavated, and the subsurface soil conditions were examined by sampling as well as downhole logging by geotechnical engineers in a large-diameter boring. The fill consisted primarily of two types of soils, a clayey sand (soil FV-1) and a sandy clay (soil FV-2). The results of classification and compaction (ASTM D1557) tests for each soil, an average of three series of tests, are summarized in Table 1:

Table 1 – Soil Properties

	Soil FV-1	Soil FV-2
Percent gravel (+4)	15	10
Percent fines (-200)	48	56
Liquid limit	40	37
Plasticity index	17	16
Specific gravity, G_s	2.68	2.69
Unified classification	SC	CL
Maximum dry density, pcf	116.0	114.2
Optimum water content, %	14.3	16.0

Although sample FV-1 classified as clayey sand (SC) and FV-2 as clay of low plasticity (CL), this difference in classification was due only to the fact that FV-1 had 5% more gravel than FV-2. Classification of the fraction passing the No. 4 sieve for both soils was CL, with FV-1 having a higher liquid limit and a higher plasticity index than FV-2. Consequently, because the swell/compression tests were performed on the fraction passing the No. 4 sieve, FV-1 showed somewhat higher expansion than FV-2.

Triaxial Tests

Specimens for triaxial swell/compression tests were compacted in a 1.4-inch diameter by 3.0-in. high 4-way split mold. Soil density was controlled by using a tamper with a Harvard miniature compacter foot, and compacting each specimen in ten equal layers. The weight, height (H_o) and diameter (D_o) of each specimen were measured after it was taken out of the split mold. The compacted specimen was then assembled in a triaxial cell using dry porous stones, dry filter strips and two thin membranes.

Geonor triaxial cells with Geonor anisotropic loading accessories were used. The system compliance of each triaxial cell for various anisotropic stress conditions was measured in advance and used in the data analysis. The amount of system compliance correction was significant, varied slightly from cell to cell, and depended on the cell pressure.

After assembling the compacted sample in the triaxial cell, radial and axial stresses were applied in small increments, keeping the ratio of total radial stress to axial stress (K) a constant value, until the target stress condition was established. The sample was then allowed to reach equilibrium before water was introduced.

The elastic axial and radial deformations during this initial loading were measured and the new sample height (H_1) and diameter (D_1) were calculated for use in computing the subsequent swell/compression strains caused by wetting. The elastic change in specimen height under applied axial stress before wetting was measured directly with dial gauge; the change in diameter was calculated, based on both isotropic and deviatoric components of stress and Poisson's ratio, and checked against direct measurements on a number of specimens that were subjected to load-unload cycles and taken out of the cell for direct height and diameter measurements.

Water was introduced to triaxial specimens after de-airing the drainage line; the sample gradually absorbed water from an open reservoir with no back pressure, as the object was soaking by percolation (as in the field), not full saturation by backpressuring. The volume of water absorbed by the sample was continuously monitored. The process of soaking typically took several days until the sample reached equilibrium and no more water was absorbed. At this point, all samples were nearly saturated, with degrees of saturation typically being higher than 95%.

It was not feasible to measure the change in the sample diameter by monitoring the volume of water in the triaxial chamber because Geonor cells are designed to operate with oil on top of water, and oil leaks slowly and continuously around the loading piston. Also, four units of electrical lateral clip strain gauges were used, but turned out to have unstable readings over time. Thus, because the only important value of the diameter after soaking was the final diameter, it was measured directly at the end of the test when the sample was taken out of the triaxial cell. Correction was made for the small undrained elastic rebound that occurred as the axial load and the cell pressure were removed before the sample was taken out of the cell. This was done by measuring the axial rebound strain directly in two steps, for the deviatoric component of the stress removal and the isotropic component of the stress removal. Using these data and the soil Poisson's ratio after wetting, the amount of radial elastic rebound was calculated, and subtracted from the final diameter measured (D_3) to determine the final diameter inside the cell (D_2). The average of three diameter measurements, at mid-height and at quarter points, was used.

All triaxial swell/compression tests were run by wetting-after-loading procedure, i.e., each sample was wetted only under one set of anisotropic stresses representing one single point in the fill. This necessitated testing a large number of samples, but was considered to be the only reliable way of duplicating the post-construction wetting process that takes place in the field. Trial runs showed that multiple load tests (loading-after-wetting) (Lawton et al., 1991) did not produce the same results as wetting-after-loading tests.

Results of Triaxial Swell/Compression Tests

Five series of tests were run on soil FV-1, and seven series on soil FV-2. A typical set of results is plotted in Figure 1. In this figure, the horizontal axes show the major (vertical) stress under which soaking took place, and the vertical axes show the corresponding vertical and radial swell/compression strains that resulted from soaking. Three different K values of 0.5, 0.7 and 1.0 (ratio of radial total stress to vertical total stress) were used to allow interpolation for other values in that range.

The test results in Figure 1 are for samples compacted at a water content of 15% and a relative compaction of 90%. Additional tests were run for samples compacted at a water content of 15% and a relative compaction of 95%; also at a water content of 16.6% and a relative compaction of 93.5%. The results provided adequate data for application to field conditions for soil FV-1. Similar tests, in seven series at water contents varying from 16.0% to 18.1% and relative compactions varying from 90% to 95%, were performed on soil FV-2.

The data from Figure 1 have been replotted in Figure 2 depicting volumetric strain versus the mean principal stress. It can be seen from Figure 2 that the relationship between mean principal stress and volumetric strain caused by wetting seems to be independent of the principal stress ratio, K. This indicates that material anisotropy, if any, did not have a significant influence on the swell/compression behavior of soil FV-1. As can be seen in Figure 3, the same was true for soil FV-2.

Finite-Element Analysis of Slope Deformation Caused by Wetting

The swell/compression strains measured in triaxial tests were used in a finite-element analysis to compute slope deformation caused by wetting. The finite element program for swell/hydrocompression (SWELL) was developed as an extension of a previously developed finite element program, SATURN (Sweet, 1978), and is described in detail by Sweet (1991). SWELL is based on a material model that simulates the swell/compression phenomenon by the following incremental constitutive relations:

$$d\sigma_x = (B - 2/3\,G)\,(d\varepsilon_z - de_z - de_y) + (B + 4/3\,G)\,(d\varepsilon_x - de_x) \quad (1)$$

$$d\sigma_z = (B - 2/3\,G)\,(d\varepsilon_x - de_x - de_y) + (B + 4/3\,G)\,(d\varepsilon_z - de_z) \quad (2)$$

$$d\sigma_{xz} = 2\,G\,d\varepsilon_{xz} \quad (3)$$

where
σ_{ij} = stress tensor
ε_{ij} = strain tensor
e_i = swell/compression strain components
B = bulk modulus
G = shear modulus

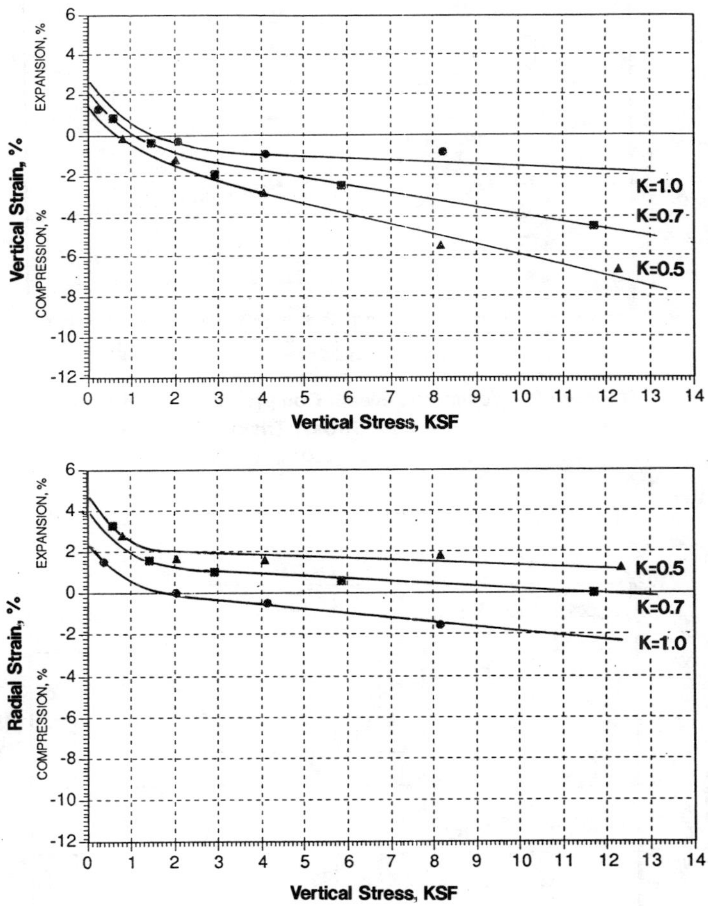

Figure 1 - Results of Triaxial Swell/Compression Tests on Soil FV-1, w=15.0%, RC=90%

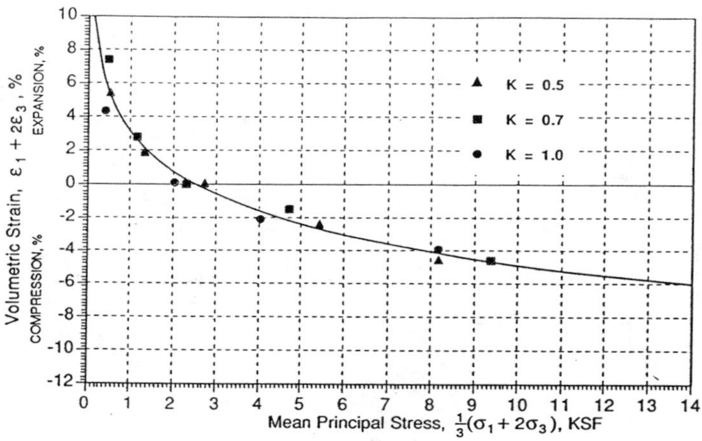

Figure 2 - Volumetric Swell/Compression Strain Versus Mean Principal Stress, Triaxial Tests on Soil FV-1, w=15.0%, RC=90%

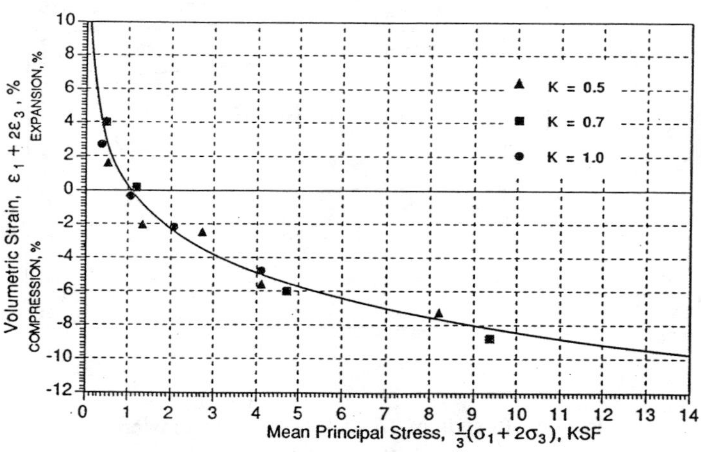

Figure 3 - Volumetric Swell/Compression Strain Versus Mean Principal Stress, Triaxial Tests on Soil FV-2, w=16.0%, RC=90%

A plane-strain condition is assumed for the slope, with the x-coordinate in the lateral direction, the z-coordinate in the vertical direction, and the y-coordinate perpendicular to the x-y plane. For this plane-strain condition, total strain in the y-direction is zero, but swell/compression strain in this direction is not zero.

The two-step coupling of the gravitation stress field with swell/compression strains is similar to the solution proposed by Fredlund (1985) and used by de Miranda (1988), except that it uses a total stress approach utilizing swell/compression data as a function of in-situ vertical and horizontal total stresses.

The first step in SWELL's solution is to estimate the in-situ gravitational state of stress using the as-compacted soil density, bulk modulus and shear modulus. As mentioned earlier, because of the residual compaction prestress, lateral stresses cannot be estimated accurately even though the vertical stresses can be calculated with confidence. In SWELL, only the vertical stress component is used directly; the lateral components are taken to be K-times the vertical component, with K varying from top to bottom of the soil profile as a function of depth and estimated compaction prestress.

The estimated gravitational stresses are then used as initial condition for various elements in the finite-element mesh to calculate the swell/compression strains caused by wetting. The wetting region of the soil profile is prescribed and swell/compression calculations are made for that region. In SWELL, calculations are made by a computer in an incremental fashion using equations 1, 2 and 3, and by an iterative process a compatible solution is completed.

The soil swell/compression properties are input as a function of in-situ vertical stress and lateral pressure coefficient, K, in the form of families of curves similar to those in Figure 2. Of course, in terms of moisture content, density and degree of compaction, these data should represent the as-built condition of the fill. Because the solution is dependent on the estimated values of the in-situ lateral pressure coefficient, K, in a typical problem, a parametric study can be made by multiple runs with different combinations of K, and a range of reasonable computed values could be selected based on the judgment of the engineer and the available data regarding actual deformations in the field.

Example of Application

Figure 4 represents a simplified soil profile for a 50-ft high slope at Fairway Vistas residential development in San Diego, California. The five soil layers shown in the profile consist of two types of soil, FV-1 and FV-2, with different placement conditions. The as-built conditions of moisture content, dry density and relative compaction (RC) applicable to each layer were determined from the grading reports for the project (San Diego Soils Engineering, 1981) and are summarized in Table 2.

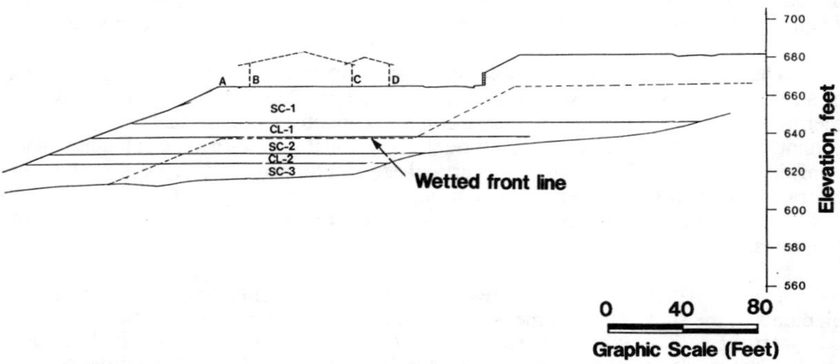

Figure 4 - Simplified Soil Profile Used in Finite-Element Analysis

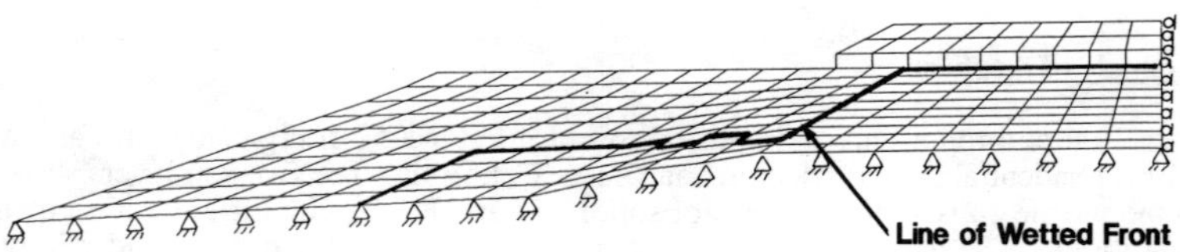

Figure 5 - Finite-Element Mesh and Location of Line of Wetted Front

Table 2 – As-Compacted Conditions for Fairway Vistas Slope

Soil Layer	As-compacted Moist.Content %	As-compacted Dry Density pcf	As-compacted Unit Weight pcf	Relative Compaction %
FV-1a	14.6	108.7	124.6	92.5
FV-2a	16.6	107.0	124.8	93.4
FV-1b	16.7	106.4	124.2	93.5
FV-2b	17.6	106.3	125.0	94.8
FV-1c	20.0	103.4	124.1	93.0

The swell/compression design curves for these five soil layers were developed by interpolation from the results of twelve series of triaxial tests previously mentioned for soils FV-1 and FV-2. In developing the design curves, allowance was made for the presence of 15% gravel in soil FV-1 and 10% gravel in soil FV-2.

The line of wetted front delineating the soaked and unsoaked regions was selected based on the moisture content and degree of saturation data obtained from the forensic subsurface investigations in December 1990 (Shepardson Engineering Associates, Inc., 1991). The finite element mesh and the position of this line are shown in Figure 5.

Computations were made using the soil input data, but varying the as-built lateral pressure coefficient, K from 1.0 at the top of the soil profile to 0.7 at the bottom. Also, runs were made with and without a building uniform stress of 150 psf, and for the present condition of line of wetted front as well as complete wetted condition.

A typical result of the finite element analysis in graphical form is shown in Figure 6; the deformations in this figure have a magnification factor of 10. The vertical and horizontal deformations computed at several key points on the ground surface, at locations B, C and D in Figure 4, are listed below in Table 3.

Table 3 – Computed Ground Surface Deformations

Bldg. Stress	Wetting Condition	Vertical Deformation Inches			Relative Horizontal Deformation Inches		
psf		B	C	D	B-A	C-B	D-B
0	Partial	6.5	5.7	5.7	-1.6	-0.7	-1.5
0	Complete	6.3	6.2	6.4	-1.5	-0.5	-1.2
150	Partial	6.3	5.1	5.3	-1.6	-0.3	-0.9
150	Complete	6.1	5.6	5.9	-1.5	-0.2	-0.5

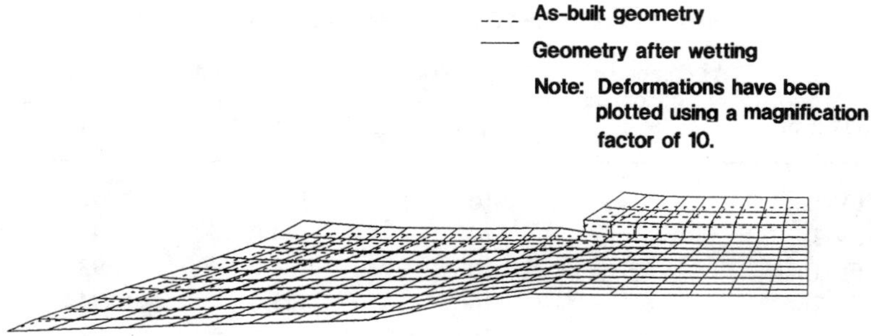

Figure 6 - Typical Pattern of Deformation Caused by Wetting

In Table 3, all positive values of vertical deformation indicate ground heave. The current condition of wetting (as of December 1990) is referred to as "partial" wetting condition; "complete" wetting refers to a situation where the rest of the fill is soaked too.

Because of a relatively uniform fill thickness under the building area in this case (Figure 4), the computed differential heave in the building area (points B and C) is only 1.2 inches for the current condition of wetting, i.e. a 1.2-inch tilt in the direction of the slope was computed. For complete wetting condition, a slight (0.2 inch) reduction in ground heave was computed for point B. This will occur because of additional settlement of the deep layers when they become wet. In contrast, on the other side of the building (point C), an additional heave of 0.5 inches will occur because in this area the fill is both thinner and more confined. This will actually reduce the current slab tilt of 1.2 inches (elevation difference between B and C) to 0.5 inches.

With respect to horizontal deformations, the minus sign indicates a deformation in the direction of the slope. The differential horizontal deformations indicate stretching between B and A (a distance of 16.2 feet in the patio), C and B (a distance of 55 feet across the building), and D and B (a distance of 75 feet across the building including the garage). It is further noted that with respect to lateral deformation, additional wetting does not seem to worsen the present conditions for this particular case.

Although the computed values are reported to 0.1 inch in Table 3, this is not to imply that such a fine degree of accuracy can be expected of the analysis. The trends of deformation, however, are well indicative of the actual field conditions. Although there was not sufficient field data to verify the estimated total ground heave values, there was good agreement for differential vertical and lateral deformations. Floor-level surveys of the building slabs for several units in the vicinity of the analyzed cross-section showed tilts toward the slope between 0.7 inch and 1.3 inch. The amount of lateral stretching toward the slope in the building area, as evidenced by cumulative widths of cracks in the slab parallel to the slope, was between 0.4 inch and 0.9 inch, well in the range of relative horizontal deformation between D and B for partial wetting in Table 3. The measured lateral stretching in the joints of the hardscaping in the rear patios was about 0.7 to 1.1 inch, which is indicative of part of the actual total horizontal displacement between B and A, and is in good agreement with the estimated values in Table 3.

Summary and Conclusions

Post-construction wetting by precipitation and heavy irrigation can cause heave or settlement of compacted fills, and lateral deformation of fill slopes. The response of a compacted fill slope to wetting can be modeled by anisotropically loaded triaxial tests performed on samples compacted to the initial fill placement conditions and then wetted. The measured vertical and radial strains caused by wetting can be used as input parameters in a finite-element analysis of slope to estimate long-term heave, settlement, and lateral deformation as the wetted front advances through the fill. The analytical method proposed in this paper seems to provide a satisfactory solution for a very complex problem, and can be used as a design tool to estimate the long-term behavior of confined structural fills, slopes and embankments. Field instruments for long-term monitoring of heave/settlement and lateral deformation of fills should be installed at every opportunity to provide engineers with data to further confirm the applicability of this procedure.

Acknowledgments

The laboratory tests were carried out at San Diego State University with the aid of graduate assistants Steve Fitzwilliams and Tan Doan. Financial support for this project was provided by Benton Engineering Inc., Buchanan-Rahilly Inc., Geocon Inc., Group Delta Consultants Inc., San Diego Soils Engineering, Moore & Taber, Shepardson Engineering Associates Inc., and Sukut Construction. Their support is gratefully acknowledged.

Appendix I – References

Brandon, T.L., Duncan, J.M. and Gardner, W.S. (1990). "Hydrocompression Settlement of Deep Fills." *Journal of Geotechnical Engineering*, ASCE, 116 (10) 1536-1548.

Duncan, J.M. and Seed R.B. (1986). "Compaction-Induced Earth Pressures under K_0-Condition." *Journal of Geotechnical Engineering,* ASCE, 112 (1) 1-22.

Duncan, J.M., Williams, G.W., Sehn, A.L. and Seed, R.B. (1991). "Estimation Earth Pressures Due to Compaction." *Journal of Geotechnical Engineering,* ASCE, 117 (12) 1833-1847.

Fredlund, D.G. (1985). "Soil Mechanics Principles that Embrace Unsaturated Soils." *Proceed., 11th International Conference on Soil Mechanics and Foundation Engineering,* San Francisco, 2, 465-472.

de Miranda, A.N. (1988). "Behavior of Small Earth Dams During Initial Wetting." Ph.D. Dissertation, Department of Civil Engineering, Colorado State University, Fort Collins, Colorado.

Justo, J.L., Delgado, A. and Luiz, J. (1984). "The Influence of Stress-path in the Collapse-Swelling of Soils at the Laboratory." *Proceed. of the Fifth International Conference on Expansive Soils*, Adelaide, South Australia, 21-23 May 1984, 67-71.

Lawton, E.C., Fragaszy, R.J. and Hardcastle, J.H. (1989). "Collapse of Compacted Clayey Sand." *Journal of Geotechnical Engineering,* ASCE, 115 (9) 1252-1267.

Lawton, E.C., Fragaszy, R.J. and Hardcastle, J.H. (1991). "Stress Ratio Effects on Collapse of Compacted Clayey Sand." *Journal of Geotechnical Engineering,* ASCE, 117 (5) 714-730.

Lloret, A., Alonso, E.E. and Gens, A. (1986). "Undrained Loading and Consolidation Analysis for Unsaturated Soils." *Proceed. of the European Conference on Numerical Methods in Geomechanics,* Stuttgart, Germany, Vol. II.

Maswoswe, J. (1985). "Stress Path for a Compacted Soil During Collapse Due to Wetting." Ph.D. thesis, Imperial College, London University.

Noorany, I. and Stanley, J. (1990). "Swell and Hydrocompression Behavior of Compacted Soils: Test Data." Department of Civil Engineering, San Diego State University, San Diego, California.

Nwabuokei, S.O. and Lovell, C.W. (1986). "Compressibility and Settlement of Compacted Fills. Consolidation of Soils: Testing and Evaluation." ASTM STP 892, R-N. Yong and F.C. Townsend, eds., Philadelphia, PA, 184-202.

San Diego Soils Engineering, Inc. (1981). Various reports of grading for Bernardo Heights, Rancho Bernardo, San Diego, California.

Schreiner, H.D. (1988). "Volume Change of Compacted Highly Plastic African Clays." Ph.D. thesis, Imperial College, London University.

Shepardson Engineering Associates, Inc. (1991). Report of Field and Laboratory Test Results, Fairway Vistas Fill Analysis. Rancho Bernardo, California." February 5, 1991.

Sweet, J.A. (1978). "SATURN – A Multi-Dimensional Two-Phase Computer Program Which Treats the Non-Linear Behavior of Continna Using the Finite Element Approach." Report No. JSA-79-016, Joel Sweet & Assoc., California.

Sweet, J.A. (1991). "SWELL – A Finite-Element Program that Predicts the Deformation and Stress State Resulting from the Swelling or Compression of Soils." Coast Analytics, Inc., California.

Appendix II – Conversion to SI Units

To convert	To	Multiply by
ft	m	0.3048
in.	cm	2.54
lb	kg	0.4536
psf	kPa	0.04788
pcf	Mg/m^3	0.016

LONG TERM BEHAVIOR OF URBAN FILL EMBANKMENTS

J. David Rogers[1], M. ASCE

Abstract

Some aspects of the author's experience with long-term performance of urban fill embankments in California are briefly summarized. Analysis of field performance through insitu monitoring and the comparison of recently-derived laboratory data with that taken at the time of fill placement are compared. These comparisons suggests that many of the design assumptions incorporated into the design of urban embankments were insufficient to prevent long-term problems with settlement, slope stability and seepage. Conclusions and recommendations for future urban embankment design are presented at the end of the article.

Introduction

Since the late 1950's, large urban earthwork projects have been constructed on the coastal hills of California's largest suburban areas, principally the San Diego, Los Angeles basin, and the San Francisco Bay areas. Initially modest in size, some of these projects now involve as much as 35 million cubic yards (26.8 million m^3). Most of this earthwork is accomplished to create level pads for hillside residential development.

At the time of their construction, no substantive baseline of experience existed from which empirical estimations of long-term behavior could have been made. However, in the 30+ years since these embankments were constructed, the geotechnical criteria governing their emplacement has essentially remained unchanged. Most estimates of "long-term" settlement have been limited to gross calculations of primary consolidation, based upon short-term, one dimensional consolidation tests.

[1]Principal, Rogers/Pacific, Inc., 396 Civic Drive, Pleasant Hill, California 94523-1921

Most cases involving differential settlement have occurred within 5 to 15 years of placement, following periods of extranormal precipitation. In most instances, it is the deeper sections of fill (more than 35 feet, or 10.67 m) that have consolidated, after initial saturation. This phenomena appears ascribable to hydrocompression, or the tendency of a soil to compress when wetted under moderate levels of confinement (Brandon et al. 1990). Other researchers have shown that long-term strength of compacted clayey soils may decrease substantially, due to repeated cycles of wetting and drying (Stark and Duncan, 1990).

In most cases, shallow groundwater was found to be "trapped" within embankments, fed by overwatering, broken utilities, underlying aquifers, or partially failed subdrain collection systems (Noorany 1990). Brandon et al. (1990) reported that domestic landscape watering accounted for an equivalent precipitation of 70 inches annually in an area naturally recieving only 8 inches a year.

The performance of the deeper residential embankments has been noticeably poor with respect to differential settlement/swell, subdrainage and slope stability. During the past decade, the author has participated in numerous forensic studies undertaken to evaluate the causes and mechanisms of earth movement within these fills. This article summarizes some of the collected data, taken over a wide geographical area.

Effects of Long-Term Saturation

Within a few months or years of placement, embankment soils will begin to swell or consolidate, depending on effective stress, soil plasticity, and sources of moisture to which the soils are exposed. Figure 1 presents a standard moisture-density relationship, or "compaction curve". Most soils are compacted between 75% and 85% of saturation in order to achieve optimum density for the least amount of compactive effort. With the passage of time, embankment soils will eventually become saturated. Upon saturation, soil mixtures containing clay can be expected to swell under low levels of confinement (Figure 2).

The amount of swell generated by the soil is a function of effective confinement and the moisture content at the time of initial placement (Seed and Chan, 1959). The most dramatic alterations appear to occur in clayey soils initially compacted dry of optimum moisture content, a common occurrence in late summer months or drought years.

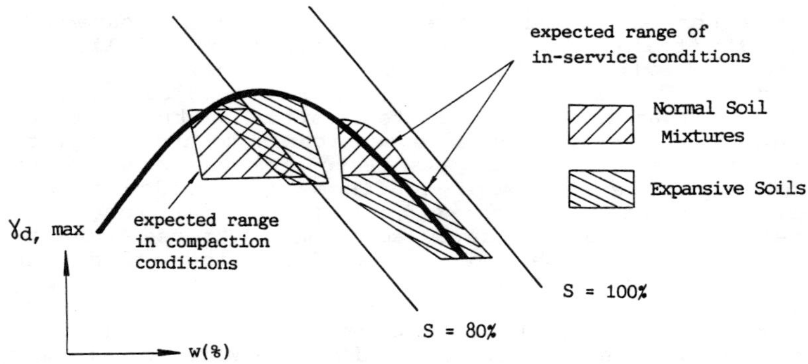

Figure 1 - *Comparison between dry density and water content at time of initial placement with in-service conditions.*

Figure 3 presents the laboratory-derived swell travel paths for San Ramon silty clay, compacted in accordance with ASTM D-1557-78 (Modified Proctor) at varying initial moisture levels, and allowed to swell under low levels of confinement. Samples of this soil were found to swell as much as 50% when initially compacted to 90% of the Modified Proctor standard at moisture levels dry of optimum. The increased swelling tendency of clay mixtures compacted dry of optimum moisture has traditionally been ascribed to flocculation of the clay platelets (Seed, Woodward and Lundgren, 1962), as opposed to the dispersed condition achieved when the soil is compacted wet of optimum moisture content. Regardless of these physiochemical effects, the soils' ability to swell as additional water is introduced means that the bulk density of such material decreases as the mixture inevitably becomes saturated. For this reason, comparisons of present-day insitu density with similar determinations made at the time of construction cannot be strictly compared at low to moderate levels of confinement as a method to evaluate adequacy of the original compactive effort.

Clayey soils under moderate confinement (more than 35 feet, or 10.7 meters, of cover) will rapidly undergo primary consolidation (as the overburden load compresses the air voids and expels interstitial pore water). But, area experience also suggests that further settlement occurs after a cycle of saturation, likely due to hydrocompression (Brandon et al. 1990). In many instances, additional settlement has been observed in conjunction with a lowering of groundwater within such embankments (such as that experienced during periods of drought). Hydrocompression appears driven by initial saturation (Brandon et al. 1990), while secondary consolidation appears to be influenced by the hydrologic regimen acting within the embankment.

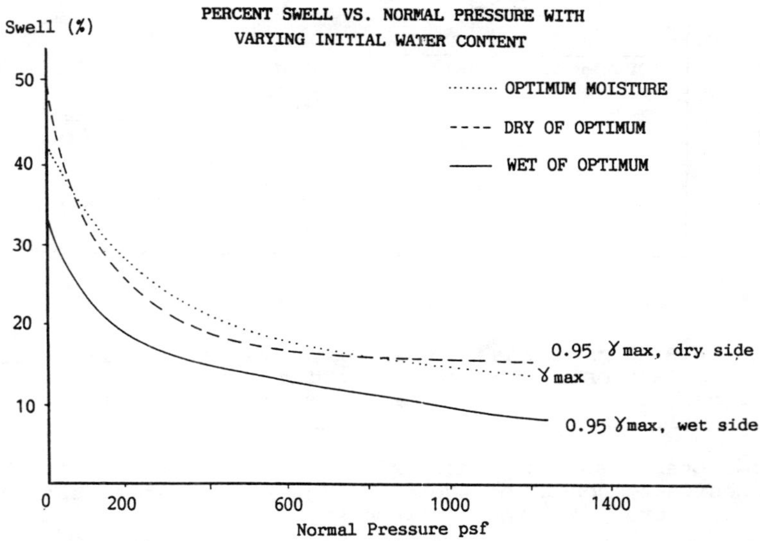

Figure 2 - *Variation of swell pressure with water content for sandy clayey silt compacted to 95% and 100% of the Modified Proctor (ASTM D-1557) standard. Note the variance in swell between samples compacted dry and wet of optimum moisture.*

Because embankments are compacted in layers of heterogeneous consistency, certain zones will become saturated and affected by these phenomena before other, nearby zones, thereby necessitating a broad range of predicted settlement values.

Long-Term Strength Loss

At lower levels of confinement (generally, less than 1 tsf, or 96 kPa), soil mixtures with any appreciable amount of clay can be expected to swell. Through such swelling, their density is decreased, with a corresponding effect on shear strength, as presented in Figure 4. Stark and Duncan (1990) presented similar data in their analysis of long-term strength degradation of the colluvial foundation soils beneath a portion of California's San Luis Dam that failed in 1981. Stark and Duncan's research revealed that native clay soils, which have be

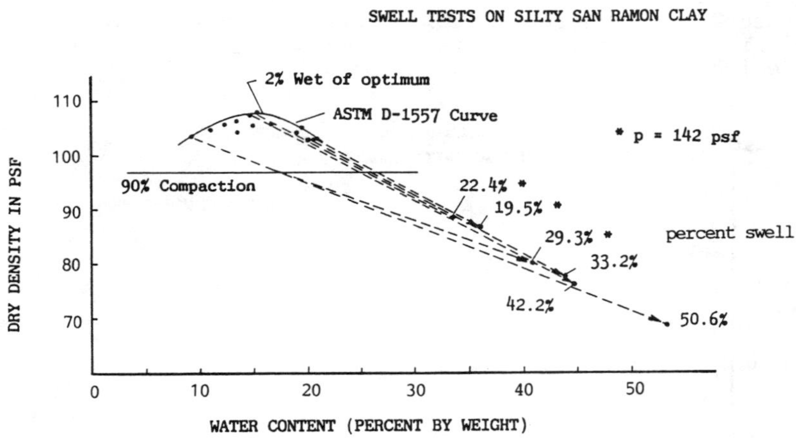

Figure 3 - *Swell characteristics of San Ramon silty clay as overlain on a standard compaction curve (ASTM D-1557). Although initially compacted to maximum achievable density, all samples swelled, some as much as 50%.*

come overconsolidated through desiccation, will lose the cohesive component of their strength upon initial saturation, thereby causing a marked strength loss.

This discovery was in contrast to the accepted premise that mechanically overconsolidated (compacted) clays will not lose their cohesive strength, even though some softening is expected, in proportion to the number of shrink-swell cycles the soil is subjected.

Secondary Consolidation Effects

For many years, geotechnical practitioners have recognized that measured settlements often exceed theoretical predictions based on primary consolidation. By the 1960's, Terzaghi and Peck (1967) concluded that a "secondary time effect" was likely due to the structural realignment of mineral grains. In the early 1970's, Mesri (1973) began to explore the various mechanisms of secondary consolidation, demonstrating its occurance under constant effective stress, well after the excess pore pressures associated with primary consolidation have dissipated. Depending on the size of an embankment, its degree of drainage, and the hydraulic regimen; secondary consolidation may not initiate for a number of years, or even decades, following placement. In subsequent studies, Mesri and Godlewski (1977) demonstrated that secondary

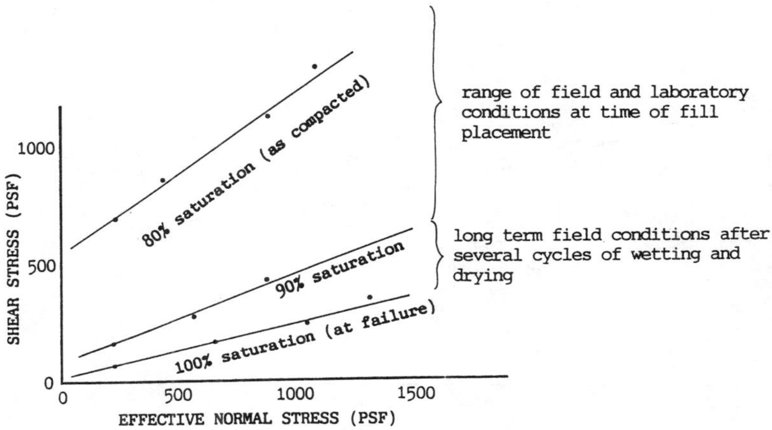

Figure 4 - *Comparison between direct shear test results on San Ramon silty clay for both initial and long-term conditions.*

consolidation index appears related to clay mineralogy, which is physiochemically tied to the natural water content of the clay minerals in the soil mixture.

An example of secondary consolidation test data in both sandy and clayey soils derived from the Pleistocene-age Merced formation are presented in Figure 5. Homes constructed upon the sandy soil mixture underwent marked amounts of differential settlement in the first decade after placement, following above-average rainfall in the late 1960's. Settlement since that time, however, appears to be occurring at a diminished rate, well below that experienced in the first decade.

Differential Settlement of Sloping Hillside Fills

One of the most troubling aspects of California's urban fills has been the employment of combination cut/fill pads (Figure 6). Level lot pads were popular with builders because they minimized construction costs. But, over the long term, such lots inevitably creep and settle, towards the sloping face. Early on, sidehill embankments were weakly keyed into the underlying strata, if at all. Such slopes were termed "sliver fills", because of their decreasing downslope thickness appears sliver-like in cross section. In "sliver fills", the

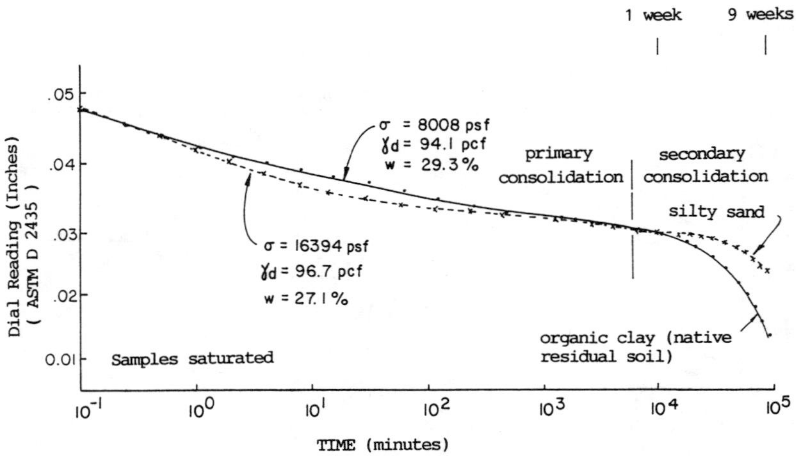

Figure 5 - *Time rate of settlement for sandy and clayey soils taken from the Merced formation. By extending the conventional consolidation test past a week, dramatic levels of secondary consolidation can be appreciated.*

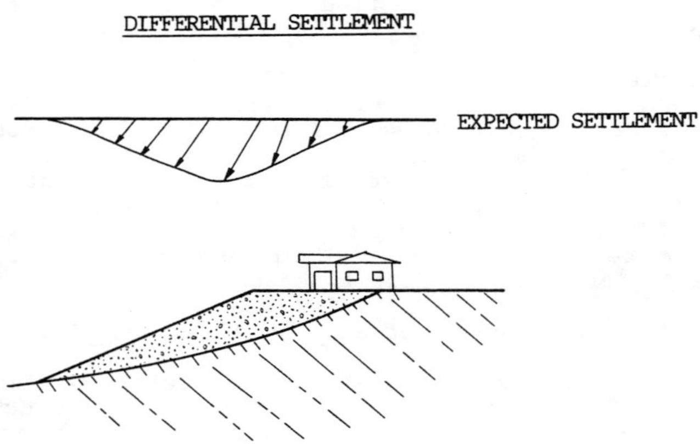

Figure 6 - *Section view through a cut/fill in an urban embankment. The horizontal component of settlement does the greatest damage to residential structures.*

brow of the embankment usually represents the thickest section of fill, and thereby, that area which can be expected to foster the maximum settlement (Figure 6).

If we consider the trajectories of effective stress within a transversely anisotropic clay fill (Figure 7, upper), the influence of the embankment's sloping face can be appreciated. In this case, the effects of fill layering and creep over a non-yielding foundation are modeled, incorporating capillary tension effects within the upper, non-saturated zone. As the embankment is allowed to progressively strain (to simulate time), the effective stress vectors become increasingly skewed in the horizontal plane. Embankment settlement will follow the trajectories of maximum effective stress, which in the example presented, are so inclined as to engender a significant horizontal component of motion (Figure 7, middle).

It is this small, but important, differential component of horizontal motion, which serves to pull apart the framing of a house and/or cause extension of brittle foundation elements. If the embankment is properly keyed into the underlying ground (Figure 7, lower), the same phenomena can occur, but at somewhat retarded levels.

Differential Settlement Due to Overconsolidation and Swell

Most geotechnical practitioners in California have assumed that desiccated native soils are suitably strong because of their overconsolidated nature (caused by their drying out). Based on such a premise, many engineers discounted future consolidation of these soils, if the effective overburden pressure was less than the preconsolidation pressure engendered by the desiccation. Pre-consolidation values range between 800 and 12,000 psf (38 to 573 KPa), depending on soil plasticity and mineralogy.

When such materials are moisture conditioned and reworked by compaction equipment, the soil is overconsolidated to a new level, commensurate with the compactive input energy (Duncan et al. 1990). Beneath a shallow portion of a valley fill embankment (positions "A" and "B" in Figure 8), the effective confinement (geostatic stress) may be less than the preconsolidation pressure. But, out in the deeper sections of the fill (positions "C" and "D" in Figure 8), the soil feels much greater load, and may be consolidating along the virgin compression curve for that material. The deeper material can be expected to densify almost immediately, depending on drainage conditions.

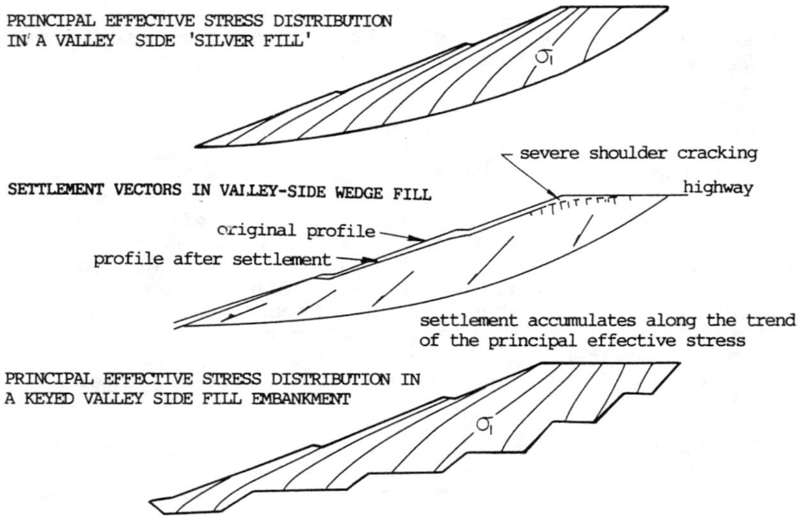

Figure 7 - *Maximum effective stress trajectories and expected settlement vectors in old side-hill embankments comprised of highly aniostropic clay fill.*

But, subsurface monitoring of large urban fills over the past decade suggests that the deeper portions of valley fills may take decades to saturate initially, depending on the permeability of individual compaction lifts. Once saturated, however, the deep areas will retain their moisture, even through periods of significant drought. If water levels rise in the embankment, hydrocompression will likely occur in those portions of the embankment subject to the greatest effective stress. The threashold level between swell and hydrocompression (shown in Figure 8) varies according to the plasticity of the soil mixture. In their study of sandy soil mixtures in the San Diego area, Brandon et al. (1990) reported a threashold of around 1000 psf (48 kPa), wehile the author's experience in the San Francisco Bay area has ranged between 2,000 and 5000 psf (96 to 240 kPa); increasing in porportion to soil plasticity. In this manner, swell or hydrocompression is simply a function of three factors: (1) soil plasticity; (2) effective confinement; (3) water content at time of placement; and (4) the initial dry density of the soil mixture. Cycling of entrained groundwater will engender load cycles similar to those performed in the laboratory (Figure 8, upper). Through successive cycles of saturation (buoyance) and shrinkage (densification), the soil appears to progressively consolidate through a combination of hydrocompression and secondary consolidation.

By alternating the moisture levels within a clayey soil mixture, the "memory" of its initial compaction is gradually erased. This process can be physically appreciated in the gradual smoothing over of compaction lifts recorded on successive cone penetrometer soundings of deep fills, in the years following placement (Figure 9). We now know from the work of Stark and Duncan (1990) and Noorany (1990) that such a mechanism is an explainable and demonstrable occurrence.

Swell/consolidation phenomena appear to have occurred in both side-hill and valley fills in the drought years immediately following the above-average rainfalls of the early 1980's along California's coastline. In most instances, the swelling appears to have been promoted by landscape watering, while the settlement reflected more regional trends, due to weather pattern changes.

Engineers must appreciate that the long-term settlements we can expect in situations such as that depicted above are exceedingly complex and most difficult to predict with any great deal of accuracy. The big unknowns in attempting such an assessment will be the inherent variability of moisture conditions and soil types during construction and the hydrologic regime acting within the embankment during its service life. In addition, the number and severity of saturation/drying cycles may also play a significant future role in embankment behavior.

Perched Groundwater in Fills

When the urban embankments in California were constructed, most engineers assumed that subdrains incorporated into the fills would serve to drain them over the long haul. But, embankment soils are compacted in near-horizontal lifts which serve to make them much more permeable in the plane parallel to the compaction lift, as compared to the near-vertical permeability, normal to the plane of compaction (most geotechnical engineers assume a 5:1 variance, or a horizontal permeability roughly five times that of the vertical permeability; Daniel, 1989). As a consequence, water tends to become trapped within the fills, as shown in Figure 10.

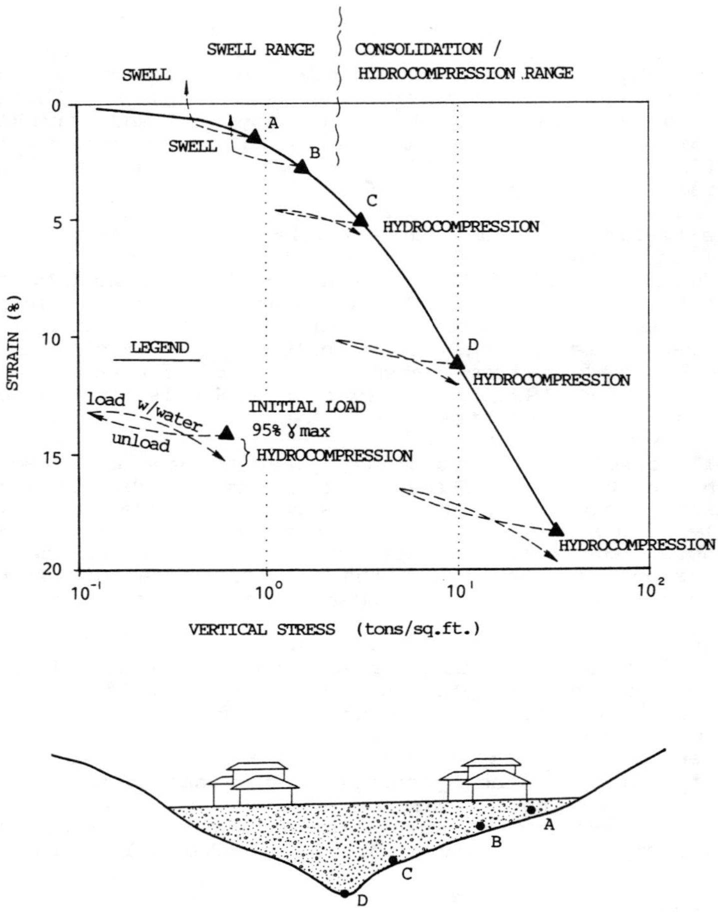

Figure 8 - *Comparison of stress conditions fostering swell and hydrocompression in a valley fill. The threashold between swelling and consolidation behavior is a function of soil plasticity, initial density and water content.*

Many engineers also forget that soil pores must fill to near-saturation before hydraulic pressures can develop to "push" groundwater down, towards the underlying subdrains. Installation of discrete subdrains beneath an embankment is not a tacit guarantee that the embankment won't become saturated.

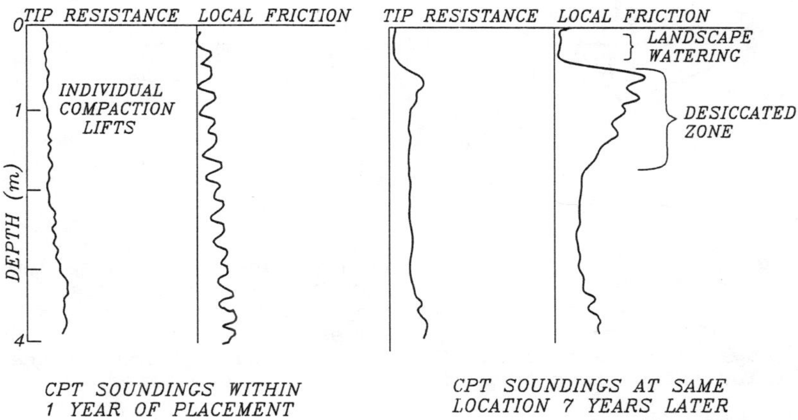

Figure 9 - *Comparison of portions of cone penetrometer soundings of a silty clay in an urban fill in the Blackhawk area of San Francisco Bay, taken 7 years apart. Note how the initial effects of compaction in lifts are erased due to moisture change and desiccation effects.*

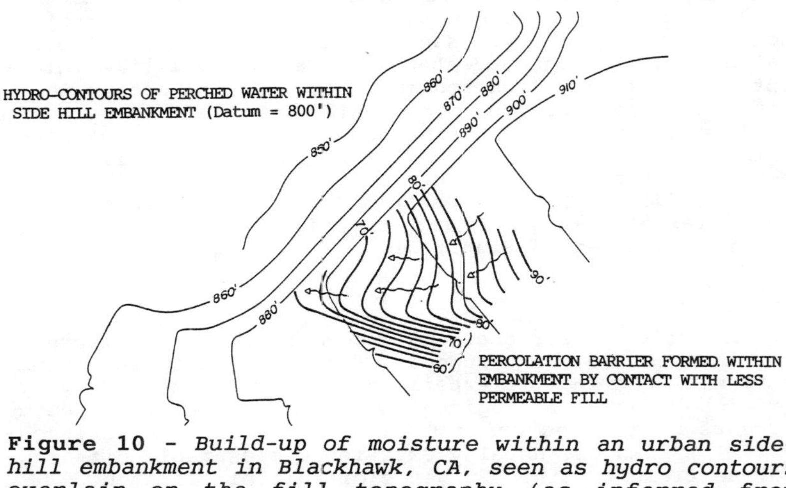

Figure 10 - *Build-up of moisture within an urban side-hill embankment in Blackhawk, CA, seen as hydro contours overlain on the fill topography (as inferred from piezometers). In this instance, a percolation barrier appears to have formed within the embankment, caused by adjoining materials with contrasting hydraulic conductivity. This variance was likely caused by a change in borrow areas for the fill.*

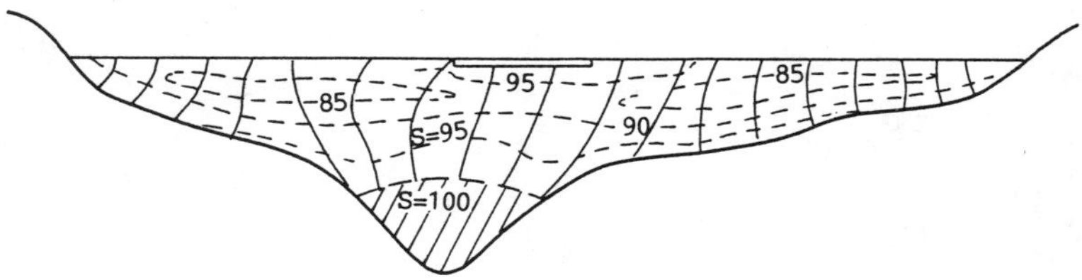

Figure 11 - *Variation of percent saturation in a valley fill in San Bruno, CA. The deepest section infilled an old ravine and was experiencing settlement. Note how most of the fill has not yet reached saturation, even though over 20 years old. The subdrain placed in the ravine bottom is either disfunctional or overtaxed.*

Another problem inherent in sorting out primary/secondary consolidation and swell is the presence of subterranean water and the seepage paths such water followed prior to placement of the embankment. Groundwater will tend to collect and flow in preferential "conduits" that have been formed over thousands of years in the geomorphic development of natural terrain (Wilson and Dietrich, 1987). Recent experiences with groundwater instrumentation of both urban and sanitary landfills has revealed that pre-existing groundwater flow paths are little interrupted by the earthwork accomplished upon the native terrain. As a consequence, groundwater will tend to collect and concentrate in old drainages (Figure 11). These same drainages act as "sumps", or low points in the fill which tend to collect any near-surface water that finds its way into the fill. Zones of saturation and partial saturation invariably occur within the fills. A very typical case is presented in Figure 11.

Conclusions

The historic performance of urban fills in California over the past 35 years suggests that most of these embankments will continue to experience movement, but likely at diminished levels. Regional experience and survey data suggests that homes built upon more than a 30% variance in fill thickness, usually experience measurable differential settlement within 25 years of initial placement.

The California experience also suggests that urban fills are periodically subjected to large moisture fluctuations, due to landscape watering, septic fields, precipitation and shallow groundwater fed from underlying bedrock. Underdrainage incorporated into such fills has not served to completely drain embankments. As a profession, we must collectively appreciate that the embankment's designers will be incapable of controlling future moisture fluctuations within urban fills.

In designing for long-term performance, geotechnical professionals should assume that embankment soils will become saturated and utilize appropriate laboratory data to replicate such conditions. The present practice of utilizing strength data generated from tests on partially-saturated samples (usually at field moisture contents) would appear inadequate for estimation of long-term behavior.

The long-term performance of embankments also appears to be influenced by the cycling of moisture within the embankments. Mechanically compacted soils will shrink, swell, creep, and consolidate as a function of effective confinement and moisture level fluctuations within the embankment. Stark and Duncan (1990) have shown that the shrink/swell effect is even more marked in native soils.

In a long-term sense, it would appear that the construction of urban embankments should incorporate four key components:

1. Thorough subexcavation of embankment foundations, including complete removal of surficial soils, colluvium, and weathered rock which shows evidence of down-slope creep and dilation. Such a premise is presented schematically in Figure 12;

2. The recognition that embankment soils will eventually become saturated, most often within discrete lifts, or zones of the fill. Soil strength parameters must incorporate considerations of plasticity, swell and creep taken as an aggregate whole. Long-term strengths will generally be much less than short-term, partially saturated values, and final slope inclinations should reflect long-term, saturated conditions (recognizing that the outer crust of the embankment may become desiccated);

3. The recognition that moisture levels can be expected to fluctuate within an embankment, in a cyclic fashion. Seepage collection and underdrainage beneath the fill and along natural

drainage swales must be emplaced with adequate capacity and margin of error to encourage long-term drainage of the embankment; and

4. On the basis of performance, long-term differential settlement engendered by primary and secondary consolidation, hydrocompression, and shrink/swell/creep mechanisms can be avoided by restricting fill thickness variance beneath proposed structures to a differential of no more than 15%, taken vertically (Figure 12). In most instances, such a limitation can be accommodated through simple overexcavation in those areas immediately beneath structures.

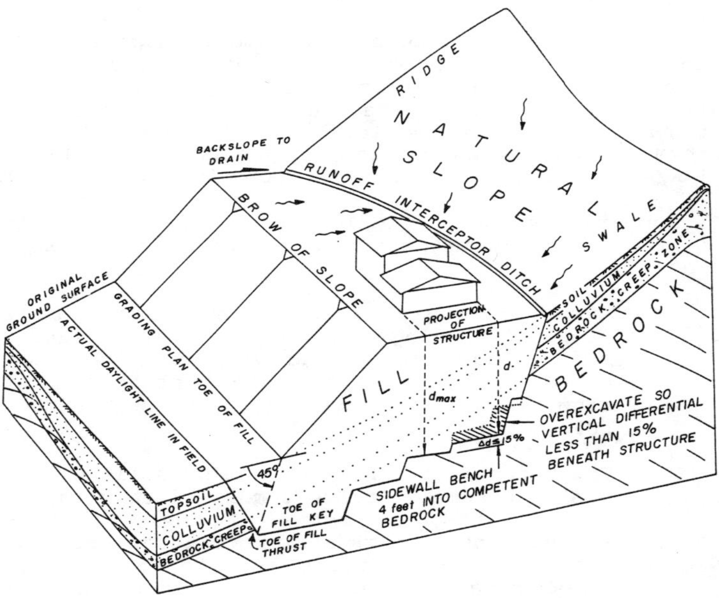

Figure 12 - *Recommendation for toe-of-fill key and sidewall bench keys in a sidehill fill. Stripping of topsoils, colluvium and creeping bedrock would serve to retard the concentration of distortion noted in embankment foundations (taken from Scullin, 1983).*

References

Brandon, T.L., Duncan, J.M., and Gardner, W.S. (1990). "Hydrocompression Settlement of Deep Fills", *Journal of Geotechnical Engineering*, American Society of Civil Engineers, v. 116, n. 10, October, pp. 1536-1548.

Daniel, D.E. (1989). In Situ Hydraulic Conductivity Tests for Compacted Clay, *Journal of Geotechnical Engineering*, American Society of Civil Engineers, v. 115, n. 9, September, pp. 1205-1226.

Duncan, J.M., Williams, G.W., Sehn, A.L., and Seed, R.B. (1991). "Estimation of Earth Pressures Due to Compaction", *Journal of Geotechnical Engineering*, American Society of Civil Engineers, v.117, n.12, December, pp. 1833-1847.

Mesri, G. (1973). "Coefficient of Secondary Compression", *Journal of the Soil Mechanics and Foundations Division*, American Society of Civil Engineers, v. 99, n. SM 1, pp. 123-137.

Mesri, G. and Godlewski, P.M. (1977). "Time and Stress-Compressibility Interrelationship", *Journal of Geotechnical Engineering*, American Society of Civil Engineers, v. 103, n. GT 5, pp. 417-430.

Noorany, Iraj (1990). "Swell and Hydrocompression Behavior of Compacted Soils", *unpublished test data*, San Diego State University, 90 p.

Seed, H.B. and Chan, C.K. (1959). "Undrained Strength of Compacted Clays After Soaking", *Journal of the Soil Mechanics and Foundations Division*, American Society of Civil Engineers, v. 85, n. SM-6, December, pp. 31-48.

Seed, H.B., Woodward, R.J., and Lundgren, R. (1962). "Prediction of Swelling Potential for Compacted Clays", *Journal of the Soil Mechanics and Foundations Division*, American Society of Civil Engineers: v. 88, n. SM 3, June, pp. 53-89.

Stark, T.D. and Duncan, J.M. (1990). "Mechanisms of Strength Loss in Stiff Clays", *Journal of Geotechnical Engineering*, American Society of Civil Engineers, v. 117, n. 1, January, Paper 25421, pp. 139-154.

Terzaghi, K. and Peck, R.B. (1967). *Soil Mechanics in Engineering Practice, Second Edition,* John Wiley & Sons, 729 p.

Wilson, C.J., and Dietrich, W.E. (1987). "The Contribution of Bedrock Groundwater Flow to Storm Runoff and High Pore Pressure Deveolpment in Hollows", *Proceedings of the International Symposium on Erosion and Sedimentation in the Pacific Rim*, 3-7 August 1987, Corvallis, OR.

SLOPE REMEDIATION

By Manfred R. Hausmann, Member, ASCE[1]

ABSTRACT: In the engineer's view the causes of slope failures can be attributed to changes in the slope geometry or external loading, changes in the internal state of total stress, changes in pore pressure and seepage, and/or changes in the shear strength of the soil. Strategies in dealing with slope instability are avoidance (in the planning stage), prevention (through good land management), acceptance (design structures which tolerate movement) or remediation (stop or reverse the destabilizing process). Stabilizing forces can be introduced externally by geometric measures, removing and replacing weak materials, structural barriers and surface coatings, including vegetation. Improving the internal resistance against sliding is achieved by ground modification, which may be hydraulic (dewatering), by inclusions and confinement (reinforcement), chemical (admixtures and grouting), physical (heating and freezing) or mechanical (compaction). Slope remediation has benefitted from improvements in the recognition and mapping of instabilities, land management practices, landslide monitoring, and from innovations in soil reinforcement and biotechnical stabilization. In addition, the introduction of risk management systems is now helping to find acceptable solutions to slope instability problems.

SLOPE FAILURES - THE ENGINEER'S VIEW

Slope failures and therefore also slope remediation may be associated with natural slopes or engineered earth structures such as dams, embankments, excavations and retaining walls; slope failures can also

---

[1]School of Civil Engineering, University of Technology, Sydney, Box 123, Broadway 2007, Australia

be observed in open cut mining and stockpiling operations. Compared with the economic loss caused by other geotechnical hazards, slope failures are said to be only second to the damage caused by swelling soils. However, in terms of the possible magnitude of individual events, slope failures and associated stabilization works should probably be ranked highest on the list of challenges to the geotechnical engineer and engineering geologist.

Slope failures may involve mass movement or surface erosion. Soil erosion, characterized by Gray and Leiser (1982) as the "removal of surface layers of soil by the agencies of wind, water and ice" has in the past been the domain of the soil scientist and agricultural engineer as much as that of a civil engineer. The scope and magnitude of soil erosion caused by human activities around the world has, however, escalated to such a degree, that soil conservation will be a major issue in the 21st century. In many instances engineering solutions will have to be applied to problems which in the past could be corrected by proper land use practices. Since many of the factors contributing to soil erosion may lead directly or indirectly also to mass movement, the measures taken to prevent and control erosion are often also beneficial for the treatment of more deep seated failures.

Mass-movement is defined by Gray and Leiser (1982) as the "downward and outward movement of slope-forming materials - natural rock, soils, artificial fills, or combination of these materials". In the engineer's view, mass movement occurs where the shear stress in the material exceeds its shear strength. Therefore, any process which leads to higher stress levels or which reduces the shear strength of the soil or rock may lead to instability. Conversely, remedial measures are aimed either at reducing the magnitude of shear stresses in existing or potential failure zones, or at increasing the strength of the soil or rock mass subject to movement.

An infinite variety of failure modes are possible, dictated by the topography, soil and rock properties, drainage patterns and other factors. The primary cause or causes of slope movement are not necessarily evident from its physical appearance or the sequence of events. Nevertheless, for the design of cost efficient prevention and control measures it is not only essential to know the factors contributing to a failure, but also the geometric configuration of the moving soil or rock mass.

Not all types of mass movement fit the failure models which are most commonly relied upon by engineers, such as translational and rotational slides. Neither are all slides amenable to the traditional limit equilibrium methods of analysis. Rock falls and topples, lateral spreads and flows, as distinguished by Varnes (1958), and slope movements occurring progressively or made complex by highly variable materials often cannot be analysed accurately. Nevertheless, even if an absolute safety factor cannot be determined reliably, it may still be possible to assess the relative improvement obtained by remedial measures, such as changes in the slope geometry and a reduction in the pore water pressures.

CHANGES WHICH LEAD TO SLOPE FAILURES

In the literature, the majority of the most devastating landslides are attributed either to extreme rainstorms or earthquakes (e.g. see Crozier 1986). These may well be the primary causes of a failure. A geotechnical engineer, however, would more likely interpret the failure in terms of the impaired state of equilibrium of the soil mass, such as due to additional external forces acting on the soil mass, an increase in the internal pore pressure, or a reduction in the shear strength of the soil.

For a slope to fail suddenly or move periodically over a long time, changes must be taking place either in the external environment or within the soil or rock mass itself. Terzaghi (1950) already distinguished between external and internal changes in stability conditions which cause mass movement.

For the purpose of discussing remedial measures based on halting, reversing or counteracting destabilizing influences, the author finds it useful to group factors leading to slope failures into the following categories:

1. Changes in slope geometry (height or steepness) or external loading (permanent or transient)
2. Changes in the internal state of (total) stress
3. Changes in pore pressure and seepage
4. Changes in shear strength

These changes are further detailed in Table 1. Each of these changes can be the primary factor of slope failure or may come about as a secondary effect. For example, an earthquake force may cause an

TABLE 1. Changes leading to slope instability, and counter measures

Change in	Examples of destabilizing events	Examples of remedial measures
Slope geometry, external loading	Constructed excavations in slopes. Cuts and fills for highways or railways. Toe erosion by streamflow or wave action. Structural loads on slope (buildings, abutments, walls). Water, snow or ice loading. Traffic loading, vibrations, earthquakes. Drawdown in reservoirs.	Provide toe support, remove material from head of slope. Reduce steepness, make slope longer. Build berms to prevent shallow failures. Prevent accumulation of water, snow and ice. Redirect traffic, remove sources of vibrations. Slow down changes in reservoir levels. Provide structural support by walls founded on strong strata. Use anchorage to introduce stabilizing forces and/or change the geometry of the potential mass movement.
Internal state of (total) stress	Removal of lateral constraints in overconsolidated soils (may lead to a reduction in horizontal stresses). Formation of cracks and fissures due to environmental influences. Structural and stress changes due to subsidence in mining or in karst areas. Underground excavations, tunnelling. Internal erosion (piping). Movement along faults. Stress changes due to creep. Ice formation and root growth in cracks, swelling soils and rocks.	Reduce effect of environmental influences leading to cracks and fissures (vegetation may help). Prevent root wedging, fill in cracks accumulating water. Tensile restraint through passive anchors (e.g. soil nails) may be mobilized as movement progresses. Use of tensioned anchoring systems (incl.tensioned rock bolts) may increase frictional resistance along failure planes.
Pore pressure and seepage	Fluctuating ground water levels. Increased seepage due to rainfall, broken pipes, changing soil permeability. Dissipation of negative pore pressures. Excess pore water or air pressures due to external dynamic loading.	Reduce ground water levels through gravity drainage or pumping from wells. Collect and divert surface runoff. Repair conduits leaking water into the moving soil mass. Prevent access of water to unsaturated ground by installing impermeable barriers and appropriate drainage systems. Where possible decrease levels of saturation and increase air permeability (vegetation may help).
Shear strength	Loss of strength due to physico-chemical effects (leaching, weathering, freeze-thaw, wetting and drying), possibly resulting in changed soil structure (reduced interparticle attractive forces, breaking of cementitious bonds). Remoulding of sensitive clays. Increased pore pressures or collapse of soil structure may lead to liquefaction, fluidization.	Remove and replace weak soils with higher strength materials. Improve in-situ material through compaction, consolidation, admixtures, grouting, heat treatment, or reinforcement. Reduce effects of weathering. Reinforce through vegetation.

increase in internal total stress which is transferred into high pore pressure, which in turn may induce liquefaction (a total loss of shear strength).

In any one incident of failure, a combination of the above causes may be identified, even if they are initiated by a single action. For example, the removal of trees, bushes and other vegetation from a slope will cause a change in the evapotranspiration regime of the soil, leading to higher water contents and pore pressures; the runoff from rainfall will intensify and cause erosion, changing the slope geometry; and the effect of root reinforcement will be lost with time, resulting in a loss of shear strength along potential failure planes.

STRATEGIES IN DEALING WITH SLOPE INSTABILITY

Before resorting to expensive remedial action, it is appropriate to review whether it is possible to avoid the problem of slope instability, to prevent it, or to live with it and accept the risks involved. These alternative strategies are discussed below.

Avoiding the risk of instability

The problem of land instability in the construction of highways, dams, reservoirs and other human endeavours may be avoided by relocating these works or activities to a stable area, preferably in the planning stage. This action requires access to detailed maps indicating landslide hazards.

According to Walker et al (1987) regional maps (1:1 million to 1:10000) of landslide distribution are available in the U.S., Japan, and a number of other countries. Large scale "landslide inventory maps" or "maps of landslide deposits" are usually based on the interpretation of aerial photographs and may not indicate the type of landslide, age or rate of movement. For detailed planning, reference to a site specific "slope stability map" (say 1:1000) showing all geomorphic features may be needed.

Ideally, landslide maps should give sufficient information for assessing the risk involved for a specific project. The landslide risk will depend not only on the probability of instability occurring, but also on the potential damage incurred.

In absence of maps dedicated to landslide evaluation, more general maps produced by terrain evaluation (Grant 1974) or terrain analysis

(Way 1978) may yield sufficient information for the preliminary planning of civil engineering projects.

Identification of potential mass movement is also part of the Urban Land Capability classification system developed by the Soil Conservation Service of New South Wales (Hannam and Hicks 1980, Charman and Murphy, 1990). This classification system is based on data collected covering climate, geology, landforms, slopes, hydrology and soil types, with specific reference to expansive, dispersive and saline soils as well as existing and potential mass movement. "Urban capability maps" may be produced on a scale of 1:50000 for strategic planning, down to subdivision plans (1:2000) or plans showing details of proposed works to control mass movement, erosion and runoff.

Given the risk involved for a civil engineering structure in an area of potential instability, economic analysis will assist in making the decision whether to move elsewhere, live with the risk or undertake remedial measures.

Protecting land subject to mass movement

Poorly planned engineering construction as well as agricultural activities can accelerate the incidence of mass movement in form of rotational slides, slumping, earthflows and other phenomena. The Soil Conservation Service of New South Wales has embarked on a program of mapping areas prone to mass movement hazard and declaring it "protected land", the same status as given to areas which need habitat or flaura and fauna protection. The objective is to "limit the potential for slope failure rather than try to repair the damage after the event". Construction of access tracks, tree clearing and farm dam construction can all be the cause of mass movement which in turn may lead to increased surface erosion and siltation of waterways. The management guidelines for protected land issued by the Soil Conservation Service include the following advice:

- Retain all trees and regrowth on mass movement prone land and the immediate surrounding areas
- Do not construct farm dams without geotechnical advice
- Plan all farm activities involving earth disturbance such as road, track, fence and firebreak construction with the advice of the Soil Conservation Service
- Plant deep rooted tree species on areas with highest risk of mass movement, such as seepage areas, gullies and below existing dams

- Do not disturb existing slope failures unless geotechnical advice has been obtained

The documents prepared by the Soil Conservation Service highlight the role of trees in reinforcing the ground through roots and in limiting saturated soil conditions.

Mapping of protected lands increases the public awareness of potential mass movement problems. Involving members of the community in promoting good land management practices is the objective of another government initiative in Australia: the formation of Landcare groups. Good land management not only helps to prevent slope instability but also improves the water quality in lakes, lagoons and estuaries.

Living with slope instability

The detrimental effect of soil mass movement on human endeavours can be reduced or avoided by

- Inducing a landslide or rockfall in a controlled risk situation, e.g. with explosives
- Bridging over or tunneling under a landslide
- Diverting the sliding soil mass (for example by building a reflective shell around a bridge pier) or deflecting falling rocks
- Redesigning the endangered structure so that it can tolerate large foundation movement
- Building structures strong enough to resist high earth pressures or impact forces from mass movement
- Adopting a risk management system

Inducing a slope failure as a safety measure may have merit in open cut mining operations. Rather than securing an unstable soil or rock mass with elaborate cable anchor systems or other structural support, it may be more economical to cart the fallen material away after failure, since very high capacity earth moving equipment is readily available.

In general, partial removal of debris accumulating at the toe of a slide is not a practice which enhances slope stability, although perpetrated by many a road maintenance gang. However, as explained by Baker and Marshall (1958), there are exceptional circumstances where undercutting of the toe may bring about "relative" stability. Cases quoted involve so-called "piston" slides (a form of lateral spread) in the State of Washington, and the formation of benches to allow for additional slope movement without damaging structures further downhill. The latter is similar in principle to the formation of benches on rock slopes, designed to neutralize falling rocks.

Digressing slightly, it is worthwhile to mention that initiating snow avalanches using mortars and artillery has been practiced in Switzerland and Canada.

Huder (1990) explains how the bridge piers of a major viaduct of the N13 highway near Mesocco in Switzerland were built within 4.2 m diameter open vertical shafts fully penetrating the unstable surface layer. This design also allowed to significantly improve the drainage of the slope. Veder (1981) describes similar solutions to bridge foundations in slopes which are subject to creep. Where the protective shell around the pier cannot be designed to take up the full earth pressure due to slope movement, it can at least provide sufficient space to allow creep movement for many years before the bridge foundation is endangered.

Roads in mountainous country can be protected against rockfalls by galleries whose roofs deflect the falling rock (Descoeudres 1988) or earth covered shelters which can absorb some of the kinetic energy of the falling rocks (Kirsten et al 1986).

It is possible that the mass movement is slow enough and occurs on such a large scale that damage to individual buildings is minimal and regular maintenance is sufficient to keep services and transportation infrastructure going. Together with a string of other buildings, the church in the village of Peiden in south eastern Switzerland has steadily moved downhill; over a period of 100 years it moved a total distance of 20 m, without visible damage (Huder 1990). Significant total and differential displacements in this region occur over more than two square kilometers. Typically, in this area, individual buildings are small and rigidly constructed, retaining walls and other engineering structures are of the flexible type, and roads need periodic repairs.

Larson et al (1987) reported on the mitigating measures for a high pressure gas pipe line in the Rocky Mountains which was threatened by landslides over a distance of several miles. Amongst the evasive techniques used were embedding the gas line below the moving soil mass (defined by inclinometer readings), and supporting the line on a timber structure above ground, designed in such a way as to allow relative movement. These measures were complemented with surface water diversion and subsoil drainage. In at least one location rapid relief of high tensile stresses in the pipe line could be achieved by excavating a trench parallel and uphill of the line. Landslide activity was monitored by strain gauges installed on the pipe line itself or by borehole inclinometers.

In the case of rockfalls, boulders rolling downhill, earth flows or debris slides it may be possible to protect vital buildings or services by constructing defenses in form of retaining walls or embankments. Their function would be to absorb the energy of the moving earth and to deflect it towards an area where damage is minimal. Impact loads by landslides have been studied by Loewenstein and Clemence (1984).

Defensive installations on rockfall prone slopes may include hanging nets and chains, wire mesh fences, berms and energy absorbing catchpits and gravel beds.

Remote monitoring of movements and pore pressures may allow extra precautions to be taken or emergency procedures to be initiated when increased landslide activity is indicated, for example during and immediately after an intensive rainstorm. This approach can be part of a risk or safety management system, such as operated by the State Rail Authority (SRA) of New South Wales.

Within the framework of statistical decision theory, the "risk" involved in a slope failure depends on the probability of it occurring within a given time as well as its consequences, such as loss of life or damage to property. Table 2 gives an example of a simplistic way of assessing geotechnical risk for railway track engineering. This example is similar to the current practice of the State Rail Authority of New South Wales. Depending on the risk category, safety measures such as track closures and speed restrictions may be introduced, while rail movement surveys, detailed geotechnical monitoring or remedial works are undertaken. Risk categories are also considered when determining the priority of remediation projects competing for limited funds.

The adoption of risk categories by the SRA is not unique. Depending on the environment in which they are operating, other organisations may need a similar, possibly more elaborate system to standardise their response to slope instability. The Hong Kong Geotechnical Control Office differentiates between risk-to-life and economic risk (Geotechnical Manual for Slopes 1981). Whether the risk is judged to be negligible, low, or high, determines the safety factor to be chosen in the design of new slopes or remedial measures.The SRA has taken a risk management approach in dealing with slope instability near the Tickhole tunnel along the Sydney-Newcastle line. At this cutting the track is threatened by the lateral movement of a conglomerate block, which is some 30 m high and 100 m long. The stop-start sliding, apparently associated with the build-up of hydrostatic pressures after extended periods of rain, can result in a track displacement in the order of 100 mm. Catastrophic failure was considered unlikely, but the possible movement could lead to the derailment of high speed commuter trains. To ensure track safety in the short term, an early warning system has been installed, involving the monitoring of ground heave, lateral movement (using borehole inclinometers), rainfall and

TABLE 2. Example of Risk Assessment Scheme

1 - highest risk, 5 - lowest risk

Consequence of event affecting the track	Probability of event affecting the track (on time frame)		
	High Event is anticipated in the near future	Moderate Event is possible and is expected in the long term	Low Event is possible but not expected
Extreme - loss of life expected - extensive damage and disruption	1	2	3
Severe - loss of life is possible - appreciable damage and disruption	2	3	4
Moderate - loss of life or serious injury not expected - minor damage to structures and facilities	3	4	5

Examples:

Risk Cat.1 (very high risk): Slope failure is occuring or is imminent, requiring temporary or permanent track closure. Remedial action required prior to re-opening track to normal traffic.

Risk Cat.2 (high risk): High embankment constructed on a side fill. Mechanism is dentified as a rapid slide which could occur in the long term, taking out the track. Early warning system advisable.

Risk Cat.3 (moderate risk): Cutting in weathered rock. Potential failure onto the track may cause train to derail or plough into debris. Surveillance program required.

Risk Cat.4 (low risk): An embankment constructed on a gently sloping ground which is experiencing instability of the slope face. May affect access road or embankment shoulder, but not the track geometry.

Risk Cat.5 (very low risk): Recent remedial works have stabilized the slope and are being monitored for effectiveness.

pore pressures. The Slope Indicator Company's (SINCO) Intelligent Data Acquisition (IDA) system controls the readings, can send alarm messages to railway staff and activates signals stopping all trains if critical values are exceeded.

The risk management approach has been greatly enhanced by the advances made in the automatic monitoring of slope movement, pore pressure, soil moisture and other geotechnical parameters. As stated by Hinckley et al (1987), portable dataloggers, capable of processing and storing measurement on site, are now readily available and fairly inexpensive. They can be programmed to make logic decisions, such as setting off alarms. Remote communication via hard wire, modem or radio is technically and economically feasible, even communication via satellite.

The risk management approach has also benefited from a gradual formalization of risk assessment and the establishment of decision making frameworks with respect to the danger of landslides. Einstein (1988) has proposed such a probabilistic framework which allows not only to quantify the risk of living with the danger of a landslide, but also provides a basis for optimizing counter measures.

Engineered slope remediation

Engineered slope remediation is aimed at *stopping or reversing the destabilizing process*. Table 1 gives examples of remedial measures which can be taken in response to the four basic causes of instability. Arresting the movement and/or increasing the factor of safety of a slope is achieved by a variety of structural and geotechnical methods.

Techniques which introduce stabilizing forces or barriers which are essentially external to the sliding soil or rock mass include:

- **Geometric measures**. Examples: Create toe berm keyed into strong strata, remove material from the upper portion of the slope or decrease overall steepness of constructed slopes. The inclusion of berms on rock slopes may not only make construction easier, but it may also be useful for arresting rockfalls and installing surface drains . Special types of failures, such as toppling failures in foliated rock, may need detailed studies of the effect of the foliation dip on slope stability (Orr et al 1991).
- **Remove and replace** the unstable material. This approach could be seen as an extension of geometric measures. The replacement soil

may be imported granular fill or may be the original soil improved by admixtures or reinforcement and drainage facilities.

- **Structural barriers**: Rigid walls and piles made of timber, steel or concrete intersect the unstable soil mass, possibly keyed or anchored back into a stable strata.
- **Permeable or impermeable surface coatings**. Essentially to prevent erosion or saturation of the endangered soil mass. Examples: Sprayed concrete or bitumen, geosynthetics, vegetation.

Processes which improve the internal resistance against mass movement may be grouped under the heading of *ground modification* (Hausmann, 1990). Specifically, a slope may be stabilized by:

- **Hydraulic modification**. In sandy and gravelly soils, gravity drainage or dewatering by wells reduces pore pressures and seepage forces; external capture and redirection of surface runoff helps this process. In fine-grained soils an increase in shear strength can be achieved by a reduction in water content by consolidation, induced by electroosmosis or preloading (with or without vertical drains). The latter would only be feasible in special circumstances, such as prior to making a cut or excavation.
- **Modification by inclusions and confinement** (reinforcement). Soil reinforcement with all sorts of tensile elements, soil nailing, rock bolting, various methods of anchoring as well as forming gravity walls by confining local soils by gabions and cribwalls could be included in this category.
- **Chemical modification**. This covers modification with admixtures such as cement, lime and bitumen. For example, weak material is removed, mixed with a stabilizing agent and used to rebuild the slope. Alternatively, deep mixing, the formation of lime columns or grouting techniques could be employed.
- **Physical modification**. Temporary stabilization can be achieved by ground freezing. In-situ soil heating has also been employed in the past, but may have undesirable aspect with respect to energy use and environmental impact.
- **Mechanical modification** (compaction). Shallow compaction usually plays a part in the changing of the slope geometry and in the remove and replace procedure. Deep compaction could be considered before making cuts and excavations.

Engineers generally do not rely on one single method for stabilizing a slope, but choose a combination of techniques, most likely proceeding in stages. Unless confronted with an immediate high risk problem, low cost methods are tried first. In most cases initial works are a combination of geometric measures and surface treatment, particularly the control of runoff. If this is found inadequate, either by analysis or experience, subsoil drainage and reinforced soil type retaining structures are investigated. Anchored walls and piling may be considered next. Constructing a viaduct or tunnel bypassing the unstable area is likely to be the costliest alternative.

Table 3 illustrates the range of remedial measures recently executed or currently planned by the SRA within 100 km of Sydney (Christie, 1991). All these projects are located in two of the four main areas of New South Wales which are best known for rainfall induced landsliding (Blong and Eyles 1989). In preparing this table, only the major method of stabilization was identified. It should be noted that control of surface water is incorporated in all projects and that subsoil drainage layers or slots are part of the design wherever extra fill is placed. Slope remediation work was given higher priority within the general maintenance program of the SRA after two people were killed in a 1988 landslide apparently triggered by a blocked drain along the railway track.

TABLE 3. Slope remediation along railway tracks north and south of Sydney.
(Christie 1991, State Rail Authority of New South Wales)

Remedial measure	Number of projects		Total
	Embankment	Cutting	
Altering geometry:			
benching, flattening		5	
toe berm, buttress	6		14
toe key	3		
Gravity walls (gabions etc.)	2	5	7
Subsurface drainage:			
gravity drains	4		6
eductor wells	2		
Rockface treatment:			
removal+ rock bolting		2	6
shotcrete, meshing		4	
Piles (single or contiguous)	2	1	3
Soil nailing		1	1

Railway systems are particularly vulnerable to slope instability. This is due to a number of factors: the geometric restriction imposed on the original line design, the age of the cuts and fills (often extremely steep and affected by weathering) and the high risk of derailment resulting from only very minor movement of the track.

PRINCIPAL METHODS OF SLOPE REMEDIATION

Each of the basic engineering methods of slope stabilization which were introduced in the previous section will be further discussed below. It is not attempted to give a balanced review of alternative solutions, as far as their economic significance, general applicability, or level of technology required is concerned. Rather, in terms of the detail given and illustrations presented, emphasis is placed on the more recently developed slope remediation techniques.

Changing the geometry and composition of a slope

Removing material from the head of the slide reduces the driving force in a slide mass and placing fill at the toe increases the resisting force. If stabilization by weight displacement of this kind achieves a safety factor of 1.3, this is usually considered adequate (Sembenelli 1988). The economics of such a solution may depend on the availability of and access for heavy earth moving equipment, the cost of imported fill material and the additional land resumptions required. The large size of the earthworks usually required may represent an adverse environmental impact.

Stabilization of a landslide by excavating some 60 000 m^3 of material from its head is described by Jubien and Abbott (1989). Since this remedial work has been completed, the Canadian National Railway track which traverses the slide area has shown a dramatic decrease in subsidence and misalignment.

Where a certain grade has to be maintained, consideration could be given to replacing material at the top of a slide by light weight substitutes: fly ash (γ_{dry} =1.2 to 1.9 t/m^3), blast furnace slag (γ_{dry} = 1.5 to 1.9 t/m^3), sawdust and woodchips (say 0.6 t/m^3), polystyrene blocks (0.02 to 0.03 t/m^3) or "pneusol" (0.6 to 0.8 t/m^3). Pneusol is described in more detail later.

Fig. 1 shows a cross section of the rockfill toe berm created to stabilize the Hue Hue Road landslide near Wyong, N.S.W. (Fell et al, 1987).

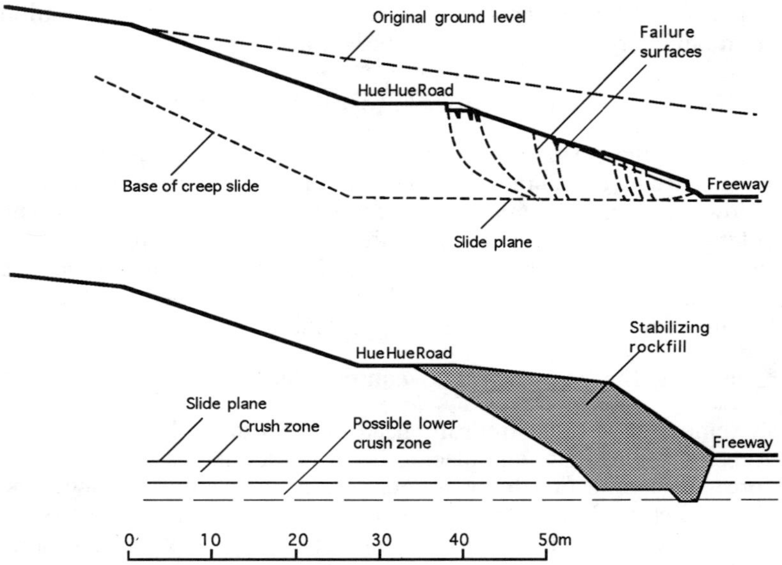

Figure 1. Stabilizing the Hue Hue Road landslide (Sketch based on Fell et al, 1987)

Control of surface water and seepage

Control of water on the surface and within the ground is an integral part of most slope remediation programs. Reducing the water content and decreasing the pore pressure in a soil mass increases the shear resistance along potential failure planes and decreases the total weight of the soil. Lowering the rate of seepage reduces the destabilizing forces and the risk of erosion and/or piping.

Surface runoff is usually collected in V- or U-shaped concrete lined channels or semi-circular corrugated steel pipe sections and diverted away from the slide prone soil mass. These channels should be strategically placed at the head of the slope and along berms. Their design should be based on local hydrological data. The engineering detailing of surface water collection systems should provide for minimizing future maintenance with respect to blockages and relative displacement due to slide movements.

The sealing of cracks in the soil or rock subject to mass movement, surface coatings (shotcrete or bitumen), as well as the establishment of vegetation help to reduce water infiltration. Temporary placement of geomembrane covers on the ground surface may reduce the risk of movement during the construction of drains and berms.

The principal methods of *subsoil drainage* are:

- Gravity flow slots cut into the unstable soil mass and filled with gravel and leading to a common sump or ditch.
- Horizontal drains (boreholes acting as relief wells, or drainage layers installed during fill placement)
- Vertical holes or slots connecting with a horizontal drain, tunnel or natural aquifer lower down which acts as a drainage layer
- Pump-out wells (well points, eductor wells or submersible pumps)

Natural slopes are rarely homogeneous enough to allow reliable subsoil drainage design according to simple principles of dewatering by slots and wells. The secret of success lies in a good understanding of the geological structure and choosing a drainage system layout which maximizes the probability of intersecting the major water bearing layers.

Lew and Graham (1988) show how sand filled trench drains (or counterfort drains) helped to stabilize a clayey riverbank in Winnipeg, Canada. Typically, the trenches were 5 m deep, 5 to 15 m apart, and resulted in a drawdown of about 2 m, improving the safety factor by some 15 %. Another 15 % improvement was attributed to the reinforcing effect.

Classic applications of horizontal drains using innovative drilling techniques are described by Jedlicka (1986). In a typical application in Czechoslovakia, six 80 m to 140 m long boreholes were fanned out from several surface location near the toe of a large slide area, achieving a drop in ground water level of 2 to 8 m. The initial yield of the drainage system was 300 to 400 l/min; it reduced to an average of 10 to 40 l/min some two months after the installation of the drains.

Safeguarding a 17th century town sitting on an ancient landslide in central Italy was the objective of a 600 m long drainage tunnel located at an average of 40 m below ground level (Bianco 1988). A total of 6000 m of vertical and inclined drains were drilled from the tunnel into the failure zone above.

A similar solution helped to control a slow debris slide which occurred during commissioning trials for the Fisher Hydro-Electric Scheme in Tasmania (Paterson et al 1975). The slide was caused by water leaking from a pressure tunnel into the interface between sandstone bedrock and relatively impermeable talus. A 400 m long Y-shaped adit was driven in the sandstone to the area below the head of the slide. A drainage curtain consisting of a series of vertical holes (150 to 115 mm diameter) was then drilled from the surface into the adit. Additional holes were then drilled upwards as well as downwards from the adit. After refilling the tunnel/penstock system the flow into the adit rose to 410 l/min, which was enough to lower the piezometric pressures at the base of the talus, effectively stabilizing the slide.

Another solution to deep drainage of slopes is to sink a series of large diameter (say 1.5 m) holes from the surface and interconnect them at depth through horizontal boreholes providing gravity discharge. Such a solution was chosen by Collotta et al (1988) for the stabilization of a highway embankment stretching across an unstable slope.

Drainage wells in combination with vacuum-assisted pump-out wells were used for the dewatering of a trial pit for an open-cut mining development in South Australia (see Hausmann, 1990). Depressurization of the artesian aquifers above and below the coal seam permitted side slopes at angles between 45 and 60°.

Dewatering with eductor wells is part of the solution to the stabilization of the railway track at Coledale, south of Sydney (Elliott 1991). A line of 19 wells were drilled along the railway line, transverse to the direction of the slip, at 10 m centres. The 100 mm diameter wells were drilled through embankment fill, colluvium and talus into residual clays, shale and sandstone. Eductors work on the Venturi principle. A 15 kW pump draws water from a 10,000 liter tank to supply drive water to the eductors. The water discharging from the nozzle into the Venturi throat sucks groundwater from the well into the return pipe leading back into the pump tank. The overflow of the pump tank represents the groundwater removed from the wells. Mean dry weather discharge is about 30 l/min, but the maximum discharge capacity is 420 l/min.

The advantage of the Coledale pump-out system is that it could be installed quickly without interrupting the train traffic. Installation costs were only $ 100,000. Annual continuous running costs are estimated to be $ 12,000. The next stage of the stabilization works involves the construction of a rockfill berm and buttress keyed into the residual soil, likely going to cost well in excess of $ 3 Mill. Improved gravity drainage may then make the pump-out system superfluous.

Rigid structural barriers

Conventional walls

Rigid concrete or masonry gravity or cantilever walls may be used to flatten oversteep slopes, as necessitated in the construction of roads, railway tracks and for buildings on slopes. They may also serve to reduce cut and fill volumes of newly created earth structures. Due to their inability to tolerate much movement, conventional rigid retaining structures are now rarely designed for slope stabilization. They have been replaced by elemental wall systems based on the principle of soil reinforcement (e.g. Reinforced Earth) or soil confinement (crib walls and gabions), as discussed under the heading of "Flexible support systems".

If conventional walls fully penetrate an unstable soil mass and can be founded on a strong strata, they could theoretically be used to stabilize an existing landslide. However, the excavation necessary to build such a wall might well accelerate the slope movement. If a rigid barrier is needed, it makes much more sense to consider installing a row of single or contiguous piles, possibly anchored back into a solid strata. In a cut and fill situation the necessary excavation can then proceed in stages, with anchors being installed at progressively lower levels.

Piles and piers

A row of cast-in-situ concrete piles (900 mm diameter) at 2.25 m intervals were used by the SRA to stabilize 100 m of track near Stanwell park, south of Sydney. The piles are some 10 m deep and socketed into weathered rock. They were installed in December 1989 and are still waiting for anchors to tie them back into solid rock. In the present condition they helped to arrest movement along a known failure plane at a depth of about 6 m, but the safety against a possible deep seated failure is not as yet guaranteed.

Lippoman and Gudehus (1989) describe the stabilization of a 8 m deep, 150 m wide creeping clay slope by means of a total of 188 reinforced concrete piles as "dowelling". These vertical piles were 900 mm in diameter, 11 m long and arranged in three rows across the slide. The top of each pile was 2 m below the surface, so as not to be visible.

Three rows of mixed-in-place columns helped to stabilize the sideslopes of an embankment of the Paris-Bordeaux railway line (Cartier et al 1989). The admixture consisted of lime (45%) and cement (55%), making up about 18% of the final column material.

The problem of designing piles and piers for slope stabilization has received considerable attention (Fukuoka 1977, Viggiani 1981, Nethero 1982, Oakland and Chameau 1989) but their actual application seems to be reserved for high risk situation where high costs are justified. For a case study on large diameter piers used to stabilize a deep cut in an unstable slope, see Brandl (1988).

Anchored walls

The design of anchored walls with sloping backfill is covered in most geotechnical textbooks, although the various methods of analysis and alternative definitions of the safety factor still manage to arouse controversy (e.g. Anderson et al 1983 and discussion by Schnabel 1984, also see Carpenter 1988).

For rigid slope support walls or piles, prestressed (active) friction anchors are most common. The anchor transfers the applied force into a stable soil or rock zone via the grouted fixed anchor length. The free anchor length, characterized by a constant tension force may be grouted after prestressing for the purpose of corrosion protection.

Rockbolting and soil nailing are considered method of ground reinforcement and are discussed under the heading of flexible support structures.

Once an anchored wall is in place, it is necessary to check whether a new failure mechanism could develop which would result in "overtopping" the wall. As quoted by Leventhal and Mostyn (1987), Stone (1987) developed a method of analysing the need for additional walls in an infinite slope situation. Assuming a passive wedge of soil is pushed over the wall by an unstable active block further uphill, the analysis yields the force acting on the wall and the spacing of additional support structures.

Flexible support structures

Concepts and economics

Flexible support structures in this context refers to retaining wall systems of elemental construction based on the principle of reinforcement by inclusions or confinement (Hausmann 1990). Typical for a reinforcing "inclusion" is that stress transfer between the reinforcement and the surrounding ground occurs along the full length of the reinforcing element and the tensile stress varies along the element. Reinforcement by confinement, as in gabions and crib walls, forms gravity structures by enclosing soil or rock by structural elements able to resist tensile forces. Gabion walls can be transformed into a internally reinforced soil structures by attaching horizontal layers of mesh embedded in the backfill (in Australia called Terramesh).

Rockbolting and soil nailing or dowelling are methods of "in-situ reinforcement", in contrast to "constructed reinforced wall systems", such as Reinforced Earth. Soil nailing and dowelling are a form of passive reinforcement, that means no prestressing is involved. Rockbolt action can be passive or active. Rockbolts can be considered a method of reinforcement, rather than just anchoring, where its main purpose is not transferring the load into a stable zone, but confining a jointed and fragmented rock mass in such a way as to mobilize the frictional strength between individual blocks.

Reinforced soil walls can be used as structural units for slope stabilization, just like any other gravity wall. They have the advantage of being more flexible than reinforced concrete retaining structures, thus being able to accomodate considerable settlement or slope movement before becoming unserviceable.

Reinforced soil techniques allow the formation of vertical or steeply inclined benches or toe fills; this may help to stabilize a slope subject to shallow slides. Reinforced soil walls may be constructed with good imported fill or with local material, possibly improved with admixtures.

Metal reinforced wall systems generally put more stringent conditions on the suitability of fill materials than systems using synthetic elements. Standard design procedures for metal strip or mesh reinforced walls, such as Reinforced Earth or VSL Retained Earth generally only consider free draining cohesionless backfill. Synthetic strips, grids and textiles can be used with all kinds of soil, provided no physical damage occurs during their installation.

Besides of having the technical advantage of being flexible and not needing elaborate foundations, reinforced soil retaining systems are generally cheaper than conventional concrete or masonry walls, except at small heights. Geotextile walls are likely to be cheaper than walls with other types of reinforcement, but their permanency and aesthetics may be questioned. Given the task of building walls along an 850 m long stretch of road in mountainous terrain in Hawaii, Castelli and Munfakh (1986) compared wall designs with gabions, Reinforced Earth and geotextiles. The geotextile wall, designed with a 40 mm thick wire mesh gunite facing, attracted the lowest bid. A cost comparison of various versions of geogrid reinforced slopes with other retaining systems for a project in Malaysia was performed by Toh et al (1986). In this case, a 45° to 75° slope reinforced by Tensar geogrids proved the most economic proposition for heights in excess of about 6 m.

Analysis of reinforced soil walls and slopes

Reinforcement can readily be designed to guarantee internal stability, basically by ensuring that neither rupture of the reinforcement, nor slippage between the reinforcement and the fill material occurs. The latter may involve reinforcement pull-out, or block sliding on the reinforcement-soil interface. Global slope stability is improved where critical potential slip planes of the unreinforced slope are intersected.

For vertical reinforced soil walls with metal (="inextensible") reinforcement, analysis is usually based on the 1978 Reinforced Earth (RE) procedure (e.g. see Hausmann, 1990) or minor variations thereof. The design can be adapted for a sloping face, say at an angle of 45 to 50°. However, according to Boyd (1991) it can be shown that, in terms of global stability, a sloping RE structure may be relatively less stable than a vertical structure of the same block width.

For reinforced slopes in general, a number of different procedures of conventional limit equilibrium analysis are available, suitable for cohesionless as well as cohesive soils. Failure geometries assumed include the infinite slope (Ingold 1982), a two part wedge mechanism (e.g. Jewell 1990), circular arcs (e.g. Ruegger 1986) and log spirals (Leschinsky and Reinschmidt 1985). Each of these references contains design charts.

The most recent design charts for steep reinforced slopes, defined as having slope angles from 30° to 90°, are those of Jewell (1990). These new charts represent a revision of earlier work, generally giving a less conservative design and introducing some new concepts such as bond allowance and load-shedding allowance. The charts have been prepared for different pore pressure coefficients, which relate the pore pressure u to the overburden pressure γz ($r_u = u/\gamma z$). They yield the required length of reinforcement and the minimum required reinforcement force at all levels of the structure. The design friction angle ϕ_d' is set equal to the critical state shearing resistance ϕ_{cs}', implying c'= 0.

For many practical problems of slope reinforcement the various published design charts are often oriented towards specific products and limited with respect to geometry and design parameters. As demonstrated by Greenwood (1990) conventional limit equilibrium analysis by the method of slices can be successfully adapted to incorporate the effect of the reinforcement, either yielding a factor of safety in terms of a moment equilibrium (for circular failure surfaces) or horizontal force equilibrium (for non circular or wedge type mechanisms).

In slope remediation work soil strength parameters may be estimated from a back analysis of the failed slope. The reinforcement magnitude T, a tensile force, introduced into the stability equation is determined by the design strength of the reinforcement or its pull-out resistance calculated for the particular depth and effective length of embedment, whichever is lower.

The stability analysis of a slope repair option is already valuable even if it indicates only the relative improvement gained, rather than an accurate value for the absolute safety factor achieved. A "sensitivity analysis" of a slope with respect to parameters such as soil strength, pore pressure and reinforcement density develops an understanding of the problem in hand.

With respect to absolute safety factors, Greenwood (1990) indicates that where conservative design parameters have been chosen, it is quite common to work to safety factors in the range from 1.1 to 1.3; a higher value may, however, be appropriate for high risk situations.

Metal strip and mesh reinforced walls

The first U.S. Reinforced Earth wall on a highway project was built on Route 39 near Los Angeles, California, in September, 1971. This wall was actually built across a debris slide which closed the original road. At roadway elevation, the slide was 120 m wide; it extended some 270 m above and more than 140 m below the road, at a slope between 1 to 1.5 : 1. The natural slope below the slide area is about 3:1 (Chang, Forsyth and Smith, 1974).

Another early Reinforced Earth project, well documented by Royster (1974), also represents a slide repair, this time involving 200 m of roadway reconstruction along the Interstate 40 Tennessee.

The first of the above RE projects used metal face panels, the second the now standard concrete cruciform units. Many projects of similar nature have now been completed.

Fig. 2 shows the reinstatement of the Lawrence Hargrave Drive south of Sydney closed by a landslide during the heavy rainfalls of 1988. The structure, designed by Longmac Associates of North Sydney consists of two Terramesh walls with a combined height of 13 m. Slag from the nearby Port Kembla steelworks was used as backfill in the reinforcement zone. A geotextile acted as a filter layer behind the gabion face. Enkamat, hay mulch and topsoil helped to establish vegetation on the slag and earth fill surfaces.

Geogrid and geotextile walls

Greenwood (1990) describes the use of geogrids for the repair of highway embankments in South East England. Many of these embankments have been built in clay which tend to show shallow slips due to softening of the clay some 5 to 6 years after construction. The repair begins with excavating the softened material and benching the remaining slope. The slope is then rebuilt, at an angle of say 30° to 35°, with layers of geogrid reinforcement at typically 750 mm vertical spacing. The excavated clay may be reused as backfill, possibly with the addition of lime for improved compactibility.

The repair of a large landslide in San Pablo, California, reported by Bonaparte and Margason (1984), is said to be this state's first use of geogrids in slope remediation. Reoccurring slide movement extending over an area of some 50000 m^2 (12 acres) was accelerated by heavy winter rains in 1981/82 and 1982/83. To save the road leading across the headscarp, and adjacent houses, emergency stabilization was initiated, followed by permanent repairs. Temporary stabilization was achieved by driving 63 steel H-piles along the high side of the road. The piles were driven to a depth of 10 m, which represented twice the depth of the lowest of the known multiple failure planes. This provided temporary support of the sliding soil mass and, after installation of tiebacks in critical places, allowed the construction of a geogrid reinforced soil prism keyed into hard claystone. This reinforced soil structure was designed to stand up even after further slope movement would remove downhill earth support. A drainage system complemented this more permanent slide repair (Fig. 3).

A typical slope rehabilitation design using geogrid as practiced in Malaysia is shown in Fig. 4 (Courtesy Netlon Pty.Ltd). The slide material is excavated, the slope benched back, a drainage system is installed and the slope is reformed using horizontal layers of geogrid reinforcement.

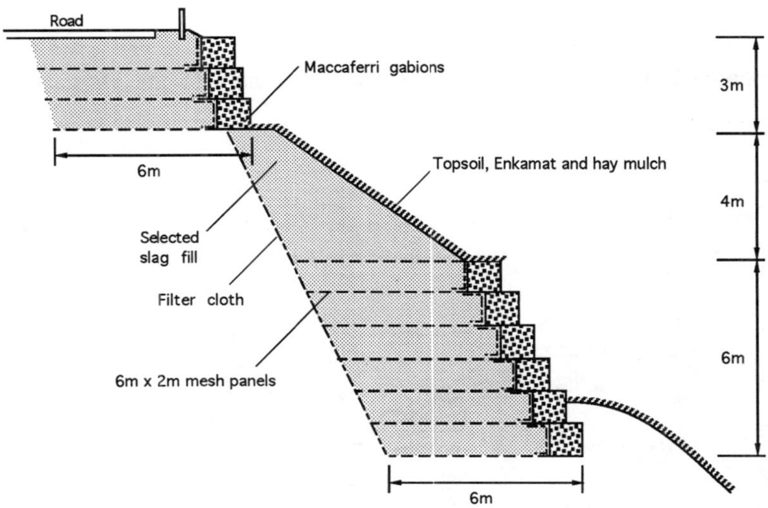

Figure 2. Terramesh wall at Lawrence Hargrave Drive

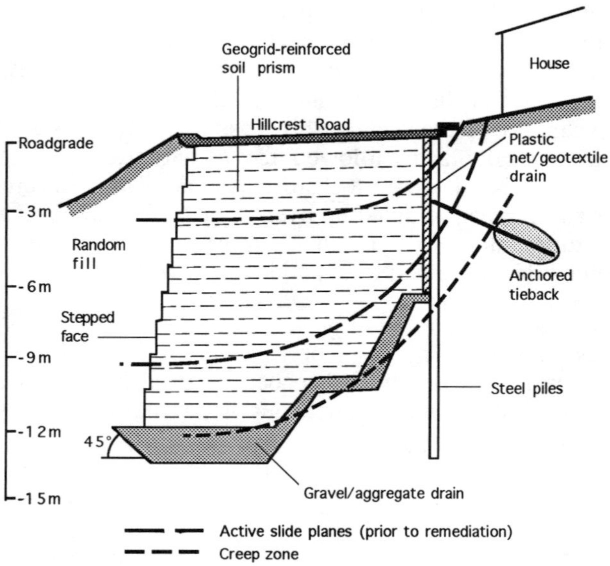

Figure 3. Geogrid reinforced slide repair (Bonaparte and Margason 1984)

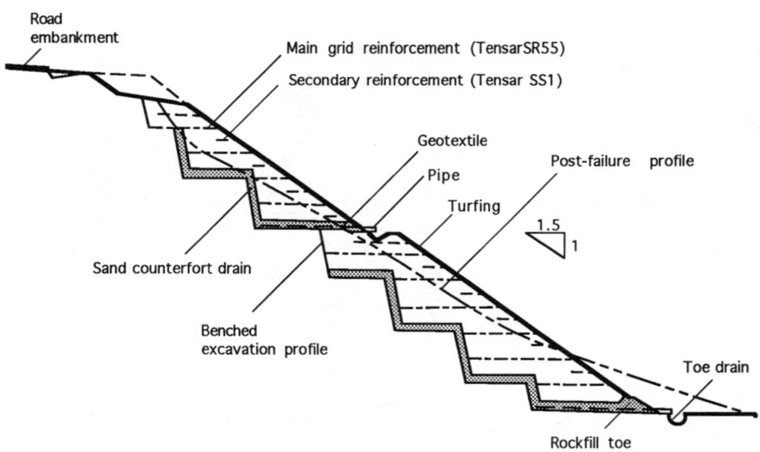

Figure 4. Slope rehabilitation in Malaysia (Information provided by Netlon Pty.Ltd.)

Fig. 5 shows the scheme adopted by the Road Construction Authority of Victoria (Flintoff 1988) to stabilize a failed embankment in the Strzelecki ranges (Grand Ridge Road). The first stage involved building an access ramp and inserting timber power poles into prebored holes to form a key and platform on which to construct the second stage, a geotextile wall. The fabric employed was a woven polypropylene with a tensile strength of 18 to 20 kN/m, placed at 300 mm intervals, lapped over at the face. For UV-protection the face was sprayed with hot bitumen, but this technique is not recommended since it softened the fabric and did not achieve complete penetrating cover.

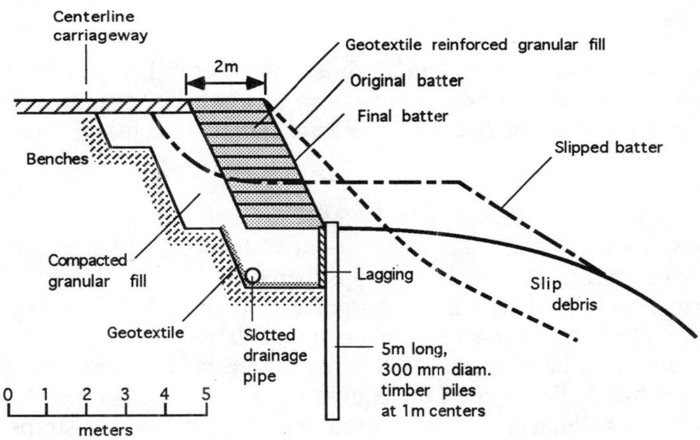

Figure 5. Grand Ridge Rd. - Slip repair with geotextile reinforced fill (Flintoff 1988)

Pneusol

The use of automobile tires as high strength tensile inclusions in a soil mass was one of the ideas stimulated by Vidal's invention of Reinforced Earth (Vidal 1969) and the need to find solutions for the disposal and recycling of the ever growing mountain of discarded tires. According to Forsyth and Egan (1976) the California Department of Transportation was charged in 1973 to study the tyre disposal problem. Interestingly, the first practical application was the repair of a sidehill fill slipout along Calif-236 north of Santa Cruz. After removing the debris to below the slip surface, a drainage system was installed and the embankment rebuilt with tire sidewall mat reinforcement at 0.6 m vertical intervals. The individual sidewall mats were interconnected by steel clips and lapped over the edge of the embankment, thus providing erosion protection. Reinforcement allowed the embankment to be constructed with a side slope of 0.5:1, rather than the conventional 1.5:1, thus saving some 70000 m^3 of fill.

In France, research into the use discarded tires dates back to about the same time as in California. As reported by Long (1990) French engineers have now built more than 200 structures with tire reinforcement in France, and 12 in Algeria. Not only that, but this construction technique has now been given a name: Pneusol (or tiresoil), a registered trade mark.

Research carried out by the Laboratoire Central des Ponts et Chaussées looked at reinforcement in form of whole tires, sidewalls, or treads placed on edge or flattened. Here are some of the major findings (Long 1990):

- Pull-out resistance varied from 19.5 kN for a single sidewall mat to 68 kN for a group of treads. (Pull-out tests were conducted with single tire elements as well as groups thereof.)
- The unit weight of trial fills incorporating tires varied from 6 to 8 kN/m^3. (This makes it a very light weight fill!)
- The stress-strain modulus of the reinforced soil is less, but the strength higher than that of the equivalent unreinforced soil.
- A variety of connectors can be used: Synthetic ropes and straps or metal hooks. (A layer of geotextile could also provide interconnection between tires, apart from separating soil/tire layers).

One quarter of the French projects have used Pneusol for slope remediation and erosion protection (Fig. 6). Other applications have been: building retaining structures, providing a light weight fill, reducing earth pressures on structures, absorbing energy (including that of avalanches), creating arching (load distribution) above buried culverts, and improving trafficability of the ground for military vehicles.

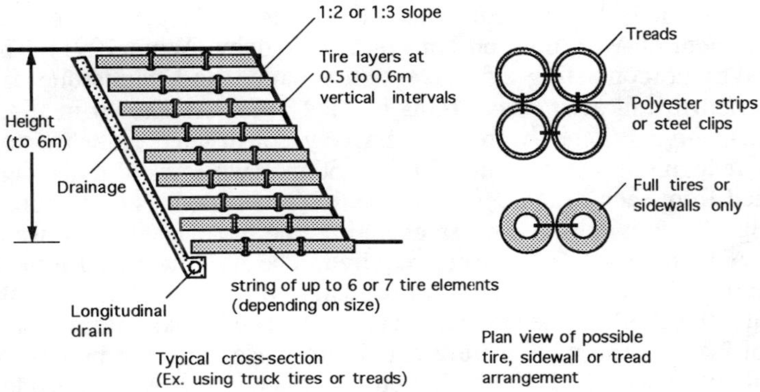

Fig. 6. Pneusol

In-situ reinforcement: rock bolting, soil nailing, dowelling

Rock bolting is frequently used to stabilize rock slopes. Rock bolts typically are 15-30 mm diameter steel bars which are inserted into boreholes and fixed in position by wedges or grout. They may be tensioned by tightening a nut against the face plate. Grouting after installation is essentially for corrosion protection.

Rockbolts have two main purposes: To pin or nail well-defined blocks or slabs of rock onto a more stable formation, or to form a new structural entity out of jointed rock by applying compressive stresses. In both cases it is attempted to mobilize the inherent shear strength along existing joints by confining and/or stressing the rock mass. Rockbolts therefore can do more than transfer a surface load to a deeper rock zone; by preserving the integrity of a jointed and fractured mass of rock, they act as a form of reinforcement rather than anchorage.

Soil nails typically are 20 to 30 mm diameter steel bars providing primarily tensile reinforcement in a soil mass. They are generally installed in boreholes and bonded into place with low pressure grout. Soil nails allow the in-situ formation of retaining structures acting similarly to a Reinforced Earth or geogrid wall. Recent investigations (Pedley et al 1990) seem to indicate that in the analysis of steep slopes and excavations stabilised by conventional soil nailing the potential benefit of bending stiffness is best ignored.

The soil nailing solution was chosen to stabilize a freeway embankment at Mt.White, 60 km north of Sydney (Won, 1991). The embankment consisting of clayey silty sands and sandstone fill threatened to fail after heavy rains in June 1989. As shown in Fig.7 four 2 m high benches were progressively constructed from the top down, reducing in length from 120 m to 60 m. Two rows of nails were installed from each bench before excavation proceeded. The horizontal spacing of the nails was 2 m, staggered at one meter centres between rows. A total of 400 nails were required. The nails were 20 mm in diameter, 12 m long, inclined downwards at 10°; they had an ultimate capacity of 125 kN, which compared to a calculated maximum working load of 70 kN. The nails were grouted into place from the bottom of the hole up. Each nail had a bolted on face plate which was embedded in the 150 mm thick mesh reinforced shotcrete facing. Six of the nails were instrumented with strain gauges. The maximum tension measured since the completion of the wall in December 1990 has been less than 25 kN. As an experiment, six so-called HOPE (Highly Oriented Poly-Ethylene) bars were also installed on this project. Monitoring is proceeding.

The term "soil dowels" is now increasingly used for in-situ reinforcing elements which are relatively stiff and have diameters ranging from 70 to 200 mm. They may be steel pipes driven into the ground or cast-in-situ concrete "micropiles" or "root piles". Characteristically they are installed in a slope so as to intersect actual or potential failure planes, where they have to primarily resist significant shear forces (say 20 to 100 kN per dowel). In a typical root pile application for slope stabilization the dowels are oriented at a variety of angles and therefore may also have to resist compressive and tensile stresses.

To use the term rock dowel for all untensioned rock bolts is not recommended; it should only be used where the primary purpose is to resist shear, rather than tension or compression.

An application of root piles for the stabilization of a landslide in Pennsylvania has been described by Dash (1984). This project involved the installation of 458, 125 mm diameter cast-in-place concrete piles forming an underground "wall" across the landslide.

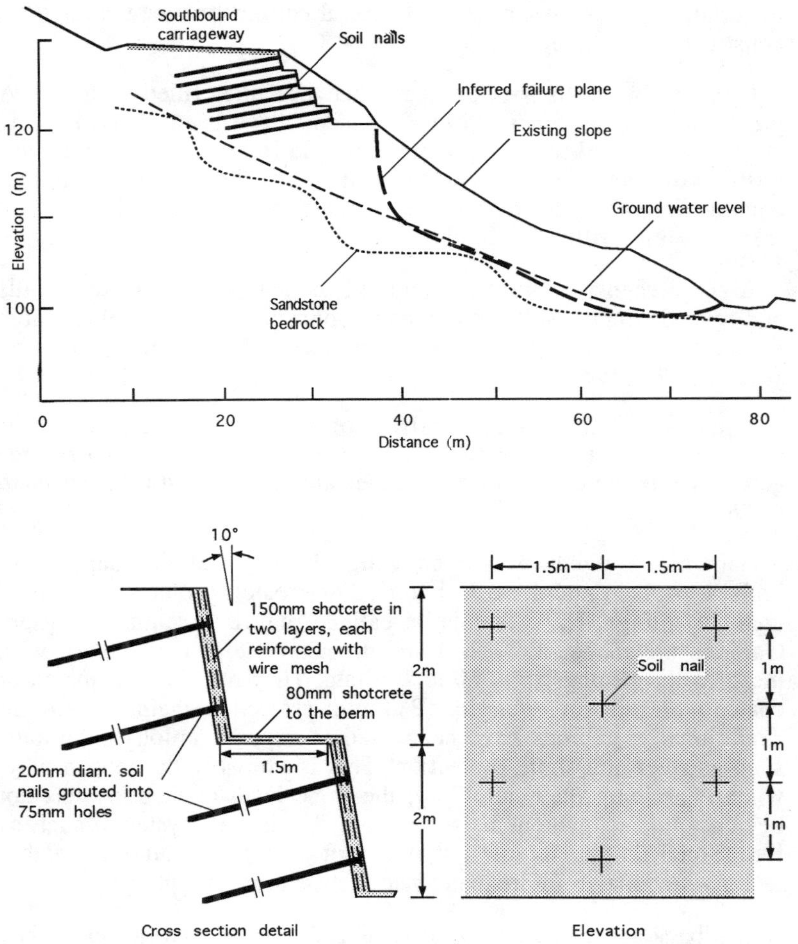

Figure 7. Soil nailing at Mt.White (Won 1991)

Reinforcement by confinement: gabions, crib and bin walls

Reinforcement by inclusions is conceptually not as far away from reinforcement by confinement as one might think. In both cases elements able to resist tensile stresses are combined with a particulate material, together forming new structural entities or a new composite construction medium.

Crib and bin walls are box like lattice structures filled with soil. A great variety of different crib wall systems have been devised, using steel, concrete and timber elements, either in form of beams or frames. Crib walls have the advantages of high flexibility, elemental construction, aesthetically pleasing appearence and permeability (with free draining infill).

A comprehensive long-term study of the performance of crib walls and related structures, including their application in slope stabilization, has been carried out in Austria, and its results have been reported by Brandl (1980, 1982, 1984, 1986).

What is different in the analysis of crib walls as against other reinforced soil structures is the existence of so-called silo-pressures acting within the cells. Basic principles are summarized by Hausmann (1990).

Gabions are wire baskets filled with rock. A typical unit may be 1 m high, 1 m deep and 1 or 2 m long. For erosion protection, such as channel linings, large thin units called "mattresses" may be used. Gabions are made of 2- to 3-mm-diameter welded or woven wire mesh, with openings from 60 to 100 mm. They may be galvanized or coated with polyvinyl chloride (PVC) for protection against corrosion. Traditionally, gabions have been used in very hazardous situations, such as river and shore protection. This is probably the reason why, when examining their durability, the same level of scrutiny has not been applied to gabions as to reinforced soil wall systems, such as Reinforced Earth. The low initial cost of gabion structures is another factor which makes future maintenance costs more acceptable.

Sandbags are another form of reinforcement by confinement. They have many uses for temporary erosion protection and emergency slope repair.

The close relationship between reinforcement by inclusions and confinement, as well as anchorage, becomes apparent when examining some of the innovative elemental retaining wall systems which have appeared on the market in recent years: NEW (Neue Ebenseer Wall), American Geotech walls, Anchored Earth, Hilfiker Welded Wire Wall - just to name a few (Hausmann 1989).

Fiber reinforcement: Texsol

Texsol, another French invention, is granular soil reinforced with continuous synthetic fiber. Although only making up 0.1 to 0.2 % by weight of the natural soil, one cubic meter of mix may contain several tens of kilometers of fiber, creating a cohesion in the order of 100 to 300 kPa. With Texsol, oversteep slopes, say at 60° to 70° to the horizontal can be constructed. Texsol therefore has the potential to be used for the reinstatement of unstable slopes (LCPC 1990).

Vegetation

The principles of stabilizing earth structures with vegetation is now well established and is covered in books by Gray and Leiser (1982), Bache and MacAskill (1984) and Coppin and Richards (1990). The latter also propose a modification of Greenwood's slope stability equation which takes the effect of vegetation into account by considering parameters such as:

u_V = decrease in pore-water pressure due to evapotranspiration by vegetation at slip surface, kPa
c'_R = enhanced effective soil cohesion due to root matrix reinforcement by vegetation along slip surface, kPa
S_W = surcharge due to vegetation, kN/m
D = wind loading force parallel to slope, kN/m
T = tensile root force acting at base of slice, kN/m
θ = assumed angle between roots and slip surface

Note that wind loading D as well as surcharge S_W can be destabilizing effects. The exact values of the above parameters will depend on the type of vegetation, particularly the strength and depth of the root system. The increase in shear strength of soils due to tree roots alone can be in the order of 2 to 10 kPa or more.

Lee (1985) also adapted a conventional method of stability analysis to include the effect of vegetation. In Lee's approach, soil suction due to evapotranspiration is represented in terms of an increased effective cohesion, rather than a decrease in pore pressure as assumed above.

Sotir and Gray (1989) describe the stabilization of a highway embankment using layers of live brush, fascines and stakes. In some areas where there appeared to be a need for additional reinforcement, Tensar geogrids were placed beneath the brushlayers.

Vegetation may be an integral part of reinforced soil structures where face "panels" consist of a combination of wire or geogrid mesh and/or geotextiles. The TEXTOMUR-system developed in Switzerland (Ruegger, 1986) allows over-steep (say 60°-) slopes to be constructed using *L*-shaped wire mesh units at the face and horizontal non-woven geotextiles reinforcement in the backfill. An open-pored "vegetation fabric" inside the retaining mesh protects the fill against erosion, acts as a carrier for Hydroseed and supports the growing plants.

Other methods of ground modification

Hydraulic modification and modification by inclusions and confinement are the principal categories of ground improvement methods relied upon for slope stabilization. Other techniques have been used, essentially for increasing the shear strength in the ground, but are not widespread: mechanical modification (compaction), physical and chemical treatment (heating and freezing, stabilization with admixtures), grouting, and consolidation by preloading and electroosmosis (both of which could be classified as a form of hydraulic modification). Case studies involving these methods are rare in the landslide or slope stability literature. One reason for their absence is that they are best applied as a preventative measure, such as for ground treatment before an excavation is made which may bring about slope instability.

Compaction should be part of the engineering applied to any reshaping of a slope. Stabilization with admixtures (lime, cement) may be included in rebuilding a failed slope with local materials. Densification and admixtures do increase the shear strength of the soil, but may also reduce its permeability. Care has to be taken to make appropriate improvement in the drainage provisions.

Blasting of rock materials may contribute to their stability by improving drainage at the same time as causing densification, but this method is controversial, to say the least (see Baker and Marshall, 1958). Blasting may of course also initiate further sliding.

Grouting is a valid method in many situation where the foundation strength has to be improved. However, caution is advised in slope remediation, since grouting may also produce a temporary increase in the pore pressure which could aggravate movement. Pore pressure may also be increased where existing drainage paths are blocked. Permeable grouts are available but this technology is not well known.

The chemical stabilization of an active landslide using columns of quicklime was described by Handy and Williams (1967) as quoted by Gedney and Weber (1978).

Pressureless chemical grouting was part of the techniques used to stabilize a slope in a valley near Interlaken, Switzerland (Rieder and Ischi, 1988). Improved drainage and nailed wall supports were other works carried out in this project.

Thermal treatment of soils has been applied to slope stabilization (reviewed by Hausmann, 1990) but this method would only have application in exceptional circumstances due to the high energy requirements and undesirable environmental impacts.

Freezing (either naturally occuring or artificial) has potential for temporary slope stabilization, e.g. during the construction of supporting structures.

Consolidation and the accompanying gain in shear strength is a beneficial side effect of surcharging the lower portion of a sliding mass of clayey soil. Although additional fill further up the slope may be destabilizing in the short term, it could be beneficial in the long term, as a build-up in effective stresses increases frictional resistance. This is provided the fill is below the so-called neutral line (see Bromhead 1986, quoting Hutchinson, 1977).

Electrokinetic dewatering and stabilization, reviewed by Hausmann (1990), remains an intriguing phenomenon. Active dewatering through electroosmosis would only be considered for slope remediation in exceptional circumstances, partly due to the energy costs involved. However, Veder (1981), who was an astute observer of electroosmotic effects, reported astounding stabilizing effects when metal rods (15- to 25-mm diameter, at 3- to 8-m intervals) were driven through the unstable zone in several active landslides. He concluded that rather than just providing reinforcement and possibly drainage, these rods acted as short-circuit conductors between soil layers of unequal potentials. This stopped electroosmotic flow of water into the contact zone (= failure plane), which was responsible for high pore pressures leading to instability. Veder further speculated that the success of soil nailing and even rock bolting may be partly due to eliminating electroosmotic phenomena.

COMPARING SLOPE REMEDIATION OPTIONS

It is difficult to map out a generally valid approach to slope remediation, because of the many different forms slope instability can take. Topography, geology and hydrology change from case to case and even a detailed investigation cannot eliminate all uncertainty. The economic and social risks associated with a landslide hazard may be assessed differently by the particular groups of people affected. The risk of a landslide is itself not a static measure. With steadily growing traffic along a nation's transportation routes and with the ever wider urbanization of the landscape, the potential damage caused by mass movement is also increasing. The corporate liability of owners of transportation facilities rises in sympathy, as does the personal responsibility of the design engineer. All these developments seem to require increasing conservatism in slope remediation, yet the economy of the country may not allow the necessary expenditure. This situation is only resolved by accepting higher risks - or better risk management.

The two examples following illustrate how different options for slope remediation can be assessed. Both relate to instabilities affecting the railway track between Sydney and Wollongong in the south.

The first case concerns the Coalcliff area where mass movement has been recorded as far back as 1890. Considerable track distortion became visible in 1988 during an extremely wet period. Apparently contributing to the land instability has been extensive coal mining below the general area. The consultants, Longmac Associates Pty.Ltd, prepared a remedial options table, reproduced in Table 4. The first three options represent total drainage systems, stabilizing the embankment as well as the slope it sits on. Options D to G stabilize the embankment and track by isolating it from the lower slope area. The various options are ranked according to criteria of technical merit, ease of construction, maintenance costs, capital costs, disruption to traffic and environmental impact. The necessity for major earthworks is considered to be detracting from a solution's environmental desirability. As advised by the consultants, the actual solution will have to be chosen by the SRA on the basis of general policy, available resources, cost benefit and the future plans for the railway.

TABLE 4. Summary of options for Coalcliff slope remediation
(State Rail Authority of New South Wales*)

Option	Description	Ranking (1 - most beneficial, 7 - least beneficial)					
		Technical merit	Ease of con-struction	Mainte-nance costs	Capital costs	Disrup-tion to traffic	Environ-mental impact
A	Slot drains, extended under sidings	1	4	5	2	7	7
B	Slot drains with drainage drives under sidings	5	6	6	5	3	4
C	As per A, with stability berm	2	2	3	4	4	6
D	Eductor wells	7	1	7	1	1	1
E	Berm + partial slot drains	3	3	4	3	5	5
F	Buttress earthworks	4	7	2	6	6	3
G	Pier wall	6	5	1	7	2	2

* Based on a report prepared by Longmac & Associates Pty.Ltd)

The second case concerns the Coledale embankment which failed in 1988 with the loss of two lives. This site also has a long history of instability and repair work. Some eight factors were identified which contributed to the failure, but it was a blocked culvert which triggered the disaster when it caused water to overflow and saturate the embankment. Immediate remedial works involved reconstructing the embankment with improved culverts and the installation of eductor wells, which are closely monitored. Long term remediation options considered by Golder Associates Pty. Ltd. (Sydney) were: Complete rebuilding of the embankment, constructing a new culvert system, building a gravity wall, installing contiguous piles, and providing full length drainage trenches. Even building a viaduct bridging the unstable area was investigated. Another option involving rockfill berms, partially keyed into the residual soil, was proposed by SMEC (Snowy Mountains Engineering Corporation Ltd). Each of these options was assessed in the following terms:

- time required for design and construction
- maintenance of rail services
- risk
- expected life of works

- estimated cost
- land resumption required
- ease of construction

Final decisions with regard to the permanent rehabilitation of this area have not yet been made.

CONCLUSIONS

The bulk of the literature on landslides and slope stability is concerned with the description of the many and varied forms mass movement and erosion can take, and the equally diverse long list of methods of analysis which are available to explain and predict this behaviour. Papers and books relating the success or otherwise of engineered slope stabilization are in the minority, although many ingenious solutions to slope instability problems have been found and are still being developed. Landmark contributions to landslide control and slope design can be found in the reports and proceedings edited by Eckel (1958), the ASCE Soil Mechanics and Foundation Division (Stability and Performance of Slopes and Embankments 1966), Schuster and Krizek (1979), Reeves (1982) and Walker and Fell (1987).

Strategies in dealing with slope instability can be grouped into avoidance, prevention, acceptance, and remediation. Engineered slope remediation seeks to introduce external stabilizing forces or attempts to improve the soil through ground modification, either hydraulically, by inclusions and confinement, chemically, physically or mechanically.

When confronted with existing or potential slope instability the engineer now benefits from progress being made in the following areas:

- Recognition and mapping of instabilities
- Improved land management practices, including legislation protecting unstable land
- Automation of geotechnical performance monitoring
- Innovations in soil reinforcement techniques and anchorage systems
- Developments in biotechnical methods of slope protection
- Introduction of risk management systems

The engineer's work is still hampered by the fact that the hazard of land instability, just like the possibility of foundation failure, is not as readily accepted by the community as other risks. This puts a heavy burden on the organisations and individual engineers involved in infrastructure design and maintenance, because decisions will always have to be made in an environment of uncertainty, despite all the progress made in the last 25 years. To paraphrase Ralph Peck (1966), it is still possible that nature will outwit us on occasions.

ACKNOWLEDGEMENTS

The cooperation of the State Rail Authority of N.S.W., especially the help received from David Christie, Principal Geotechnical Engineer, is gratefully acknowledged. Valuable information was also received from Greg W. Won (Geotechnical Group of the Road and Traffic Authority of N.S.W.), Mike Dobie (Netlon Ltd. Malaysia), Ian Hannam (Soil Conservation Service of N.S.W.) and John Elliot (Cumec Pty.Ltd.). Prof. Harry Poulos, Phillip Pells (Coffey Partners International) and Paul Hewett (Arup Geotechnics) made valuable suggestions after reviewing the draft manuscript. The author is also indebted to the librarians of the University of Technology, Sydney, for their assistance in finding important reference material.

APPENDIX. REFERENCES.

Anderson, W.F., Hanna T.H., and Abdel-Malek, M.N. (1983). "Overall stability of anchored retaining walls." *Journal of the Geotechnical Engineering Division*, ASCE, GT 109 (11), 1416-1433.Bache, D.H., and MacAskill, I.A. (1984). *Vegetation in civil and landscape engineering*. Granada, London.

Baker, R.F., and Marshall, H.C.(1958). "Control and Correction". Chapter 8 in "Landslides and Engineering Practice", E.B. Eckel, Editor. *Highway Research Record*, Special Report 29, NAS-NRC Publication 544, 150-188.

Bianco, B., (1988). "Landslide stabilization by means of drainage tunnels in central Italy." *Proc. Fifth International Symposium on Landslides*, (Bonnard, C., Editor) 2, 857-862.

Blong, R.J., and Eyles, G.O. (1989). "Landslides; Extent and economic significance in Australia, New Zealand and Papua New Guinea. *Proc. 28th International Geological Congress: Symposium on Landslides*, Washington. Balkema, Rotterdam (Brabb E.E. and Harrod, B.L., editors).

Bonaparte, R., and Margason, E. (1984). "Repair of landslides in the San Francisco Bay area". Proc. Conf. on *Polymer Grid Reinforcement*, London. Thomas Telford, London.

Boyd, M.S. (1991). "Elements of Reinforced Soil Wall Systems with particular reference to the design and application of Inextensible (Metallic) Reinforcement Systems". Lecture presented at the Ground Modification Seminar No.1 on "Soil Reinforcement", April/May 1991, University of Technology, Sydney.

Brandl, H. (1980). "Tragverhalten und Dimensionierung von Raumgitterstützmauern (Krainerwänden)." Bundesministerium für Bauten und Technik, *Strassenforschung*, Heft 141, Wien.

Brandl, H. (1982). "Raumgitter-Stützmauern (Krainerwänden)." Bundesministerium für Bauten und Technik, *Strassenforschung*, Heft 208, Wien.

Brandl, H. (1984). "Schadensfälle an Raumgitterstützmauern." Bundesministerium für Bauten und Technik, *Strassenforschung*, Heft 251, part 2, Wien.

Brandl, H. (1986). "Stützmauersystem 'NEW' und andere Konstruktionen nach dem Boden-Anker-Verbundsprinzip" Bundesministerium für Bauten und Technik, *Strassenforschung*, Heft 280, Wien.

Brandl, H. (1988). "Stabilization of deep cuts in unstable slopes." *Proc. Fifth International Symposium on Landslides*, (Bonnard, C., Editor) 2, 867-872.

Bromhead, E.N. (1986). *The stability of slopes*. Surrey University Press (U.K.), Chapman and Hall (U.S.A.).

Carpenter, J.R. (1988). "Analysis of tieback slopes and walls using STABL5 and PCSTABL5." *Transportation Research Record* 1191, 176-187.

Cartier, G., Waschkowski, E., Deniau, A., and Dupeuple, P. (1989). "Renforcement des sols par colonnes compactées mélangées en place". *Proc. of the Twelfth International Conference on Soil Mechanics and Foundation Engineering*, Rio de Janeiro. Balkema, Rotterdam, Vol. 2, 1237-1238.

Castelli, R., and Munfakh, G. (1986). "Geotextile walls in mountainous terrain". *Proc. Third International Conference on Geotextiles*, Vienna, 2, 459-463.

Chang, J., Forsyth, R., and Smith, T. (1974). "Reinforced Earth Highway Embankment - Road 39". California Dept. of Transport, Rep. No. CA.DOT. TL 2115-9-7437.

Charman, P.E.V., and Murphy, B.W., (ed.) (1990). *Soils, their properties and management* - A soil conservation handbook for New

South Wales. Sydney University Press/Oxford University Press Australia.

Christie, D. (1991). *Personal Communication.*. Principal Geotechnical Engineer, State Rail Authority of New South Wales, 9-13 Unwins Bridge Rd., Sydenham, N.S.W. 2044, Australia.

Collotta, T., Manassero, V., and Moretti, P. C. (1988). "An advanced technology in deep drainage of slopes." *Proc. Fifth International Symposium on Landslides*, (Bonnard, C., Editor) 2, 887-892.

Coppin, N.J., and Richards, I.G. (1990). *Use of vegetation in civil engineering*. Butterworths, London.

Dash, U. (1984). "New methods of slope stabilization". Proc. of the International Conference on In-Situ Soil and Rock Reinforcement, Paris, 249-255.

Descoeudres, F. (1988). "Special lecture: Stabilizing methods in rock slopes against sliding, toppling or rock falls." *Proc. Fifth International Symposium on Landslides*, (Bonnard, C., Editor) 2, 821-828.

E.B. Eckel, Editor (1958). "Landslides and Engineering Practice." *Highway Research Record*, Special Report 29, NAS-NRC Publication 544.

Einstein, H.H. (1988). "Special lecture: Landslide risk assessment procedure." *Proc. Fifth International Symposium on Landslides*, (Bonnard, C., Editor) 2, 1075-1090.

Elliott, J., (1991). *Personal communication*. Cumec Pty.Ltd., 66 Gipps Street, Birchgrove, N.S.W. 2041, Australia.

Fell, R, MacGregor, J.P., Williams, J., and Searle, P. (1987). "The Hue Hue Road landslide, Wyong." *Proc. of an extension course on "Soil Slope Instability and Stabilisation"*, Sydney. Walker, B. and Fell, R., Editors, Balkema, Rotterdam, 315-324.

Flintoff, W.T. (1988). "The repair of landslips on mountain roads in Victoria." *Proc. 14th Australian Road Research Board (ARRB) Conference*, Canberra. ARRB, Vermont South, Victoria, Australia.

Forsyth, R.A., and Egan, J.P. (1976). "Use of waste materials in embankment construction." *Trans. Res. Record* No. 593, PP3-8.

Fukuoka, M. (1977). "The effects of horizontal loads on piles due to landslides." *Proc. 9th Int. Conf. on Soil Mechanics and Foundations*, Tokyo, Specialty Session 10.

Gedney, D.S., and Weber, G. Jr. (1978). "Design and Construction of Soil Slopes." Chapter 8 in "Landslides, Analysis and Control", Schuster, R.L., and Krizek, R.J., Editors. Transportation Research Board, National Academy of Sciences, Washington, Special Report 176, 172-191.

Geotechnical manual for slopes. (1984), 2nd edition. Geotechnical Control Office, Engineering Development Department, Kowloon, Hong Kong.

Grant, K. (1974). "The PUCE Programme for Terrain Evaluation for Engineering Purposes. II-Procedures for Terrain Classification". *Div. of Applied Geomechanics, CSIRO, Australia, Technical Paper No.15.*

Gray, D.H., and Leiser, A.T. (1982). *Biotechnical slope protection and erosion control.* Van Nostrand Reinhold Company, New York, N.Y.

Greenwood, J.R. (1990). "Design approach for slope repairs and embankment widening."*Proc., Conf. on Reinforced Embankments, Theory and Practice in the British Isles*, Cambridge University, Sept.1989, Thomas Telford, London, 51-61.

Hannam, I.D., and Hicks, R.W. (1980). "Soil conservation and urban land use planning."*J. of the Soil Conservation Service of N.S.W.*, July, 36(3), 134-145.

Hausmann, M.R. (1990). *Engineering principles of ground modification.* McGraw-Hill, New York.

Hinckley, A.L., Tanner, B.D., and Campbell, E.C. (1987). "Automation of monitoring of geotechnical instrumentation". *Transp.Research Record* 1119, 26-29.

Huder, J. (1990). "Geotechnik des Bündnerschiefers". *Mitteilungen des Instituts für Grundbau und Bodenmechanik*, Eidgenössische Technische Hochschule Zürich, Nr. 139, 3-17.

Ingold, T. (1982). "An analytical study of geotextile reinforced embankments."*Proc. of Second International Conference on Geotextiles*, Las Vegas, August, 3, 683-688.

Jedlicka, M. (1986). "Stabilisierung durch Horizontalbohrungen." *Tiefbau, Ingenieurbau, Strassenbau*, (28) 1, 22-26.

Jewell, R.A. (1990). "Revised design charts for steep reinforced slopes."*Proc., Conf. on Reinforced Embankments, Theory and Practice in the British Isles*, Cambridge University, Sept.1989, Thomas Telford, London, 1-30.

Jubien, W.E., and Abbott, B.C. (1989). "A landslide stabilized by an unloading excavation". *Proc. of the Twelfth International Conference on Soil Mechanics and Foundation Engineering*, Rio de Janeiro. Balkema, Rotterdam, Vol. 3, 1569-1572.

Kirsten, H.A.D., Steffen, O.K.H. and Stacey, T.R. (1986). "Discussion, Session 7: Rockfall protection methods." *Conf. on Rock engineering and excavation in an urban environment*, February 1986, Hong Kong. The Institution of Mining and Metallurgy, London, 492-495.

Larsen, K., Bukovansky, M., and O'Sadnick, D. (1987). "Mitigating measures for lines buried in unstable slopes." *Pipe Line Industry*, Oct., 67(4), 22-25.

LCPC (Laboratoire Central des Ponts et Chaussées, 1990). *TEXSOL ouvrages de Soutènement*. Guide technique LCPC SETRA. LCPC Service IST-Publications, 58 boulevard Lefebvre 75732 Paris Cedex 15, France.

Lee, I.W.Y. (1985). "A review of vegetative slope stabilisation". *Journal of the Hong Kong Institution of Engineering*, 3 (7), 9-21.

Leschinsky, D., and Reinschmidt, A.J. (1985). "Stability of membrane reinforced slopes."*J. Geotech. Engrg.*, ASCE, 111(11), 1285-1300.

Leventhal, A.R., and Mostyn, G.R. (1987). "Slope stabilization techniques and their application." *Proc. of an extension course on "Soil Slope Instability and Stabilisation"*, Sydney. Walker, B. and Fell, R., Editors, Balkema, Rotterdam, 183-230.

Lew, K.V., and Graham, J. (1988). "Riverbank stabilization by drains in plastic clay." *Proc. Fifth International Symposium on Landslides*, (Bonnard, C., Editor) 2, 939-944.

Lippomann, R., and Gudehus, G. (1989). "Dowelled clay slopes: Recent examples". *Proc. of the Twelfth International Conference on Soil Mechanics and Foundation Engineering*, Rio de Janeiro. Balkema, Rotterdam, Vol. 2, 1269-1271.

Loewenstein, R.M., and Clemence, S.P. (1984). "Study of impact loads on retaining walls." *Proc. Fourth Australian-New Zealand Conf. on Geomechanics*, Perth, (1) 39-43.

Long, N.T. (1990). "The Pneusol." Laboratoire Central des Ponts et Chaussées, Paris. *Études et recherches des laboratoire des ponts et chaussées*, série géotechnique GT 44.

Hausmann, M.R., and Ring G.J. (1989). "Developments in Elemental Retaining Systems", *Proc. Fifth National Conference on Local Government Engineering*, Sydney, 1989, 164-168.

Nethero, M.F. (1982). "Slide control by drilled pier walls". In "Application of walls to landslide control problems", Proc. ASCE National Convention, Las Vegas, Nevada, April 29, 1982, Reeves, R.B. (Editor) 61-76.

Oakland, M.W., and Chameau, J. (1989). "Analysis of drilled piers used for slope stabilization." *Transportation Research Record* 1219, 21-32.

Orr, C.M., Swindells, C.F., and Windsor, C.R. (1991). "Open pit toppling failures: Experience versus analysis". *Computer Methods and Advances in Geomechanics*, Beer, Booker & Carter (eds) Balkema, Rotterdam, 505-510.

Paterson, S.J., Hale, G.E.A., and Ikin, D.B. (1975). "Stabilizing a landslide above Fisher penstock, Tasmania." *Proc. Second Australia-New Zealand Conference on Geomechanics*, Brisbane. The Institution of Engineers, Australia, National Conference Publication No.75/4, 314-318.

Peck, R.B. (1966). "Stability of natural slopes." *Stability and performance of slopes and embankments* (1966). Proc. of a Conference held at Berkeley, California, Aug. 22-26, 1966. ASCE, Soil Mechanics and Foundations Division, 437-451.

Pedley, M.J., Jewell, R.A., and Milligan, G.W.E. (1990). "A large scale experimental study of soil-reinforcement interaction." *Ground Engineering* July/August, 44-50 (Part 1), and September, 45-49 (Part 2).

Reeves, R.B. (1982). "Application of walls to landslide control problems", Proc. ASCE National Convention, Las Vegas, Nevada, April 29.

Rieder, U., and Ischi, H. (1988). "Investigation and stabilization of unstable slopes in the Lombach Valley near Interlaken, Switzerland." *Proc. Fifth International Symposium on Landslides*, (Bonnard, C., Editor) 2, 979-981.

Royster, D.L., (1974). "Construction of a Reinforced Earth fill along Interstate 40 in Tennessee." A paper presented at the 25th annual Highway Geology Symposium May 24, 1974, at Raleigh, North Carolina.

Ruegger, R. (1986). "Geotextile reinforced soil structures on which vegetation can be established."*Proc. of Third International Conference on Geotextiles*, Vienna, April, 2, 453-458.

Schnabel, H. Jr. (1984). "Discussion: Overall stability of anchored retaining walls." *Journal of the Geotechnical Engineering Division*, ASCE, GT 110 (12), 1817-1818.

Schuster, R.L., and Krizek, R.J., Editors (1978). "Landslides, Analysis and Control". Transportation Research Board, National Academy of Sciences, Washington, Special Report 176.

Sembenelli, P. (1988). "General report: Stabilization and drainage." *Proc. Fifth International Symposium on Landslides*, (Bonnard, C., Editor) 2, 813-819.

Sotir, R.B., and Gray, D.H. (1989). "Fill slope repair using soil bioengineering systems". *Public Works*, 120 (13), 37-40.

Stability and performance of slopes and embankments (1966). Proc. of a Conference held at Berkely, California, Aug. 22-26, 1966. ASCE, Soil Mechanics and Foundations Division.

Terzaghi, K. (1950). "Mechanism of landslides."*Geological Society of America, Berkey Volume*, 83-123.

Toh, C.T., Chee, S.K. and Ting, W.H. (1986). "Design, construction and performance of a geogrid reinforced high slope and unreinforced fill slopes." *Proc. IEM-JSSMFE Joint Symp. on Geotechnical Problems*, Kuala Lumpur, 90-111.

Varnes, D.J. (1958). "Landslide types and processes." Landslides and engineering practice, ed. by Eckel, E.B., *HRB Spec. Rept. 29*, NAS-NRC, Washington, D.C., 12-33.

Veder, C. (1981). *Landslides and their stabilization.* Springer Verlag, New York.

Vidal, H. (1969). "The principle of Reinforce Earth." HRB, *Highway Research Record* 282, 1-16.

Viggiani, C. (1981). "Ultimate lateral load on piles used to stabilize landslides." *Proc. 10th Int. Conf. on Soil Mechanics and Foundations*, Stockholm, 555-560.

Walker, B., Dale, M., Fell, R., Jeffrey, R., Leventhal, A., McMahon, M., Mostyn, G., and Phillips, A. (1985). "Geotechnical risks associated with hillside development." *Australian Geomechanics News*, No. 10, 29-35.

Way, D.S. (1978). *Terrain Analysis.* Dowden, Hutchinson & Ross, Inc., Stroudsburg, Pa., and MacGraw-Hill Book Company, New York, N.Y.

Won, G.W. (1991). *Personal communication.* Geotechnical Group, Materials Section, Roads and Traffic Authority, 28 Ennis Road, Milsons Point, N.S.W. 2061.

USE OF DRILLED SHAFTS IN STABILIZING A SLOPE

By Lymon C. Reese[1], Honorary Member, ASCE,
Shin-Tower Wang[2], Member, ASCE, and
Jeffrey L. Fouse[3], Member, ASCE

ABSTRACT

A rational method is presented for the analysis of drilled shafts that are used for improving the stability of an earthen slope. The method includes procedures for estimating the loads due to earth pressures from a potentially-sliding mass of soil; procedures for assessing the resistance of the soil below the possible surface of sliding; procedures for estimating the response of the drilled shaft, both above and below the estimated sliding surface; and procedures for estimating the factor of safety of a slope that is reinforced by drilled shafts. The paper is aimed at presenting a complete analysis for the stability of slopes with drilled shafts in place. The benefits of using the method are elaborated in the paper, and rationality and convenience are indicated that were not previously available.

INTRODUCTION

Remediation of a slope in distress can be accomplished sometimes by drainage, but the use of piles is an option when more substantial measures are necessary. Drilled shafts or bored piles at various spacing have been used successfully to prevent sliding, but generally accepted guidelines for the analysis of such a problem have been unavailable. This paper makes a contribution to the development of a method of analysis. For simplicity of expression, the broad term "pile" will be employed when referring to drilled shafts or bored piles. Assuming that a slope-stability method is at hand that can utilize the resisting forces from a single pile or a group of piles, the principal problems to be solved are: the distribution of earth pressures along the portion of a pile above the sliding surface, the response of the portion of the pile below the sliding surface, and the capacity of a reinforced-concrete section to resist a bending moment. Each of these problems is discussed.

---

[1] Nasser I. Al Rashid Chair Emeritus, Dept. of Civ. Engrg., The Univ.of Texas at Austin, Austin, Tex., Telephone: 512/458-1128; Fax: 512/467-1384.

[2] Project Engr., Lymon C. Reese & Associates, P. O. Box 180348, Austin, Tex. 78718.

[3] Senior Project Engr., Woodward Clyde Consultants, St. Louis, Missouri.

The rational method presented herein includes: (1) procedures for estimating the loads due to earth pressures from a potentially-sliding mass of soil; (2) procedures for assessing the resistance of the soil below the possible surface of sliding; (3) procedures for estimating the response of the pile, both above and below the estimated sliding surface; and (4) procedures for estimating the factor of safety of a slope that is reinforced by piles. While the discussion about a method for predicting the factor of safety of a slope against sliding is limited, mention is made of the necessity of including the influence of the piles in finding the potential surface of sliding.

ANALYTICAL PROCEDURE

A view of a pile embedded in a slope is shown in Fig. 1a where the depth to the critical surface of sliding is given by the symbol h_p. The distributed forces from the sliding soil are shown by the arrows, parallel to the slope, in Fig. 1b. The resultant of the horizontal components of the forces from the sliding soil is given by the symbol F_s. The boundary conditions for the portion of the pile in the stable soil are given in Fig. 1c as a shear P_t and a moment M_t. The portion of the pile below the critical surface of sliding is caused to deflect laterally by P_t and M_t and the resisting forces from the soil are shown in the lower portion of Fig. 1b. Making use of the assumptions shown in the next paragraph, the behavior of the pile can be found by the procedures for piles under lateral loading.

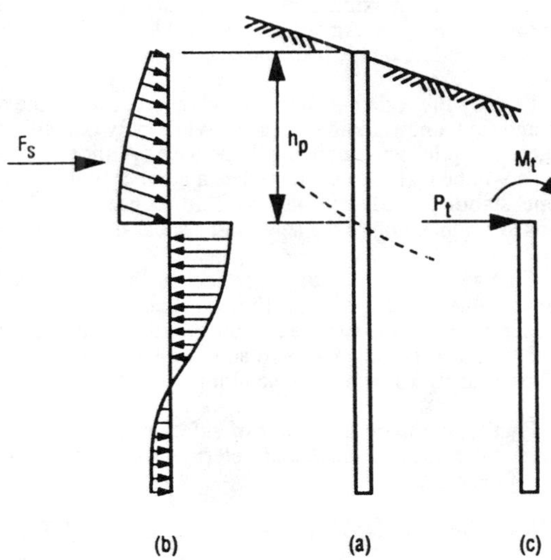

Figure 1. Forces from Soil Against a Pile in a Sliding Mass.

The principles of limit equilibrium are usually employed in analyzing the stability of a slope. Therefore, a consistent assumption is that the sliding soil has moved a sufficient amount that the ultimate resistance from the soil has developed against the pile. If one considers the force against a pile from a wedge of soil with a sloping surface, the force parallel to the soil surface is larger than that if the surface were horizontal. Studies of driving force from soils above the sliding surface will be given later.

The discussion above leads to the following step-by-step procedure of analysis:

1. With an appropriate computer code, find the factor of safety against sliding of the particular slope to be stabilized by the use of piles.

2. At a position where a pile is to be placed in a slope, tabulate the relevant properties of the soil as a function of depth.

3. Select a pile with a given diameter and bending stiffness and find its ultimate bending moment.

4. Assume that the surface of sliding is the same as found in Step 1, then compute the ultimate resistance of the soil against the pile at depths above the sliding surface. Employ the ultimate resistance as the distributed earth pressure on the pile, as shown in Fig. 1b, and analyze the pile using the p-y curve method.

5. Compare the maximum bending moment found from Step 4 with the ultimate value found in Step 3. An adjustment of the geometry of the pile may or may not be made, depending on the results of the comparison.

6. Employ the resisting shear P_t and moment M_t in the computer code used in Step 1 and find a new sliding surface. While only one pile is shown in the illustration, a group of piles will undoubtedly be used. A slice of appropriate width through the slope will be taken to accommodate a number of piles. Some computer codes for slope stability analyses may be able to use directly the profile of distributed loads in computation of the new most critical sliding surface.

7. Change the depth h_p to the critical surface of sliding, employed in Step 4, get new values of M_t and P_t (Fig. 2), and repeat the analyses until agreement is found between that surface and the resisting forces for the piles. Also, the geometry of the piles should have been adjusted so that the maximum bending moment found in the analyses is close to the ultimate bending moment of the pile.

8. Finally, compare the factor of safety of the slope with no piles in place with that for the piles in place and determine whether or not the increase justifies the use of the piles.

COMPUTATION OF MAXIMUM BENDING MOMENT IN A PILE

A physical model for the laterally-loaded pile is shown in Fig. 3. A pile is shown in the figure with loadings at its top. The soil has been replaced with a series of mechanisms that show in concept the response of the soil. The soil reaction p (force per unit length along the pile) is a function of lateral deflection y and depth x and is defined by the curves.

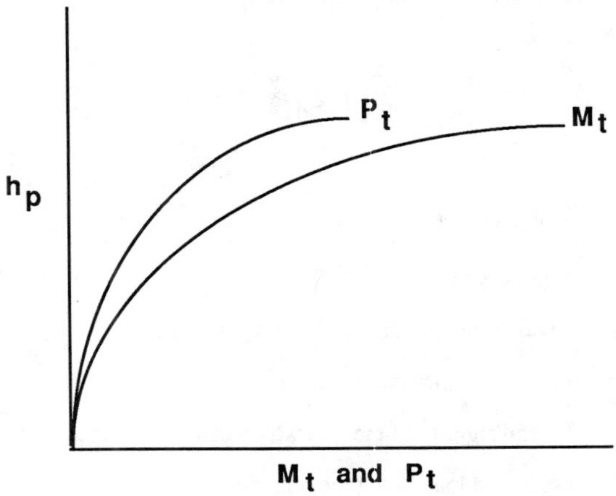

Figure 2. Boundary Conditions for Bottom Part of Pile as a Function of the Depth of the Critical Sliding Surface.

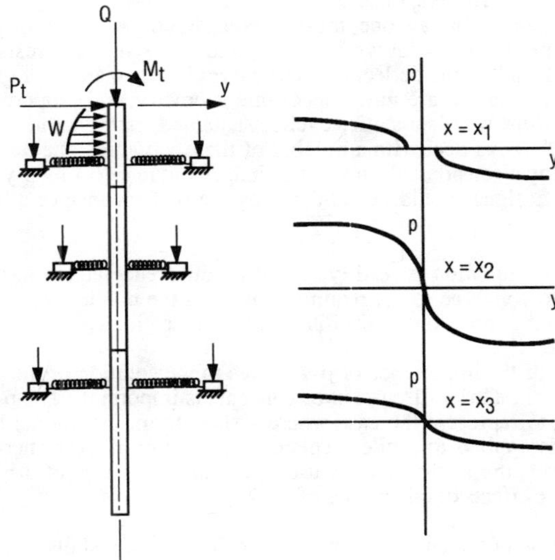

Figure 3. Model for Pile under Lateral Loading with p-y Curves.

The behavior of the pile under lateral loading can be obtained by solving the following differential equation (Hetenyi, 1946)

$$EI\frac{d^4y}{dx^4} + Q\frac{d^2y}{dx^2} - p + W = 0$$

where

Q = axial load on the pile,

x = length coordinate,

y = lateral deflection of the pile at a point x,

p = soil reaction per unit length,

EI = flexural rigidity of the pile which can vary with x, and

W = distributed load along the pile.

A number of authors have presented expressions for p-y curves for clays and sands under both static and cyclic loading (Matlock, 1970; Reese et al, 1975; Welch and Reese, 1972; Nyman, 1980; Murchison and O'Neill, 1984; Gazioglu and O'Neill, 1984). The response of the soil to sustained loading, which could readily occur in stabilizing a slope, must be considered. A family of p-y curves, obtained by experiment, is shown in Fig 4, and gives the soil resistance as a function of depth and of pile deflection. The curves in Fig. 4 are for a specific case but, in general, p-y curves are functions of pile geometry, soil properties, and the methods of loading whether static, cyclic, sustained, or dynamic. While the references noted above deal with a number of these variables, the designer must frequently employ a considerable amount of judgement in selecting p-y curves. In important jobs, designs should be validated by the performance of a field test or tests.

The above equation is readily solved by difference-equation techniques, using iteration to converge to the required points on the nonlinear p-y curves. A solution satisfies the conditions of equilibrium and compatibility.

Because of the importance of p-y curves to the solution of the problem of piles in sliding soil, a brief discussion of concepts is important. The p-y curve in Fig. 5 is meant to represent the case where a short-term, monotonic loading, or static loading, is applied to a pile. This case is seldom if ever encountered in practice; however, the static curve is useful because as a sort of a baseline for demonstrating the effects of other types of loading.

The portion of the curve in Fig. 5a from Points a to b shows that the value of p is increasing at a decreasing rate with increasing deflection y. This behavior undoubtedly is reflecting the nonlinear portion of the in situ stress-strain curve. The straight-line, horizontal portion of the p-y curve in Fig. 5a implies that the in situ soil is behaving plastically with no loss of shear strength with increasing strain.

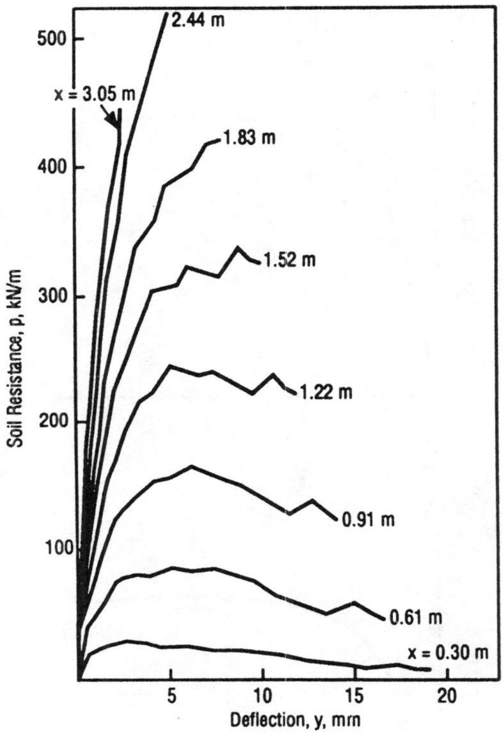

Figure 4. p-y Curves Developed from Static-Load Test on 641 mm Diameter Pile (after Reese, et al, 1975).

With that assumption, some analytical models can be used to compute the ultimate resistance p_u as a function of pile dimensions and soil properties.

Fig. 5b shows that there is an increasing deflection with the sustained loading. The decreasing value of p implies that the resistance is shifted to other elements of soil along the pile as the deflection occurs at some particular point. The effect of sustained loading is negligible for granular soils and for heavily overconsolidated clays, but p-y curves must be modified if the soil below the sliding surface is a soft clay, for example. The geotechnical engineer must deal with such modifications on a case-by-case basis.

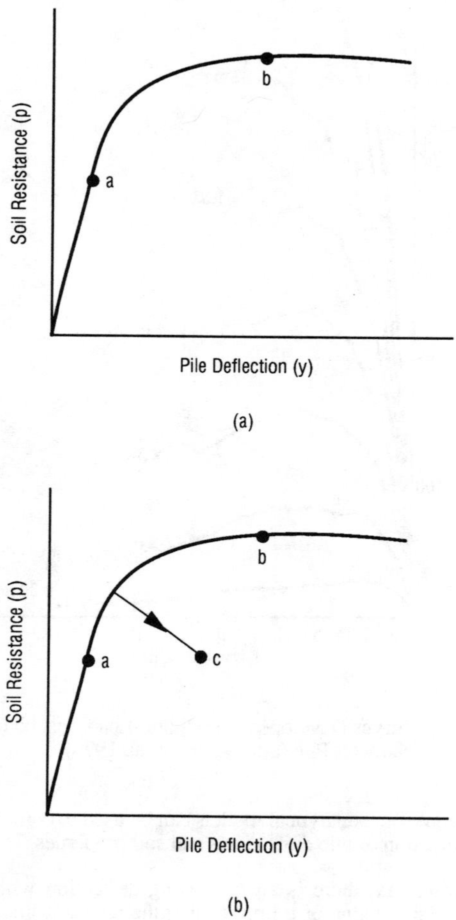

Figure 5. Conceptual p-y Curves.

COMPUTATION OF THE LATERAL THRUST FROM MOVING SOIL TO STABILIZING DRILLED SHAFTS AND RESPONSE OF DRILLED SHAFTS TO LATERAL LOADING.

There are two cases where the lateral thrust from moving soil and response of a pile can be computed in a relatively straight forward manner: (1) when the piles are side-by-side with no spacing between (a wall), and (2) when the piles are so widely spaced that no interaction occurs. With regard to thrust or loading, in the first case Coulumb's equations for computing passive earth-pressure on retaining walls have been well-accepted in practice. In the second case, the concept of the failure soil wedge (Figs. 6 & 7) presented by Reese (1958) and Reese et al (1974) can be used to compute the lateral thrust on each drilled shaft.

That complications arise, with regard to both loading and response, can be seen by referring to Fig. 8. The overlapping of failure wedges is shown, which gives rise to complicated equations.

In the general case of piles in a group, with respect to both loading and response, the edge effect and the in-line effect must be considered. With respect to thrust or loading, research has been limited and the studies on response can be used to some advantage. With regard to response, the studies have been found in some technical literature and will be discussed in the following paragraphs.

Experimental studies conducted by Prakash (1962), Cox et al (1984), Wang and Reese (1986), and Lieng (1988) have included loading tests on side-by-side piles. The factor for reducing P is defined as the ratio of the averaged capacity of individual piles in a group to the pile capacity of a single pile. The P-reduction versus the pile spacing is in terms of S/b (S is the center to center spacing, b is the pile diameter) are presented in Fig. 9. It clearly shows that the reduction is negligible if the spacing is greater than 3 times the pile diameter. According to the analytical study by Wang and Reese (1986), the reduction for ultimate resistance at a shallow depth is near 0.5 when piles are in contact edge to edge. While the scatter in the results in Fig. 9 is significant, the curve in the figure represents at present the best estimate of the reduction factor as a function of pile spacing regardless of the soil type.

When both edge effects and in-line effects must be considered, a modification factor similar to f_m, recommended by Brown et al (1987), is needed to reduce the soil resistance. The result is similar to that in Fig. 10.

In summary, the relative position of each pile in a group is considered and a value of f_m is found for each individual pile. Both edge effect and in-line effect must be considered. The, p-y curves, modified as necessary, can be used to compute the behavior of each of the piles.

CASE STUDIES AND EXAMPLE COMPUTATION

Case Studies

Fukuoka (1977) described a field experiment that was performed by Fukumoto at the landslide at Higashi-tono in the Niigata Prefecture. A pile, instrumented with strain gages, was installed in the slide which continued to move

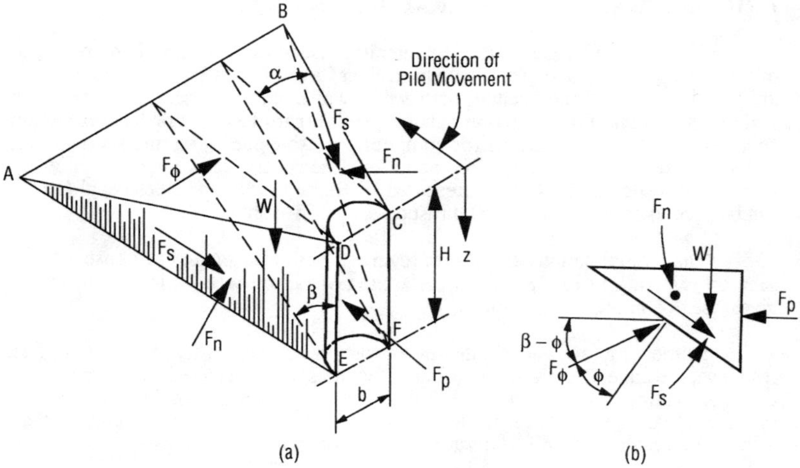

Figure 6. Assumed Passive Wedge-Type Failure for Sand
(a) Shape of Wedge; (b) Forces acting on Wedge.

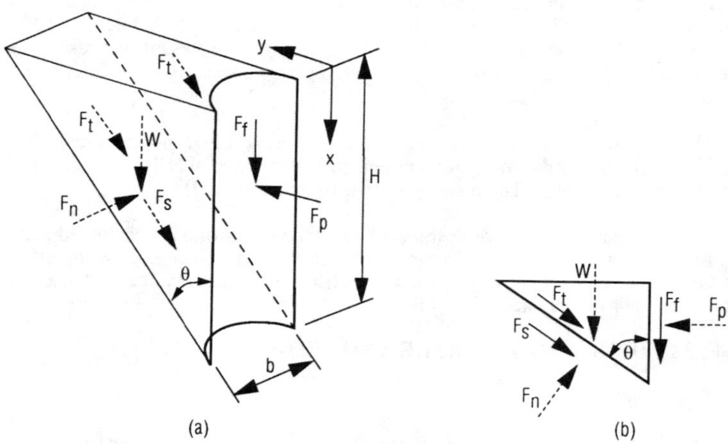

Figure 7. Assumed Passive Wedge-Type Failure for Clay
(a) Shape of Wedge; (b) Forces acting on Wedge.

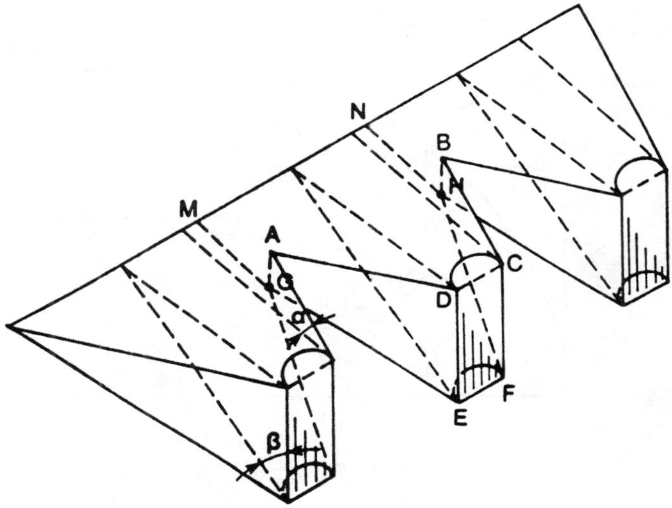

Figure 8. General View of Failure Wedge for Side-by-Side Piles in Sand.

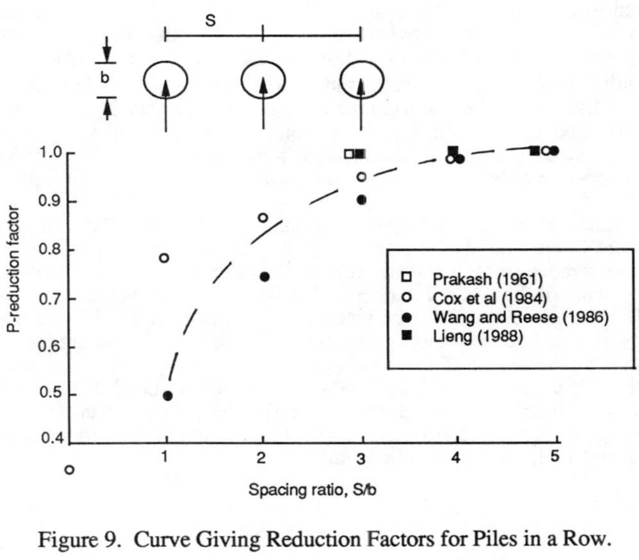

Figure 9. Curve Giving Reduction Factors for Piles in a Row.

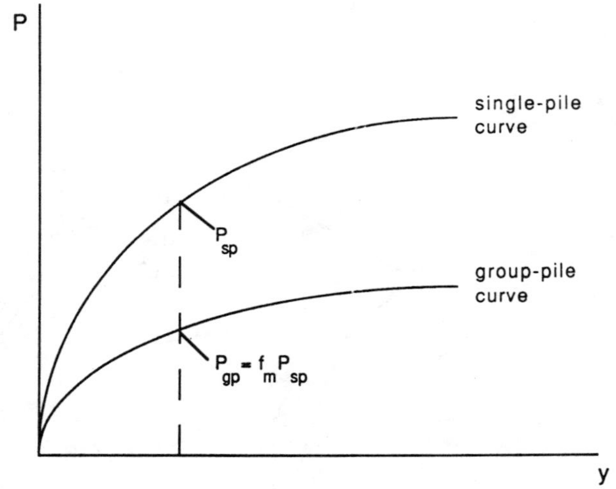

Figure 10. Definition of f_m.

at a slow rate. The sliding soil was mudstone and the N-value from the Standard Penetration Test near the sliding surface was found to be 20. The pile was 22 m in length, had an outer diameter of 406.4 mm, and had a wall thickness of 12.7 millimeters. The bending moment in the pile increased rapidly after installation and appeared to have reached the maximum value after being in place about three months. The strain gages showed the maximum bending moment to occur at a depth of about 10 m below the ground surface and to be about 220 kN-meters. The maximum bending stress in the pile, thus, was about 1.5 x 10^5 kN/m^2, a value that shows the loading on the pile from the sliding soil to be very low. Therefore, it was concluded that the driving force from the moving soil was far from its maximum value. The positive conclusion from this field test is that the bending-moment curve given by Fukuoka had the general shape that would be expected.

At another site at the Higashi-tono landslide, Fukuoka (1977) described an experiment where a number of steel-pipe piles were used in a sliding soil. Some of them were removed after a considerable period of time and found to have failed in bending. One of them had a diameter of 318.5 mm and a wall thickness of 10.3 mm. The collapse moment for the pipe was computed to be 241 kN-m. Assuming a triangular distribution of earth pressure on the pile from the sliding mass of soil, which had a thickness of 5 m, the undrained shear strength that was required to cause the pile to fail was 10.7 kPa. The author merely stated that the soil had a value of N_{SPT} from the Standard Penetration Test that was less than 10. That value of N_{SPT} probably reflects an undrained shear strength that encompasses the computed strength to cause the pile to fail.

Example Computations

The example that was selected for analysis is shown in Fig. 11. The slope exists along the bank of a river where sudden drawdown is possible. Slides had been observed along the river at numerous places and it was desirable to stabilize the slope to allow a bridge to be constructed.

The undrained analysis for the sudden-drawdown case was made based on the Spencer's method, and the factor of safety was found to be 1.06, a value that is in reasonable agreement with observations. Plainly, some method of design and construction would be necessary in order for bridge piers to be placed at the site. The method described herein was employed to select sizes and spacing of drilled shafts that could be used to achieve stability.

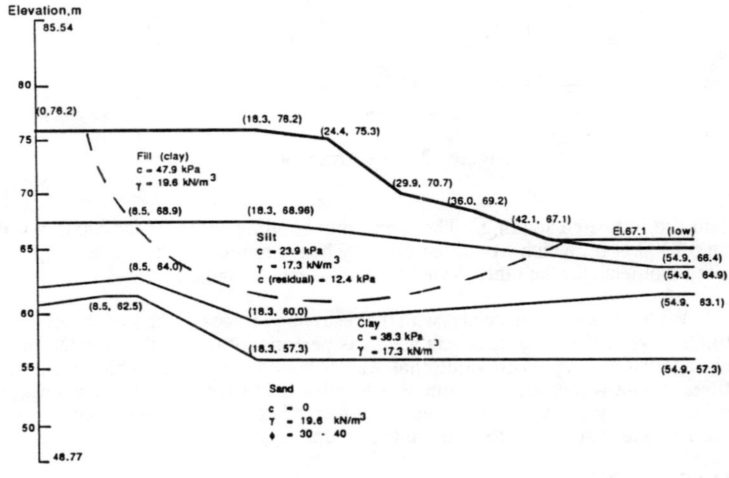

Figure 11. Failure Surface for Low-Water Level.

Figure 12 shows a preliminary design, but not shown in the figure is the distance along the river for which the slope was to be stabilized. Drilled shafts were selected that were 915 mm in diameter and that penetrated below the sliding surface, as shown in the figure. Further, as shown in the figure, it was found that the tops of the shafts had to be restrained with anchors in stable soil. The use of the anchors was required because of the depth of the slide. The results of the analysis, for each of the groups perpendicular to the river, gave the following loads at the top of the drilled shafts: Shafts 1, 2, and 3, +1,090 kN; Shaft 4, -1,310 kN; and Shaft 5, -1,690 kN. The member connecting the tops of the 5 piles would be designed to

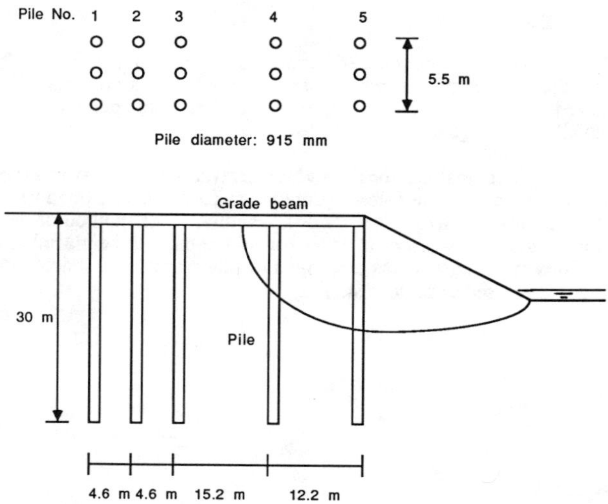

Figure 12. Pile Arrangement.

sustain the indicated loading. The maximum bending moment for Shaft 5 was about 6,250 kN-m, which would require very heavy reinforcement. The computed bending moments for the other drilled shafts was much smaller.

With the piles in place and with the restraining forces of the piles against the sliding soil (Fig. 13), a second analysis was performed to find the new factor of safety against sliding. The value that was obtained was 1.82. This result was sufficient to show that the technique was feasible. However, in a practical design, a series of analyses would have been performed to find the most economical geometries and spacings of the piles in the group.

CONCLUSIONS

The results predicted by the proposed design method are compared with results from available full-scale experiments. The case studies yield information on the applicability of the proposed method of analysis.

A complete analysis for the stability of slopes with drilled shafts in place is presented. The method of analysis is considered to be practical and can be implemented by engineers by using readily available methods of analysis. The benefits of using the method is that rationality and convenience are indicated that have not been previously available.

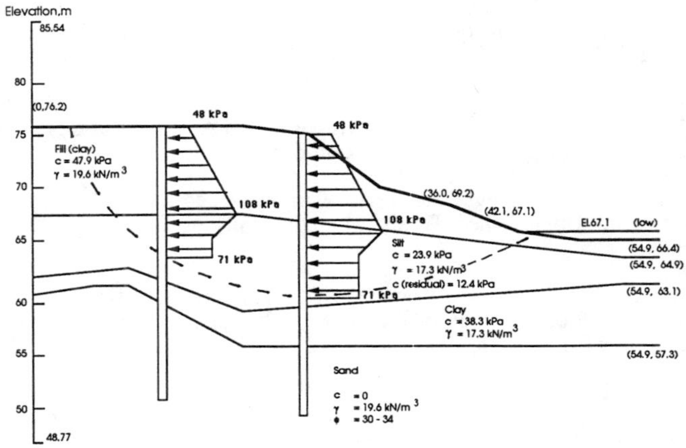

Figure 13. Load Distribution on Piles 4 and 5.

REFERENCES

Brown, D. A., Reese, L. C., and O'Neill, M. W. (1987). "Cyclic Lateral Loading of a Large-Scale Pile Group," *Journal of the Geotechnical Engineering Division,* ASCE, 113(11).

Cox, W. R., Dixon, D. A., and Murphy, B. S. (1984). "Lateral Load Tests of 25.4 mm Diameter Piles in Very Soft Clay in Side-by-Side and In-Line Groups," *Laterally Loaded Deep Foundations: Analysis and Performance,* ASTM, SPT835.

Fukuoka, M. (1977). "The Effects of Horizontal Loads on Piles Due to Landslides," *Proceedings,* Special Session 10, 9th International Conference, International Society of Soil Mechanics and Foundation Engineering, Tokyo, Japan, 1-16.

Gazioglu, S. M., and O'Neill, M. W. (1984). "Evaluation of p-y Relationships in Cohesive Soil," *Proceedings,* Symposium on Analysis and Design of Pile Foundations, ASCE, San Francisco, 192-213.

Hetenyi, M. (1946). Beams on Elastic Foundation, "The University of Michigan Press, Ann Arbor.

Lieng, J. T. (1988). "Behavior of Laterally Loaded Piles in Sand - Large Scale Model Tests," Ph.D. Thesis, Department of Civil Engineering, Norwegian Institute of Technology, 206 pages.

Matlock, H. (1970). "Correlations for Design of Laterally Loaded Piles in Soft Clay," *Proceedings,* Second Annual Offshore Technology Conference, Vol. 1, Paper 1204, Houston, Texas, 577-594.

Murchison, J. M., and O'Neill, M. W. (1984). "Evaluation of p-y Relationships in Cohesionless Soils," *Proceedings*, Symposium on Analysis and Design of Pile Foundations, ASCE, San Francisco, California, 174-191.

Nyman, K. J. (1980). "Field Load Tests of Instrumented Drilled Shafts in Coral Limestone," Unpublished Thesis, The University of Texas at Austin, May, 1980.

Reese, L. C., Cox, W. R., and Koop, F. D. (1975). "Field Testing and Analysis of Laterally Loaded Piles in Stiff Clay," *Proceedings,* Seventh Offshore Technology Conference, Vol. II, Paper 2312, Houston, Texas, 672-690.

Reese, L. C., Cox, W. R., and Koop, F. D. (1974). "Analysis of Laterally Loaded Piles in Sand" *Proceedings,* Fifth Offshore Technology Conference, Vol. II, Paper 2080, Houston, Texas, 473-485.

Reese, L. C. (1958). "Discussion of Soil Modulus for Laterally Loaded Piles," by McClelland, B. and Focht, J. A., Jr., *Transactions,* ASCE, Vol. 123.

Wang, S-T. and Reese, L. C. (1986). "Study of Design Method for Vertical Drilled Shaft Retaining Walls," *Research Report 415-2F,* Center for Transportation Research, Bureau of Engineering Research, The University of Texas at Austin.

Welch, R. C. and Reese, L. C. (1972). "Laterally Loaded Behavior of Drilled Shafts," *Research Report No. 3-5-65-89,* conducted for the Texas Highway Department, in cooperation with the U. S. Department of Transportation, Federal Highway Administration, Center for Highway Research, The University of Texas at Austin.

Prakash, S. (1962). "Behavior of Pile Groups Subjected to Lateral Loads," Unpublished Dissertation, University of Illinois.

Slope Stabilization Using In-Situ Earth Reinforcements

Seth L. Pearlman,[1] Member ASCE
Bradley D. Campbell[2] and
James L. Withiam,[2] Member ASCE

Abstract

A method to control the movement of marginally stable soil slopes is described which relies on the resistance mobilized from a pattern of vertical and near vertical reinforcing elements (Pin Piles) installed across a slide plane. The measured performance of several walls is summarized and used to evaluate an analytical procedure developed for preliminary design. The procedure encompasses the interactive response of the soil-pile system, and is presented in the form of charts which can be used for preliminary design. Recommendations are also provided regarding procedures for final design.

Introduction

A general overview of known applications of Type "A" INSERT (In-Situ Earth Reinforcement Technique) walls for slope stabilization in the U.S. is presented. Reference to these structures as walls is made because the structures are used in applications where more traditional approaches may also be considered (e.g., soil buttresses, anchored soldier pile walls or conventional gravity retaining walls). The use of these traditional systems may be precluded by site constraints or cost. The Type "A" INSERT wall system is the least disruptive to the site, and may be constructed on extremely steep slopes using relatively small, lightweight drilling equipment.

---

(1) Chief Design Engineer, Nicholson Construction Company, Bridgeville, Pennsylvania, 15017.

(2) Project Engineer and Principal, respectively, D'Appolonia Engineering Division of Ground Technology, Inc., Monroeville, Pennsylvania, 15146.

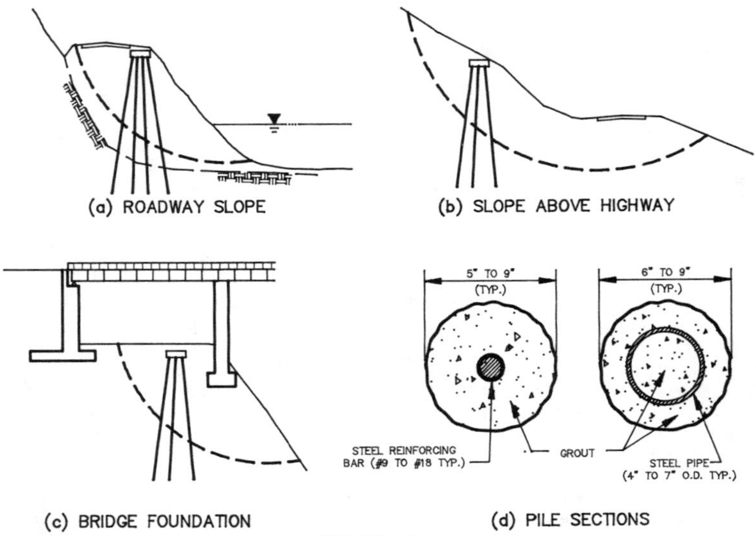

FIGURE 1
TYPICAL CONFIGURATIONS AND APPLICATIONS
FOR TYPE "A" INSERT WALLS

Typical Applications and Wall Construction

Typical configurations and applications for Type "A" INSERT walls are shown in Figure 1. They consist of vertical and near-vertical Pin Piles (small diameter drilled and grouted reinforcing elements) which are placed in specific patterns and extended below the known or postulated slide plane. The pile elements are typically extended to the surface and connected together with a reinforced concrete cap located at or just below the ground surface. A Pin Pile consists of a centralized reinforcing bar or steel pipe which is grouted into the ground using a neat cement grout. Diameters typically range from five to nine inches (13 to 23 cm). Primary resistance is provided by lateral movement above and below the slide plane which is manifest by bending in the piles. For relatively shallow slide planes (i.e., within about 20 feet (6 m) of the ground surface), the cap beam provides added stiffness. For deeper slides, wall designs are being evaluated which do not require a cap beam or full extension of the pile reinforcement to the ground surface. Details regarding the construction of Type "A" INSERT walls are presented in Bruce (1992).

Case History Examples

As summarized in Tables 1 and 2, seven well documented and instrumented case history examples of Type "A" INSERT walls and similar systems have been reported in the literature. (Other case histories have been reported, but the information presented in them was insufficient for the purposes of the work described herein.) Table 1 provides general background information for each project regarding location, construction date, ground conditions, slope geometry and depth from ground surface to the slide plane along the centerline of the wall. Details regarding the geometry of the wall and pile elements, measured lateral displacements and general remarks are presented in Table 2. In Table 2 it is noteworthy that the density of piles has reduced since the earliest installations from greater than two piles per foot to less than one pile per foot for more recently constructed walls. However, this reduction has not noticeably affected the performance of the walls with regard to limiting slope movements after wall construction. This observation suggests, and subsequent analyses later demonstrate, that the walls may have considerable reserve resistance if all possible soil-structure interaction modes are considered in design.

The information presented in the tables was used to evaluate the applicability of the proposed procedure for preliminary design of Type "A" INSERT walls. Project Nos. 6 and 7 are highlighted below as they are generally similar in construction to the other walls, are well instrumented, and use a lower density of piles to control slope movements.

Project No. 6: State Route 4023, Armstrong County, Pennsylvania. Portions of the two-lane roadway were constructed on a slope adjacent the Allegheny River. An approximately 250-foot long section of the road, and the railroad tracks located upslope, were experiencing damage caused by slope movements toward the river. A monitoring program initiated by the Pennsylvania Department of Transportation (PADOT) indicated a slip-plane was located about 25 to 35 feet (8 m to 11 m) below the road and that the slope was moving at a rate of up to 0.7 inches (2 cm) per month.

PADOT designed a repair using prestressed rock anchors and tangent caissons extending into competent rock. The earth pressures used for the design were based on the results of stability analyses, for which the soil along the slip-plane was assigned a residual angle of friction. This design provided a minimum factor of safety with regard to the overall slope stability equal to 1.5 and 1.2 for the normal and rapid drawdown

TABLE 1

SUMMARY OF IN-SITU EARTH WALL CASE HISTORIES

NO.	PROJECT NAME, LOCATION AND REFERENCE	CONSTRUCTION DATE	GROUND CONDITIONS	SLOPE GEOMETRY UPSLOPE/DOWNSLOPE	DEPTH TO SLIDE AT WALL (ft)
1	Forest Highway No. 7 Mendicino National Forest, CA Walkinshaw (1977); Palmerton (1984)	1977	Micaceous phyllite (ϕ = 13°, c = 500 psf) and schistose bedrock	2H:1V/2H:1V[1]	55
2	Route 23A Catskill State Park, NY Murray (1984); Palmerton (1984)	1977	Moist, very dense glacial till with boulders (ϕ = 30°, c = 0) and shale bedrock	2H:1V/2H:1.3V	10 - 26
3	PA-306 Monessen, PA Dash and Jovino (1980)	1979	Random fill and colluvium (ϕ = 17°, c = 100 psf, γ = 134 pcf) and weak shale	2.5H:1V/4H:1V	20
4	L.R. 69 Armstrong County, PA Dash (1987); Earth, Inc. (1986)	1985	Random fill, wet sandy clay and hard clay (ϕ = 30°, c = 0, γ = 120 pcf; ϕ = 17° along slide plane)	Level/1.2H:1V	34
5	Glady-Durbin Road No. 44 Randolph County, WV NCC[2] project files	1987	Sandy clay with rock fragments (ϕ = 30°, c = 0, γ = 120 pcf) and sandstone bedrock	Level/1H:1V	16 - 40
6	S.R. 4023 Armstrong County, PA NCC project files	1988	Random fill, stiff colluvial clay with rock fragments (ϕ = 17°, c = 0, γ = 125 pcf) and weathered claystone/ competent sandstone	Level/2H:1V	23 - 35
7	Blue Heron Road Big South Fork River, KY COE[3] project files; NCC project files	1989	Medium stiff/stiff silty clay and shale bedrock (ϕ = 19°, c = 0, γ = 125 pcf)	Level/1.5H:1V	25

[1] 2H:1V - 2 Horizontal to 1 Vertical; [2] NCC - Nicholson Construction Company; [3] COE - U.S. Army Corps of Engineers

1 ft = 0.31 m; 1 psf = 47.9 N/m^2; 1 pcf = 159 N/m^3

TABLE 2

SUMMARY OF IN-SITU EARTH WALL CASE HISTORIES
WALL GEOMETRY AND PERFORMANCE

NO.	CAP BEAM GEOMETRY (LxWxT) (ft)	PILE DIAMETER(3) (in.)	REBAR SIZE (#)	PILE DENSITY (pile/L.F.)	PILE INCLINATION (° from vertical)	MAXIMUM LENGTH (ft)	MAXIMUM EMBEDMENT BELOW SLIDE (ft)	LATERAL DISPLACEMENT (in.)	REMARKS
1	310x6x3 50x5x3	5 4.5/rock	9	2.33	19 to -19[1]	69 53	8	N.M.[4]	Cap constructed before pile construction.
2	250x11x1.75	4	10	2.80	15 to -15	50	24 - 30	4 D.C.[5] 0.3 A.C[6]	Cap constructed before pile construction.
3	200x6x2.5	5	9	2.25	19 to -19	45	9	0.1 - 0.8 A.C.	Failure of downslope after construction; little wall movement.
4	310x6x2.5	6	11/14	1.33	8 to -8	50	15	0.1 A.C.	Anchors inclined at 40° to vertical installed for rapid drawdown case.
5	115x5x3 and 50x5x3	5.5/soil 4.5/rock	11/14/18	1.00	16 to -2	37[2] 53[3]	8	N.M.	Cap constructed before pile construction.
6	122x4.6x3[2] 122x4.6x3[3]	5.5/soil 4.5/rock	11/14/18	1.25[2] 0.80[3]	26 to 0	60	17[2] 24[3]	1.0 D.C. 0.7 A.C.	Cap constructed after pile construction.
7	34x5x3	5.5	11/14	0.75	19 to -5	35	10 - 15	0.6 D.C. 0.3 A.C.	Cap constructed after pile construction.

[1] Upslope is + angle; [2] Wall A; [3] Wall B; [4] N.M. - Not measured; [5] D.C. - During construction; [6] A.C. - After construction.

1 ft = 0.31 m; 1 in = 2.54 cm.

conditions, respectively. A post-bid alternative based on a Type "A" INSERT wall design was accepted by PADOT with a resultant savings of about $1M compared to the lowest bid for the anchored caisson wall design.

In general, the wall consists of four rows of Pin Piles extending across the slip-plane and into competent rock as idealized in Figure 1(a). The wall is comprised of two equal length sections designated as Wall A and Wall B. Wall A contains a higher density of piles than Wall B, because the top of weathered rock dips to a lower elevation in the area of Wall A.

Performance monitoring consisted of: (1) strain gages on reinforcement of the outermost pile rows; (2) inclinometers through and upslope from the wall; (3) fixed-end extensometers to measure average strain along the reinforcements; and (4) surface survey monuments. Due to space limitations, only the results of the slope inclinometer data for Wall A are described herein.

The maximum movement at select inclinometer locations along the length of the wall is plotted as a function of time in Figure 2. The data for inclinometers located relatively close to and within the wall indicate that up to 1.5 inches (4 cm) of horizontal movement occurred during construction, and up to 0.4 inches (1.0 cm) of movement occurred in the year following completion of construction. Inclinometer SI1W, located approximately 60 feet (18 m) upslope from the wall, exhibited approximately 1.2 inches (3.10 cm) of additional movement in the first year after the completion of construction and 0.34 inches (0.9 cm) of movement in the second year following the end of construction.

In general, the inclinometer data shown in Figure 2 indicate that the wall has significantly slowed the slope movements at the site from a rate of approximately eight inches per year (20 cm/yr) to a rate of 0.3 in/yr (0.8 cm/yr). Furthermore, the rate of movement is decreasing with time. The data also show that the stabilizing effects of the relatively flexible pile elements become greater with increasing movement of the wall. If movements were to continue and eventually reach an unacceptable magnitude, additional piles could be installed to arrest slope movements.

Project No. 7: Blue Heron Road, Big South Fork River, Kentucky. This project, located in southeastern Kentucky, was initiated in early 1989 when personnel from the U.S. Army Corps of Engineers (COE), Nashville District, observed a moving slope downhill from a bridge abutment and above a land pier supporting an historic railroad bridge. These movements opened a gap between

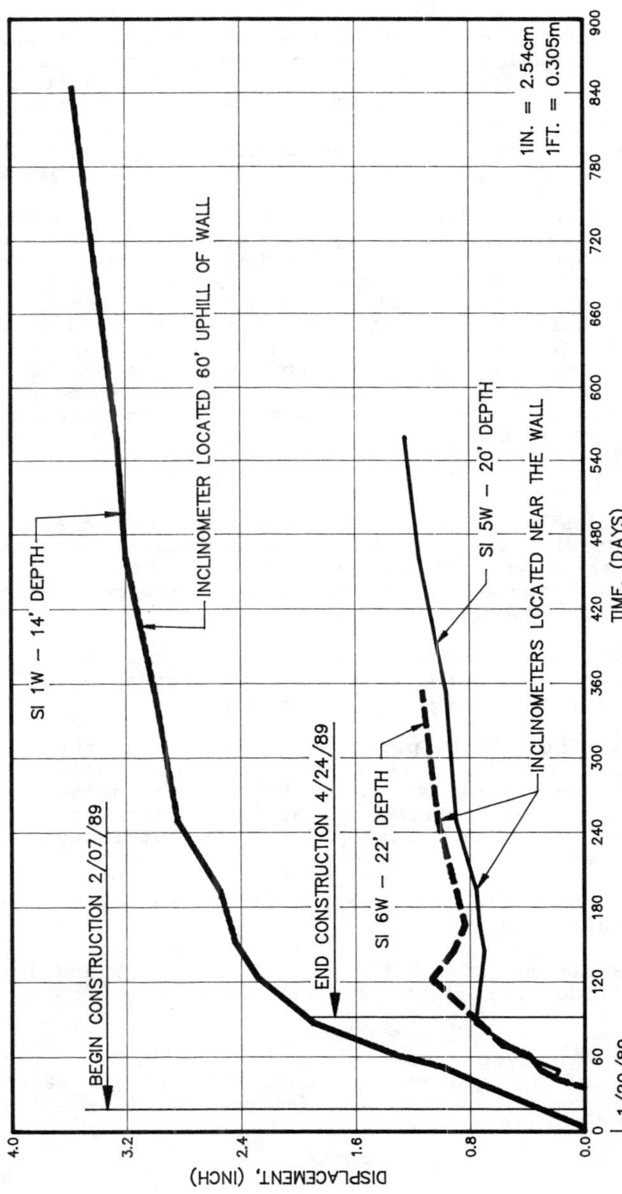

STA. 22+12 WALL A

FIGURE 2

MAXIMUM MOVEMENT AT SELECT INCLINOMETER LOCATIONS WITH TIME
STATE ROUTE 4023, ARMSTRONG COUNTY, PENNSYLVANIA

the steel base plate and the concrete pier which threatened the stability of the structure.

The results of a subsurface exploration are summarized in Table 1. Monitoring of inclinometer casings indicated that slope movements were occurring near the top of rock. The COE accepted a contractor-designed Type "A" INSERT wall similar to that shown in Figure 1(c) to arrest future slope movements and protect the bridge pier. The wall geometry is presented in Table 2.

Based on the results of survey data, the bridge pier experienced approximately 0.6 inches (1.5 cm) of horizontal movement during the 10 days it took to construct the wall; an additional 0.3 inches (0.8 cm) of movement occurred within five days after completion of the wall. Subsequent surveys indicate that movement of the pier has stopped. Inclinometer measurements within the wall and slightly uphill of the wall also show that 0.3 inches (0.8 cm) of deflection occurred after completion of the wall and that the slope movements were subsequently stopped by the wall.

Original Type "A" INSERT Wall Design Procedure

The designs for the two case history examples described previously, as well as those for Project Nos. 4 and 5, were developed assuming the wall behaved as a gravity retaining wall. In summary, the procedure involved the following:

- Determining the pressure acting on the back of the wall based on slope stability analysis or earth pressure theories;
- Assuming enough piles are provided in the cross section of the wall to retain soil between the piles, check the wall for sliding and overturning stability in conformance with the design of gravity retaining walls;
- Provide sufficient shear elements to resist sliding; and
- Sum moments about the toe of the wall and provide sufficient tension elements to resist overturning.

Basic to the procedure was an assumption that the pile density needed to achieve adequate factors of safety against sliding and overturning failure would be sufficient to stop slope movements.

New Procedure for Preliminary Design

Based on an evaluation of wall performance data, it was evident that Type "A" INSERT walls were not behaving as a gravity wall. For example, wall movements seemed to be confined to a relatively thin and localized zone along the slide plane and additional slope movements were occurring after wall construction. Therefore, a new procedure was developed for preliminary design of Type "A" INSERT walls to better model the behavior of this relatively flexible slope stabilization system. The theoretical basis for the procedure has been verified by comparison with back analyses of the instrumented walls. In general, the design procedure involves the following:

- Conducting stability analyses to determine the increase in resistance along a potential or existing slip surface required to provide an adequate factor of safety.
- Checking the potential for structural failure of the piles due to loading from the moving soil mass.
- Checking the potential for plastic failure (i.e., flow of soil around the pile).

Typically, movement of marginally-stable, non-creeping soil slopes occurs within a relatively thin zone which is subjected to large shear strains. Generally, the materials above and below the zone of failure experience negligible shear strains. The purpose of the Pin Piles is to connect the moving zone (above the slip surface) to the nonmoving zone (below the slip surface), and thus increase the sliding resistance along the slip-plane.

Because the Pin Piles are relatively flexible, the maximum bending moments in the piles tend to develop relatively close to the slip-plane. Fukuoka (1977) devised a theory to evaluate the bending moments which develop in a pile oriented perpendicular to a slip-plane assuming an uniform velocity distribution of the soil above the slip-plane.

Figure 3 is a chart for preliminary design of Type "A" INSERT walls. The chart was developed using the method described in Fukuoka (1977) and considers four sizes of pile elements. It should be noted that the ultimate horizontal resistance designated in Figure 3 is either the load which causes yield stresses to develop on the outer edges of the steel pipes (i.e., Pile Types 1, 2 and 3) or the load which causes crushing of

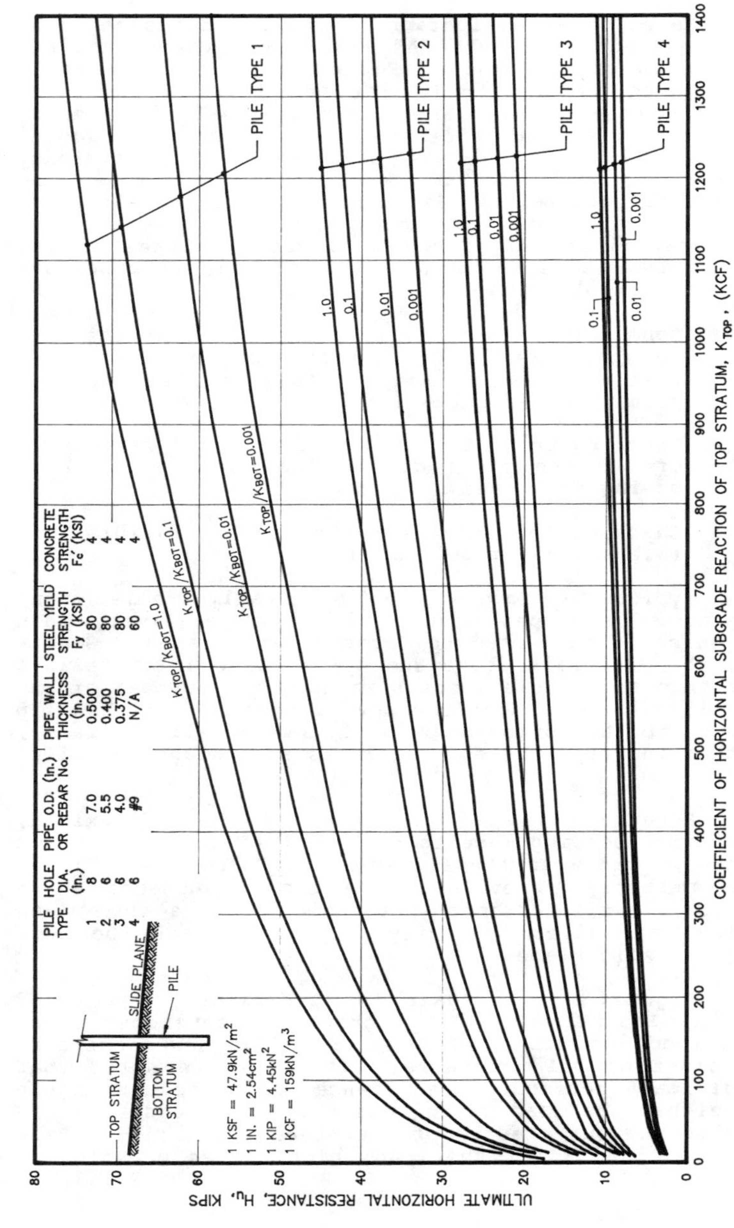

FIGURE 3

PRELIMINARY DESIGN CHART FOR ULTIMATE HORIZONTAL RESISTANCE OF PILES

the concrete surrounding the centralized reinforcing bar (i.e., Pile Type 4).

Figure 3 shows that the ultimate horizontal load resistance of the piles is a function of the coefficient of subgrade reaction K_s of the soil or rock above and below the slip-plane. The K_s of the soil also has a significant effect on the amount of horizontal movement required before the pile reaches its ultimate horizontal resistance. Typical deflections and bending stresses along a pile are shown in Figure 4.

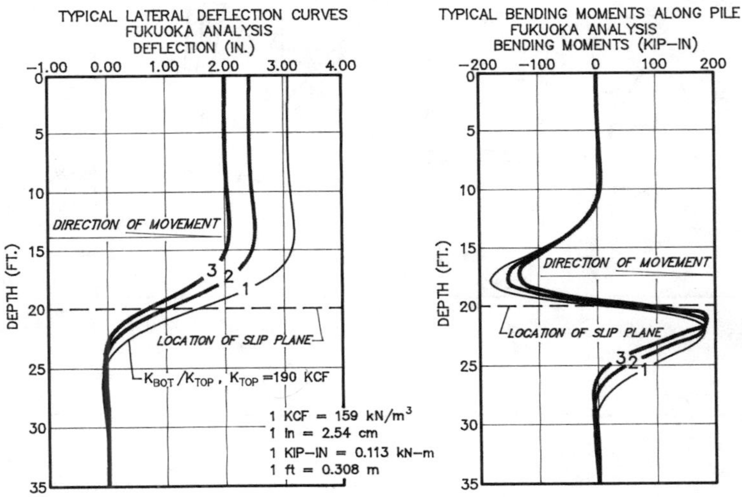

FIGURE 4
TYPICAL RESULTS OF ANALYSES OF TYPE 3 PILE(1)
(1) SEE FIGURE 3 FOR DIMENSIONS

Plastic failure of soil around the piles is analyzed using a procedure developed by Ito and Matsui (1975). The method is based on the fundamental consideration that soil deformation is restricted to a plane strain condition. Typically, this type of failure occurs if the soil above the slide plane is relatively soft and the piles are stiff and spaced far apart. This is usually not the case with relatively flexible Pin Piles, but may govern when stiffer pipe elements are employed. Based on the theory proposed by Ito and Matsui (1975), the predicted results for various pile spacings and soil conditions are plotted in Figure 5.

190 kcf => 110 pci

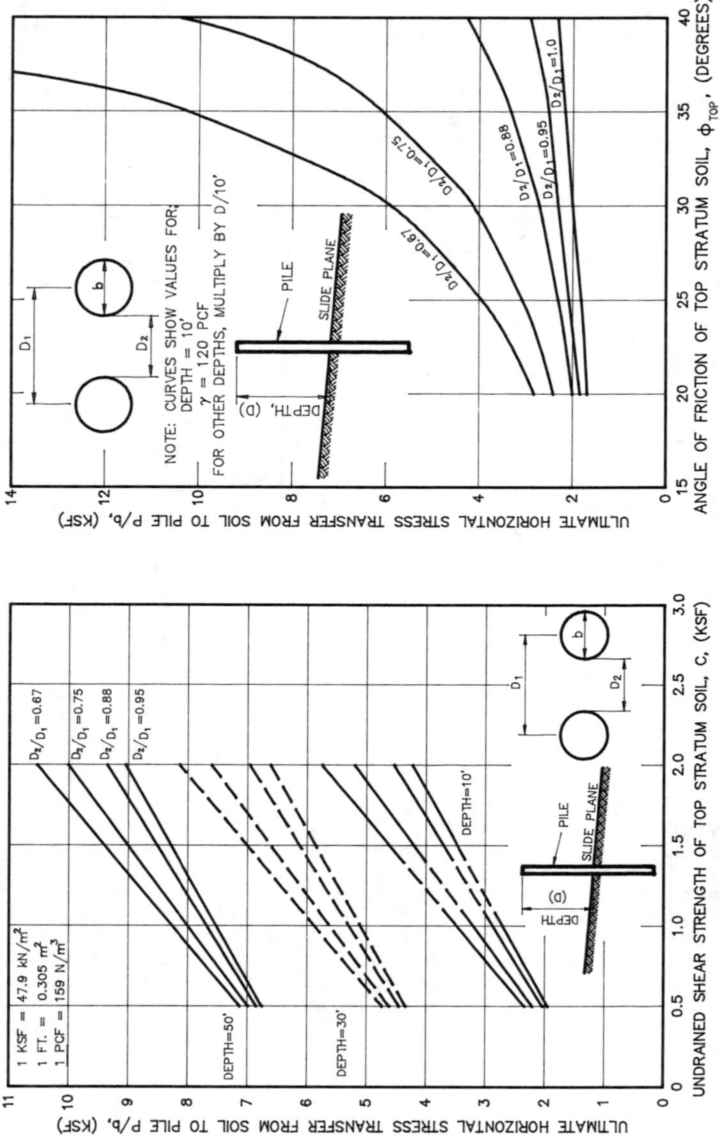

FIGURE 5

ULTIMATE STRESS TRANSFER FROM SOIL TO PILE VS. SHEAR STRENGTH OF SOIL

Considerations for Final Design

The procedure for preliminary design is useful in providing a preliminary estimate of the pile density and type of piles which are feasible for a particular application. The procedure is conservative for wall cross-sections which use battered piles. Battering the Pin Piles with respect to the slide plane, and/or direction of slope movement, tends to mobilize the axial resistance of the Pin Piles. Since the piles are typically small in diameter, their surface area to cross-sectional area ratio is relatively large. Hence they are very efficient at mobilizing skin friction, and typically have much higher axial capacity than lateral capacity.

The Type "A" INSERT walls reviewed herein have used a centralized reinforcing bar in a grout-filled hole. As shown on Figure 3, piles constructed using a centralized #9 reinforcing bar (e.g., Type 4 Piles) have significantly less resistance to horizontal loading than the piles installed with pipe reinforcement. However, once the concrete crushes, the reinforcing bar will provide additional resistance to horizontal loading by the development of tensile forces in the bar across the slide plane. A Type "A" INSERT wall designed assuming the Pin Piles behave in this manner would therefore be more economical than a wall designed using the procedure proposed for preliminary design. However, the additional movement of the slope needed to reach this condition may be intolerable in some cases.

Back Analysis of Case Histories

The two case histories described previously were back analyzed to check the recommended procedure. A more thorough consideration of the procedure will require better information regarding the strength properties of materials controlling slope behavior than was available for the projects described herein.

State Route 4023, Armstrong County, Pennsylvania. As shown in Figure 2, this Type "A" INSERT wall has decreased the rate of movement along the slide plane from approximately eight inches/year (20 cm/yr) to approximately 0.3 inches per year (0.8 cm/yr). The location of the slip-plane was estimated based on inclinometer data, and the driving forces of the slide were calculated to be approximately 132 kips/lf (1.9 MN/m).

Using equations developed by Leinenkugel (1976) and considered by Winter, et al. (1983) for the stabilization of creeping clay slopes, the loads acting on the Type "A" INSERT wall can be estimated. The basic assumption (Winter, et al., 1983) is that the mobilized shear stress along the failure plane equals the shear strength associated with an initial strain rate. The placement of the Pin Piles across the zone of movement reduces the stresses in the soil along the slide plane, causing the slope to move at a slower rate, thereby reducing the mobilized soil shear resistance along the slide plane. Consequently, equilibrium dictates that a decrease in the mobilized shear resistance in the soil along the slide plane must result in an increase of the resisting forces provided by the piles. Using this approach and the actual measured rates of movement, the resisting force along the slide plane provided by the Type "A" INSERT wall is equal to about 13 percent of the driving load, or 17 k/lf (250 kN/m).

Using the procedure for preliminary design, the maximum resistance provided by the wall was calculated to be approximately 7.5 kips/lf (110 kN/m), or 5.7 percent of the driving load. A more detailed analysis using the GROUP 1 program (Reese, et al., 1987), and assuming that the rock located below the slip plane acts as a pile cap, indicates that a maximum resistance of 17.5 kips/lf (260 kN/m), or 13 percent of the driving load, can be mobilized. This load is reasonably close to the back-calculated load imposed on the wall based on the measured decrease in the strain rate.

The field data suggest that the Pin Piles are at the limit of resistance as given in the procedure for preliminary design (i.e., the grout surrounding the reinforcement is crushing). Further slope movements should result in the Pin Piles acting in tension which would provide additional capacity and ultimately lead to cessation of slope movement.

<u>Project No. 6, Blue Heron Road, Big South Fork River, Kentucky</u>. Based upon the monitoring data collected at this site the Type "A" INSERT wall has effectively stopped the slope movements after moving approximately 0.3 inches subsequent to completion of the wall. Back-analysis of this wall indicates that the total driving load of the creeping slope is approximately equal to 45 kips/lf (660 kN/m).

The maximum resistance provided by the wall is calculated using the procedure for preliminary design to be about 4.6 kips/lf (67 kN/m) or 10 percent of the driving load. Based upon this back analysis, a 10 percent increase in the stability of the slope, as predicted by the procedure for preliminary design, is sufficient to stop the slope movement.

Summary and Conclusions

The paper presents an analytical procedure for preliminary design of Type "A" INSERT walls based on the bending resistance of Pin Piles across a slide plane and the plastic flow of soil between the piles. Selected case history summaries of Type "A" INSERT walls constructed in the U.S. are presented. Based on evaluation of the design procedure using information from these projects, the following is concluded:

- Back-analyses of two case histories indicate that the procedure for preliminary design is conservative for Pin Piles with a centralized reinforcing bar in a grout-filled hole.

- If the procedure for preliminary design is used and 10 percent of the driving load is provided as additional resistance, a stable slope is created with only minor additional slope movement.

- More instrumented installations of Type "A" INSERT walls are needed to further the knowledge base and improve the design procedure.

- If properly designed, Type "A" INSERT walls are effective for arresting slope movements.

In closing, it should be noted that the charts presented herein are for preliminary design only to establish general requirements for pile size and spacing. Final design of Type "A" INSERT Walls requires consideration of various factors including pile batter relative to the orientation of slope movement, the depth of the slide plane relative to the stiffness of the piles, and the additional capacity of the reinforcing bars after concrete crushing occurs.

ACKNOWLEDGMENTS

The authors appreciate the use of information provided by representatives of the the Pennsylvania Department of Transportation, District 10-0, and the U.S. Army Corps of Engineers, Nashville District, which was used in the development of the case histories described in the paper. In addition, the authors acknowledge the constructive comments of the conference reviewers, and those of their colleagues at Nicholson Construction Company and D'Appolonia, in clarifying and amplifying the paper.

REFERENCES

Bruce, D. A. (1992) "Two New Specialty Geotechnical Processes For Slope Stabilization," ASCE Specialty Conference on Stability and Performance of Slopes and Embankments - II, 15p.

Dash, U. and P. Jovino (1980) "Construction of a Root-Pile Wall at Monessen, Pennsylvania," Embankment Stabilization and Soil Mechanics, Transportation Research Record 749, National Research Council, Washington, D.C., pp. 13-21.

Dash, U. (1987) Pin-Pile Wall Evaluation, FHWA-PA-86-047-84-36, Federal Highway Administration, U.S. Department of Transportation, Washington, D.C., 174p.

Earth, Inc. (1986) Performance Evaluation Report, Existing INSERT Wall, L.R. 69 - Section 255, Armstrong County, Pennsylvania, Report for NCC, 25p.

Fukuoka, M. (1977) "The Effects of Horizontal Loads on Piles Due to Landslides," Proceedings, Specialty Session 10, Ninth International Conference on Soil Mechanics and Foundation Engineering, Tokyo, Japan, pp. 27-42.

Ito T. and T. Matsui (1975) "Methods to Estimate Lateral Force Acting on Stabilizing Piles," Soils and Foundations, Vol. 15, No. 4, pp. 43-59.

Leinenkugel, H. J. (1976) "Deformations und Festigkeitsver Halten Bindiger Erdstoffe," Experimentelle Erges-Bnisse und Ihre Physikalische Deutung Veroffentil. Inst. f. Bodenmech. u. Felsmech. 66, Karlsruhe.

Murray, R. (1984) In-Place Roadway Stabilization, New York Department of Transportation, Albany, NY, 19p.

Palmerton, J. (1984) Stabilization of Moving Land Masses by Cast-in-Place Piles, Miscellaneous Paper GL-84-4, Waterways Experiment Station, Vicksburg, MS, 101p.

Reese, L., K. Awoshika, P. Lam, and S. Wang (1987) "Documentation of Computer Program, GROUP1, Analysis of a Group of Piles Subjected to Axial and Lateral Loadings," Ensoft INC., Austin, TX.

Walkinshaw, J. L. (1984) "Correction of a Landslide with a 'Reticulated Root Pile' Structure," Federal Highway Report, Region 9, San Francisco, CA.

Winter, H., W. Schwarts, and G. Gudehose (1983) "Stabilization of Clay Slopes by Piles," Proceedings 8th European Conference on Soil Mechanics and Foundation Engineerings, Helsinki, Vol. 2, pp. 545-550.

HYDRAUGERS AT THE VIA DE LAS OLAS LANDSLIDE

By W.H. Roth[1], Member ASCE, R.H.Rice[2], D.T. Liu[1], Member ASCE, J. Cobarrubias[3]

ABSTRACT

In 1988 a subsurface drainage system consisting of 32 hydraugers was installed to enhance the stability of the 780,000 cubic-foot Via de Las Olas landslide in Pacific Palisades, California. The results of a three-year monitoring program involving measurements of hydrauger discharge and corresponding changes in piezometric levels are discussed in this paper. The presented data demonstrate the effectiveness of the hydraugers through measurable quantities of both discharge and reduction in piezometric levels. The feasibility of dewatering low-permeability siltstone/sandstone bedrock formations by hydraugers has been a controversial issue for some time. This case history should lend additional credence to the concept of dewatering the seemingly "impervious" formations of the coastal bluffs of southern California by small-diameter horizontal drains - as long as these formations are fractured. Because the latter is also a condition for rendering a "rock" formation unstable in the first place, most bedrock slides are likely to benefit from this approach.

---

(1) Dames & Moore, 911 Wilshire Boulevard, Suite 700, Los Angeles, California 90017

(2) Dames & Moore, 221 Main Street, Suite 600, San Francisco, California 94105

(3) City of Los Angeles, Department of Building & Safety, 200 N. Spring Street, Room 460-A, Los Angeles, California 90012

INTRODUCTION

GENERAL BACKGROUND

The Via de las Olas landslide in Pacific Palisades is typical of the wide-spread stability problems affecting the coastal bluffs of southern California. Located adjacent to a controversial proposed oil drilling site near a popular beach, the implications of possibly reactivating the large slide mass through oil drilling activities played a major role in a nearly 25-year public campaign against oil production at this location. In 1988, after the installation of an extensive drainage system to enhance the stability of the slide area, oil drilling permits were finally obtained. However, coastal oil drilling was banned by a public referendum shortly thereafter and the project was abandoned.

Regardless of the oil drilling issue, the drainage system of the Via de las Olas landslide has been maintained and monitored the past three years. It provides much needed protection against further deterioration of this sprawling slide area whose advancing head scarp has begun to encroach upon the residential area overlooking the coastal bluffs. Continuous monitoring of hydrauger flow rates and piezometric levels over a period of three years has established a valuable data base on the effectiveness of subsurface drainage in the low-permeability clay and siltstone formations typical of the coastal bluffs of southern California.

SITE DESCRIPTION AND HISTORY

A plot plan of the slide area is shown in Figure 1, and a typical cross section is presented in Figure 2. The slide is about 800 feet square in plan dimensions, with a height of approximately 240 feet from the toe to the top of the head scarp. The main head scarp ranges in height from 50 feet in its eastern part, to as much as 130 feet to the west. Reported slope movements at the site date back to 1890, and intermittent slides have been noted since the 1920's, when residential development of the upper mesa began. The last major slope failure occurred in 1958, when approximately 780,000 cubic yards of slide debris forced the relocation of Pacific Coast Highway 200 feet towards the ocean.

Numerous investigations have been performed since the 1958 slide episode. The most extensive studies, however, were performed between 1983 and 1988, in the course of appealing an earlier denial of permit for oil drilling at this site. In this period extensive geologic mapping and exploratory drilling was performed. A total of 21 borings between 40 and 200 feet deep were drilled using dry auger and continuous coring techniques, some with oriented cores. In addition, extensive geotechnical instrumentation was installed and continuously monitored as described elsewhere in this paper.

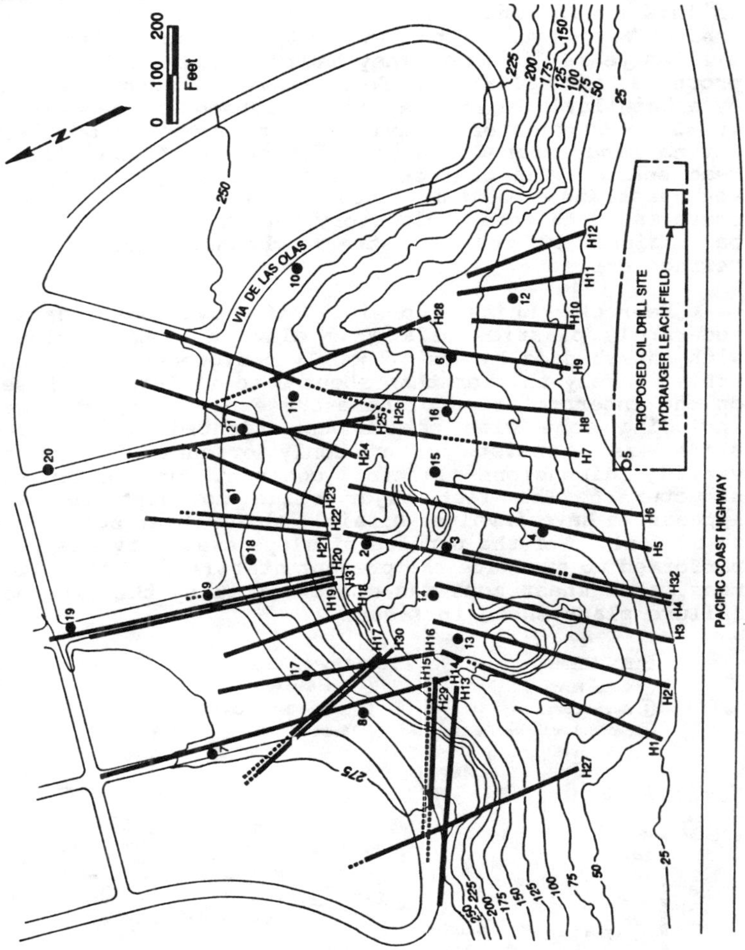

Figure 1. Planview with Boring Locations and Hydraugers (the dotted lines denote sand and gravel zones encountered by the hydraugers).

GEOLOGIC CONDITIONS

STRATIGRAPHY

Failure occurred within a sequence of thinly bedded claystones, siltstones, and weakly cemented fine sands of the Modelo formation which constitute the "bedrock" in this area. The bedrock units, also referred to as shales, are of marine origin and Upper Miocene (about 10 million years) in age. They have been planed by marine processes and uplifted to form a wave-cut terrace capped by a thin layer of marine sands. Subsequent deposition of 30 to 50 feet of alluvial deposits on top of these marine sands shaped today's mesa north of the slide's head scarp. The bedrock contact slopes gently toward southeast and, based on visual observation of the head scarp and data from exploratory borings, is believed to be incised by several stream channels filled with terrace deposits.

Bedrock units consist of overconsolidated, moderately plastic, dark gray clays and silty clays, with interbedded layers of poorly cemented fine-grained sands, rarely thicker than about 6 to 12 inches. Based on the interpretation of numerous exploratory borings, including some with continuous oriented coring, the slide mass consists of randomly oriented blocks of varying dimensions lacking coherent or continuous structure. The last major slide movement in 1958 appears to have involved a fairly thick shear zone of 30 to 50 feet. For the purpose of slope stability analysis performed to evaluate the present stability of the slide mass, this shear zone was approximated by the distinct failure plane shown in Figure 2.

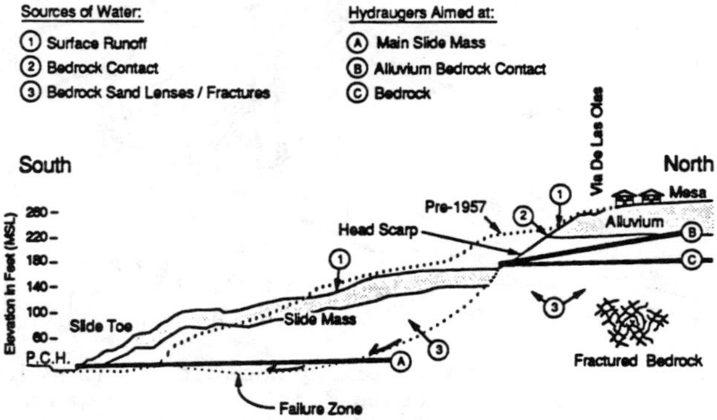

Figure 2. Cross Section with Hydraugers

SLOPE STABILITY

Water is the primary factor affecting slope stability. The increase in unit weight and loss in strength experienced by fractured shales as a result of water saturation, makes these bedrock units particularly susceptible to landsliding. In contrast, bedrock material which is not exposed to saturation by water can form near-vertical bluffs of more than 200 feet in height.

Based on back calculations of the 1958 slope failure, it was concluded that the main slide mass in its present configuration is stable with a comfortable margin of safety. By shifting its weight from the top of the slope to its toe, the stability of the slide mass has improved considerably compared to pre-1958 slope conditions. Hence, future large-scale earth movements are far less likely to occur at this site than at other coastal bluffs between Santa Monica and Malibu, which have not yet failed. Existing slide stability is further evidenced by the fact that no deep-seated slope movements have been detected since the 1958 failure. This, in spite of the extremely wet storm seasons of 1968/69, 1977/78, and 1982/83, which triggered large numbers of major slides in similar geologic formations along the coast. Rainfall amounts of these recent wet seasons were higher than the 1957/58 season which triggered the 1958 slide.

In contrast to the main slide mass, the stability of the rather steep head scarp was evaluated and found to be only marginally stable. Incipient slide movement is evidenced by an extensive crack pattern in the street pavement of Via de las Olas along the crest of the scarp. In addition to possible future damage of adjacent homes on the upper mesa, a slope failure in the scarp area was also seen as a potential threat to the stability of the main slide mass. The debris from a scarp failure loaded onto the head of the main slide mass could considerably reduce the safety factor of the latter. It was, therefore, recommended to enhance the stability of both the main slide mass and the head scarp through the installation of open surface drains and a subsurface drainage system consisting of hydraugers. The purpose of the drainage system is to intercept the active ground water flow before it enters the main slide mass.

GROUND WATER REGIME

The upper mesa extending north towards the Santa Monica Mountains was concluded to constitute a major recharge area for ground water entering the slide. As indicated in the cross section of Figure 2, there are three water sources which pose potential problems for the site: (1) surface runoff, (2) ground water migrating along the alluvium/bedrock contact at the head scarp, and (3) ground water which approaches the slide mass essentially from below its failure zone, after entering fracture systems in the bedrock below the terrace deposits of the upper mesa.

Recharge in the upper mesa occurs as a result of precipitation during the months between November and April and from irrigation during the summer months. Much of the ground water flowing in the alluvium is apparently concentrated in shallow channels incised into the bedrock surface. Water has been observed discharging from these bedrock channels at the western cliffs and in the central portion of the headscarp following periods of extended rainfall. Established water-loving vegetation evidences these seepage areas.

Deep wells located in the bedrock of the upper mesa experience distinct fluctuations apparently related to rainfall. In contrast, measurements taken over the last 20 years indicate that the ground water levels within the slide mass do not significantly fluctuate with rainfall. In fact, piezometric data and water chemistry indicate that the ground water in the slide mass may be isolated from the surrounding ground water regime. It is likely that this isolation may be due to the shear zone of the slide forming a hydraulic barrier. The migrating ground water from the upper mesa, which probably played a sig-nificant role in triggering the 1958 failure, now appears to find its way around the slide. Concentrated seepage at the slide toe, observed near the western and eastern slide boundaries, supports this interpretation.

DRAINAGE SYSTEM

The drainage system consists of two components: (1) surface drainage (open channel drains), and (2) subsurface drainage (hydraugers). Only the latter is discussed in this paper. The subsurface drainage system consists of horizontal drains (hydraugers), a leachfield at the toe of the slide, and a hydrauger manifold line connecting the hydraugers to the leachfield. The leachfield is designed to return ground water produced by the hydraugers back into the lower ground water regime below the toe of the landslide.

HYDRAUGER INSTALLATION

The basic layout of the installed hydraugers is shown in the plan view and cross section of Figures 1 and 2, respectively. The hydraugers in the main slide mass (A) were designed to drain both perched water within the slide mass and ground water which is migrating through the bedrock underlying the landslide shear zone. At the head scarp, there are two types of hydraugers: (B) inclined hydraugers penetrating and draining the alluvium/bedrock contact of the upper mesa; and (C) near-horizontal hydraugers draining the fractured bedrock underlying the terrace deposits. Type (B) includes hydraugers tapping into the alluvial channels of the western portion of the upper terrace where some of the recharge water is believed to be concentrated.

The hydraugers consist of 1.5-inch diameter slotted PVC pipe inserted in approximately 3-inch diameter holes, which were generally uncased. The holes were drilled through a 20-foot long, 5-inch diameter steel casing grouted in place prior to drilling the deep 3-inch hole. Thirty-two hydraugers were installed, varying in length from 250 to 790 feet at inclinations of 2 to 15 degrees from the horizontal. The total length of installed drainage pipe was 14,455 feet.

Drilling was performed with a rotary-wash horizontal drill rig mounted on a crawler tractor. An attempt was made to drill some of the borings with compressed air, in order to identify saturated zones; however, due to the clayey nature of the bedrock formation, drilling progress was hampered by continuous plugging of the bit. All hydrauger borings were logged by recording the drilling resistance encountered and observations of the drill cuttings. In general, drilling in the main slide area encountered medium and, occasionally, soft resistance, indicative of slide debris. In contrast, drilling in the head scarp was hard, except for four hydraugers, H13, H16, H23, and H30, which encountered moderate resistance. Sand and gravel zones which were large enough to be recognized in the drill cuttings are shown in the hydrauger layout of Figure 1 as dotted lines.

INSTRUMENTATION

Extensive geotechnical instrumentation has been installed within and adjacent to the landslide mass for the monitoring of potential slope movements and ground water conditions. Except for shallow sloughing, none of the slope indicators and extensometers have shown signs of earth movements which could be interpreted as a reactivation of the main slide mass.

Piezometric levels were collected from an array of instruments which include open-standpipe wells and pore pressure gages. With the exception of a few "average" wells with full-depth screened zones, the standpipes have short isolated screened intervals targeting specific sandy beds within the bedrock, or the alluvium/bedrock contact of the upper mesa. The pore pressure gages consist either of pneumatic or electric-strain-gage type instruments. The locations of the observation wells and pore pressure gages are shown in plan view in Figure 1. A schematic cross section depicting the three typical ground water monitoring devices, (1) average wells, (2) isolated wells, and (3) pore pressure gages, is shown in Figure 3.

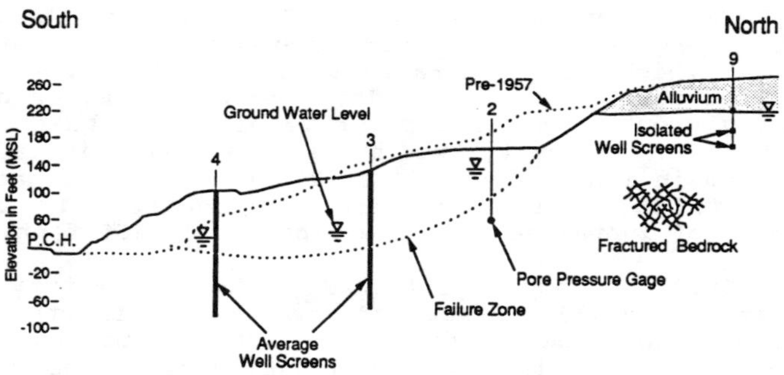

Figure 3. Cross Section with Observation Wells and Pore Pressure Gages

Plans for a fully automated data acquisition system, including computerized warning devices for early detection of threshold ground water levels and rates of slope movement, were abandoned when the oil drilling project was halted in 1988. Nevertheless, the instruments were monitored manually on a rather frequent basis until summer of 1991.

DISCUSSION OF MONITORING DATA

In order to evaluate the effectiveness of the installed subsurface drainage system, hydrauger flow rates and ground water data were recorded for several years. Furthermore, in an attempt to assess the hydraulic connections between alluvium, bedrock, and the slide mass, ground water samples were chemically analyzed and attempt was made to group them by source.

However, the results of the chemical analysis were inconclusive, probably because of the mixing of water types occurring within the contorted slide mass and fractured bedrock formations.

HYDRAUGER DISCHARGE

Plots of hydrauger flow rates and cumulative discharge volumes versus time are presented in Figures 4 and 5, together with a bar graph showing local rainfall data in inches per week. The hydraugers are grouped by the local ground water regions which they tap: (1) the upper-mesa alluvium, (2) the bedrock north of the head scarp, and (3) the main slide mass, respectively. The bottom plot of Figure 5 presents the sum total of all data. Table 1 presents the total discharge volumes accumulated since installation for each of the 31 hydraugers.

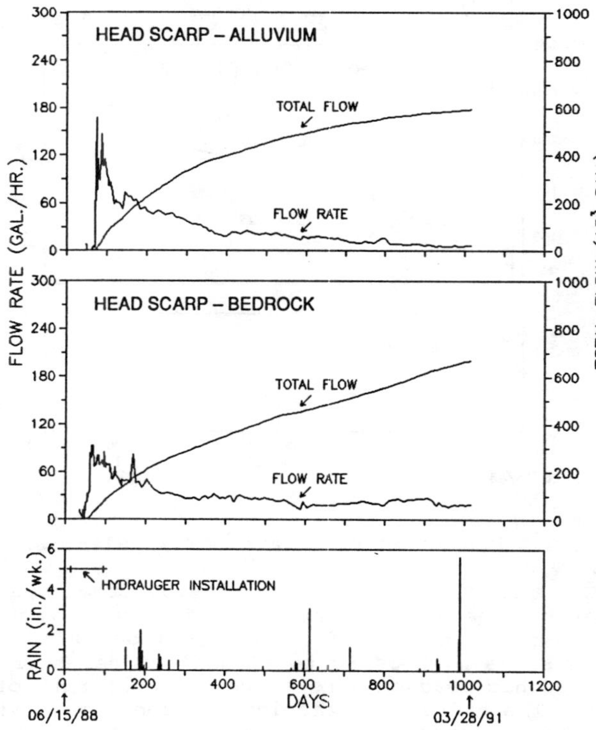

Figure 4. Hydrauger Flow Rates, Head Scarp

Hydrauger H12 which is located at the mouth of a ravine east of the slide area, was excluded from the plotted data, because of its atypically high, and occasionally erratic, flow rates. This hydrauger is thought to have some direct hydraulic connection with surface runoff and/or a leaking storm drain or sewer transversing this ravine.

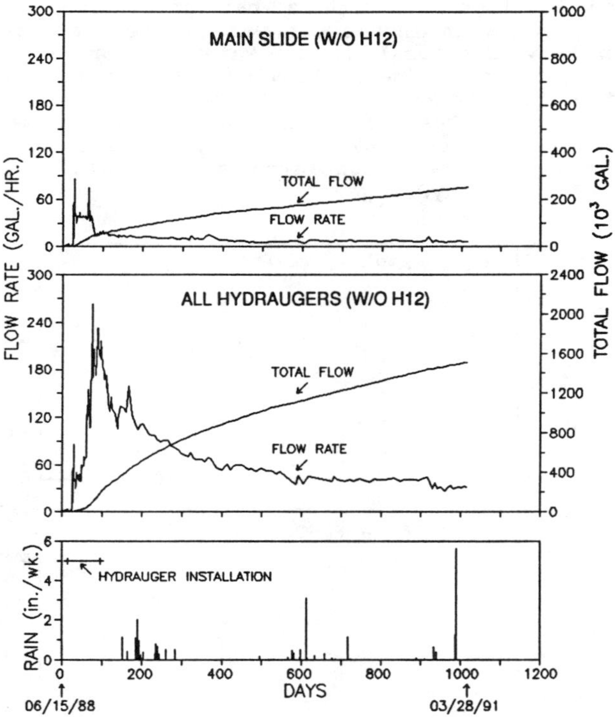

Figure 5. Hydrauger Flow Rates, Main Slide and Total

Of the three groups analyzed, the hydraugers dewatering the bedrock at the head scarp appear to produce most consistently. Starting out with initial flow rates as high as 60 to 90 gallons per hour (gph), discharge has reached a relatively stable rate of 25 to 30 gph. The hydraugers tapping the upper mesa alluvium above the headscarp initially produced at significantly higher rates of 90 to 150 gph, but declined to 25 to 30 gph in the first year, and are presently as low as 5 to 10 gph. Finally, the main slide hydraugers, with the exception of two brief 80 gph peaks in the beginning, only produced about 40 gph for the first 2 months. After that they sharply declined to about 15 gph, and have reached about 10 gph today.

TABLE 1. Total Hydrauger Discharge from 1988 to 1991

Hydrauger #	Total Flow (Gallons)	Hydrauger #	Total Flow (Gallons)	Hydrauger #	Total Flow (Gallons)
1	12999	12	440834	23	42368
2	29617	13	234413	24	35809
3	65714	14	59502	25	31545
4	43910	15	79872	26	4375
5	23005	16	20391	27	178850
6	1432	17	143	28	2772
7	9200	18	252	29	428995
8	69772	19	17292	30	37191
9	90	20	29	31	34378
10	0	21	62501	32	5924
11	2487	22	23144		

The entire hydrauger monitoring period reported herein, of course, coincided with a severe regional drought. Therefore, little insight can be obtained about the interrelationship of rainfall, regional ground water recharge, hydrauger flow rates, and piezometric levels. Notwithstanding this situation, the hydrauger flow rates do not appear to be much affected by rainfall. If anything, it almost seems as if flow rates in the past had decreased in response to rainfall.

The gradual decrease of hydrauger flow rates over time may be related to decreases in permeability of the soil envelope around the hydrauger. This phenomena has been noted in falling-head permeability tests in vertical piezometers, repeated over periods of time. It would not be easy to check this condition in the upward-inclined hydraugers, though air has been used as a testing fluid for in situ permeabilities of soils (Blight, 1971).

PIEZOMETRIC LEVELS

The recorded piezometric levels versus time for the periods before and after the installation of hydraugers are shown in Figures 6 and 7. Only data from those instruments which experienced detectable and consistent changes in response to hydrauger installation are shown. The dotted, dashed, and solid-line plots represent three different types of instruments, (1) pore pressure gages,

(2) average wells, and (3) isolated wells, as discussed are presented in Table 2.

The head scarp piezometric levels were essentially constant before the installation of the hydraugers, but had shown temporary peaks in response to seasonal rainfall (Figure 6). During hydrauger installation many of these piezometric levels experienced small temporary above. Installation characteristics of these instruments spikes (Figure 7), probably caused by the injection of drilling fluid in the installation process. After these initial spikes, most of the bedrock piezometric levels decreased, over a period of about 1 to 2 months, reaching steady levels up to 15 feet lower than before the hydraugers were installed. In addition, the seasonal peaks typically experienced before the hydraugers, appear to have all but disappeared in the second monitoring period. This behavior, however, must be evaluated in light of the lower rainfall experienced throughout this last period.

The atypical behavior of Gage 7D is likely due to malfunctioning of this instrument which, after some fluctuations beginning at approximately Day 680, broke down completely and was abandoned. Of the observation wells located at the bedrock-alluvium interface of the upper mesa, only 9B experienced a relatively small decrease of 2 feet. The remaining wells in the alluviumdid not show any significant changes; their piezometric levels are not shown here.

TABLE 2. Installation Data of Wells and Pore Pressure Gages

Well #	Location	Type	Installation Date	Ground Surface Elevation (Ft. M.S.L.)	Screen/ Gage Elevation (Ft. M.S.L.)
2	Main slide	Pore pressure gage	03-Jun-83	168	73
3	Main slide	Average well	1967	152	147 to -8
4	Main slide	Average well	03-Apr-83	100	95 to -60
7A	Head scarp	Isolated well	24-Oct-83	274	162 to 160
7C	Head scarp	Isolated well	28-Oct-83	274	192 to 190
7D	Head scarp	Pore pressure gage	01-Sep-87	274	161
9A	Head scarp	Isolated well	19-Oct-83	262	192 to 190
9B	Head scarp	Isolated well	04-Nov-83	262	223 to 221
9C	Head scarp	Isolated well	04-Nov-83	262	166 to 164
9D	Head scarp	Pore pressure gage	02-Sep-87	262	166
12	Main slide	Isolated well	17-Nov-83	100	3 to 1
13	Main slide	Isolated well	27-Aug-87	141	42 to 22
14	Main slide	Isolated well	01-Sep-87	142	43 to 23
19	Head scarp	Pore pressure gage	14-Sep-87	275	78
20	Head scarp	Pore pressure gage	11-Dec-87	270	171
21	Head scarp	Pore pressure gage	09-Dec-87	252	151

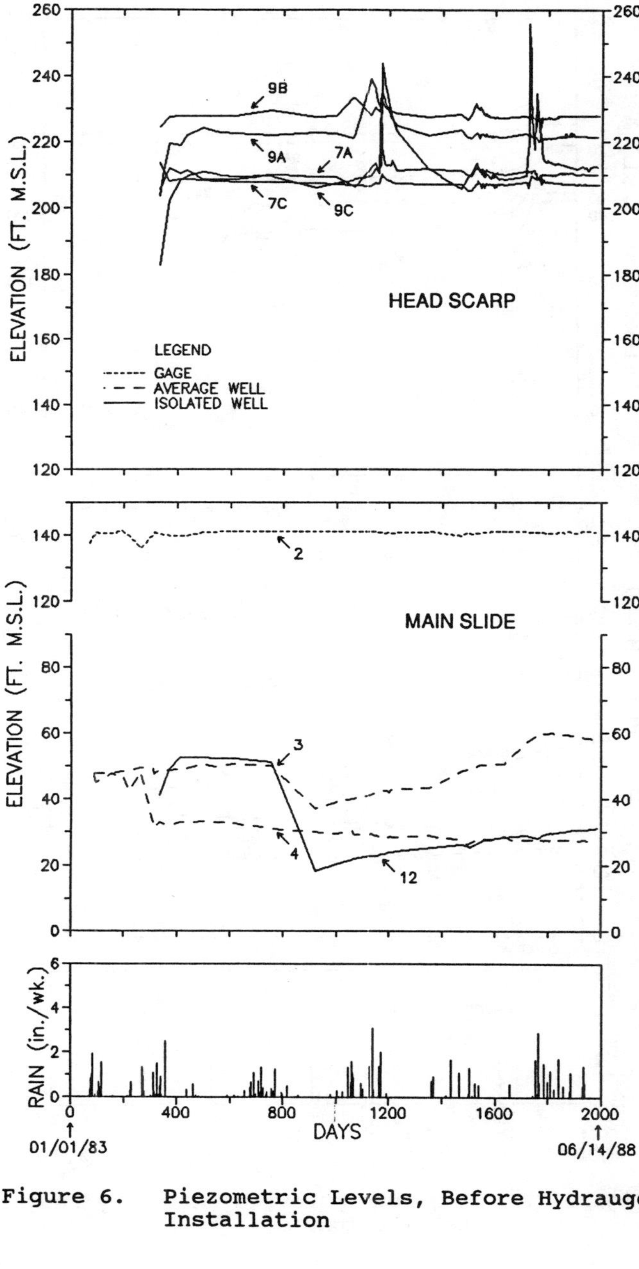

Figure 6. Piezometric Levels, Before Hydrauger Installation

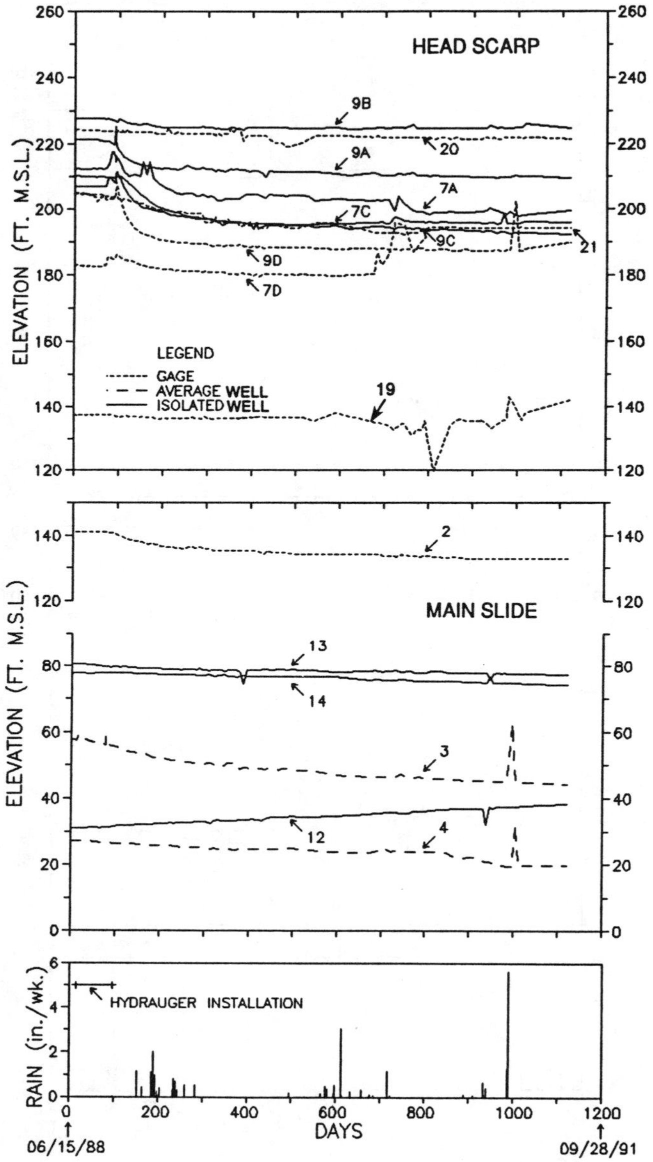

Figure 7. Piezometric Levels, After Hydrauger Installation

In the region of the main slide mass many of the observation wells remained essentially unaffected by the hydraugers. The data from those instruments which did experience changes are presented in the bottom half of Figure 7. The most significant decreases in piezometric levels, about 15 and 8 feet, occurred at Locations 3 and 4, respectively. In addition to these two average-well type instruments, the pore pressure gage at Location 2 showed a decrease of about 5 feet.

The steadily rising piezometric level at Location 12 represents a notable exception from the general trend of constant or decreasing levels across the site. It is situated near the mouth of the same ravine which is thought to be affected by external water influx causing the excess flow rates of Hydrauger H12 as discussed above.

CONCLUSION

Several years of monitoring the drainage discharge and corresponding changes in the ground water regime of the Via de las Olas slide have established a useful data base to evaluate the effectiveness of subsurface drainage in the low-permeability clay- and siltstone formations typical of the coastal bluffs of southern California.

In evaluating the effectiveness of hydraugers, it is important to focus on their main purpose of reducing the pore water pressures within the saturated soil or rock materials. Depending on whether the permeability of these materials is low or high, the pressure reduction is accompanied by small or large volumes of flow, respectively. Hence, in low-permeability materials, such as the bedrock formations of the Via de las Olas slide, even hydraugers which are merely dripping are likely to be effective. Also, the absence of demonstratively dropping piezometric levels is not necessarily a sign of nonfunctioning hydraugers. In the highly contorted and fractured bedrock formations, some of the observation wells may be hydraulically isolated from nearby hydraugers tapping into independent ground water regimes.

Notwithstanding the above, the data presented herein demonstrate the effectiveness of the installed hydraugers through measurable quantities of both discharge and reduction in piezometric levels. Among practitioners dealing with the enormous landslide problems of southern California's coastal bluffs, there have long existed two different schools of thought on this subject. Specifically, the feasibility of dewatering bedrock slides by hydraugers has been a controversial issue. The case history presented herein should lend additional credence to the concept of dewatering these seemingly "impervious" formations of the coastal bluffs by small-diameter horizontal drains - as long as these formations are fractured. Because the latter is also a condition for rendering a "rock" formation instable in the first place, most bedrock slides are unlikely to benefit from this approach.

APPENDIX

REFERENCES

Blight, Geoffrey E., "Flow of Air Through Soils," Journal of the Soil Mechanics and Foundations Division, ASCE, Vol. 97, No. SM4, Proc. Paper 8026, April, 1971, pp. 607-624.

SI-UNITS CONVERSION

1 in = 0.0254 m;
1 ft = 0.3048 m;
1 cubic yard = 0.7646 m^3
1 gallon per hour = 3.785 liter per hour

STABILIZATION OF TABLACHACA DAM LANDSLIDE

By

Richard A. Millet [1] M. ASCE, Gil M. Lawton [2] M. ASCE,
Pedro C. Repetto [3] M. ASCE, Vinod K. Garga [4] M. ASCE

ABSTRACT

The Tablachaca Dam Landslide exists above the dam of a major hydroelectric power project in Peru. When movement threatened the dam, an emergency comprehensive program of multiple remedial measures designed to act together was developed and successfully implemented. This paper describes the background of the landslide and presents details regarding several of the emergency stabilization measures.

INTRODUCTION

Tablachaca Dam is a 72 m high concrete gravity-arch dam on the Mantaro River in Peru, located approximately 300 km from Lima (Figure 1). Its objective is to divert water from the Mantaro River through a 20 km long pressure tunnel to the Santiago Antunez de Mayolo hydroelectric plant (798 MW). From there water is conveyed through another tunnel to a second power plant (216 MW). These two power plants are owned by the government company ELECTROPERU and provide as much as 50% of the electrical capacity for central Peru. Construction of the dam was completed in 1972.

In September 1972, the reservoir was first filled from river level at about Elevation 2,640 m to a maximum elevation of 2,695 m. During the period when the reservoir was rising, movement was observed in part of an ancient slide on the

---

1. Executive Vice President, Woodward-Clyde Consultants, Denver, CO 80237
2. Senior Project Engineer, Woodward-Clyde Consultants, Oakland, CA 94607
3. Associate, Golder Associates, Mt. Laurel, NJ 08054
4. Associate Professor, Department of Civil Engineering, University of Ottawa; Ottawa, Canada; Geotechnical Consultant to ELECTROPERU, Lima, Peru

Figure 1. PROJECT LOCATION

right abutment immediately upstream from the dam. Because of this event, the reservoir was lowered and subsequent movements of the slide were monitored during a period from October 1972 to December 1972. However, the movement of the slide did not stop and between December 1972 and January 1973, a volume of about 5,000 m^3 of debris slid into the reservoir.

A program of sporadic measurements of the slide movement continued until 1979 when visible evidence of slide activity was again apparent. In December 1979, a series of investigations and measurements were undertaken to develop information about the extent and depth of the landslide. The work included geodetic triangulation techniques to record the slide movement, and boreholes to explore the depth, geologic characteristics of the slide, groundwater, and to provide data on the location of subsurface slip surfaces by monitoring distortion of casings.

By the end of the rainy season of November 1981 to April 1982, there was serious concern that a very large active slide mass was in danger of sliding into the reservoir. Inspection of the survey data for surface displacement of several monuments located along the faster moving downstream side of the slide showed accelerating rates of displacement from year to year. The active slide surface area measured about 300 m along the reservoir bank by 350 m up the slope.

Description of Landslide

The landslide begins immediately upstream from the dam and it extends from elevation 2660 m to 3200 m, with an average inclination of 37° to 40°. The landslide consists of two parts, designated as lower and upper part. The lower part was the most active, with a volume of about 3 million m^3 and extends from approximately elevations 2660 m to 2850 m. The upper part has a volume of about 10 million m^3, and extends up to elevation 3200 m. It initially was inferred to be inactive. The slide mass is essentially a melange of fractured, black phyllite and brown quartzitic phyllite slabs and fragments in a matrix of gravel, sand and silt-sized detritus, overriding black phyllite in the upstream three-quarters of length of the landslide and overriding quartzitic phyllite in the downstream quarter.

Development of Emergency Works

Due to the urgency of the situation at the end of the 1981-82 rainy season, ELECTROPERU, with the assistance of advisors and specialists, retained Woodward-Clyde Consultants, together with Peruvian partner, Michelena, Repetto y Asociados (WCC/MR), in April 1982, to design the emergency stabilization works for the slide. A program that would permit the immediate start of the works in the shortest possible time was required; its first priority was to immediately improve the slide stability.

In preparing that plan, it was considered necessary that the emergency works meet the following basic criteria:

- The stabilization works had to be susceptible to field modifications during construction as more information was to be obtained as a result of an investigation and instrumentation program to be implemented concurrent with construction.

- The type of stabilization works had to be such that they could be implemented in the shortest possible time.

- Since the stabilizing effect of the works would be gradual and, therefore, the landslide would likely continue to move during construction, the selected works had to be capable of deforming with the slope without being damaged.

- The emergency stabilization works had to be such as to lend themselves to being incorporated into a more definitive solution in the future.

- The stabilization works had to be such as to not compromise the operation of the reservoir and dam (sediment trap, low level outlets used to flush the high sediment load, spillway gates, etc., as shown in Figure 2).

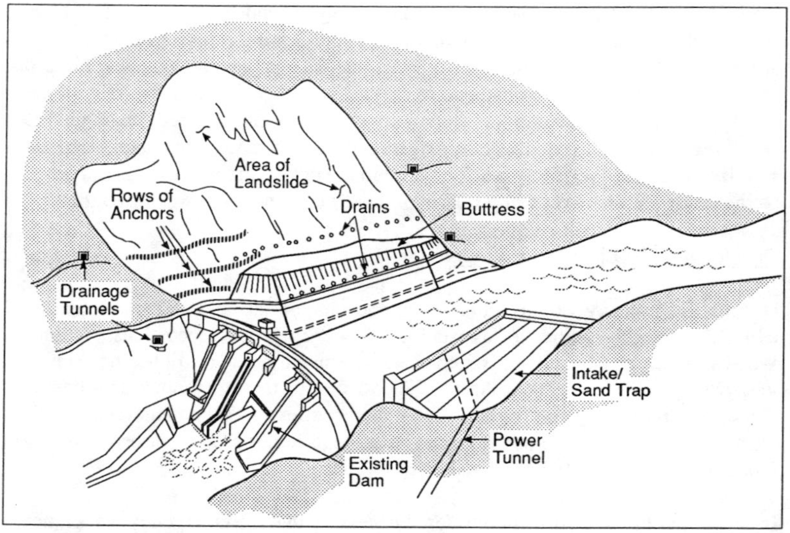

Figure 2. OVERVIEW OF EMERGENCY WORKS

The emergency stabilization works program was developed based on the above requirements and was a comprehensive program of multiple remedial measures designed to act together. The emergency stabilization works constructed include:

- Construction of a fill buttress at the toe of the slide, including foundation improvement using compacted gravel columns;

- Prestressed rock anchors in the slide area close to the dam (where a buttress fill would impinge on the low level sluice gates);

- Drainage tunnels excavated below the slide and radial drains installed from the tunnels;

- Horizontal drains installed from the slope surface; and

- Excavation of a rock protrusion in the river upstream from the landslide (so that the main hydraulic flow path of the river would not impinge against the base of the landslide).

Actually, considering such factors as ease of construction, cost and schedule, the buttress was found to be superior to any other solution; however, in the area immediately adjacent to the dam abutment, the buttress had to be very small so as not to interfere with the low level sluice outlets, and in that area, it needed to be supplemented by 120-ton working load rock anchors.

The emergency works also included the implementation of geotechnical instrumentation already planned by ELECTROPERU, and the design and implementation of additional geotechnical instrumentation to investigate the slide mechanisms and the effectiveness of the stabilization works. A schedule of all the completed work is shown in Table 1.

TABLE 1

SCHEDULE OF STABILIZATION WORK

Geotechnical Field Investigations	April 1982 to September 1984
Instrumentation	April 1982 to September 1984
Buttress Construction	
Investigation of Reservoir Sediments	June to August 1982
Reservoir Sediment Improvement	August 1982 to March 1983
Buttress Construction	August 1982 to September 1983
Rock Anchors	August 1982 to September 1984
Tunnels and Radial Drains	April 1982 to September 1984
Horizontal Drains	April 1982 to September 1984
River Rock Protrusion	
Hydraulic Model	July 1982 to February 1983
Protrusion Excavation	February 1983 to July 1983

A more-detailed description of the landslide characteristics and stabilization works has been presented elsewhere (Repetto, 1985).

SCOPE OF PAPER

This paper discusses the general setting of the landslide and then focuses on several aspects of the design and construction of the stabilization measures mentioned above. A background of the stability analyses also is discussed. Significant issues addressed include the buttress design and construction, including the improvement of the foundation which consists of river-deposited low density silty sediments in an active seismic environment; and the need for a comprehensive combination of stabilization measures for the landslide, compared to many landslides where only one or two methods may be required for stabilization.

SLOPE STABILITY ANALYSES

Static Stability Analyses

A very limited period of time was available for design, preparation of construction specifications, construction bidding and contractor selection, and starting the actual construction of the emergency works to be performed before the following wet season. Therefore, basic stability analyses were required to support the design work immediately after the project was authorized by ELECTROPERU in April, 1982. During the period of the basic analyses, no data were available on soil types and properties of the slide mass and the reservoir sediments. Thus, the basic analyses conducted prior to July 1982 were primarily to assess the relative effect on the stability of the active slide of all slope-stabilizing schemes contemplated. These schemes included toe buttress, rock anchors, surface drainage, subsurface drainage using horizontal drains and drainage tunnels, and unloading of upper slopes.

Cases Analyzed. A series of static slope stability analyses were conducted to examine the stability of the most active zone of landsliding incorporating various emergency remedial schemes being implemented (WCC/MR, 1983a).

For the active zone of the landslide, the following analyses were performed:

(a) Back-calculation of the shear strength parameters for an estimated overall slip surface. Two slightly different sections were analyzed. A typical section is shown in Figure 3. In both cross-sections the elevation of the

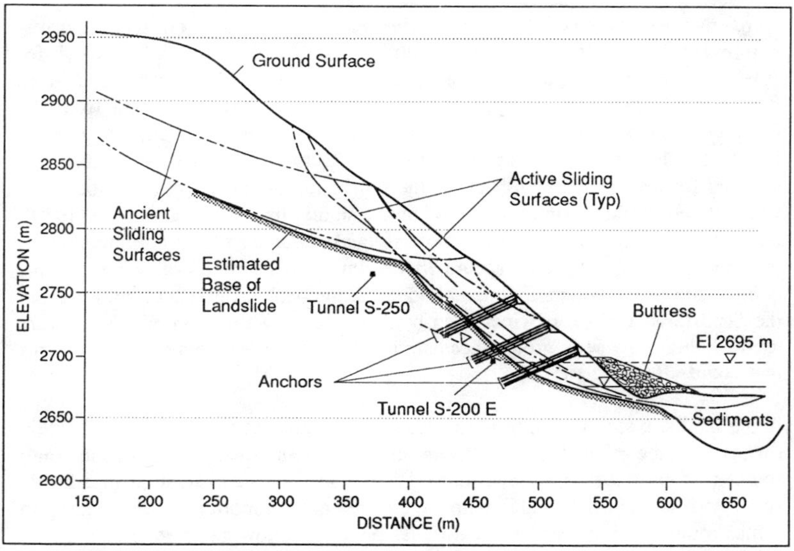

Figure 3. SCHEMATIC CROSS SECTION OF LANDSLIDE

slide surface at the toe was given at El. 2665 (Hutchinson, 1982). In these analyses, the back-calculated shear strength parameters correspond to a factor of safety of about one in the dry season with the reservoir water level at about Elevation 2676 m (36 m above the river level). Using this method, a value of ø = 37° was calculated.

(b) Using the back-calculated strength parameter, analyses were made to examine the effects of buttress and rock anchor remedial measures being implemented as part of the emergency work.

(c) Analyses were made to examine the effect of pore water pressures including rapid drawdown conditions on the stability of the active slide.

(d) Analyses were made to examine the effect on the stability of the active slide if the reservoir sediments were to be removed.

Seismic Ground Motions

The Tablachaca landslide is in a moderate to high seismically active zone. Therefore, a study was performed to develop ground motions for use in a seismic stability analysis of the landslide after completion of the emergency works. Site peak accelerations on rock were estimated using modifications of the acceleration attenuation curves presented by Seed and Idriss (1982). Because these curves were developed based on data for shallow crustal earthquakes (i.e. such as strike-slip events in California), they involve greater uncertainty when used for subduction zone earthquakes in Peru. The peak accelerations from Seed-Idriss curves were compared with the peak acceleration recorded in Lima from the October 3, 1974 earthquake (M_s 7.8) that occurred on the subduction zone plate interface offshore Peru. The data from the October 3, 1974 earthquake was selected for this comparison because the source rupture area appears to be fairly well defined for this earthquake, based on the studies by Dwey and Spence (1979), and thus the closest distance from the earthquake source to the recording stations in Lima appeared to be fairly accurately known for this earthquake. Based on this comparison, for subduction zone earthquakes, the peak ground acceleration from the Seed-Idriss curves was increased by 25 percent. These revised curves were used to obtain the peak ground acceleration for the two subduction zone sources that could affect the site.

A probabilistic seismic exposure analysis was conducted to develop information for the design earthquake. Local (intraplate faults) and distant (subduction zone) sources of earthquakes were evaluated. For design and evaluation of seismic stability of the landslide stabilization at the site, ground motions corresponding to a maximum credible earthquake (MCE) on the shallow plate interface of the subduction zone were selected on the basis that this earthquake source is the most likely source of strong ground shaking at the site during the life (50-year return period) of the project. For the MCE (M_w 9.0, M_s 8.5) on the shallow plate interface of the subduction zone (95 km to the dam), the estimated peak rock acceleration is 0.15 g and the estimated duration of strong shaking is 50 seconds. This level of shaking is greater than a comparable 50-year return on the local fault (WCC/MR, 1983b).

In selecting the design earthquake for the design of the remedial works, a 50-year return period on a local fault was selected, in consultation with ELECTROPERU. This return period was based on the economic balance of the remediation costs compared to the likely consequence of the landslide which would be the permanent blockage of the intakes for the power tunnels to the power plant. Because of the low reservoir capacity, there is a relatively low risk for a large rapid breach.

Seismic Stability

The consequences of earthquake shaking on the Tablachaca slide after the full effects of long-term remedial measures have taken place were evaluated in terms of seismically-induced displacements of the main slide and the buttress. The effects of the drainage system behind the main slide were also evaluated for one of the main slide sections with the reservoir level at el. 2695 m. The seismically-induced displacements were estimated using a relationship based on accelerograms recorded in Lima, Peru, and the relationships developed by Makdisi and Seed (1977), respectively. Based on all the analyses and evaluations performed, the following conclusions were reached:

(1) As long as the saturated part of the shear zone within the main slide is kept below about el 2680 m (40 m above the river elevation) and as long as earthquake shaking does not increase the excess pore water pressure (pwp) in the shear zone within the main slide by more than approximately 30 percent of the corresponding effective overburden pressure, the computations showed that the likely consequence of earthquake shaking is a downward displacement of about 2 m or less. If the water level in the slide rises above el 2680 m or the excess pwp is greater than 30%, then the movement may be greater than 2 m.

(2) If the main slide has a well-developed shear zone consisting of soils that sheared significantly in the past, then the likelihood of the excess pwp in the saturated shear zone exceeding 30 percent of the corresponding effective overburden pressure because of the postulated shaking conditions is low.

(3) Because of the first conclusion, an effective and sustained full drainage system all along, and particularly toward the bottom of the slide, was considered to be the most critical element as far as the seismic response of the slide is concerned.

(4) In addition, the densities estimated for the buttress foundation sediments indicate that these sediments would liquefy during the design earthquake. Therefore, improvement of these reservoir sediments was required to develop a suitable foundation for the buttress.

FOUNDATION STABILIZATION FOR BUTTRESS

The buttress is a 467,000 m^3 (611,000 yd^3) earth and rockfill embankment built at the toe of the landslide within the reservoir. It acts as a counterweight and prevents erosion of the lower slope by the river and the reservoir. It is believed that erosion was a major factor in the destabilization of the slide. In sizing the buttress, two contradictory requirements were faced: producing the maximum resisting force to the landslide, while at the same time keeping the interference of the embankment within the reservoir to a manageable level. The buttress designed and built at Tablachaca invades about half of the reservoir width, reducing the available space between its toe and the sediment trap opposite to only 60 m (200 ft). A hydraulic model was developed and showed that the reservoir outlet works could be operated as required, even with the buttress in place.

The buttress was constructed from August 1982 through October 1983 and is 300 m (1,000 ft) long, 50 m (164 ft) high, and 22 m (72 ft) wide at its crest. The exterior slopes vary from 1.6:1 to 1.4:1 (horizontal to vertical). The internal geometry of the buttress is shown in Figure 4. Since the part of the buttress comprised between its base (el. 2672 m) and maximum reservoir level (el. 2695 m) would not be compacted due to schedule constraints and would be subjected to immersion, it was constructed with screened, plus 1.5 in. material (Zones IA, IB and II), so that it would drain freely and thus avoid susceptibility to liquefaction. The buttress lower platform was constructed first to an elevation 2m above minimum reservoir level, this platform was constructed by dumping plus 1.5 inch screened gravels (Zone IA) in the river and then filling the inside with screened, minus 1.5 in. material (Zone IB). Sediment improvement was made from this platform. As sediment improvement progressed, construction of the buttress continued simultaneously on the improved areas. Above elevation 2695 m to elevation 2,720, the buttress consist of Zone IV (minus 1.5 in. compacted). Above El. 2,695 m to El. 2,720 m. the buttress consists of Zone IV (minimum 1.5-in.) material compacted with vibratory roller. The outside slopes of the buttress below the maximum reservoir level, as well as its downstream portion are covered with riprap for erosion protection (Zone III).

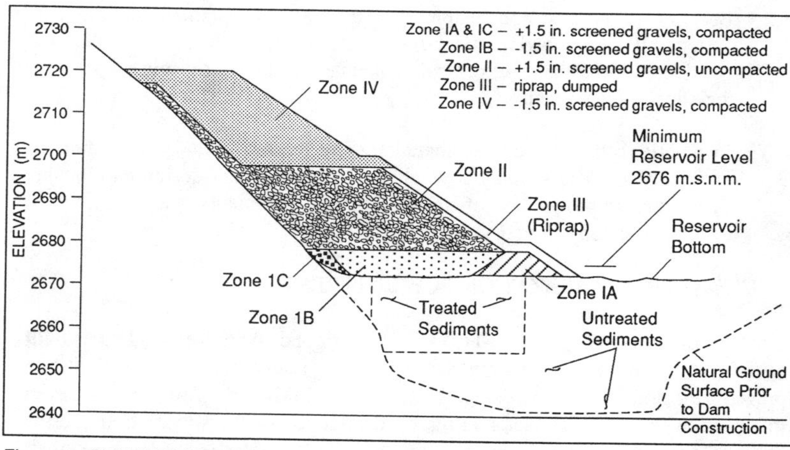

Figure 4. BUTTRESS SECTION

The major consideration, however, in the construction of the buttress was that it is founded on the reservoir sediments, which had been deposited since the dam was constructed. In April-July 1982, when the design of the preliminary stabilization works, including the buttress, was underway, it had been assumed that these sediments would be sandy and silty and in a loose state and, as a consequence, would be susceptible to liquefaction; therefore, it was also assumed that they would require densification. At that time, the plan was to densify the sediments using vibroflotation from a barge before constructing the buttress.

Character of Reservoir Sediments

In June and July 1982, the stratigraphy and consistency of the reservoir sediments were investigated by means of borings, standard penetration tests and dynamic cone soundings made overwater from a platform with jacked-up legs. The overwater investigation disclosed the presence of much fine-grained sediments intercalated within the sand, which led to a reconsideration of the vibroflotation method and eventually to deciding to improve the sediments using compacted gravel columns.

The sediments were found to be some 35 m (115 ft) thick near the center of the old riverbed. Their stratigraphy is erratic due to their deposition regimen between 1972 and 1982 after the dam was constructed. Generally, they consist of medium to fine sand (SP) and silty sand (SM) with some gravel (GP-GM). Layers of lenses of cohesive soil were found, forming a complex horizon, which was called the "clay layer" at a depth of 5 m to 12 m (16 ft to 40 ft) below the underwater surface of the sediments. The thickness of the clay layer increases from the downstream to upstream portion of the reservoir, and from the right back to the center of the reservoir.

Sediment Improvement

It was concluded that the sediment improvement method had to be capable of increasing the relative density of the sandy sediments and at the same time increasing the shear strength of the cohesive sediments of the "clay-layer" and intercalated fine-grained sediments elsewhere within the sandy soil. It was also concluded that vibroflotation considered earlier in the project was not appropriate. After a review of available methods, including blasting, displacement piles, excavation and replacement, compacted gravel columns were selected as they could densify the sandy soil by displacement and vibration, as well as accelerating consolidation of the clays by drainage (Davidson and Perez).

Due to the project schedule at that time (August 1982), it became preferable to eliminate the use of barges and to perform the sediment improvement from the fill platform which was to constitute the base of the buttress. Also, for reasons of schedule and budget, it was decided to postpone sediment improvement outside the buttress footprint to a later phase of the project.

Two types of compacted gravel columns were considered, the impact driven casing method (Franki) and the vibroreplacement method (Vibroflotation

Company). The Franki method consisted of driving a steel casing to depth through the impact of a hammer on a plug of gravel at the tip of the casing, driving out the plug and an additional amount of gravel to form a dense bulb of gravel at the base, and then driving gravel out of the casing as the casing is extracted. The gravel filled the cavity created by the advancing casing at it displaced the soil. The degree of compaction was controlled by the volume of gravel compacted into the column and the spacing between columns. The vibroreplacement method created a cavity for the vibroflot by liquefying vibration and water injection. Gravel was deposited from the surface into the cavity as the vibroflot was removed. The amount of densification was controlled by the amount of energy imparted to the soil by the vibroflot and the spacing of adjacent columns.

Before the start of the production gravel columns, prototype tests for both the Franki and Vibroreplacement methods were conducted to evaluate their effectiveness.

Franki Gravel Column Test

The objectives of the Franki gravel column test were to:

- confirm that no widespread liquefaction of the platform sediments would be caused by installation of the compacted gravel column;

- evaluate the effectiveness of the equipment in penetrating the perimeter dike;

- evaluate the effectiveness of compacted gravel columns in densifying the granular sediments and increasing the shear strength of the cohesive sediments in the "clay layer", and

- measure the time required for dissipation of excess pore pressure generated during column installation.

The compacted gravel columns were installed using a Franki Model XIII machine which permitted installation of a 21.5 m (70 ft) long casing and the formation of a bulb of gravel approximately 1 m below the base of the casing. A spacing of 2.4 m (8 ft) between columns on an equilateral triangular pattern was selected based on experience. The steel casing had 52 cm (20.5 in.) outside diameter and 47 cm (18.5 in.) inside diameter. The casing was advanced by driving a bulb of gravel in the tip of the casing with a 4-ton hammer drop of 2.4 m (8 ft). Once refusal or the maximum length of casing was reached, the gravel plug was extruded from the casing by driving while the casing was simultaneously extracted by cable tension. Gravel was added and compacted into the casing cavity with the hammer at a rate of 2.5 times the casing volume as the casing was raised. This method produced a gravel column with an average diameter of 75 cm (30 in) to

80 cm (31.5 in.). For several columns, four times the casing volume of gravel was compacted into the soil producing a 100 cm (39 in.) diameter column.

Comparison of Before and After Densities

Standard penetration resistances measured before and after sediment improvement with a gravel volume 2.5 times the casing volume, showed approximately 100 percent increase in the penetration resistance of the granular sediments and moderate improvement of the cohesive sediments. The information also indicated that the densification was sufficient to prevent widespread liquefaction during the design earthquake.

Production Franki Gravel Column Improvement

Production Franki gravel columns were installed using three Franki machines. Because of the emergency nature of the work, all three machines were airlifted from Brazil to Peru in Peruvian Air Force C-130 Hercules transport planes, and then transported by truck to the site. Disassembled, each of these 40-ton machines barely fit in the airplane. Each Franki machine required a nine-man crew to operate the equipment. Typical Franki production was five 22 m-long gravel columns per machine in two good 12-hour shifts. Control of verticality and location was good. The quality of the gravel of column could be easily controlled by counting the number of buckets of gravel placed and blows of the hammer for each meter of penetration. The Franki method was very amenable to a design specification which required specific measurable construction quality control elements. The Franki machines started on the eastern end of the platform adjacent to the most active zone of the slide. Because of the problems discussed below of the vibroflot in working through the + 1.5 inch material, the area of the perimeter dike was treated primarily by Franki gravel columns. The Franki production rate reflects that it had to treat the most difficult areas of the platform. A total of 882 gravel columns were installed by this technique.

Vibroreplacement Method

During production Franki gravel column installation, Vibroflotation Company USA was requested to conduct a prototype test program similar to that conducted by Franki. It was hoped that the much faster vibroreplacement method could be used in addition to the Franki method to shorten the schedule required for platform sediment improvement to enable buttress construction to be completed sooner.

The principal objectives of the vibroreplacement test were to:

- verify the degree of densification of the granular sediments achieved by vibroreplacement;

- demonstrate the capability of the vibroreplacement equipment in penetrating the material in the perimeter dike; and

- check the vibroreplacement production rate.

The gravel columns were constructed by advancing the vibroflot suspended from a crane into the soil. The vibroflot penetrated the soil by the combined effect of vibration, dead weight and water injection which liquefied the soil adjacent to the probe. When the full length of the vibroflot was reached (22 m, 72 ft depth), gravel was introduced into the annular space around the vibrating probe from the surface. The gravel was worked downward and compacted by successively raising and lowering the vibrating probe, while more gravel was added. Eventually the vibroflot was removed from the gravel filled opening it created in the soil. The design gravel consumption was 0.5 m^3/linear m (0.2 yd^3/ft) while the actual consumption was 0.8 m^3/linear m (0.3 yd^3/ft).

The effectiveness of the vibroreplacement gravel columns was verified by standard penetration resistances measured in borings before and after improvement. A substantial increase in blow count was observed in the granular sediments and moderate improvement was measured in the cohesive sediments. Vibroreplacement effectively decreased the susceptibility of the sediments to liquefaction.

Vibroreplacement equipment had great difficulty installing the columns below the + 1.5 inch material because of the formation of a crater several meters in diameter. Some loosening of the upper 2 m (6.5 ft) of the platform was observed. Penetration of the vibroflot through the cohesive sediments was difficult and slow, because the vibroflot produced a cavity nearly the same size as the probe and this small annular space impeded passage of both the gravel and the probe in these zones.

Production Vibroreplacement Gravel Column Improvement

Vibroreplacement gravel columns were installed with a single vibroflot and support equipment. Since Franki was already working in the eastern portion of the platform, vibroreplacement started in the western end of the platform. Vibroreplacement was substantially faster than Franki in areas of limited clay and Zone I fill, installing fifteen 22-m long gravel columns in two 12-hr shifts. However, whenever clay was encountered, the vibroflot became stuck; in fact one probe was left permanently in the platform. The full pulling capacity of the crane was required to move the vibroflot in and out of the cohesive sediments. The vibroreplacement equipment required a crew of only three to four people plus the loader to place the gravel. Vibroreplacement had some difficulty in maintaining verticality, and surface location of the columns was not particularly accurate. A total of 701 gravel columns were installed by this technique.

Sediment Improvement Summary

The summary of all post-improvement normalized standard penetration resistances confirm the reduced liquefaction susceptibility of the improved sediments. Approximately 130,000 m^3 of sediments were treated during the emergency stabilization program from August 1982 to March 1983 at a cost of approximately $3 million.

DRAINAGE TUNNELS

The need for drainage of the landslide mass and reduction of pore water pressures acting at the base of the landslide was clearly shown in the results of the static and dynamic stability analyses. Movement of the landslide was observed to accelerate during the rainy season. This acceleration was thought to be caused by an increase in the weight of the slide from saturation of the slide mass and by an increase in pore pressure on the slide plane caused by infiltration of water. Therefore, a system of tunnels and radial drains was planned to drain water from the landslide to improve overall slope stability under static and dynamic conditions.

Two drainage tunnels and transverse drainage galleries totalling about 1527 m in length were excavated in the rock behind the landslide. The excavated tunnel diameter was approximately 2.2 m wide by 2.4 m high. The lower tunnel (S-200 at el. 2700 m) extended completely around the slide mass from a portal downstream from the dam to an upstream portal. An upper tunnel (S-250 at el. 2750 m) extended under the more active part of the landslide but did not daylight at the upstream end. The tunnels were started from portals located in rock outcrops outside the slide area. It was required that the tunnels be located in bedrock, beneath the landslide. However, bedrock locations were not known when construction was started. Therefore, the alignment had to be changed numerous times when the base of the landslide was intersected. In order to increase drainage, transverse drainage galleries were excavated at different locations from the tunnels within bedrock and the slide mass. A total of 190 radial drains (3290 m total length) were drilled from inside the tunnels.

Tunnel support initially consisted of timber cut from local eucalyptus trees due to schedule constraints. Some of these timbers were so green that they sprouted new growth in the tunnel. Permanent support consisted generally of steel sets and lagging, with a pre-cast concrete invert.

The tunnels encountered a variety of rock and groundwater conditions. At the surface of sliding, clayey zones of crushed and broken materials were observed in the tunnels. The clayey gouge zones were typically several meters thick and of very low permeability and are believed to act as an aquiclude. The mass of sliding materials on the other hand consists of broken rock materials with large voids, as evidenced by the loss of drilling fluid and by the voids encountered in the borings made from the surface of the slide for installation of instrumentation.

Substantial areas of water were encountered in the graphitic phylite, particularly in crushed and faulted zones in the rock. Two separate sources are believed to be present: 1) regional seepage within the phyllite behind the slide, and 2) water within the slide mass which infiltrates from rainfall and which may be perched on top of the gouge zone(s) as it drains toward the reservoir. As part of the emergency stabilization measures, horizontal drains consisting of perforated pipe wrapped in filter fabric were installed from the surface of the slide to intercept this infiltrated "surface water".

Effects of Tunnels on Stability

Measuring flow from the tunnels was hampered by water seepage through the tunnel invert. Flow loss was estimated to be 25 to 50 percent of the intercepted groundwater. Flows were measured to be on the order of 10 gpm at the entrance of Tunnel S-200. The tunnels and radial drains acted to lower the water table and reduce pore pressures acting on the base of the slide. In terms of the cost of stabilization measures, the cost of the tunnels was low (approx. $100/m) and rapidly contributed to increasing the factor of safety of the landslide.

CONCLUSIONS

(1) In a major landslide, multiple remedial measures, any of which often are sufficient to stabilize smaller slopes, are likely to be necessary to stabilize the landslide. Multiple measures were required because of the magnitude of the landslide, the limited working area at the dam, and the limited time available.

(2) Evaluation of all the techniques used in the stabilization indicate that: (1) internal drainage of the landslide provided by both radial drains associated with tunnels and surface horizontal drains, and; (2) erosion protection of the reservoir slope by the buttress may be a basis for improved long-term stability.

(3) Short-term stability can be developed by such means as the buttress and rock anchors.

(4) Gravel columns are very effective in densifying and improving loose sediments.

To date, significant movement of the landslide has not occurred since completion of the emergency works.

ACKNOWLEDGEMENTS

The writers are indebted to ELECTROPERU for their support and contribution in performing the work described in this paper, and to Woodward-Clyde Consultants for assistance in preparing the paper. Special thanks go to Mr. Richard Davidson of WCC for his input regarding treatment of the foundation sediments.

REFERENCES

Davidson, R.R. and Perez, J.P. Treatment of Recent Sediments using Compacted Gravel Columns, Tablachaca Landslide, Peru, Unpublished.

Dwey, J.W. and Spence, W. (1979). "Seismic Gaps and Source Zones of Recent Large Earthquakes in Coastal Peru," in Pure and Applied Geophysics, (Vol. 11), Birhauser Verlag, Basel.

Hutchinson, J.N. (1982). Embalse Tablachaca, Report on Derrumbe No. 5, Report to ELECTROPERU, Lima, April.

Makdisi, F.I. and Seed, H.B. (1977). "A Simplified Procedure for Estimating Earthquake-Induced Deformations in Dams and Embankments," Report No. UCB/EERC-77/19, Earthquake Engineering Research Center, College Engineering, University of California, Berkeley, August.

Repetto, P.C. (1985). "The Tablachaca Dam Slide No. 5 Problem," in Proceedings of the Eleventh International Conference on Soil Mechanics and Foundation Engineering," Vol. 5, San Francisco.

Seed, H.B. and Idriss, I.M. (1982). "Ground Motions and Soil Liquefaction During Earthquakes," Earthquake Engineering Research Institute.

Woodward-Clyde Consultants/Michelena-Repetto y Asociados (1982). Design Memorandum, Buttress Embankment, Emergency Works for Static Stability, Tablachaca Derrumbe No. 5.

Woodward-Clyde Consultants/Michelena-Repetto y Asociados (1983a). "Static Slope Stability of Tablachaca Derrumbe No. 5."

Woodward-Clyde Consultants/Michelena-Repetto y Asociados (1983b). Selection of Ground Motions for Tablachaca Derrumbe No. 5 Stabilization.

SLIDE STABILIZATION WITH STONE-FILL TRENCHES

George L. Sills[1] M.ASCE and
Robert L. Fleming, Jr.,[2] M.ASCE

ABSTRACT: The U. S. Army Engineer District, Vicksburg developed a different method of slide stabilization using gravel-fill or stone-fill trenches. A number of slides have been successfully stabilized with this method. The design, construction, and performance monitoring of one of these repairs is discussed, along with, a history of the slide, description of the soils, and mechanism of failure. The applications of this method of slide stabilization are also evaluated.

INTRODUCTION

The U.S. Army Engineer District, Vicksburg has successfully used gravel-fill or stone-fill trenches to stabilize a number of slides. In 1982, a long shallow translatory slide in a medium to stiff highly plastic clay was stabilized with a technique using trenches excavated below the slip plane and partially filled with a washed gravel aggregate (Wardlaw 1984). The slide which had closed one lane of a county road and had destroyed a timber bridge, has remained essentially stable since it was stabilized in 1982. Only a small amount of strain cracking in the pavement has been evident. Since that time a number of other slides have been successfully stabilized using washed

---

1 - Geotechnical Specialist, P.E., U.S. Army Engineer District, Vicksburg, 3515 I-20 Frontage Road, Vicksburg, MS 39180-5191

2 - Chief, Foundation and Materials Branch, P.E., U.S. Army Engineer District, Vicksburg, 3515 I-20 Frontage Road, Vicksburg, MS 39180-5191

gravel and more recently quarry run graded stone. Two of the more recent applications involving the use of stone-fill trenches to stabilize slides were on the inlet channel at John H. Overton Lock and Dam on the Red River Waterway and on the Ouachita River at Rilla, Louisiana. The Rilla slide was located 9 miles downstream of Monroe, Louisiana, at approximate River Mile 166 on the Ouachita River and is shown in plan on Figure 1. The slide had involved a portion of the Ouachita River Levee and was considered to be an endangerment to the flood control system. The repairs were made under emergency conditions in December 1988.

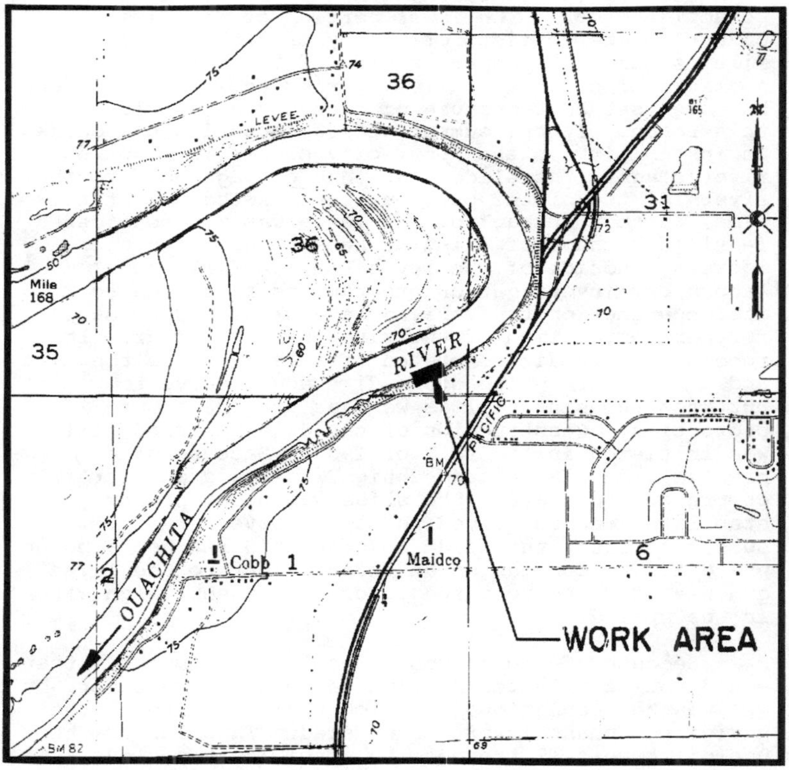

Figure 1. Site Location

A method of repair, not materially affected by the weather conditions, that could be completed prior to the next high water season was selected. In addition, the repair method could not extend into the navigation portion of the channel.

This paper summarizes the nature and history of the Rilla Slide, the design of the stabilization measures, construction of the stone-filled trenches and the results of post construction monitoring.

BACKGROUND

The slide location, in the outside channel slope of a meander of the Ouachita River, has been subjected to active scour. The lower portion of the river channel slope consists of highly erodible sands and silty sands. The upper portion of the channel slope consists of clays and silts. Prior to 1983, the levee had been set back because of bank caving problems in this reach. In the summer of 1983 scouring had caused an instability in the upper clay slope. During low river stages, the slide scarp had propagated into the riverside toe of the previous levee setback. At this time, further setback of the levee was not considered feasible because of existing houses and structures located landside of the levee toe. A field and laboratory investigation program was initiated to design a section to stabilize the slide and to stop further degradation of the river bank by scour. The recommended repair consisted of a large stone toe dike and a sand berm. The stone dike would serve to buttress the sand fill as well as provide scour protection. Construction of the dike and sand fill was initiated in the fall of 1984. Because of problems, such as extended high water and an extremely compressed work area, the slide was never properly stabilized and continued to slowly move. By the summer of 1988, the slide movement had reached a point where it was considered an endangerment to the levee and needed to be repaired prior to the next high water in the spring.

Because of these problems only those methods that would not be affected by fluctuations of the river, by wet weather conditions or which would not impact the navigation channel were considered. This led to the use of stone-fill trenches to stabilize the slide. Because of the emergency nature of the project and problems with the previous construction, the specifications were very specific in terms of the order and the compressed timing of the work. Construction was initiated on December 7, 1988 and was completed on December 19, 1988. As of the time of this paper in December 1991, there has been no evidence of further movements along the slip plane.

SLIDE ASSESSMENT

Geology. The area is located in the western part of the Mississippi Alluvial Valley in a region known as the Ouachita lowland. Fleetwood (1969) presented a detailed study of the area geology. The alluvial deposits in the area consist of an upper fine grained unit called the topstratum and a lower coarse grained unit called the substratum. The alluvial topstratum deposits of the Ouachita River Valley are classified according to depositional environment. The topstratum at the Rilla site, located 9 miles downstream of Monroe, Louisiana, at approximate River Mile 166 on the Ouachita River, consists of a 2- to 5-foot thick sequence of natural levee silt (ML) which overlies a 20- to 25-foot thick deposit of medium to very stiff backswamp clay (CL and CH). Natural levee deposits form broad low ridges that flank both sides of the river and are a result of overbank deposition during periods of flooding. Backswamp deposits consist of fine grained sediments laid down in broad shallow basins within the floodplain during major periods of stream flooding. The topstratum deposits rest directly on a 40- to 50-foot thick layer of highly erodible substratum sand (SP and SM) with a 2- to 5-foot gravel strata at the base. These alluvial deposits in turn lie directly on the Tertiary age material that consists of a hard, dark gray clay (CH).

Exploration Program. The investigation of the soil conditions at the site consisted of three undisturbed borings taken in September 1983 and two undisturbed borings taken in June 1988. The borings indicated a silt layer of the natural levee deposit and a backswamp clay deposit consisting of a layer of very stiff clay (CL) overlying a medium to very stiff clay (CH). The generalized soil and slide profile is presented in Figure 2. The borings which were made through the existing levee are shown in plan on Figure 3.

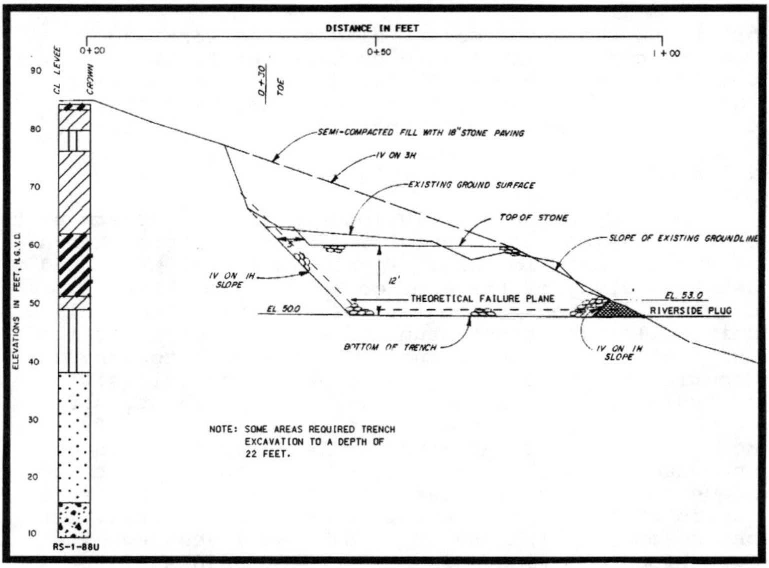

Figure 2. Typical Section of Slide Area

<u>Laboratory Testing</u>. Laboratory testing consisted of water content determinations, Atterberg limit tests, and unconfined compression (UC) tests. The UC tests indicated the shear strengths range from a cohesion of 250 psf to 780 psf with an average value of 500 psf. The water contents varied from a low of 21 to a high of 42. It should be noted that the slickensided nature of the backswamp clay help to contribute to the lower UC values.

<u>Conditions Contributing to Slide</u>. The site investigations in 1983 and 1988, cross sections of the area taken on regular intervals and familiarity with the construction problems associated with the 1984-85 repair led to the conclusions that the initial triggering mechanism for the slide was the steepening of the river bank due to erosion of the silty sand substratum. This in turn caused sliding to occur in the backswamp clays. The initial attempt to stabilize the slide using the stone dike and sand berm was to

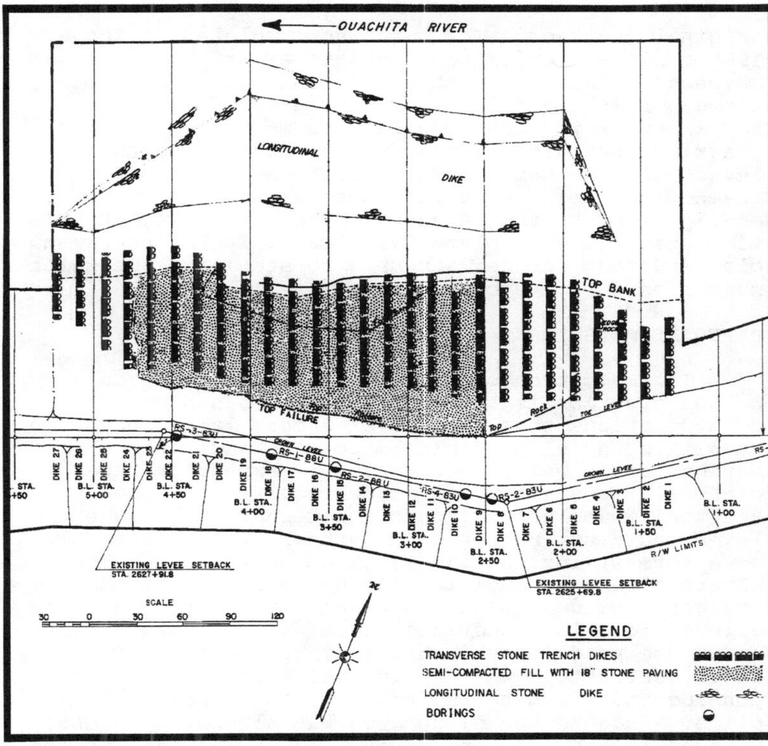

Figure 3. Site Plan

consist of building a stone dike to elevation 62 and rebuilding the slope to a 1V on 3H with sand fill. The sand fill would be protected from scour with engineering fabric and riprap. In November 1984, high water conditions existed at the construction site. The contractor chose to place the sand fill prior to completion of the stone dike. The weight of the additional sand fill caused the slide to experience further movement in the backswamp clay. The contract was modified to ensure the stone dike was completed before any more sand fill was placed. The modification also allowed placement of the riprap in other than dry conditions which was precluded in the original contract. The work was completed in January 1985 and the section continued to experience some movement. This movement was assumed to be occurring along a remolded slip plane that existed in the backswamp clay. The movement was more than strain cracking, but did not seem to be endangering the levee. In the summer of 1988, the movement had reached the point that it was considered to be an endangerment to the levee. It was also clear at this time that the slip plane was located in the backswamp clay and that further attempts to stabilize the slide must plan for this condition.

REMEDIAL MEASURES

The selection of stone-fill trenches as the method of stabilization at Rilla was based on considerations that the use of a berm large enough to stabilize the slope would be both a hinderance to navigation as well as be extremely expensive because of the large quantity of stone required. Slope flattening was not a feasible alternative because the levee could not be set back further. This left only some form of insitu stabilization as an acceptable alternative. An additional consideration was the emergency nature of the work and the short time frame allowed for plans and specifications. A method that could use readily available equipment and be bid by local contractors was highly desirable. A first consideration was to use trenches to stabilize the slide. Because the slide area was already protected with 18 inches of stone paving, it was decided to use this stone to fill the trenches. Quarry run stone would further provide a material whose strength was less susceptible to degradation due to contamination from clay and silt fines. Quarry run stone would be much larger and more angular than rounded, pit run washed gravel. The concept of stone-fill trenches is similar to stone column replacement and is applicable where digging of the trenches is within the capability of the digging equipment and conditions. In this case the maximum trench depth required to extend the trench below the bottom of the backswamp clay was 22 feet. This depth is below the theoretical depth an open trench will stand in this type of material. The depth of cut could be reduced by leveling or removing material from the top of the slide zone where possible. However, it was decided to minimize the double handling of material and modify the construction technique by only opening a small portion of the trench before backfilling.

Design of remedial measures consisted of detailed analyses of the existing conditions to determine a strength acting upon the inferred slip surface at the base of the backswamp clay. Figure 2 represents the cross section of the slide, with the boring log plot, and the location of the inferred slip surface which was used in the analyses.

A computer program utilizing a wedge method was selected to analyze the slip surface, assuming the sliding was to be in a state of limiting equilibrium. This program was written by Cheek (1975). Even though no slope indicator data was available to define the slip surface, assuming the slip plane at the bottom of the clay layer was considered conservative. These analyses also assumed a phreatic profile of 5-feet below the ground surface.

Drained strength parameters of $\phi' = 16^{\circ}$ c' = 0 psf were backfigured as an average strength acting on the slip surface producing a safety factor of unity.

Once the strength acting on the slip surface was determined, it was necessary to determine the average strength required to provide a factor of safety of 1.25 for the repaired slope. Analyses indicated that an average drained shear strength of $\phi' = 20^{\circ}$ and c' = o psf would provide the shear strength required to yield the required factor of safety.

It was decided that the existing riprap on the slope would be used in the trenches. The shear strength was estimated for this material as $\phi' = 40^{0}$ and c' = 0 psf. The trenches were to be excavated to the bottom of the backswamp clay and filled with a minimum of 12 feet of rock. Semicompacted fill would be placed on top of the repair to restore the section back to the original section.

Assuming that a 2.5-foot-wide bucket would be used to excavate the trenches, and with the shear strength parameters for the slip surface and stone-fill known, the next step was to determine a trench spacing that would provide the composite shear strength required to raise the computed factor of safety to 1.25. To determine the trench spacing, an equation was developed using a weighted average principle. In the equation the ϕ' of the existing soil mass is multiplied by the distance between trenches. This term is then added to the ϕ' of the trench backfill material multiplied by the width of the trench. These terms are equal to the ϕ' of the composite section multiplied by the total distance between the trench plus the width of the trench. The Sills/Fleming equation used is as follows:

$$\phi'_r \times S + \phi'_f \times W = \phi'_a (S+W) \quad (1)$$

where

ϕ'_r = the average ϕ' acting on the slide mass
S = distance between trenches
$\phi'_f = \phi'$ of the trench backfill material
W = width of trench
ϕ'_a = the average ϕ' of the composite section

Since the trench spacing is generally measured center to center

$$S+W = T_s \quad (2)$$

where
T_s = trench spacing measured center to center

Therefore equation (1) can be rewritten

$$\phi'_r \times S + \phi'_f \times W = \phi'_a \times T_s \quad (3)$$

Solving equation 3 it was determined that a center to center trench spacing of 15 feet would provide the required average strength parameters of $\phi' = 20°$ and c' = 0 psf. It should be noted that since arching of the soil should occur between the stone-fill trenches the actual factor of safety should be greater than 1.25. A typical design section is shown on Figure 4.

CONSTRUCTION

Construction of the stone-fill trench stabilization was accomplished by contract for the U.S. Army Engineer District, Vicksburg, in December 1988. The contract required that 27 transverse stone-fill trenches (Figure 3) be constructed beginning downstream and proceeding upstream. The work area was within the Columbia Lock and Dam Pool which maintained a minimum river of 52 feet NGVD. Since the trenches were required to be excavated to elevation 50 a soil plug was left in place on the riverside of the trench, Figure 2. The trenches were constructed beginning at the riverside plug and proceeding landward toward the levee toe.

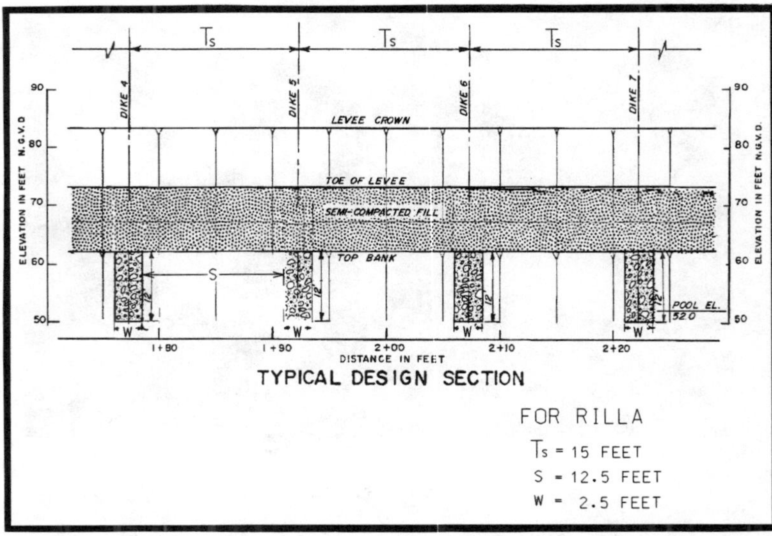

Figure 4. Typical Design Section

Prior to trench construction, existing bank paving was removed along the trench where excavated material would be placed. Trench excavation was performed using a tracked backhoe. The trenches were excavated to full depth. Excavated material was placed downstream of the trench in a manner not to overload the bank. This also ensured the excavated material was placed on a stabilized section and was not mixed with the surface stone. As the backhoe moved upslope, the existing bank paving was placed in the trench as backfill. This allowed the backhoe to both excavate and backfill the trench. The contractor was required to have less than 5 feet of unfilled trench open at any time. The backhoe moved upslope until the vertical scarp was encountered, the trench was backfilled, the backhoe turned 180° and finished the trench, digging from the top of the scarp to the previously complete portion of the trench. Figure 5 is a photo of the slide area.

Figure 5. Slide Area

This procedure was repeated for each of the 27 trenches. The soil excavated from the trenches was used to backfill the slide area to the original section. Finally, 18 inches of bank paving was added to the repair area. Figure 6 is a photo of the completed work. The construction of the repair was completed in only 12 days with a final total cost of less than $100,000.

MONITORING AND PERFORMANCE

After completion of this repair in December 1988, there has been no visual movement of this slide. During the spring of 1991, the Ouachita River reached record stages in this area. After flood stages rapidly decreased, the repaired area was still intact.

CONCLUSIONS

The stone-fill trench method of slide stabilization has a definite application in appropriate situations. These situations are shallow slides in soil masses with enough integrity to remain stable when trenches are excavated below the slip

Figure 6. Completed Repair.

plane with near vertical side slopes. This application has several definite advantages when compared to other methods of stabilization. Some of the merits are relatively low construction costs, moderate level of design complexity, and the ability to be performed with conventional, readily available, construction equipment.

The stone-fill trench method of stabilization met the basic requirements of the project, which was to stop the sliding of the unstable soil mass and to restore the levee section to pre-slide conditions.

The stabilization of this slide using stone-fill trenches serves as a prototype for other slide repairs. It provides information that can be used to evaluate and, hopefully, refine this method into a more economical and technically viable stabilization technique. One needed future refinement would be a study of the arching effect of the soil between the stone-fill trenches.

ACKNOWLEDGMENT

The authors would like to express their thanks to the U.S. Army Corps of Engineers, Vicksburg District, for undertaking this method of stabilization. The authors would also like to thank Messrs. Todd Hill, McKinley Harris, and Eddie Stewart for performing analyses and preparing figures, Mr. Stephen Lee for assisting in preparation of the geology, and Mrs. Sue Richardson and Mrs. Jaudon McKay for typing this paper.

REFERENCES

Cheek, B. and R.L. Hall, SSW028 Slope Stability Analysis (Wedge Method) Computer Program, U.S. Army Corps of Engineers, Waterways Experiment Station, Vicksburg, MS, 1975

Fleetwood, A.R., (1969), Geological Investigation of the Ouachita River Area, Lower Mississippi Valley, WES Technical Report S-69-2.

Wardlaw, E.G., L.A. Cooley, A.E. Templeton, and R.L. Fleming, Jr., Slide Stabilization With Gravel Trenches, Proc., International Conference on Case Histories in Geotechnical Engineering, St. Louis, MO, 1984, Vol II, 743-748

UNITS

1 foot = 0.305m
1 inch = 0.0254m
1 psf = 0.0479 kn/m^2

BIOTECHNICAL STABILIZATION OF CUT & FILL SLOPES

Donald H. Gray[1] , Member, ASCE, and Robbin B. Sotir[2]

Abstract

The use of tensile inclusions makes it possible to repair slope failures or to construct steepened slopes along highway right-of-ways. Live, brush-layers can be used in lieu of or in conjunction with geotextiles or geogrids for this purpose. This approach, which is termed biotechnical stabilization or soil bio-engineering, entails the use of living vegetation, primarily cut woody plant material, that is purposely arranged and imbedded in the ground to prevent surficial erosion and to arrest shallow mass wasting. The live cut stems and branches provide immediate reinforcement; secondary stabilization occurs as a result of adventitious rooting along the length of buried stems.

A soil bioengineering approach was adopted to repair a high, steep fill slope along a highway in North Carolina. Stability analyses indicated that over-steepening, seepage effects, and lack of adequate root cohesion near the surface all contributed to the problem. Repair and reconstruction of the fill slope with brushlayers corrected these problems and the embankment is now well vegetated and stable. A brushlayer buttress fill was also used to repair an unstable cut along a scenic highway in Massachusetts. Large amounts of groundwater seeped from fractures in the bedrock and through exposed soil in the cut. Scenic and environmental considerations in combination with stability analyses dictated the use of a composite drained rock and earthen brushlayer buttress fill. The rock section was placed at the bottom in order to intercept critical failure surfaces that passed through the toe of the slope. Biotechnical stabilization resulted in a satisfactory and cost effective solution; the treated slope has remained stable and it blends in naturally with its surroundings.

---

[1] Professor, Dept. of Civil Engr., Univ. of Michigan, Ann Arbor, MI, 48109
[2] Principal, Robbin B. Sotir & Assoc., 627 Cherokee St., Marietta, GA 30060

Introduction

Internal, tensile reinforcements permit over steepening of slopes along highway right-of-ways. Steepened soil slopes from 2:1 (26.6 degrees) to 0.35:1 (70 degrees) can be constructed in this fashion. A steepened slope used in lieu of slope flattening--a conventional response to instability--results in more useable land at the top and less encroachment on right-of-way at the bottom. Steepened, reinforced slopes can be used to repair slope failures and also provide an economical alternative to conventional grade separations within limited rights-of-way such as flattened slopes or vertical retaining walls (Berg *et al.*, 1990). Incorporation of tensile reinforcements in a slope also makes it possible to use on-site soils in reconstruction of failed slopes.

The material and structural components of a typical reinforced, steepened slope are shown in Figure 1. Tensile elements are placed in layers well within the fill to create a stable, unitary mass. These are termed *primary* reinforcements. *Secondary*, or surficial reinforcing elements are placed near the slope face to facilitate compaction near the edge and to minimize sloughing or face sliding. The soil at the outer edge may also be faced with erosion control mats and/or vegetated to prevent or minimize surficial erosion. External and internal drainage provisions are normally incorporated into the design as well to prevent saturation of the reinforced fill. This basic design can be used to support and stabilize both fill and cut slopes. In the former case, the reinforcements are simply incorporated into the outside edge of the embankment during its construction. In the latter, the reinforcements are inserted into a buttress fill that is laid up against an unstable cut.

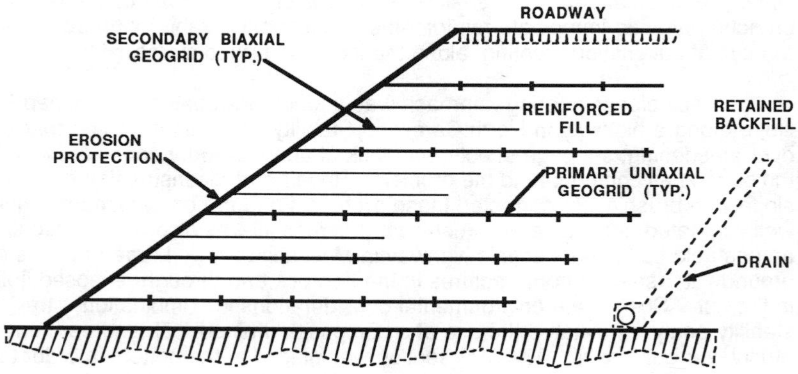

Figure 1. Material and structural components of a typical, reinforced steepened slope.

Polymer geogrids are typically used in conventional practice as the reinforcing elements in steepened slopes. High strength, extruded, uniplanar geogrids have been used as the primary reinforcements for critical highway slope and retaining wall construction. Lower strength, biaxial geogrids are normally used close to the face as secondary reinforcements. The latter are typically 3 - 6 feet long and are spaced 8 - 36 inches vertically apart as shown in Figure 1.

Tensile inclusions other than polymer geogrids can also be used to advantage to create a structurally stable composite mass. This paper describes the use of live brushlayers to stabilize both cut and fill slopes. The brush can be used alone or in conjunction with geogrids. The live brush can be substituted for the secondary geogrid reinforcement or, in some cases, actually replace both secondary and primary geogrid reinforcements. This approach is termed biotechnical stabilization or soil bioengineering. The characteristics of biotechnical slope stabilization are described in the paper together with case studies of both cut and fill stabilization using live brushlayering.

Biotechnical Stabilization

Biotechnical slope protection entails the combined or integrated use of vegetation and structural components to prevent surficial erosion and to arrest shallow mass wasting. Soil bioengineering uses living plant material itself, i.e., stems and branches, as the main structural component or reinforcement in the soil. Soil bioengineering techniques such as brushlayering and contour wattling work on this principle. Live cut stems and branches provide immediate reinforcement; secondary stabilization occurs as a result of adventitious rooting which develops along the length of buried stems. In addition to its reinforcing role woody vegetation used in biotechnical stabilization can also serve as a hydraulic drain, buttressing element, and barrier to earth movement. A classification scheme or taxonomy of different biotechnical stabilization methods is shown in Table 1.

The general principles and techniques of biotechnical slope protection and soil bioengineering have been presented elsewhere by Gray and Leiser (1982) and Schiechtl (1980). One of the earliest applications of biotechnical slope protection along a highway right-of-way was described by Kraebel (1936) who used contour wattling to stabilize steep, fill slopes along the Angeles Crest highway in southern California. Kropp (1989) describes the use of contour wattling in combination with subdrains to repair and stabilize a debris flow in California.

A biotechnical or soil bioengineering approach offers several advantages over conventional methods. These living systems are often more cost-effective (Gray *et al.*, 1980) than a purely vegetative or structural approach. They tend to utilize native, indigenous materials and are less visually intrusive in the landscape. Biotechnical methods do not cause as

TABLE 1. CLASSIFICATION OF DIFFERENT BIOTECHNICAL SLOPE PROTECTION AND EROSION CONTROL MEASURES

<table>
<tr><th colspan="2">CATEGORY</th><th>EXAMPLES</th></tr>
<tr><td colspan="2">LIVE CONSTRUCTION
Conventional Plantings</td><td>* Grass seeding
* Sodding
* Transplants</td></tr>
<tr><td rowspan="4">MIXED CONSTRUCTION</td><td>Woody plants used as reinforcements and barriers to soil movement</td><td>* Live staking
* Contour wattling
* Brush layering
* Soft gabions
* Brush mattress</td></tr>
<tr><td>Plant/ structure associations</td><td>* Breast walls with slope face plantings
* Revetments with slope face plantings
* Tiered structures with bench plantings</td></tr>
<tr><td>Woody plants grown in the frontal openings or interstices of retaining structures</td><td>* Live cribwalls
* Vegetated rock gabions
* Vegetated geogrid walls
* Vegetated breast walls</td></tr>
<tr><td>Woody plants grown in the frontal openings or instices of porous revetments</td><td>* Joint plantings
* Staked gabion matresses
* Vegetated concrete block revetments
* Vegetated cellular grids
* "Reinforced" grass</td></tr>
<tr><td colspan="2">INERT CONSTRUCTION
Conventional Structures</td><td>* Concrete gravity walls
* Cylinder pile walls
* Tie back walls</td></tr>
</table>

much site disturbance and have less severe access requirements. These characteristics make a biotechnical approach attractive in scenic corridors and environmentally sensitive areas. It is also worth noting that soil bioengineering systems function immediately as soil reinforcing units, drains, and barriers to surface erosion. In time, roots and shoots develop that further enhance stabilization. The shoots and associated foliage form a protective vegetative cover that mitigate the effects of rainfall erosion and that improve mass stability (Coppin *et al.*, 1990) by removal of excess soil moisture via transpiration. The roots permeate and indurate the soil thereby increasing its shear strength. The system grows stronger with age and encourages the natural invasion on a slope of a diverse and stable plant community.

Brushlayer System

Main Characteristics: Brushlayering is the principle soil bioengineering technique that was employed in the two slope stabilization case studies described in this paper. A brushlayer fill consists of alternating layers of earth and live, cut branches. The branches are placed in a criss cross pattern between successive lifts of compacted soil as shown schematically in Fig. 2.

The brush stems reinforce the fill in much the same manner as conventional fabric or polymer grid reinforcements, but they also act as horizontal drains in a slope. In addition, adventitious rooting along the length of buried portions of the stems provides secondary earth reinforcement. The brushlayers also act as horizontal drains and favorably modify the hydrologic regime near the face of the slope. They intercept groundwater flowing along the loose outer edge of a compacted fill, divert the flow downward, and then convey it out laterally through the brushlayer itself. Redirection of seepage flow downward in this manner results in greatly improved resistance to face sliding or sloughing (Cedergren, 1989).

Design Considerations: A brushlayer fill resembles a drained rock or earthen buttress in function; both are gravity retaining structures. The brushlayer fill, however, consists of alternating layers of earth and live, cut branches. The brush stems and branches reinforce the fill in much the same manner as conventional polymeric grid or fabric reinforcements, accordingly, the internal stability of a brushlayer fill, i.e., the resistance of the brush reinforcement layers to pullout and tensile failure can be analyzed using conventional methods (Giroud *et al.*, 1987) developed for earth slopes reinforced with geotextiles. The required vertical spacing and imbedded length of successive brush reinforcement layers are determined from the specified safety factor, allowable unit tensile strength, and interface friction properties of the reinforcement layer. The allowable unit tensile resistance for a brushlayer can be calculated from the known tensile strength of the brush stems, their average diameter, and number of stems placed per unit width (Sotir & Gray, 1989).

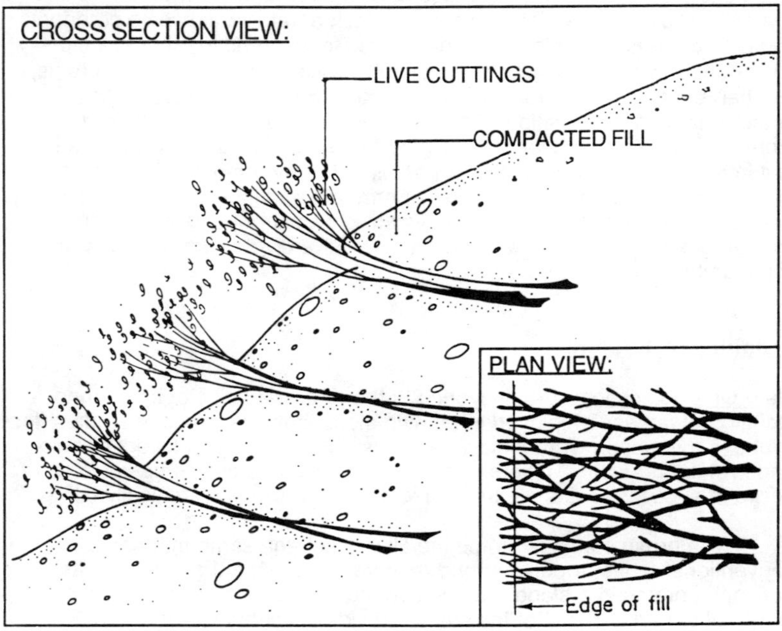

Figure 2. Schematic illustration of a brushlayer fill installation

Case Study No. 1 - Fill Slope Stabilization

Project Site: A failing highway embankment in North Carolina was selected by the North Carolina division of Highways as a test site for soil bioengineering systems. Highway embankments in North Carolina had been plagued with shallow seated slides and surficial erosion throughout the state. The demonstration site was an approximately 870-foot long fill slope on State Route 126. The embankment was constructed in 1980-81 with side slopes ranging from 1.5:1 to 2.0:1 and a maximum height of 60 feet. Shallow sloughing and erosion occurred shortly after construction and was repaired several times using conventional treatments. By the time a decision was made to repair the slope with soil bioengineering systems, sliding and erosion had progressed to the point of undermining the guard rail posts at several locations and encroaching on the pavement.

Stability Analyses: Stability analyses indicated that over steepening, seepage effects, and lack of adequate root cohesion near the face of the slope all contributed to the problems. Shallow sloughing and face sliding indicated that the failure surface was most likely located in the loose, less compacted soil near the face of the fill. Water infiltrating into the slope from the top tends to flow along this loose, more permeable layer parallel to the slope. An infinite slope analyses was used to model these slope conditions. An infinite slope analyses is appropriate for evaluating the stability of compacted fills when failure occurs primarily by shallow sloughing along a surface roughly parallel to the face of the slope. The analyses were conducted on both a 1.5:1 and 2:1 slope. Factors of safety were computed as a function of vertical depth to the sliding surface (H) and seepage direction (θ) with respect to a horizontal reference plane. The intrinsic cohesion was set low at 0.2 psi because the fill was composed primarily of granular soil or borrow. The friction angle was estimated to be approximately 35 degrees based on gradation and in-situ density of the fill.

Influence of Plant Roots: The presence of fibers (roots) provides a measure of apparent cohesion (Gray and Ohashi, 1983; Maher and Gray, 1990). This fiber or root cohesion can make a significant difference in the resistance to shallow sliding or shear displacement in sandy soils with little or no intrinsic cohesion. Actual shear tests in the laboratory and field on root/fiber permeated sands indicate a shear strength increase per unit fiber concentration ranging from 7.4 to 8.7 psi per lb of root/cu ft of soil or 3.2 to 3.7 kPa per kg of root/cu m of soil (Gray and Ohashi, 1983; Ziemer, 1981). Root concentration relationships reported in the technical literature were used to estimate likely root cohesion (c_R) as a function of depth (Riestenberg, 1983). A low to medium root concentration with depth was used in the stability analyses to ascertain the likely influence of slope vegetation on mass stability. Factor of safety is shown plotted as a function of depth and seepage direction in both the absence and presence of root reinforcement for a 1.5:1 slope and $\phi = 30$ degrees in Figure 3. In the absence of roots the factor of safety drops below unity ($F < 1$) when the seepage either parallels or emerges from the slope face at depths greater than one foot. With roots present, on the other hand, the safety factor is increased significantly near the surface and the critical sliding surface is displaced downward. The results of the stability analyses show that both seepage direction (θ) and presence of root cohesion (c_R) have a significant effect on the factor of safety. Even a small amount of root cohesion can increase the factor of safety substantially. This influence is pronounced at shallow depths where root concentrations are highest and reinforcement effects therefore greatest. Downward seepage ($\theta = 90^{o}$) greatly increases the factor of safety. This condition, in fact, yields the same factor of safety as a dry slope.

Field Installation: Soil bioengineering production work began in October of 1986. Soil bioengineering techniques tend to be labor intensive and require

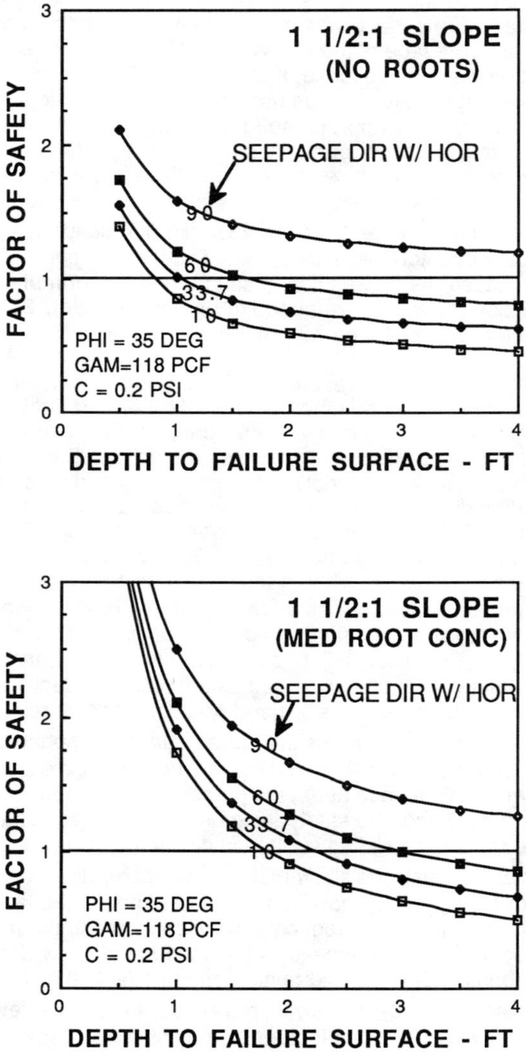

Figure 3. Factor of safety vs. depth and seepage direction for 1.5:1 hypothetical slope both with and without roots in the surface layer.

the use of hand tools during both harvesting and installation. Plant materials suitable for the live cut brush layers such as willow, alder, and dogwood were harvested off site. Harvesting related activities required approximately half of the man-hours consumed. These activities included locating harvesting sites, securing permission to harvest, improving access to the harvesting site, the actual harvesting of the plant stems, and transportation to the construction site.

Installation related activities were also quite labor intensive. A small bulldozer was used to prepare the earthen lifts on which the brushlayers were placed as shown in Figure 4. Typically, plant material delivered to the site consisted of bundles containing long live stems or brush. In addition to live stems or brush, quantities of lime, fertilizer, grass seed and jute mesh were used at the project site. Selected fill brushlayers in critical areas were reinforced, by placing a 6.5-foot wide layer of Tensar geogrid on the earthen lift beneath the brush. The large openings in the grids permit root development through the grid apertures and into the soil. These grids provided additional or supplementary reinforcement in the slope.

Views of the brush layer treated slope at various stages during and after construction are shown in Figures 4 through 6. In addition to reinforcement and favorable modification of the hydrologic regime near the face of the slope, the brushlayers acted as filters to control surficial erosion from heavy rains during and after construction. This filtering action was observed during construction when 4.5 inches of rain fell in a 3-day period. The brushlayers prevented erosion of loosely compacted fill material at the slope face. Those areas of the slope without such protection experienced soil flows and sloughing.

Case Study No. 2 - Cut Slope Stabilization

Project Site: The project site is located along Greenfield Road, just off State Route 112, in northern Massachusetts near the village of Colrain. Widening and improvement of this scenic road resulted in encroachment on an adjacent, unstable hillside which triggered cut slope failures. The slope consisted of a residual soil mantle, a silty sand, overlying a fractured quartz-mica schist bedrock. The cut was excavated back at a design slope angle of 1.5:1; the inclination of the natural slope above the cut was approximately 3:1. Cut slope heights varied in general from 20 to 60 feet. An unstable cut section extended a distance of approximately 1,800 feet adjacent the road. The vertical thickness of the soil mantle exposed above the bedrock varied from under 10 feet to the full vertical height of the cut. Slope failures were characterized by small slipouts and slumping. A substantial amount of groundwater flowed out of the cut. This water seeped out of both fractures in the underlying bedrock and through the exposed face of the soil mantle.

Figure 4. View of slope during fill brushlayer construction. N.C. State Route 126. Photo taken February 1987.

Figure 5. View of fill brushlayer after two months of stabilizing growth. N.C. State Route 126. Photo taken May 1987.

Figure 6. View of brushlayer fill one year after construction. N.C. State Route 126. Photo taken June 1988.

Results of gradation analyses showed that the native soil exposed in the cut was a silty sand, Unified Classification SM, containing 58-72 % by weight sand, 17-33 % silt, and 2-4 % clay size material. Drained triaxial tests were carried out on saturated samples of the silty sand reconstituted in the laboratory to a representative in-situ density. The effective friction angle varied narrowly from 35-36° while the cohesion intercept varied from 3.0 to 4.2 psi (20.7 to 29.0 kPa). Samples with larger clay contents exhibited the higher cohesions.

<u>Alternative Slope Treatments</u>: Two conventional structural support systems were considered for stabilization of the cut: a drained rock buttress and a gravity retaining wall. The drained rock buttress essentially consisted of a blanket of crushed rock laid up against the cut. The main design consideration in the case of a rock blanket was to determine the thickness required to provide a specified global safety factor for the cut slope. This system had been employed frequently by Massachusetts DOT for cut slope stabilization. The main objection to this system was its stark and harsh appearance which were inconsistent with the scenic nature of the highway.

The soil bioengineering alternative proposed for the site was a drained, brushlayer buttress fill. Reservations were expressed by the Massachusetts DOT engineers about the ability of an earthen, brushlayer fill to resist large shear stresses at the base or toe of the slope and to provide a desired global safety factor of 1.5. Some concern was also expressed about the possibility of a critical shear surface developing through an earthen buttress fill adjacent and parallel to a brushlayer. In view of these expressed concerns two modified brushlayer fill designs were proposed: (1) a crushed rock blanket with earthen brushlayer inclusions at periodic intervals, and (2) a crushed rock section at the base and brushlayer fill on top. The latter design was ultimately adopted; stability analyses were carried out on various configurations of this hybrid or composite system. The results of stability analyses on this composite system (see Figure 7) showed that the most critical failure surfaces passed through the basal rock section at the toe of the slope. In view of these findings a decision was made to use the composite design to stabilize the cut. An important caveat in this decision was the requirement that the earthen fill remain in a drained condition--a key assumption in the stability analyses. This requirement along with the large quantity of groundwater seeping out of the cut imposed a strict condition that a suitable filter course or drain be interposed between the earthen fill and cut face. This condition was met by placing a thin layer of crushed stone and a fabric filter cloth against the cut face during construction.

Field Installation: Successful biotechnical construction requires that harvesting and placement of live cuttings in the brush layers be carried out during the dormant season, usually November through April. The construction work at the Colrain field site began in November of 1989. Harvesting sites with suitable plant materials were located by means of an aerial survey ahead of time. Stems and branches up to three inches in diameter of willow, dogwood, alder, poplar, and viburnum shrubs were cut, bundled and transported to the project site on flatbed trucks. The cut stems and branches were laid atop successive lifts of compacted soil in a criss-cross fashion as shown schematically in Figure 2. Soil overlying each brushlayer was worked in between the branches to insure intimate contact between the brush and soil. The vertical spacing between brush layers varied from 3 to 5 feet with closer spacings used at the bottom. The length of the cut stems extended the full width of the earthen buttress. A gravel drainage course and fabric filter were placed between the fill and the cut face.

A view of the cut slope after installation of the brushlayer, buttress fill is shown in Figure 8. The appearance of the same slope some 5 months later is shown in Figure 9. By this time the brush had leafed out and adventitious rooting had commenced along the length of the imbedded stems and branches. The visual effect of the brush layers is non intrusive and pleasing to the eye; a side by side comparison of a brushlayer treatment vs. stabilization with a crushed rock blanket is shown in Figure 10. The top of the cut was treated by rounding and installation of live fascines. The latter consist of bundles of live willow cuttings which are installed in shallow trenches, staked in place, and covered with a thin layer of earth. The willow cuttings eventually root and leaf out. The live fascines provide protection against surface erosion and shallow sloughing. Part of the slope below the fascines was also stabilized against shallow sloughing and erosion by

Simplified Bishop Slope Stability Analysis

PROJECT: SLOPE REPAIR WITH COMPOSITE DRAINED ROCK AND EARTH BUTRESS

LOCATION: COLRAIN, MASSACHUSETTS

COMPLETE SLOPE CROSS SECTION

CIRCLE	X	Y	RADIUS	FS
1	32.0	185.0	145.0	1.50
2	50.0	144.0	105.0	1.46
3	65.0	152.0	101.0	1.74
4	42.0	185.0	135.0	1.83

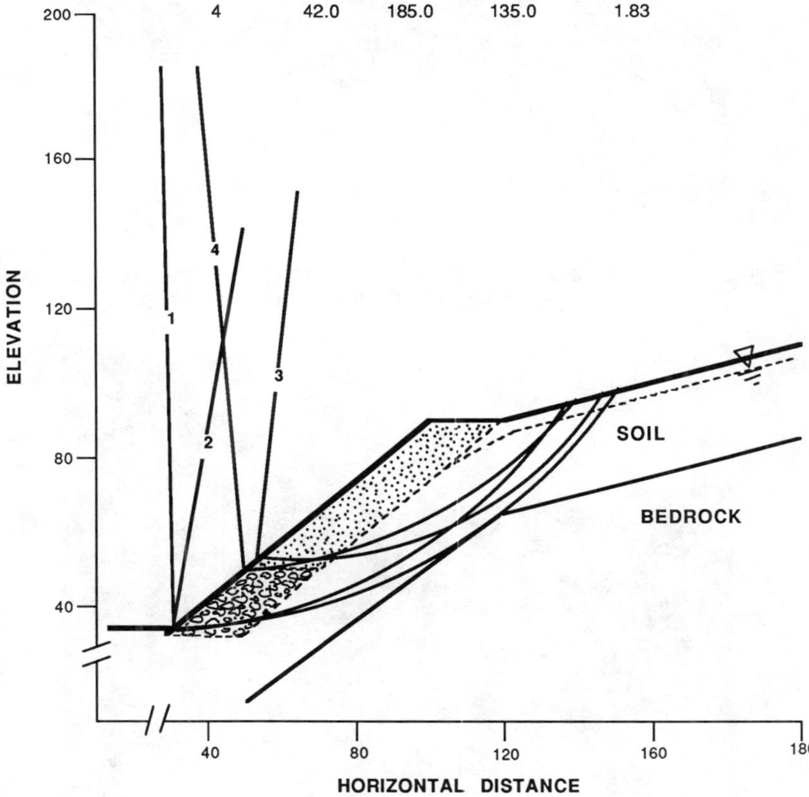

Figure 7. Factor of safety of cut slope stabilized by composite drained rock and earthen brushlayer fill.

Figure 8. View of cut slope stabilization immediately after construction of brushlayer buttress fill. Photo taken January 1990.

Figure 9. View of brushlayer fill installation seven months later. Brush layers have rooted and leafed out. Photo taken August 1990.

staking a natural geotextile tightly on the ground. The geotextile used for this purpose is known as coir netting. Coir netting is manufactured from the fibers of the coconut husks; it has a much higher durability and tensile strength than jute. Live, cut stakes were inserted through the coir netting on about 3- foot centers. These stakes also rooted and sprouted, thereby, providing additional protection to the slope.

Figure 10. Side by side comparison of slope stabilization treatment with live brush layers vs. a crushed rock blanket.

Conclusions

Soil bioengineering solutions can be used to stabilize and repair slope failures along highway right-of-ways. Live brushlayers can be used in conjunction with or in lieu of both secondary and primary polymeric reinforcements in over- steepened slopes. In addition to providing soil reinforcement, brushlayers filter soil from runoff and mitigate surficial erosion. They also act as horizontal drains and favorably modify the hydrologic regime near the face of a slope. Stems and branches of plant species must be used that root easily from cuttings such as willow, alder, and dogwood.

Soil brushlayering treatments provided an attractive and cost effective solution for stabilizing a failed cut slope along a scenic highway in Massachusetts and and an embankment fill along a state highway in North Carolina. In both cases the slope has remained stable, and vegetation has become well established on the slope face.

Appendix - References

Berg, R.R. Anderson, R.P., Race, R.J., and Curtis, V.E. (1990). Reinforced soil highway slopes. *Transportation Research Record,* No. 1288, NAS-NRC, pp. 99-108.

Cedergren, R. (1989). Seepage, Drainage and Flownets. 3rd Edition. John Wiley Sons: New York.

Coppin, N., Barker, D., and Richards, I. (1990). Use of Vegetation in Civil Engineering. Butterworths: Sevenoaks, Kent (England)

Giroud, J.P., Holtz. R. et al. (1987). Soil reinforcement design using geotextiles and geogrids, in Geotextile Testing and the Design Engineer, *ASTM Spec. Publ.* No 952, pp. 69-116.

Gray, D.H., Leiser, A.T., & White, C.A. (1980). Combined vegetative-structural slope stabilization. *Civil Engineering* (ASCE), Vol. 50, No. 1, pp. 82-85.

Gray, D.H. and Leiser, A.T. (1982). Biotechnical Slope Protection and Erosion Control. Van Nostrand Reinhold: New York, N.Y.

Gray, D.H. and Ohashi, H. (1983). Mechanics of fiber reinforcement in sand. *J. of Geotech. Engineering* (ASCE), Vol. 112, No. GT3, pp. 335-353

Kraebel, C.J. (1936). Erosion control on mountain roads. USDA *Circular No. 380,* 43 pp.

Kropp, A. (1989). Biotechnical stabilization of a debris flow scar. *Proceedings,* XX IECA Conference, Vancouver, pp. 413-429.

Maher, M. and Gray, D.H. (1990). Static response of sands reinforced with randomly distributed fibers. *J. of Geotechnical Engineering* (ASCE), Vol. 116, No. 11, pp. 1661-77,

Riestenberg, M.H. and Sovonick-Dunford, S. (1983). The role of woody vegetation in stabilizing slopes in the Cincinnati area. *Geologic Society of America Bulletin,* Vol. 94. pp. 504-518.

Schiechtl, H. (1980). Bioengineering for Land Reclamation and Conservation, The Univ. of Alberta Press, Edmonton, Canada.

Sotir, R.B. and Gray, D.H. (1989). Fill slope repair using soil bioengineering systems. *Proceedings,* XX IECA Conf., Vancouver, B.C., pp. 413-429,

Ziemer, R.R. (1981). Roots and shallow stability of forested slopes. *Intl. Assoc. Hydrol. Sciences,* Publ. No. 132, pp. 343-361.

Geosynthetic Strength - Ultimate and Serviceability Limit State Design

Fannin, R.J.[1]
Hermann, S.[2]

Abstract

The performance of an instrumented reinforced soil structure is used to establish characteristics of creep in the polymeric reinforcement. A comprehensive record of field performance is available for the four years since completion of construction in July 1987. Independent measurements of force and strain in the reinforcement are reported in detail. The field data are compared with laboratory load-strain-time data from rapid loading creep tests. Creep behaviour of the polymeric reinforcement is described. Use of the rapid loading creep data design for selection of tensile strength in ultimate and serviceability limit states design is evaluated.

Introduction

Methods of analysis used in the design of reinforced soil structures such as walls, slopes and embankments are typically limit equilibrium methods derived from conventional geotechnical analysis. Comprehensive guidelines for design of reinforced soil structures using limit equilibrium methods are reported in the US FHWA Report FHWA-RD-89-043 by Christopher et al. (1990). The guidelines for geogrid and geotextile reinforced structures address an ultimate limit state design. Limit equilibrium methods of analysis are augmented by empirical charts to establish, for example, reasonable bounds to lateral displacements. The charts are based on field data and numerical models.

---

[1]Assistant Professor, Dept. of Civil Engineering, also Harvesting and Wood Science, University of British Columbia, Vancouver, B.C. V6T 1Z4 Canada.
[2]Senior Geotechnical Engineer, Norwegian Geotechnical Institute, N-0801 Oslo 8, Norway.

It would appear codes of practice governing the in-service performance of such structures, that is a serviceability limit state rather than an ultimate limit state, are being developed from a consideration of limit strains at working stress. The approach forms the basis of the permissible tensile strains which are reported in the Technical Memorandum (Bridges) BE 3/78 (revised 1987) (DTp 1987) for retaining walls and bridge abutments in the United Kingdom. The memorandum sets out principles for the assessment of the tensile strength of materials which exhibit significant long term creep, and incorporates specific guidance for polymer based products.

The duality of approach is consistent with general principles of engineering design, but it requires careful interpretation of tensile strength for design when using extensible polymeric reinforcement such as geogrids and geotextiles.

Long-term Strength Characteristics

Geosynthetics exhibit a thermo-viscoelastic behaviour. Viscoelastic behaviour describes a response to loading that is dependent on the rate of loading. Thermo-viscoelastic behaviour describes a response that is, additionally, dependent on temperature. If the response is linear at a given temperature, then stress and strain are related by a unique value of modulus which is a function of the rate of loading. If the response is non-linear at a given temperature, the value of modulus is a function of both rate of loading and load magnitude.

Creep is the action of continued strain at constant load, and the phenomenon is well-recognized in polymeric materials. The magnitude of any creep strain is influenced by:

- type of polymer;
- manufacturing technique;
- magnitude of loading;
- temperature;
- time.

Consequently data describing the load-strain-time relationship of a geosynthetic must be used to properly evaluate long-term strength, and performance during the design life. Such data are typically presented as isochronous load-strain curves, from laboratory tension testing of unconfined samples in-air, following an approach reported by McGown et al. (1984) and further described by Jewell (1985).

The approach forms the basis of the proposed Canadian General Standards Board test method for tensile creep of rigid geogrids. The laboratory test method involves loading a series of test specimens in a rapid, smooth manner. Load is maintained throughout the test to ±1% of full load, at a controlled temperature and humidity. Deformation of each test specimen is measured. Tests are performed for at least 10,000 hrs or until failure, whichever is less. Data obtained from each test is plotted as strain against logarithm of time for each magnitude of load. A long-term design strength is selected based on a value of instability limit strain identified from a plot of strain versus logarithm of strain rate. The procedure involves extrapolation of test data from $1x10^4$ hrs, which is the test duration, to $1x10^6$ hrs, which is selected to meet requirements of a 120 year design life. There is no standard method for performing this extrapolation.

Instrumented Field Structure

A sloped (2V:1H) reinforced soil wall was constructed near Oslo in July 1987 as part of a project on soil reinforcement at the Norwegian Geotechnical Institute. The uniaxial grid reinforcement used in construction of the instrumented field trial is a high density polyethylene. During manufacture the polymer is extruded as a continuous sheet: holes are punched in the sheet, and it is drawn to configure the structure of the grid. The influence of magnitude of loading, temperature and time are examined from field measurements taken during the four years since completion of construction.

The structure is 4.8m high and 20m long, and comprises two sections, termed 'J' and 'N', each with a different arrangement and spacing of polymeric reinforcement. Tensar SR55 geogrid, a uniaxially orientated grid, was used as "primary" reinforcement, while shorter lengths of Tensar SS1, a biaxially orientated grid, were used as "intermediate" reinforcement when spacing of the layers of primary reinforcement exceeded 0.9m. Section 'J' comprises a non-uniform spacing of five layers of primary reinforcement that are 3.0m long, see Fig. 1, and Section 'N' comprises a uniform spacing of eight layers of primary reinforcement that are 2.2m long. Construction involved compacting the backfill soil in a series of layers behind discrete lightweight facing panels. The structure is founded on a competent gravelly sand foundation and seats into an existing hill slope. The backfill soil is a uniformly graded medium-to-fine sand taken from a borrow pit on site. Interpretation of direct shear tests on the soil gave a value of 38° for the angle of shearing resistance in plane strain at large displacement.

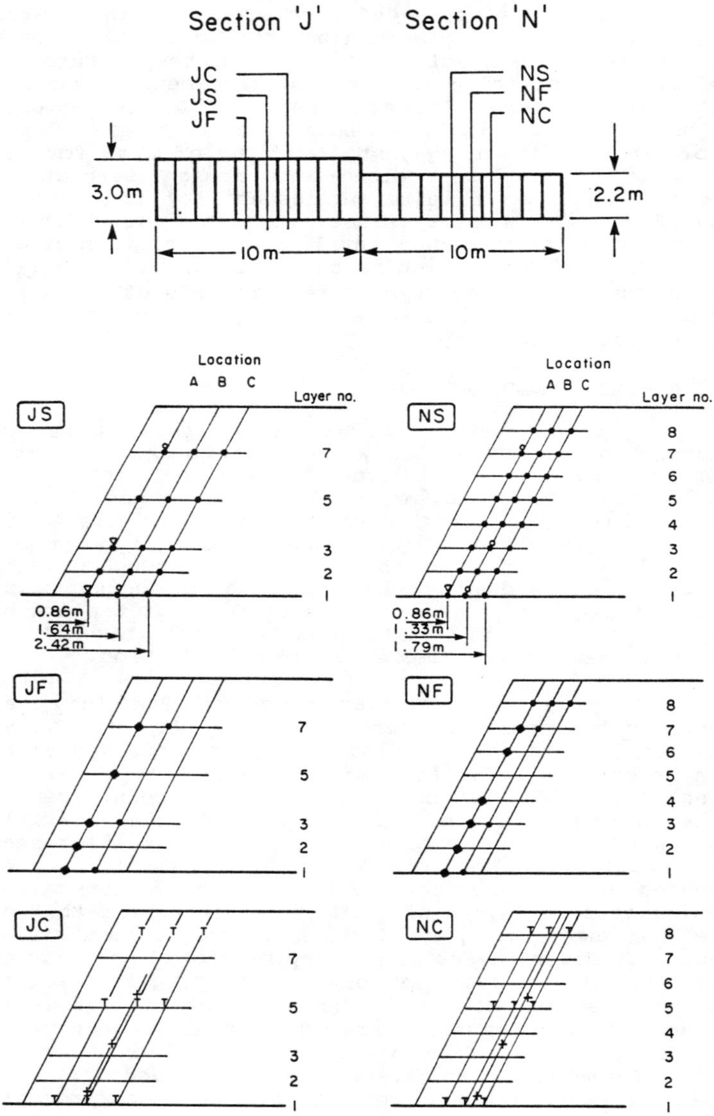

Figure 1 Instrumentation scheme (after Fannin and Hermann, 1990).

Instruments were placed during construction to monitor field performance. Measurements taken include force and strain in the reinforcement, soil strain and temperature. The uniaxial grid is manufactured to a nominal width of 1m. Force in the reinforcement is measured by cutting away a complete series of longitudinal ribs on the geogrid and installing three load cells to bridge the gap. The load cells operate on the principle of the vibrating-wire strain gauge. Strain in the reinforcement, and in the backfill soil, is deduced from the separation of pairs of Bison inductance coils. The principle of operation is that of electromagnetic coupling, whereby separation may be related to the coupling between them. Strain in the reinforcement is measured from coil pairs attached to the grid, while strain in the soil is measured from coil pairs embedded in the backfill. Dummy gauges, mounted on perspex blocks at a fixed separation, were also placed in the structure. Temperature is measured using thermistors.

Locations of instruments in the structure are shown in Fig. 1. Independent measurements of force and strain are made at location 'A' on any layer of reinforcement. The force cells and inductance coils are mounted on adjacent strips of grid reinforcement in any layer. A complete description of the instrumentation scheme is given by Fannin and Hermann (1990).

Behaviour of the structure has been examined with respect to self-weight loading, a cycle of surcharge loading, and permanent surcharge loading. Following completion of construction in July 1987 the structure was subject to self-weight loading for some four weeks. Then, in the period August to October 1987 a single load-unload cycle of surcharge was applied to the crest of the structure using water tanks that were filled to apply a maximum contact pressure of 29 kN/m^2. Finally, permanent surcharge loading was applied to the crest of the structure through placement of a 3m high berm which generated a uniform loading of 50 kN/m^2. Placement occurred over a period of three days and was completed on October 8th, 1987. Loading of the structure has remained unchanged since that time.

Measurements of force and strain allow the long-term performance of the reinforcement to be assessed with respect to our understanding of creep behaviour of such polymeric materials. A complete record of all instruments is available from completion of construction in July 1987 to May 1988, when some force cells were damaged in an electric storm. Since then force cell readings are only available for layers JF2 and JF3 in section 'J' of the structure, and layers NF3, NF5 and NF6 in section 'N'.

Field Performance

In describing field performance, and long-term behaviour of the geogrid reinforcement, emphasis is placed on measurements of force and strain taken since application of permanent surcharge loading. The force readings are believed accurate ±0.01 kN. The deduced strains are believed accurate to ±0.1%.

Influence of load magnitude

Tensile force mobilized in the reinforcement is dependent on the imposed loading. During the period of self-weight loading, the magnitude of force was in the range 1 to 2.5 kN/m. It was observed, Fannin and Hermann (1990), that smallest forces are mobilized in the base layer of reinforcement. The cycle of surcharge loading influenced the uppermost layers of Section 'N', but had little influence on the lower layers of this section or Section 'J'.

Permanent surcharge loading had a significant impact on all layers of reinforcement in the structure, with the exception of the base layers in each section. The variation of force with time during placement of the berm is shown in Fig. 2(a) and (b). Reinforcement layers refer to Fig. 1. Application of loading was not instantaneous, but occurred in stages as fill was placed in compacted lifts of 0.3m to a height of 3.0m over 3 days. The increment of force in the reinforcement was in the range 0.4 to 0.8 kN/m, causing the tensile force to be in the range 1 to 3.0 kN/m. In all layers, the mobilized force after permanent surcharge loading was the greatest experienced at that point in time.

Influence of temperature

Soil temperature is measured at nine locations in each section of the structure, giving in total eighteen measurements. They are used to correct the force cell and inductance coil outputs for temperature effects. The variation in temperature throughout the period of permanent surcharge loading is illustrated in Fig. 3. Temperature is reported as the mean value of all six measurements at location 'A'. Values are seen to vary seasonally between 0° and 16°. Most site visits led to measurements being taken on the structure in the late morning or early afternoon, so the variation does not describe the complete range of temperatures experienced in the field.

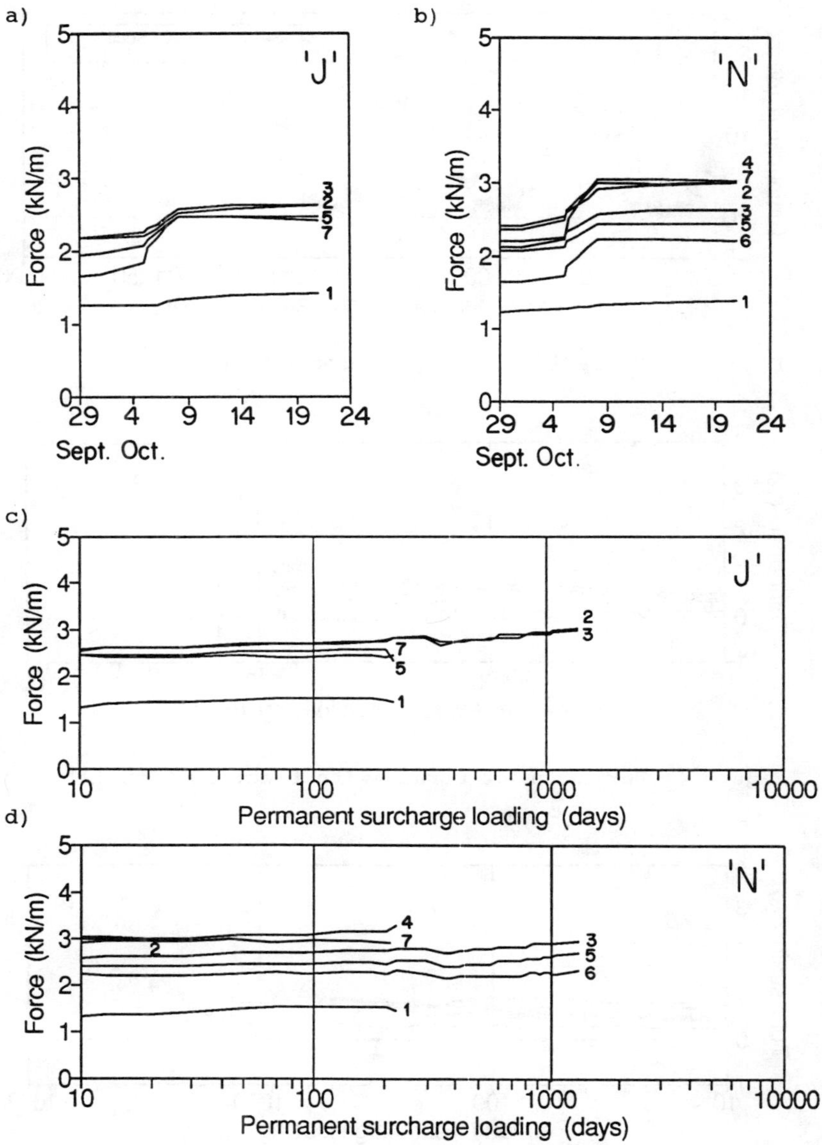

Figure 2 Variation of force with time during and after application of permanent surcharge loading.

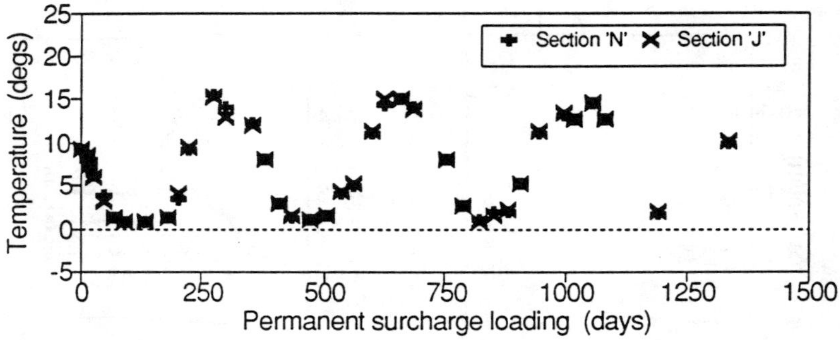

Figure 3 Variation of temperature with time.

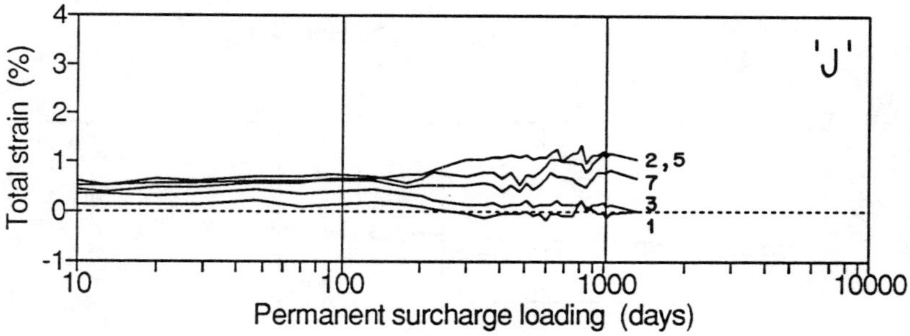

Figure 4 Variation of strain with time: Section 'J'.

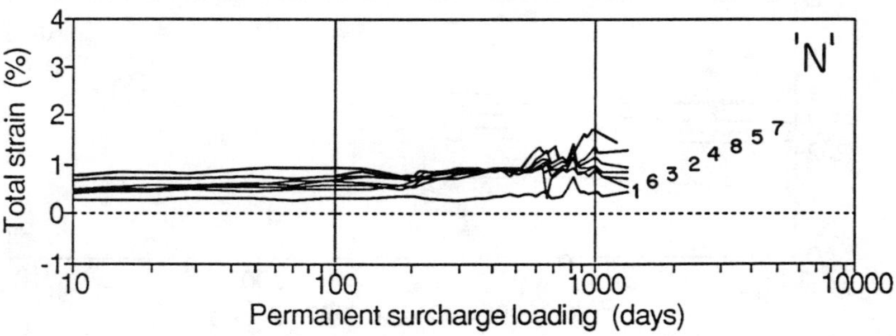

Figure 5 Variation of strain with time: Section 'N'.

Influence of time

The influence of time on measured force in the reinforcement is illustrated in Fig. 2(c) and (d) for the period of permanent surcharge loading. Elapsed time is that since completion of the berm on the crest of the sloped wall. The record describes some 1336 days, or 3.2 x 10^4 hours, of field performance. Truncated curves are a consequence of the damage caused to transducers during the electric storm. Mobilized force is seen to be nearly constant with time, both in Section 'J' and in Section 'N'. The response is not unexpected since the imposed loading is unchanging with time.

The influence of time on measured strain in the reinforcement is illustrated in Fig. 4 for Section 'J' and Fig. 5 for Section 'N'. Again elapsed time is that since completion of the berm on the crest of the sloped wall. Some variation in the magnitude of strain is evident, particularly in those layers of reinforcement in which the strain is significant, say greater than 0.5%. The general trend is one of increasing strain with time in all layers except the base layer. It is already noted that the mobilization of force in this layer in response to permanent surcharge loading was characteristically different from the other layers. The trend is apparent in Fig. 6, in which the variation of mean strain with time is shown. The value of mean strain is the average of all measurements except the base layer. The base layer has been omitted because it has not been loaded in the same manner as other layers of reinforcement, and the mobilized strain is very small.

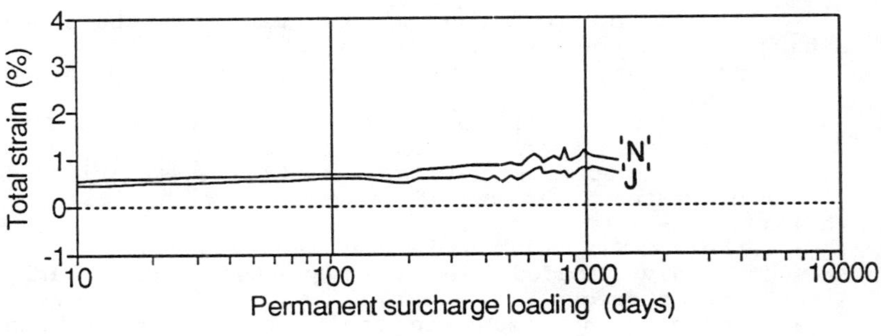

Figure 6 Variation of mean total strain with time.

Creep Behaviour of the Reinforcement

A comparison of the field measurements of force and strain at location 'A' on the reinforcement in the sloped wall shows continued strain at nearly constant load, and suggests creep is taking place. Examination of Fig. 6 indicates an increment of strain of some 0.25% to 0.4% during the period of permanent surcharge loading, with the increment being larger in Section 'N'. The force in any of these layers of reinforcement is in the range 2 to 3 kN/m, and can be characterized as nominally 2.5 kN/m.

Creep behaviour of the reinforcement is reported from laboratory testing to establish the variation of strain with time for constant load according to rapid creep test procedures. Data from such tests at 20°C and 10°C is reported in Fig. 7 from information supplied by the manufacturer. Time is measured from the application of constant load to the test specimen in the laboratory. The relationship between strain and the logarithm of time is almost linear for the smallest magnitude of loading, which is 8.8 kN/m. At this load, a strain of nearly 3% is reported 10,000 hrs after loading for test specimens at 20°C. A slightly stiffer response is observed for tests at 10°C.

Field data on creep behaviour of the reinforcement at location 'A' in the instrumented structure, taken from Fig. 6, are shown with the laboratory test data on Fig. 7(b). The field data describe mean total strain for the period between 10 days following application of permanent surcharge loading and 1336 days thereafter, which is the period illustrated in Fig. 6. It corresponds to a period between 2.4×10^3 hrs following completion of construction and 3.4×10^4 hrs thereafter, which is how data are plotted with respect to time in Fig. 7(b).

Any direct comparison of the field and laboratory data should recognize that:

- the reinforcement in the field has not been loaded instantaneously;
- the reinforcement in the field has been subject to small variations of load with time;
- the reinforcement in the field has experienced moderate variations of temperature with time;

while:

- the test specimens in the laboratory have been rapidly loaded in a way that is nearly instantaneous;
- the test specimens in the laboratory have been subject to a constant load with time;
- the test specimens in the laboratory have experienced constant temperatures during testing.

a)

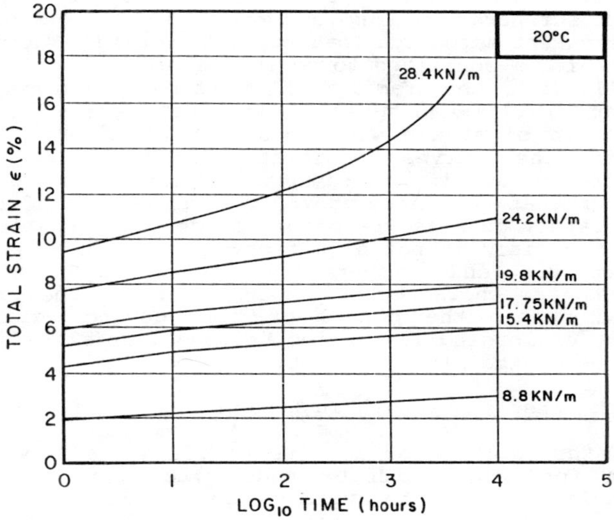

b)

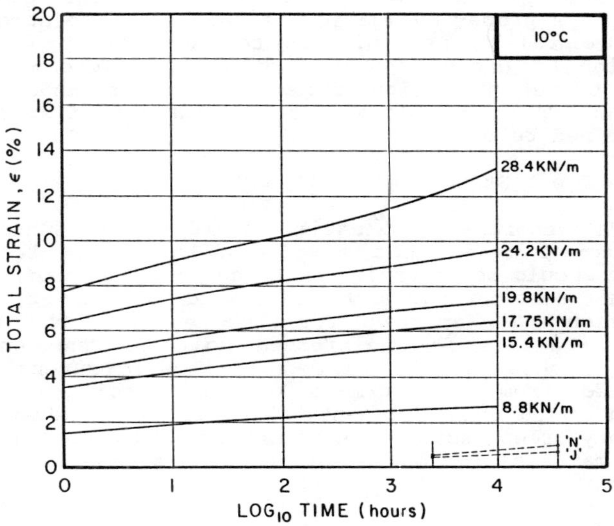

Figure 7 Load-strain-time data from laboratory testing (supplied by the manufacturer).

Agreement between field and laboratory data appears to be good. For clarity, only the data points for field measurements on October 18th, 1987 (10 days after permanent surcharge loading) and May 27th, 1991 (1336 days after permanent surcharge loading) are plotted. The data, which are for a mobilized force in the range 2 to 3 kN/m, fit well with the family of laboratory curves for different load magnitudes, and the nearly linear relationship of strain with logarithm of time compares well with other curves for relatively low loads. With regard to the qualifying remarks addressing any direct comparison that were made above, the comparison of field and laboratory data is reasonable since the variation of load in the field is not significant within the range of loads reported, and a laboratory temperature of 10°C is similar to the characteristic temperature in the field. It would appear the period of time during which an increment of loading is applied becomes less significant with elapsed time after completion of loading.

Long-term Tensile Strength in Design

Interpretation of a value of allowable long-term strength for design can be made with respect to an ultimate limit state, to avoid rupture, and with respect to a serviceability limit state, to avoid unacceptable displacements. In the first approach, partial factors of safety are applied to a value of ultimate tensile strength. A proper evaluation of this ultimate strength is fundamental to design. In the second approach, a prediction of creep behaviour of the reinforcement is made. A proper evaluation of creep strains is fundamental to design. The implications of this duality of approach are examined below.

Ultimate Limit State Design

The thermo-viscoelastic nature of geosynthetic reinforcement means that a value of allowable tensile strength should be determined with consideration of creep potential, test method and duration, and possible strength degradation. The approach forms the basis of the draft code of practice for reinforced soils in the United Kingdom, Ingold (1991). Allowable tensile strength is calculated from an estimate of the tensile rupture strength of the reinforcement, which is established from laboratory creep tests given a particular design life. A partial material factor is applied, which comprises four components:

$$T_{allow} = T_{ult}\left(\frac{1}{f_{m11}\cdot f_{m12}\cdot f_{m21}\cdot f_{m22}}\right)$$

where,

T_{ult} = tensile creep rupture strength
f_{m11} = factor of safety for material properties
f_{m12} = factor of safety for extrapolation of test data
f_{m21} = factor of safety for construction damage
f_{m22} = factor of safety for environmental conditions.

Tensile creep rupture strength must be established from appropriate load-strain-time curves. No recommendation is made in the draft code for appropriate values for these partial factors, however, Jewell and Greenwood (1988) and Ingold (1991) suggest minimum values based on practical procedures.

In the absence of sufficient long-term creep test data, the allowable tensile strength may be calculated by the following simplified expression to satisfy U.S. F.H.W.A. guidelines (after Christopher et al., 1990):

$$T_{allows} = T_{ult}\left(\frac{CRF}{FD\cdot FC\cdot FS}\right)$$

where,

T_{ult} = yield strength from the wide-width strip test
CRF = creep reduction factor
FD = factor of safety for durability
FC = factor of safety for construction damage
FS = overall factor of safety.

An empirical creep reduction factor is applied to the wide-width strength. Recommendations are made in the design guidance for values for these partial factors.

Serviceability Limit State Design

The U.K. Department of Transport Technical Memorandum BE3/78, D.Tp. (1987), sets out principles to be applied in assessing the allowable strength of materials which exhibit significant creep potential. In addition to ensuring the reinforcement does not fail in tension during the life of the structure, which is an ultimate limit state criterion, the memorandum requires that strains in the reinforcement shall not exceed a prescribed limit value during the life of the structure. The latter is a serviceability criterion. The prescribed strain limit governs a period of 120 years (the required design life), and is 1% for retaining walls and 0.5% for bridge abutments. When an incremental construction procedure and discrete facing panels are used, any strain developed during the time equivalent to the construction period is ignored. When propped, full-height facing panels are used, the time span for the strain limits should be taken as 120 years since the start of construction.

The allowable tensile strength is expressed as a permissible stress, but refers equally to an allowable strength, in which case:

$$T_{allow} = \frac{2\,T_{ck}}{F}$$

where,

T_{ck} = the characteristic load per unit width at 10°C which, when applied continuously will result in the prescribed strain limit in the designated time

F = an overall factor of safety determined for creep based on partial factors.

Tensile Strength Test Data

Application of the ultimate limit state approach and serviceability limit state approach are illustrated schematically in Fig. 8, using creep test data to establish allowable tensile strength in each case. Entering the graph for a given design life (step a), and knowing the instability limit strain against rupture for the reinforcement from manufacturer's test data (step b), establishes T_{ult} for the ultimate limit state design. Entering the graph for a given design life (step a), and knowing the construction period (step c), establishes T_{ck} for the serviceability limit state design for a particular value of limit strain $\Delta\epsilon$ (hence particular gradient of strain with logarithm of time).

A presentation of tensile test data in this manner allows a consistent evaluation of both ultimate tensile strength and creep behaviour of the reinforcement. Since the test data are typically established from laboratory experiments on unconfined specimens, validation of the load-strain-time behaviour from field structures is necessary to their use with confidence in analysis and design. The instrumented sloped reinforced soil wall has provided independent measurements of force and strain for a period of 3.4×10^4 hours, 1427 days, see Fig 7(b). The data are in good agreement with the manufacturer's data from long-term creep tests. The field response to loading is well-described by the laboratory creep tests.

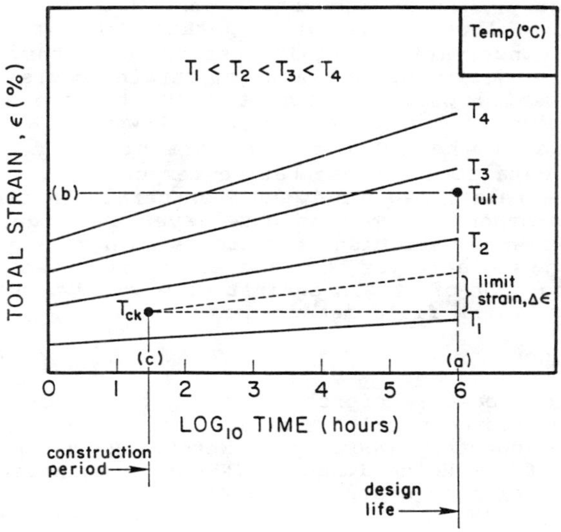

Figure 8 Determination of an allowable tensile strength from creep test data.

Conclusions

Field observations of the response to loading of polymeric grid reinforcement in a sloped reinforced wall have been used to describe mobilized long-term force and strain. The observed response has been compared with similar laboratory data on creep behaviour. The laboratory data is fundamental to a proper analysis of stability with respect to both the ultimate limit state and the serviceability limit state as defined in regulatory design guidance. A number of conclusions may be drawn from the comparison of field and laboratory data:

(1) Field measurements of force and strain near the face of the sloped wall indicate creep is occurring.

(2) Use of rapid loading creep data, from laboratory testing, to describe behaviour of the reinforcement embedded in soil is validated by the field measurements.

(3) Such rapid loading creep tests are typically performed at 20°C and 10°C. Those laboratory data at 10°C would appear to best describe the field data, which is in accordance with temperatures in the backfill of the structure typically varying in the range 0° to 16°C.

(4) The thermo-viscoelastic behaviour of the reinforcement is such that a less stiff response to loading is observed with increasing temperature.

Consequently, use of laboratory test data at 20°C will establish a lower value of T_{ult} than data at 10°C and would be conservative. Again, use of laboratory data at 20°C will suggest any increment of strain occurs during a shorter period of time than at 10°C, tending toward a conservative estimate of T_{ck}. However a correct prediction of the total strain to mobilize T_{ult} or T_{ck} requires data at a representative temperature.

(5) The relationship between mean strain in all layers of reinforcement except the base layer and logarithm of time shows some variation, but trends to a straight line. Linear extrapolation of the laboratory creep test data on a semi-log plot of strain against time would appear to be validated by the field measurements.

Acknowledgements

The sloped reinforced wall was constructed by the Norwegian Geotechnical Institute, the Norwegian Road Research Laboratory (NRRL) and Akershus Road Department. The work of Tor Helge Johansen (NRRL) in maintaining the instrumentation is gratefully acknowledged.

References

Christopher, B.R., Gill, S.A., Giroud, J.P., Juran, I., Mitchell, J.K., Schlosser, F. and Dunnicliff, J. 1990. Design and construction guidelines for reinforced soil structures - Volume I. Report No. FHWA RD-89-043, Federal Highways Administration, U.S. Department of Transportation.

Department of Transport, 1987. Technical Memorandum (Bridges) BE 3/78 (revised 1987). Reinforced earth retaining walls and bridge abutments for embankments. London (Dept. of Transport).

Fannin, R.J. and Hermann, S. (1990). Performance data for a sloped reinforced wall. Canadian Geotechnical Journal, Vol.27, pp. 676-686.

Ingold, T.S. (1991). Partial factors design of polymer reinforced soil walls. Ground Engineering, June 1991, pp. 34-38.

Jewell, R.A. (1985). Material properties for the design of geotextile reinforced soil slopes. Geotextiles and Geomembranes Vol.2, pp. 83-109.

Jewell, R.A., and Greenwood, J.H. (1988). Long term strength and safety in steep soil slopes reinforced by polymer materials. Geotextiles and Geomembranes, Vol.7, pp. 81-118.

McGown, A., Andrawes, K.Z., Yeo, K.C. and Dubois, D. (1984). The load-strain-time behaviour of Tensar geogrids. Symposium on Polymer Grid Reinforcement, ICE, London. Thomas Telford Ltd.

Finite Element Analysis of Slopes with Layer Reinforcement

By Robert M. Ebeling,[1] A.M., ASCE, John F. Peters,[1] M., ASCE and Reed L. Mosher,[1] A.M., ASCE

Abstract

A new efficient reinforcement model has been developed for use in nonlinear finite element analyses of reinforced soil structures. It allows the layout of the finite element mesh to be independent of the number of layers of reinforcement. The embedment procedure allows for the use of a coarser mesh by eliminating the restriction that reinforcement bar elements must be placed along element boundaries, significantly improving computational efficiency for nonlinear problems.

Introduction

Various types of reinforcement have been used to enhance the stability of slopes, embankments and retaining walls. The application of soil reinforcement ranges from single layers to stabilize embankments on soft soils to multiple layers for the formation of vertical walls. At the heart of the design procedures for these types of applications is a limit equilibrium method modified to include a force contribution to represent the reinforcement. The amount of reinforcement is increased as necessary to achieve the desired factor of safety against instability. These simplified design methods have criteria established to determine the appropriate limiting force in the reinforcement based on its strength, embedment length, and stiffness. The forces found from these limit equilibrium methods may not be the actual forces in the reinforcement. The development of the forces in reinforcement involves complex soil-structure interaction behavior between reinforcement and soil.

---

[1]Research Civil Engineer, US Army Engineer Waterways Experiment Station, Vicksburg, MS 39180

The finite element method provides the best available tool to evaluate the complex soil-structure interaction involved in reinforced soil problems. One technique commonly used to model the reinforcement is a one dimensional bar element sandwiched between slip or interface elements. The interface elements simulate the interaction between the soil and reinforcement. The finite element mesh for this type of representation requires the reinforcement layers to fall at element boundaries. This works well for a single layer of reinforcement, where the mesh can easily accommodate the reinforcement, but is cumbersome with closely spaced multiple layers of reinforcement or when analyzing large meshes that involve modeling the nonlinear stress-strain behavior of soils (i.e. more than 500 elements). This type of model is also inefficient when considering a number of different reinforcement schemes. Therefore, a procedure that makes the finite element mesh independent of the reinforcement layering is highly desirable when analyzing densely layered reinforcement systems.

This paper presents a new efficient reinforcement model that allows the finite element mesh to be independent of the number of layers of reinforcement under consideration. The paper describes the development of the model and its implementation in a nonlinear soil-structure interaction finite element code SOILSTRUCT (Ebeling, Peters, and Clough 1990). The application of the model to the evaluation of a reinforced soil berm to be constructed adjacent to a navigation lock demonstrates the use and versatility of the model.

Development and Implementation of Reinforcement Model

To understand the need for this new formulation used for modeling geoinclusions in an earth structure, it is worthwhile to first review the numerical procedure in which this new model is implemented.

Experience with the application of the finite element method in the analysis of earth structures has shown the importance of modeling the actual construction process as closely as possible and of including a nonlinear stress-strain soil model. One numerical procedure that has been successfully used to analyze a number of earth structures, where the results compared favorably with the measured performance, is the "backfill" analysis procedure as described in Duncan and Chang (1970). In this analysis procedure the soil is included as part of the mesh; placement of the soil fill is modeled by adding lifts of "fill" elements. The nonlinear stress-strain relationship of each soil element is modeled using an incremental construction procedure in which the loadings are applied incrementally and in combination with the adjustment of the modulus of each soil element during the course of the analysis in accordance with the stresses in the element. Application of this procedure to soil-structure interaction analyses of earth-retaining structures has led to the additional requirement that interface elements be incorporated within the finite element mesh along the backfill to structure interface. This numerical procedure has been implemented in a number of finite element computer codes, including SOILSTRUCT.

Finite Element Code SOILSTRUCT. SOILSTRUCT is a general-purpose, finite element program for two-dimensional, plane strain analysis of soil-structure interaction and soil-inclusion interaction problems. It calculates displacements and stresses due to incremental construction, excavation, dewatering and/or load application. The initial version was developed by Clough and Duncan (1969) for use in the analysis of Port Allen and Old River U-frame locks.

Three types of finite elements are used in SOILSTRUCT to represent the behavior of different materials: (a) a two-dimensional continuum (solid) element, (b) an interface element, and (c) a one-dimensional bar element.

Solid Finite Elements. The solid elements are four node subparametric, quadrilateral with an interior node (QM5) which were originally selected for SOILSTRUCT because of their ability to capture bending-like strain distributions (see Doherty, Wilson, and Taylor 1969). The constitutive relationship used for all solid elements is Hooke's law. SOILSTRUCT uses an incremental, equivalent linear method of analysis to model nonlinear material behavior. In this type of analysis, the incremental changes in stresses are related to the incremental strains through a linear relationship. This relationship is defined for each structural element by two engineering constants, the Young's modulus and the Poisson's ratio. For the soil elements, either the Young's modulus and the Poisson's ratio or the Young's modulus and the bulk modulus may be specified. The values for the modulus for each soil element are adjusted during each iteration for a given incremental load step in accordance with the stresses in each element. The nonlinear relationship between shear stress and strain of the soil elements is modeled using a hyperbolic curve (refer to Duncan and Chang 1970).

Interface Elements. The Goodman, Taylor, and Breeke (1968) four node interface element is used to allow for relative movement between different materials, such as between a soil backfill and a support wall. The properties of interface elements are defined by an interface normal stiffness and an interface shear stiffness. The nonlinear relationship between shear stress and displacement is modeled using the incremental, equivalent linear numerical procedure described previously.

Reinforcement Elements. SOILSTRUCT has two types of reinforcement elements; standard bar elements intended for use as anchors, struts and similar one-dimensional structural elements and "embedded" elements designed to model closely-spaced reinforcement layers. This section will describe the embedded element.

The embedded element is based on the idea of combining two elements, by enforcing compatibility, but expressing the deformations in terms of those for only one of the elements. In the present case, a bar element is embedded inside of a soil element with the stiffness being described in terms of the solid element forces and displacements. Each element stiffness is computed separately as follows:

$$K_e\, u_e = f_e \tag{1}$$

$$K_b\, u_b = f_b \tag{2}$$

where for the solid element K_e is the stiffness matrix, u_e are the displacements at the nodes of the soil element, K_b is the stiffness of the bar element, u_b are the displacements at the nodes of the bar. By compatibility,

$$u_b = N_b\, u_e \tag{3}$$

where N_b is the matrix of interpolation factors relating quantities at the bar nodes to the nodes of the soil element. By the principle of virtual work it is readily shown that the combined stiffness of the soil and bar is

$$(K_e + N_b^T\, K_b\, N_b)\; u_e = f_e \tag{4}$$

The increment of reinforcement strain for each step is computed from the interpolated incremental displacement and added into the total reinforcement strain. The reinforcement force is computed from the nonlinear relationship between force and strain. It is assumed that the thin reinforcement cannot resist compression. Accordingly, when the reinforcement strain becomes compressive the stiffness is set to zero. If on subsequent steps the reinforcement strain becomes tensile the stiffness is restored.

There are two advantages of using the embedded element instead of the standard bar. First, the layout of the embedded reinforcement is independent of the mesh design thus a more regular mesh can be used. Second, the assignment of reinforcement bars within each element can be automated by a straight-forward search algorithm during the preprocessing phase of the finite element code and the user only needs to supply data on the geometry of the reinforcement layers rather than the individual bars.

Examples Using Embedded Elements. The performance of the embedment procedure was evaluated through a number of simple test cases. Single elements with multiple bar segments were tested for cases where "exact" solutions could be determined. Three of these examples are shown in Figure 1 through Figure 3.

The single linear element shown in Figure 1 has 6 layers of embedded elements, spaced equidistant through the element. The nodal point constraints are as shown in the figure. The right hand face of the element is displaced uniformly in this problem, resulting in a uniform horizontal tensile strain equal to one percent ($\epsilon_x = 0.01$) throughout the element. The vertical strain and shear strain are equal to zero for this problem. The results of the finite element analysis agreed with the exact solutions summarized in Figure 1.

The problem shown in Figure 2 is the same as that shown in Figure 1, with the exception that the vertical nodal point constraints along the top two nodes are removed. In this case the uniform lateral displacement of the right hand face of the element results in both a uniform compressive vertical strain, as well as uniform horizontal tensile strain. Again, the results of the finite element analysis agreed with the exact solutions summarized in Figure 2.

The problem shown in Figure 3 tests SOILSTRUCT's ability to sense compressive strains within the reinforcement. The linear element shown in this figure has two diagonal reinforcement layers. The nodal points along the base are constrained against movement and the two top nodes are constrained against vertical movement. A uniform lateral displacement along the top results in a uniform shear strain equal to one percent ($\gamma_{xy} = 0.01$) and $\epsilon_x = \epsilon_y = 0$. In this case, two iterations are used when executing SOILSTRUCT so that the stiffness of bar number 2 was set equal to zero for the second and final iteration, after a compressive strain was computed during

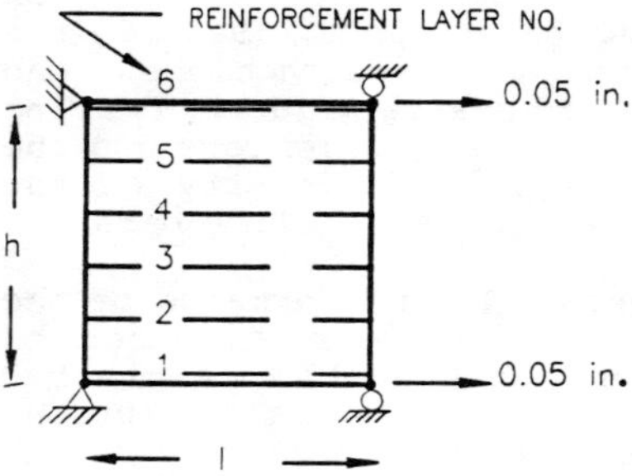

l = h = 5 in.
E = 1,000 psi
ν = 1/3
$(AE)_{bar}$ = 4,000 lbs./in.

2-Dim Element - Plane Strain

$$\begin{Bmatrix}\sigma_x\\ \sigma_y\\ \tau_{xy}\end{Bmatrix} = \begin{bmatrix} C_{11} & C_{12} & 0\\ C_{21} & C_{22} & 0\\ 0 & 0 & C_{33}\end{bmatrix}\begin{Bmatrix}\epsilon_x\\ \epsilon_y\\ \gamma_{xy}\end{Bmatrix}$$

$$C_{11} = C_{22} = \frac{E(1-\nu)}{(1+\nu)(1-2\nu)} = 1{,}500\ psi$$

$$C_{12} = C_{21} = \frac{E\,\nu}{(1+\nu)(1-2\nu)} = 750\ psi$$

$$C_{33} = \frac{E}{2(1+\nu)} = 375\ psi$$

$$\epsilon_y = \gamma_{xy} = 0$$

$$\epsilon_x = \frac{u_x}{l} = \frac{0.05\ in.}{5\ in.} = 0.01\ (tension)$$

$$\sigma_x = C_{11}\cdot\epsilon_x = \underline{\underline{15\ psi}}\ (tension)$$

Bar Elements

$$\epsilon_{bar} = \frac{u_x}{l} = \frac{0.05\ in.}{5\ in.} = 0.01\ (tension)$$

$$F_{bar} = (AE)_{bar}\cdot\epsilon_{bar} = 4{,}000 lbs/in.\cdot 0.01 = \underline{\underline{40\ lbs/in.}}$$

$$\Sigma F_{bar} = 6\ bars\cdot F_{bar} = \underline{\underline{240\ lbs/in.}}\,(tension)$$

Figure 1. One Dimensional Strain Response Of A Reinforced Linear Element (1 in. = 2.54 cm, 1 lb/in. = 0.175 kN/m, 1 lb/sq in. = 6.9 kPa).

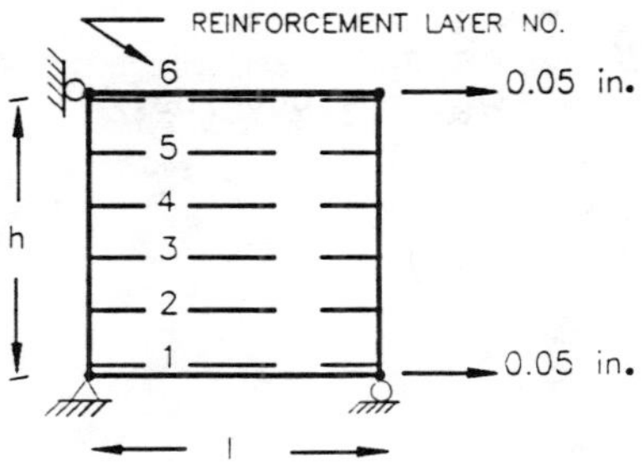

l = h = 5 in.
E = 1,000 psi
ν = 1/3
$(AE)_{bar}$ = 4,000 lbs./in.

2-Dim Element - Plane Strain

$$\gamma_{xy} = 0 \qquad \sigma_y = 0$$

$$\epsilon_x = \frac{u_x}{l} = \frac{0.05\ in.}{5\ in.} = 0.01\ (tension)$$

$$\sigma_y = 0 = C_{21} \cdot \epsilon_x + C_{22} \cdot \epsilon_y$$

$$\epsilon_y = \frac{-C_{21}}{C_{22}} \cdot \epsilon_x = -\frac{750}{1500} \cdot \epsilon_x = -\frac{1}{2} \cdot \epsilon_x = -0.005\ (compression)$$

$$u_y = \epsilon_y \cdot h = (-0.005) \cdot 5\ in. = \underline{\underline{-0.025 in.}}$$

$$\sigma_x = C_{11} \cdot \epsilon_x + C_{12} \cdot \epsilon_y$$

$$\sigma_x = 1{,}500\ psi \cdot (0.01) + 750\ psi \cdot (-0.005)$$

$$= 15\ psi - 3.75\ psi$$

$$\underline{\underline{\sigma_x = 11.25\ psi}}$$

Bar Elements

$$\epsilon_{bar} = \frac{u_x}{l} = \frac{0.05\ in.}{5\ in.} = 0.01\ (tension)$$

$$F_{bar} = (AE)_{bar} \cdot \epsilon_{bar} = 4{,}000 lbs/in. \cdot 0.01 = \underline{\underline{40\ lbs/in.}}$$

$$\Sigma F_{bar} = 6\ bars \cdot F_{bar} = \underline{\underline{240\ lbs/in.}}\ (tension)$$

Figure 2. Two Dimensional Strain Response Of A Reinforced Linear Element (1 in. = 2.54 cm, 1 lb/in. = 0.175 kN/m, 1 lb/sq in. = 6.9 kPa).

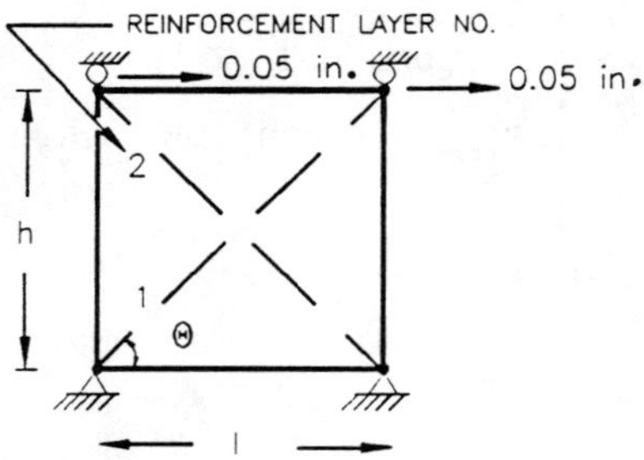

$l = h = 5$ in.
$E = 1{,}000$ psi
$\nu = 1/3$
$(AE)_{bar} = 4{,}000$ lbs./in.

<u>2-Dim Element</u> - Plane Strain

$$\epsilon_x = \epsilon_y = 0$$

$$\gamma_{xy} = \frac{u_x}{h} = \frac{0.05\ in.}{5\ in.} = 0.01$$

$$\tau_{xy} = C_{33} \cdot \gamma_{xy} = 375\ psi \cdot (0.01) = \underline{\underline{3.75\ psi}}$$

<u>Bar Element No. 1</u>

$$l_{bar} = \sqrt{l^2 + h^2} = 7.071\ in.$$

$$\tan\theta = \frac{h}{l},\ \theta = 45°$$

$$u_{bar} = u_x \cdot \cos\theta = 0.05\ in. \cdot \cos 45° = 0.3536\ in.$$

$$\epsilon_{bar} = \frac{u_{bar}}{l_{bar}} = \frac{0.03536\ in.}{7.071\ in.} = 0.005\ (tension)$$

$$F_{bar} = (AE)_{bar} \cdot \epsilon_{bar} = 4{,}000\ lbs/in. \cdot (0.005) = \underline{\underline{20\ lbs/in.}}$$

<u>Bar Element No. 2</u>

$$\epsilon_{bar} < 0\ (compression),\ therefore\ \underline{\underline{F_{bar} = 0}}$$

Figure 3. Shear Strain Response Of A Reinforced Linear Element (1 in. = 2.54 cm, 1 lb/in. = 0.175 kN/m, 1 lb/sq in. = 6.9 kPa).

the first iteration. The results of the finite element analysis agreed with the exact solutions summarized in Figure 3.

Assessment Of Pullout Potential. A limitation of the embedded element formulation is in modeling pullout. The embedded element formulation is appropriate when there is no potential for exceeding the shear transfer capacity between the soil and the reinforcement. This section describes a procedure for calculating the shear transfer without the use of special interface elements. The use of compatibility in relating u_b to u_e precludes pullout between the fabric and soil as might be modeled with conventional bars using interface (slip) elements between standard bars and soil elements. However the shear transfer between the reinforcement and soil can be computed from the gradient of bar forces thus permitting a check of potential for pullout, as shown in Figure 4 and explained in the following example.

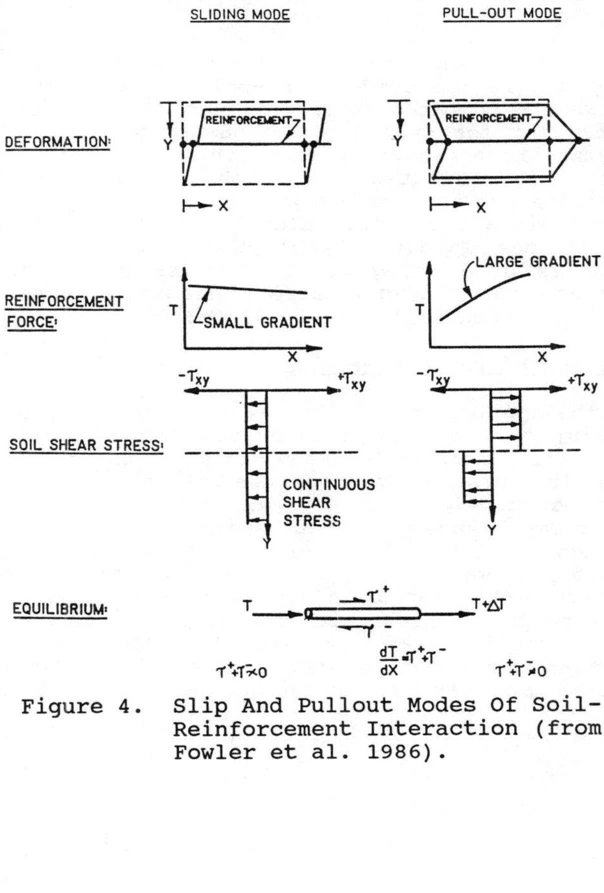

Figure 4. Slip And Pullout Modes Of Soil-Reinforcement Interaction (from Fowler et al. 1986).

Figure 5 shows the computed results using SOILSTRUCT to model a pull test of a single, 16.8 inch (6.61 cm) layer of reinforcement embedded within a 18 inch (7.09 cm) long soil container. The "stiffness" of the reinforcement, $(AE)_{bar}$, is equal to 4167 lbs per inch width (730 kN/m) for the 56 elements used to model the reinforcement. The nonlinear stiffness and strength properties assigned to the 720 soil elements corresponds to "typical" properties for dense sands (refer to Duncan et al. 1978). The initial vertical and horizontal stresses assigned to the soil are equal to 8.4 psi (58 kPa) and 4.2 psi (29 kPa), respectively. The pull force P = 30 lbs per inch width (5.26 kN/m) was applied in 120 equal load increments, using two iterations per load increment. The resulting distribution of bar force, T, and distribution of shear stress, τ, within the adjacent soil elements are shown in this figure. The magnitude of τ_{top} is computed equal to τ_{bot} but opposite in sign. The gradient of the bar force, $\Delta T/\Delta x$, is computed equal to the sum of the shear stresses in the upper and lower soil elements as shown in the figure.

In the case of embedded elements, the shear transfer may be computed as one-half the gradient of the bar force and compared to the available shear resistance when checking for pullout. It has been our experience with using the embedment element in design projects that shear transfer is not great enough to cause pullout and good results can be obtained even if more than one bar is placed within a large soil element. This is due to the fact that the layout of the reinforcement for these projects is done using force equilibrium based design procedures with a factor of safety of sufficient magnitude so as to preclude failure by pullout.

Application of the New Model

The application of the new model to the evaluation of a reinforced soil berm to be constructed adjacent to the navigation lock at Lock and Dam No. 1 on the Red River Waterway in Louisiana demonstrates the use and versatility of this new model (Ebeling et al. 1991). Lock and Dam No. 1 has experienced a serious sediment problem since its completion in 1983. The silt deposition against the riverside lock wall was significantly greater during high water than anticipated in the design of the project. During a period of high water in the Spring of 1985, silt deposition against the riverside lock wall was 14 feet higher than assumed in the design of the lock. While, the deposition of silt caused no damage to the lock, its

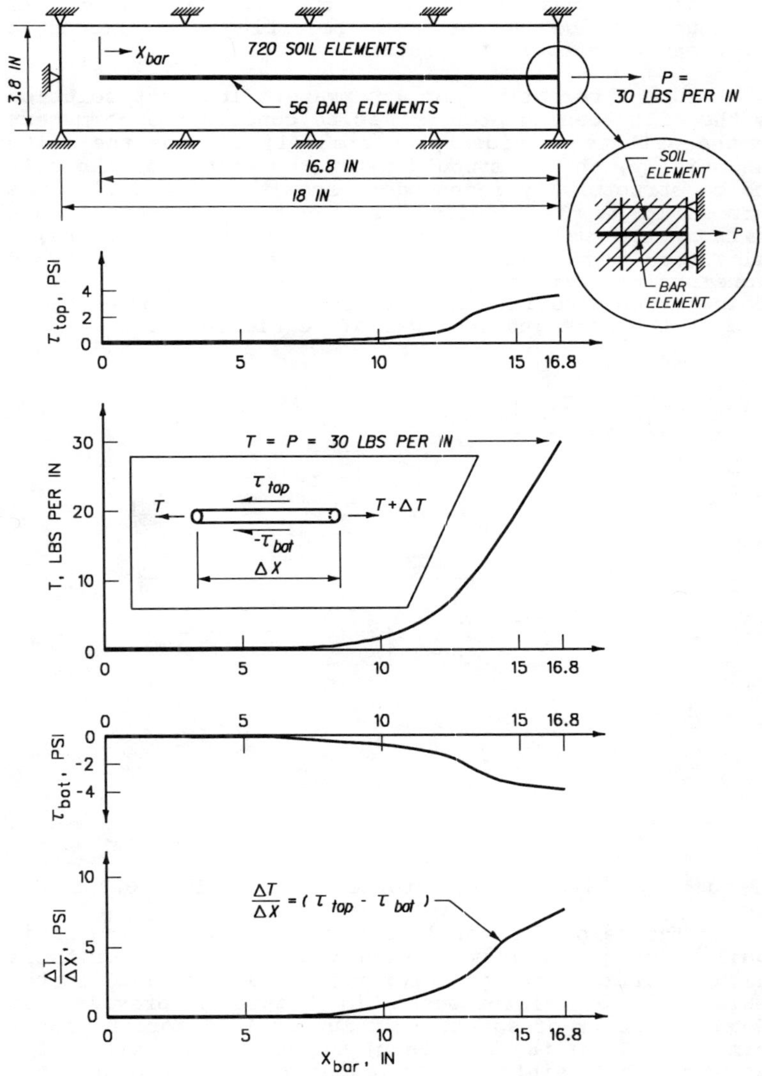

Figure 5. Pull Test Results Using SOILSTRUCT (1 in. = 2.54 cm. 1 lb/in. = 0.175 kN/m, 1 lb/sq in. = 6.9 kPA)

removal was felt to be necessary before the river receded. Since the highest level of water in the river has been less than a 10 year flood event, the silt deposition will continue to be a problem resulting in expensive maintenance costs in the future.

One alternative for a permanent low-cost solution to the silt loading problem was to construct a berm next to the wall to provide "pressure relief" from the silt. Essentially, the berm would carry the weight of the silt and be structurally independent from the lock wall. The berm would be placed with a free-standing vertical face next to the wall and therefore had to be internally reinforced with closely space reinforcing. To avoid transferring earth pressures from the berm to the wall a geoinclusion material would be placed between the vertical face of the berm and lock wall (see Figure 6).

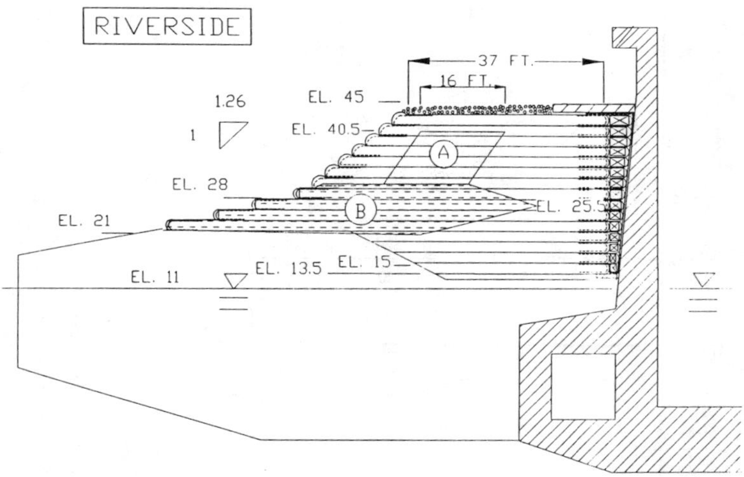

Figure 6. Riverside Reinforced Berm (1 ft = 0.305 m)

The reinforcement layout was designed using limit equilibrium procedures. Figure 7 shows the critical failure surface for the reinforced soil berm used in the design. The reinforcement in Figure 7 provides an adequate factor of safety (minimum of 1.25 for the short term and 1.5 for the long term) against slope failure for the berm. The reinforcement layout varies throughout the berm and consists of primary and secondary layers of reinforcement. Each primary layer of reinforcement extends across the entire width of the berm. Two zones of the secondary reinforcement, labeled A and B in Figure 6,

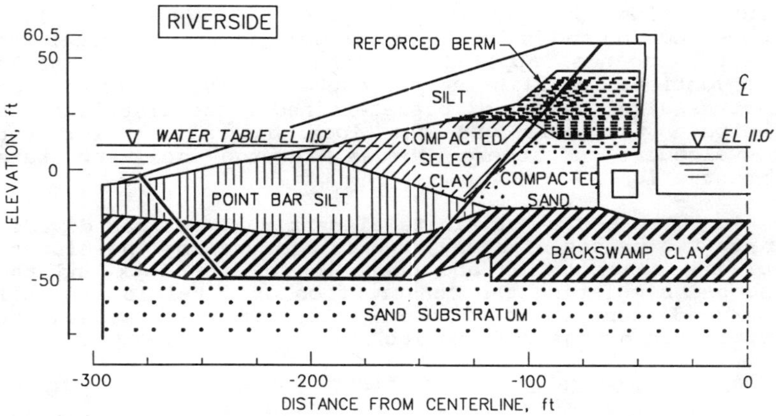

Figure 7. Critical Slip Surface Within The Reinforced Berm With Silt Loading (1 ft = 0.305 m).

provide additional reinforcement along the face of the riverside slope.

The procedures used in the design will result in a stable vertical face and riverside slope. However, the limit equilibrium procedure does not provide sufficient information with regard to the deformation of the reinforced soil berm. For the berm to protect the lock wall from silt loading, deformations must be small to avoid stress transfer through the geoinclusion material. Such detailed information on deformations is best obtained using a finite element analysis.

<u>Site</u>. The Red River Lock and Dam No. 1 is in Catahoula Parish, Louisiana, approximately 9 miles upstream from confluence of the Red and Black Rivers. The site is within the Gulf Coastal Province near the western edge of the Mississippi River Floodplain. The site has three distinct soil strata: the Point Bar Deposit, the Backswamp Deposit, and the sand substrata, as shown in Figure 7. The sand substratum is the deepest deposit and consists of mainly dense sands with some gravel present. The Point Bar Deposit consists predominantly of silts with regions of silty sand and poorly graded sands. The Backswamp Deposit is primarily overconsolidated fat clays (CH) with interbedded layers of lean clays (CL), silts, silty sands, and sands.

<u>Analysis Description</u>. SOILSTRUCT was used to analyze the behavior of the lock and the complex soil-structure

interaction by simulating the sequence of lock construction and backfilling, as well as various water and silt loadings applied to the structure. Unlike conventional equilibrium procedures, this procedure does not require the use of predetermined force distributions between the soil and the lock but allows for the development of these forces through soil-structure interaction.

The series of finite element analyses was divided into two phases of study, the first modeling the history of the construction and operation of the lock and the second simulating the construction of a reinforced soil berm adjacent to the riverside lock wall and subsequent operation of the U-frame lock.

Phase 1 Analyses. The first phase consists of an evaluation of the behavior of lock monolith no. 11, its backfill and its foundation during its construction and for three key operational load cases. The stages of construction modeled in the analyses include lowering the water table at the site, excavation, and completion of lock construction and backfilling. The computed results after completion of lock construction and for three key operational load cases for which instrumentation data is available were then analyzed and the results compared to the measured earth pressures. Good agreement was observed between the results of the finite element analyses compared with instrumentation measurements in these four cases.

Phase 2 Analyses. The second phase of the study was an extension of the first phase analyses with the simulated construction of a proposed reinforced soil berm riverside of the lock, followed by the raising of the pool levels, silt loading and subsequent lowering of the pool levels. Figure 8 depicts the model of the reinforced berm used in this series of analyses. The soil comprising the reinforced berm is assumed to be dense, well drained select sand.

The reinforcement layout used in the analysis is quite dense and varies with location in the berm. As many as six layers of reinforcement are embedded within a soil element. The finite element mesh used in the Phase 1 analyses (not shown) was modified within the riverside backfill region so as to model the reinforced soil berm. The mesh used in the Phase 2 analyses (not shown) contains 1150 elements and 1250 nodal points (2500 degrees of freedom). Use of the embedded elements within the soil elements significantly reduced the degrees of freedom for this nonlinear problem compared to those required when

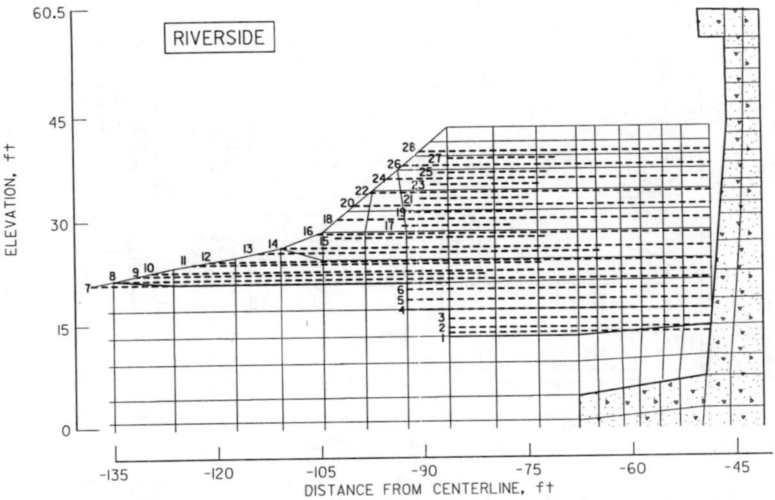

Figure 8. Layout Of 28 Layers Of Reinforcement Within The Riverside Berm Region Of The Mesh (1 ft = 0.305 m).

using traditional meshes involving standard bar elements placed along solid element boundaries.

Figure 9 shows the computed reinforcement forces in each element. These forces were relatively uniform within each element and the values shown represent average values. The distribution of forces within the embedded elements is suggestive of a Coulomb wedge near the vertical wall. It was found that the soil elements within the wedge were in a fully developed active failure condition whereby the backfill was held in place by the reinforcement layers, all of which is consistent with expectations. It is interesting that the potential failure mechanisms investigated in design (shown in Figure 7) are not evident from the distribution of fabric forces. Deformations were shown to be excessive near the top of the vertical face which called for a change in design details. Importantly, these details were not addressed by the limit equilibrium analysis.

Summary and Conclusions

A new efficient reinforcement model has been developed. It allows the layout of the finite element

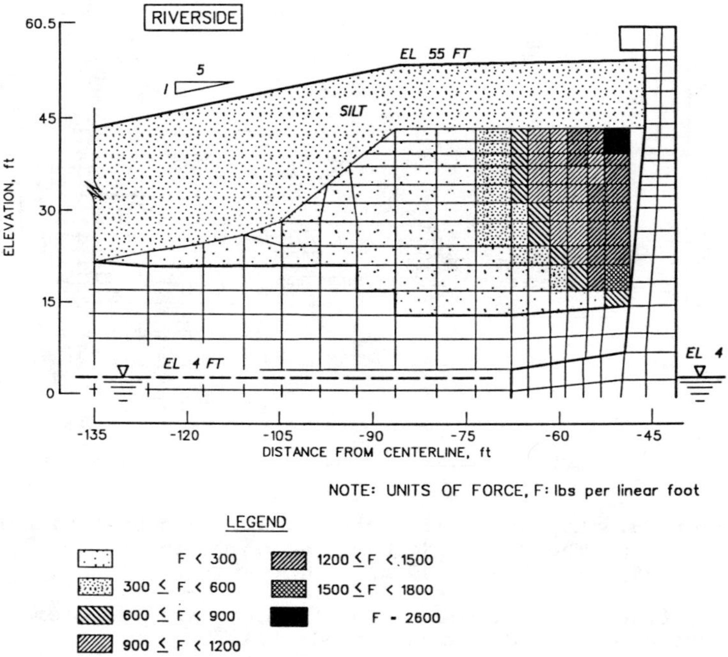

Figure 9. Variation In Reinforcement Force Within The Berm (1 lb/ft = 0.015 kN/m).

mesh to be independent of the number of layers of reinforcement, significantly reducing the required number of degrees of freedom, compared to the usual modeling procedures involving standard bar elements along solid element boundaries. A reduced number of degrees of freedom reduces the computational effort, of importance in nonlinear problems. Its use and versatility has been demonstrated in a complex large-scale finite element analysis used for the evaluation of a reinforced soil berm to be constructed adjacent to a navigation lock.

Acknowledgement

The development of the embedded element and its implementation in SOILSTRUCT was funded by the Vicksburg district of USACE. Significant technical contributions were provided during the course of the Red River Lock and Dam No. 1 evaluation by Kevin Abraham of WES and by C.C. Hamby and Ed Schilling of the Vicksburg District.

Permission was granted by the Chief of Engineers to publish this information.

Appendix I. References

Clough, G.W., and Duncan, J. M. (1969). "Finite Element Analyses of Port Allen and Old River Locks." Report No. TE-69-3 to US Army Engineer Waterways Experiment Station, Vicksburg, MS.

Doherty, W. P., Wilson, E. L., and Taylor, R. L. (1969). "Stress Analysis of Axisymetric Solids Utilizing Higher-Order Quadrilateral Finite Elements." Report No. SESM 69-3, Structural Engineering Laboratory, University of California, Berkeley, CA.

Duncan, J. M., Byrne, P., Wong, K. S., and Mabry, P. (1978). "Strength, Stress-Strain and Bulk Modulus Parameters for Finite Element Analyses of Stresses and Movements in Soil Masses." Report No. UCB/GT/78-02, College of Engineering, Office of Research Services, University of California, Berkeley, CA.

Duncan, J. M., and Chang, C. Y. (1970). "Nonlinear Analysis of Stress and Strain in Soils." J. SMFE, ASCE, 96(SM5).

Ebeling, R. M., Mosher, R. L., Peters, J. F., and Abraham, K. (1991). "Soil-Structure Interaction Study of Red River Lock and Dam No. 1 Subjected to Sediment Loading." Draft report submitted to US Army Engineer District, Vicksburg, Vicksburg, MS.

Ebeling, R. M., Peters, J. F., and Clough, G. W. (1990). Users Guide for the Incremental Construction, Soil-Structure Interaction Program SOILSTRUCT." Technical Report ITL-90-6, US Army Engineer Waterways Experiment Station, Vicksburg, MS.

Goodman, R. E., Taylor, R. L., and Breeke, T. L. (1968). "A Model for the Mechanics of Jointed Rock." J. SMFE, ASCE, 94(SM3).

Fowler, J., Leach, R. E., Peters, J. F., and Horz, R. C. (1986). "Mohincanville Reinforced Dike No. 2 Design Memorandum." Miscellaneous Paper GL-86-26, US Army Engineer Waterways Experiment Station, Vicksburg, MS.

Application of EPS for Slide Correction

Shan-Tai Yeh[1], M.ASCE and John B. Gilmore[2], M.ASCE

INTRODUCTION

Expanded Polystyrene (EPS) has been successfully used for roadbed insulation in cold regions for several years [2]. However, this material has never been used for correction of landslides in the United States although it is widely accepted in the Scandinavian countries [3]. The purpose of this paper is to illustrate the use of this material as an alternate to conventional slide repair methods, which usually involve major earth moving. Evaluation of post- construction performance is also included.

EPS is the lightest known material to be used in this type of application. The density can be as low as 1 pound per cubic foot (1 lb/ft^3) or 0.016 ton/m^3. Light weight materials used in this country have included sawdust, burned coal, and flyash which typically have densities between 30 and 60 lb/ft^3(0.48 and 0.96 ton/m^3). Because of EPS's much lower density, it has a considerable advantage over these materials.

During the spring of 1987, a section of U. S. Highway 160 between Mesa Verde National Park and the town of Durango in southwestern Colorado failed and caused the eastbound lane of this heavily traveled highway to close. The total length of the affected area was around 200 feet(61 m) and the slide covered an area of about one acre. The total amount of slide material involved was around 11,000 cubic yards (8415 cubic kiloliters).

[1]Senior Geotechnical Engineer, Colorado Department of Transportation, 4340 East Louisiana Avenue, Denver, Colorado 80222

[2]Chief Geologist, Colorado Department of Transportation, 4340 East Louisiana Avenue, Denver, Colorado 80222

GEOLOGY

U.S. 160 climbs to the highest point on Mancos Hill El. 8000 feet(2440 m) before turning north and decending. The slide area is on the east side of the hill, approximately 2000 feet(610 m) from the highest point. The highway in this area was constructed by cut and fill with the fills occupying small valleys which drained the hillside. Historically, landslides in this area have been embankment slope failures in the fills crossing these valleys. Seven slope failures have been identified in this area. Repair has generally been restoration of the slope by removal of slide debris and replacement with borrow material or sawdust, or support of the failed portion with cast-in-place or driven piles or, in a recent case, by a timber- lagged tieback wall.

Geological reconnaissance revealed that two older slides were present at the site of the problem. These two slides covered areas of approximately five and 15 acres respectively, and their age was unknown. It appeared that the two slides were fairly stable with no indication of any recent movement.

Subsurface investigation of the slide area revealed five (1.5 m) to 19(5.8 m) feet of highway embankment consisting of a mixture of glacial till and shaly clay obtained from the cut. This embankment, in turn, overlies 3(0.9 m) to 13(4.0 m) feet of natural clay which is underlain by Mancos shale, a marine deposit which often contain fossils of marine organisms. It is normally weak, easily eroded and slippery when wet. Standard penetration tests (SPT) indicated the upper undisturbed manmade and natural clay were moderately compacted and stiff, but were very soft and saturated in the slide mass. Water levels measured during the field investigation were two to three feet(0.6-0.9 m) below the surface at the slide area.

Geometry and geotechnical composition of the slide are shown in Figure 1.

STABILITY ANALYSIS

Since the water level was high, it was quite obvious that the slide was caused by rising of the water table in the embankment due to improper drainage, thus reducing the counterweight effect and subsequently causing instability of the embankment. It appears that most of the water in the embankment has resulted from the construction of the highway across the valley, which impeded both the flow of surface and subsurface water in the draw.

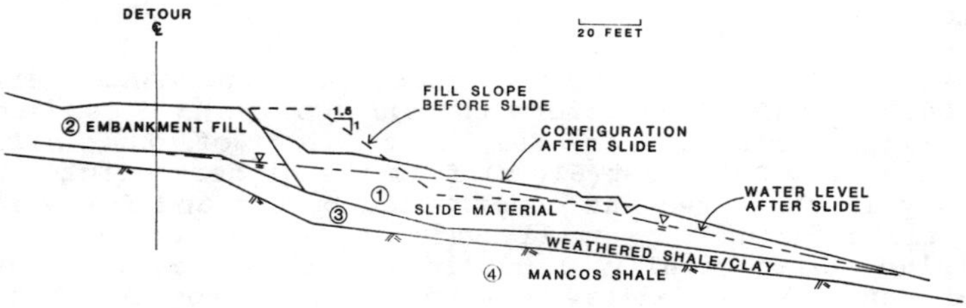

SOIL/ROCK PROPERTIES

SOIL/ROCK TYPE	TOTAL UNIT WEIGHT, PCF	MOISTURE %	L.L.	P.I.	C PSF	Ø
①	115	WET	-	-	50	15
②	115	16	35	14	500	20
③	120	14.1	36	15	1000	0
④	115	-	-	-	9000	0

FIG. 1 GEOMETRY AND CROSS SECTION THROUGH SLIDE AREA

The slip zone was found to be irregular and the Modified Janbu's method of stability analysis was used to back- calculate the soil strength. Laboratory tests were also performed on the embankment material and the upper shale to confirm the soil strength parameters. A Ø angle of 20 degrees and a cohesion of 500 psf(10,450 kN/m^2) were obtained in the laboratory for the upper embankment material. The near-surface weathered shale had a cohesion of 1000 psf(20,900 kN/m^2). A summary of the laboratory test results is also shown in Figure 1. Soil samples from the slip zone could not be obtained. Back-calculation from the slide geometry indicated a critical soil strength of 50 psf(1,045 kN/m^2) for cohesion and 15° friction angle at failure. This angle was assumed as the residual strength of the slide debris for the purpose of designing remedial measures.

DESIGN OF REMEDIAL MEASURES

Since rising of the water table in the embankment was the primary cause of the slide, it was recommended that the sources of water contributing to this rising be intercepted as the fundamental solution. An interceptor drain was installed on the uphill side along the highway ditch and the water collected in the drain was discharged well beyond the slide area.

Restoration of the embankment by completely removing the slide debris and the underlying overburden soils and founding the entire new embankment on shale was also considered. However, due to the presence of the two old slides, right-of-way limitations, and the amount of earth involved, it was decided that this option was infeasible. Two alternatives were considered practical for correction of the slide.

ALTERNATIVE 1

This involved the use of the rigid wall concept to retain the highway embankment. This alternative was considerably more expensive, but would be a permanent fix and would not be affected by future movement of the two older slides.

Based on the data collected in field investigation, it was estimated that the maximum wall height would be on the order of 35 feet(10.7 m). Approximately 7500 ft^2 (81,000 m^2) of wall would be required. Several rigid wall systems were considered, including reinforced concrete, fabric, drilled caissons and driven piles. The first two would have required excavation of the existing slide material and softer overburden to a more competent soil for foundation construction. As for the pile walls, the length of penetration required into shale would have been excessive. Considering the height of the wall and the similarity to another slide on the west side of Mancos Hill, a tieback wall system was recommended. The total cost of the project using this alternative would have been on the order of one million dollar.

ALTERNATIVE 2

Using the parameters obtained from the back-calculation for the site configuration before the slide, and the laboratory test results, the current site was found to have a marginal factor of safety against further failure. If the water table at the site could be lowered by three feet(0.9 m), the safety factor could be increased by 20%.

This assumption indicated that the slide area could also be stabilized by geometrical methods using a counterweight at the toe and lighter material at the head so that no further loading would be imposed on the drainage-improved site. The embankment could be restored to the approximate configuration shown on

Figure 2. The advantage of this alternative also included minimal site disturbance. The slide debris would not need to be removed.

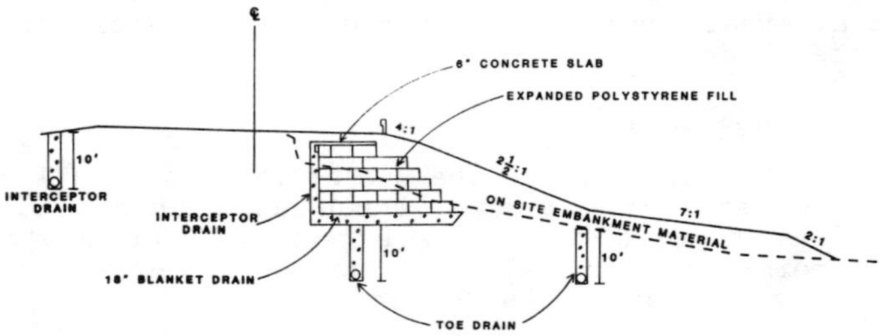

FIG. 2 DESIGNED EMBANKMENT AND DRAINS

The success of this alternative would depend heavily on the successful lowering of the water table in the slide. This could be achieved by installing two tiers of toe trench drains, and a second interceptor drain beside the first interceptor drain on the uphill side of the roadway as previously described. A cross sectional view of the recommended drains is shown on Figure 2. The estimated construction cost for this alternative was around $160,000.

Since the two older, larger slides appeared stable at the time, it was determined that Alternative 2 would be used. In order to keep the weight of the embankment to a minimum and avoid reactivation of the slides, it was also determined that extremely lightweight material would be required and expanded polystyrene (EPS) was selected. The designed bulk of the lightweight material is illustrated on Figure 2.

A tie system to prevent horizontal sliding induced by the active earth pressure from the backfill on the uphill side was considered impractical because of the need for road closure during tie installation. Special consideration was given to the design of the downhill slope soil backfill so that any active earth pressure could be resisted by it. A 1:1 backfill was constructible, but this steep a slope would not provide enough resistance to prevent horizontal sliding. A flatter downhill slope backfill could achieve the desired resistance but the thicker backfill would have caused the underlying slide material to become unstable. A compromise broken-slope design was adopted.

EPS AS GEOTECHNICAL MATERIAL

EPS is a closed cell, cellular plastic material normally used for packing or residential insulation. In the geotechnical field, it has been widely used for embankments on soft ground in the scandinavian countries [3]. The Norwegian Road Research Laboratory has been involved in more than 100 projects using EPS with great success [6]. It has been used for a variety of geotechnical purposes, including cold weather insulation beneath pavements, as embankment material to provide light loading, as wall backfill to reduce earth pressure [4], as a structural wall, and as a slide fix in conjunction with an interlayer tie system.

Strength of the EPS material depends upon the grade used and is normally equivalent to that of medium to stiff clay or silt. Load cycling tests show that it will resist unlimited cycles provided the load is kept below 80 % of the compressive strength. Slight increase of the compressive strength of the material was reported in some cases [4]. Creep studies by the manufacturers also indicated that deformations increase only very slightly under constant load if the applied load is within certain limits [6]. The material itself does not decay with time. The main problems with EPS are that it is so light that special precautions need to be taken for installation below the water table, and it is susceptible to attack by petroleum-based chemicals. Cost of the material depends on the density and shipping distance and normally ranges from \$30 to \$50 per cubic yard (\$0.04/m^3 to \$0.07/m^3).

CONSTRUCTION

The standard EPS block has dimensions of 4' x 8' x 2'(1.2m x 2.4m x0.6m). A total of 720 cubic yards(550 m^3) was designed although actual construction required 848 cubic yards(648 m^3). This translates into a total length of 108 feet(33 m) of highway alignment. Based on the design, a density of 1 pcf(0.016 ton/m^3) for the EPS would have been sufficient for pavement and traffic support. However, a density of 1.25 pcf(0.257 ton/m^3) material was chosen for added safety.

In order to protect against gasoline spills, as recommended by the Norwegian design guide [5], a 6-inch(15 cm) layer of 3000 psi(210 kg/cm^2) compressive strength concrete was designed on top of the lightweight material(Figure 2). The concrete was reinforced by 6-inch(15 cm) square 6/6 gauge welded wire fabric. This layer of concrete not only provides the required spill protection, it also helps in support of the traffic loading on top of the EPS fill. The guard rail foundation was placed directly on the second tier of EPS and was designed to be poured monolithically with the protector slab. An 18-inch sand and gravel layer was also provided between the concrete and the asphalt pavement.

To prevent possible floating of the EPS by groundwater, a blanket drain was provided and was connected with the designed toe drain. The location of the toe drains and blanket drains is illustrated in Figure 2. All filter drain material was wrapped in non-woven filter fabric. Ten-inch diameter perforated drain pipe was installed at the bottoms of the trench drains and six-inch diameter pipes spaced at 10 feet(3 m) each were installed in the blanket and interceptor drains.

The EPS blocks were staggered within each lift and blocks in alternating lifts were placed in a direction perpendicular to those above and below, as shown in Figure 3. The purpose of staggering the blocks was to prevent

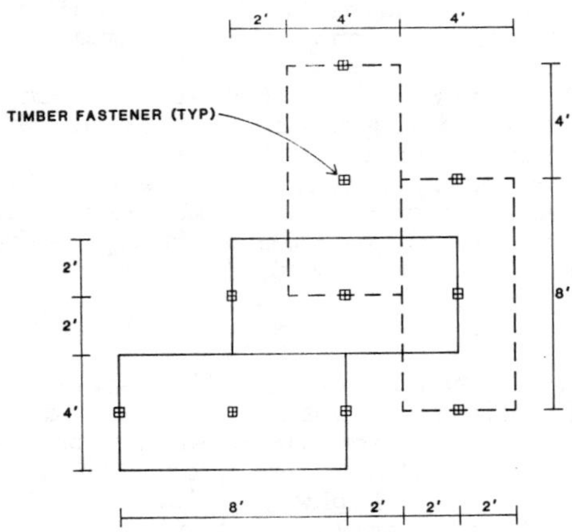

FIG. 3 BLOCK PLACEMENT AND TIMBER FASTENER DETAIL

formation of continuous joints in the same layer. To reduce the possibility of sliding between layers, double sided timber fasteners were used.

Construction of the project started in December of 1988 and the contract called for completion within two months. The project was intended to be built during the winter months so that groundwater would be at the lowest level and the ground frozen so that no further deterioration of the slope would occur.

Several problems arose during construction due to inclement winter weather conditions. Since the EPS is extremely light, containment of the blocks took some time when occasional storms passed by, and proper cover was needed each day after work to prevent the material from being blown away during the night. During the cutoff trench excavation, the temporary slope failed due to additional weight created by placement of barriers along the highway on top of the slide scarp. This additional sloughing caused the loss of the entire temporary downhill lane, posing an immediate hazard for traffic as a winter storm carrying a significant amount of snow moved into the region. A detour was made on the north side to provide two way traffic, and the detour was used for the remainder of the project.

Placement of the EPS took less than two weeks after the filter drain was placed (Figure 4). No construction difficulties were encountered other than those mentioned above.

PERFORMANCE

To monitor the performance of the EPS embankment slope, two vertical inclinometers and seven settlement monuments were designed. However, due to funding problems, only the settlement monitoring devices were installed. In order to provide at least some lateral movement data, a settlement plate with steel rod attachment extending above the ground and isolated by casing was designed.

The locations of these measurement devices are shown in Figure 5. Three of the monitoring devices were installed in the second tier of the EPS embankment, another three in the bottom tier and the last one 15 feet(4.6 m) downslope from the EPS material. Initial measurements of the devices were taken immediately after the completion of the site grading on January 19, 1989.

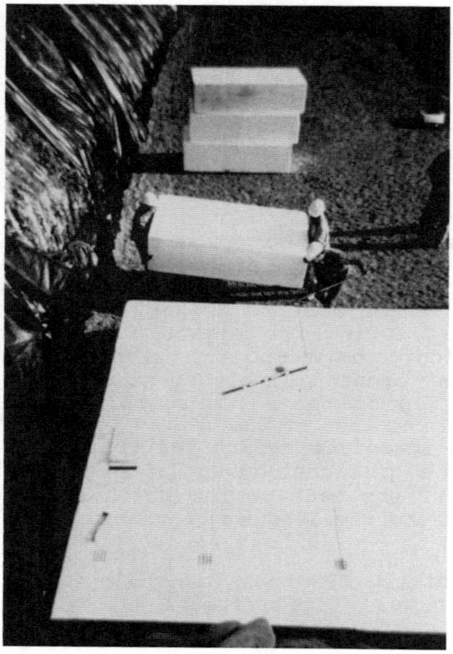

FIG. 4 EPS UNDER CONSTRUCTION

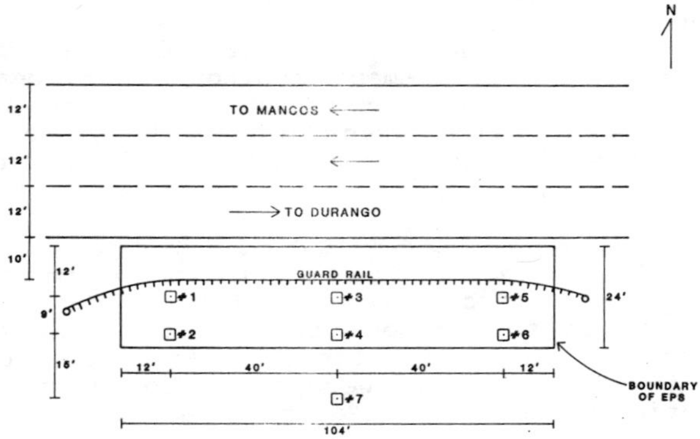

FIG. 5 LOCATION OF MONITORING DEVICES

The total vertical movements recorded at each location are shown in Figure 6 . As indicated, the EPS material compressed one to 1.5 inches(2.5 to 3.7 cm) under the initial loading from the overburden. The movement stopped a few months after project completion. As might be expected, the compression was greater at the top (Locations 1, 3, & 5) of the EPS fill than at the toe area (Locations 2, 4, &6).

The following year, 1990, was relatively dry. The record showed that little or no movement occured at the upper tier of the EPS during this period. Slight heaving, however, was recorded on the bottom tier and at Location 7, located 15 feet (4.6 m) south of the EPS. We believe that the heaving may be the result of relaxation or may merely be a surveying error.

Date	Time Months	SETTLEMENT (inches) #1	#2	#3	#4	#5	#6	#7
1-19-89	0.0	0.00	0.00	0.00	0.00	0.00	0.00	0.00
3-08-89	1.6	-1.32	-1.08	-1.08	-0.96	-0.84	-0.72	-0.24
3-16-89	1.9	-1.44	-1.32	-1.32	-0.96	-0.84	-0.96	-0.48
3-23-89	2.1	-1.56	-1.32	-1.32	-1.08	-0.96	-1.32	-0.48
3-30-89	2.3	-1.44	-1.32	-1.20	-1.08	-0.72	-1.08	-0.24
5-05-89	3.5	-1.44	-1.32	-1.44	-0.96	-0.96	-1.32	-0.60
4-04-90	14.5	-1.44	-0.48	-1.32	-0.36	-1.08	-0.72	0.12
7-15-91	30.0	-1.80	-1.44	-1.68	-0.96	-1.32	-0.96	-0.60

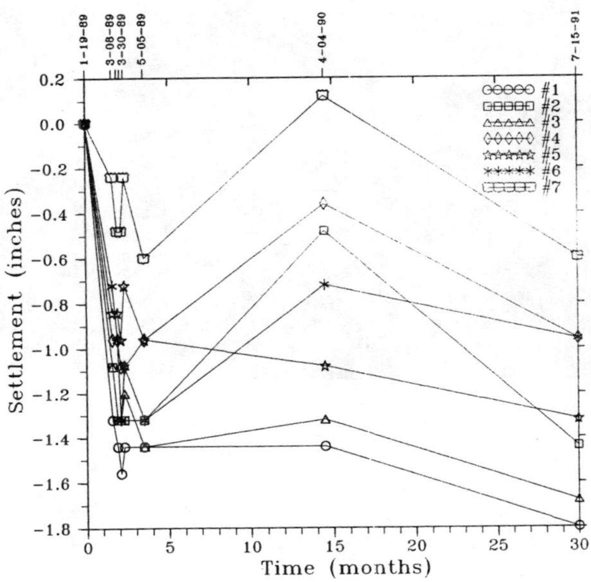

Fig. 6 – Results of the Vertical Movement

Much wetter weather was recorded this year but again the movement has been relatively insignificant. The largest movement was measured at Location 2 with less than one inch since the 1990 reading. Average downward movement of all stations was on the order of 0.5 inches(1.3 cm). The maximum movement between the 1989 and 1990 readings was less than 0.36 inches, and the average was 0.1 inches(0.2 cm). Field reconnaissance of the project site reveals no distress and the pavement is in excellent condition. The current site condition is shown on Figure 7.

Lateral movement was also recorded in all the measuring devices, but the direction of the movement was so irregular that no particular pattern could be determined.

FIG. 7 2 1/2 YEARS AFTER CONSTRUCTION

CONCLUSIONS

Based on the experience gained from the design and construction of this project, we believe much more application of EPS material can be taken advantage of. Scandinavian geotechnical engineers have used this material not only as embankment fill but also as wall backfill because of the material's low earth pressure characteristics [1]. Although the long-term behavior of the EPS, such as creep, was not well documented on this project, we believe a slightly conservative design can compensate for this deficiency. The Colorado Department of Transportation is currently conducting a research project to use this material as fill for bridge approaches. Millions of dollars have been spent in the Colorado highway system to remedy the bridge approach "bump" problem. We are hoping that EPS will bring us a permanent and economic solution.

ACKNOWLEDGEMENTS

The authors wish to extend special thanks to the engineers from CDOT's District 5 office including Keith Durphy, Dennis Young, and Larry Ruiz for the assistance in collecting the necessary construction and monitoring information. Special thanks are extended to Cheng-Kuang Su of the Materials Branch for the preparation of the Figures and giving valuable comments.

REFERENCES

1. Aaboe, R. (1985) "13 years of Experience with Expanded Polystyrene As a Lightweight Fill Material in Road Embankments", Norwegian Road Research Laboratory Report, June.

2. Ardani, A. (1987) "Frost Heave Control With Buried Insulation", Interim Research Report No. CDOH-DTD-R-89-3, Colorado Department of Highways, Denver, Colorado.

3. Flaate, K. (1987) "Superlight Material in Heavy Construction", Geotechnical News, September.

4. Frydenlund, T.E. and Aaboe, R. (1988) "Expanded Polystyrene-A Superlight Fill Material", International Geotechnical Symposium on Theory and Practice of Earth Reinforcement, Fukuoka, Japan. October.

5. Public Road Administration "Guidelines on The Use of Plastic Foam in Road Embankments", Road Research Laboratory, Norway.

6. BASF Corporation, Technical Bulletin E-3 " Strength Characteristics of EPS Thermal insulation", EPS expanded Polystyrene Insulation material.

TIMBER CRIB-FACED SOIL-NAILED RETAINING WALL

James G. Collin[1], M. ASCE, Mohammed A. Gabr[2], M. ASCE
and Alan G. MacKinnon[3], M. ASCE

Abstract

A composite retaining wall system using timber cribs as the facing material and helical anchors for soil reinforcement was used as an innovative application of soil nailing. The wall was constructed to provide access to covered boat slips for a marina expansion on the Chesapeake Bay. The wall height varied from approximately 6 to 23 feet with a 45 degree slope above the wall. The composite wall design considered global stability of the wall/slope system as well as internal and external stability of the wall.

Introduction

An existing marina located on the Chesapeake Bay was being expanded. The expansion involved the construction of a club house, swimming pool, new ships store and over 400 covered and open boat slips. In order to provide access to the covered boat slips, regrading and retaining of an existing slope was required. The design engineer for the project envisioned a conventional cantilever-type retaining wall, with a height of up to twenty feet. However, the Contractor was allowed to suggest alternative retaining systems that were cost-effective and adequate for the intended purpose.

Because regrading of the site required excavation and steepening of an existing slope, typical reinforced soil walls used in fill applications and conventional type retaining walls did not provide the most economical choice. Using these types of systems for this project would have required some temporary means of supporting the excavation during the construction of the new wall. Constructing the temporary excavation support system in addition to the permanent wall made using conventional approaches too costly.

The innovative solution to the problem included a composite retaining wall system that utilized helical anchors that were screwed into the excavated slope to provide soil reinforcement. Timber cribs, approximately two feet wide, were used as the facing material for the wall. This system is a modification to the soil nailing techniques that have been extensively used in Europe over the past twenty years.

----

[1]Vice President, Technology Development, Tensar Earth Technologies, Inc., 3000 Coporate Center Drive, Ste. 370, Morrow, GA 30260.
[2]Assistant Professor, West Virginia University, Morgantown, W.VA. 26506
[3]General Manager, Mid-Atlantic Permacrib, 10830 Guilford Rd., Annapolis Junction, MD 20701

Description of the Proposed Wall

The Permacrib Soil-Nailed retaining wall system developed for this project utilized helical anchors as the soil reinforcement. The anchors were designed for a spacing of four feet center to center, both vertically and horizontally. Each anchor, therefore, supported a tributary area of approximately sixteen feet. Two foot by two foot timber pads were used at each anchor location to add confinement to the excavated slope prior to placement of the permanent face. The permanent facing system consisted of timber cribs constructed from interlocking timber members. Each timber crib was backfilled with Number 3 stone. The reinforcement was extended from the excavation face to the inside of the timber crib and connected to the crib. A typical section of the crib-faced soil-nailed retaining wall system is shown in Figure 1. The timber crib was installed at a batter slope of 14 degrees from vertical and the reinforcement nails were installed at a slope of 14 degrees from horizontal. Thus, the reinforcement was perpendicular to the facing system, making the connection between the different components relatively straight forward (Figure 2).

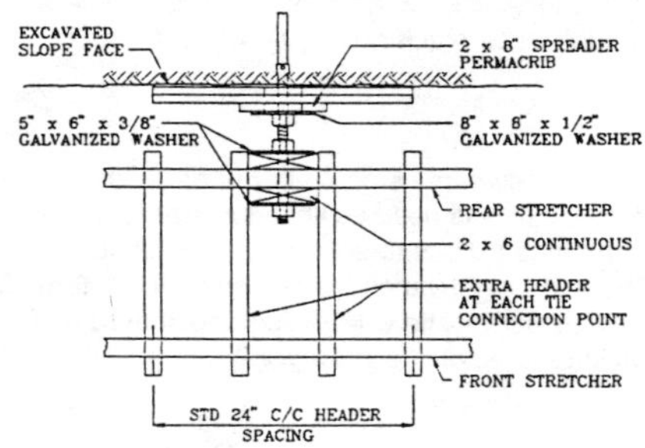

FIGURE 2
PLAN VIEW REINFORCEMENT
CRIB FACE CONNECTION

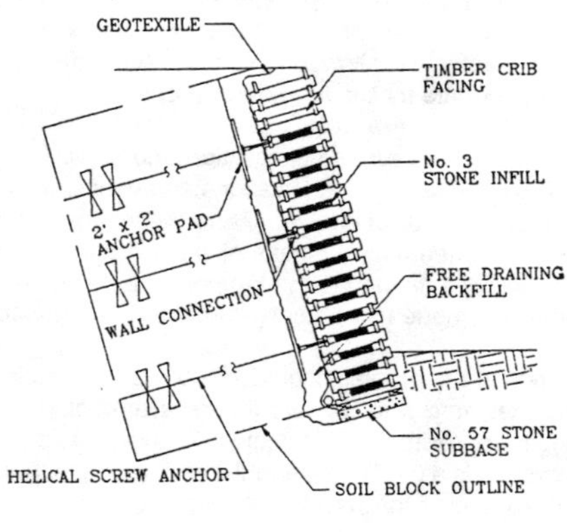

FIGURE 1
SOIL NAILED WALL SYSTEM

Soil Conditions

The site is located geologically in the Coastal Plain province. This province is generally characterized by sedimentary materials, mostly unlithified, that were transported from the uplands by river action and deposited in a succession of soil layers on the underlying metamorphic bedrock.

Six soil borings with Standard Penetration Tests (SPT), according to ASTM D-1586, were performed at the site for characterization of the in-situ soils. Data from the soil borings indicated that the site is underlain by brown to black medium sand to a depth of approximately 20 feet below the ground surface. The upper five feet of which is slightly cemented. The N values for the soil profile ranged from 5 blows per foot to 14 blows per foot. The soil was classified as SP-SM according to the Unified Soil Classification System (USCS).

For design purposes, the soil profile was idealized as having an effective friction angle of 30 degrees, zero cohesion and a total unit weight of 110 pounds per cubic foot. Although the upper 5 feet of the soil profile appeared to be cemented, no tests were conducted to investigate the nature of the cementation substance. Therefore, a zero cohesion value was used for long-term stability evaluations. No groundwater or hydrostatic conditions were detected at the site.

Analysis

There are currently several design procedures that are being used for the analysis of soil-nailed structures. All are based on limit equilibrium techniques in which the potential failure surfaces throughout the soil mass are examined. The Davis method proposed by Shen et al. (1981) and the method developed by Gassler and Gudehus (1981) consider only the tensile capacity of the reinforcement when assessing the stability of the reinforced soil mass, while the French method, developed by Schlosser (1983) also considers the effects of the shear capacity and bending stiffness of the nails on the overall stability of the in-situ reinforced soil mass (Mitchell et al., 1987).

The analysis procedures used for this project were based on the work of Shen and Gassler and Gudehus which ignore the effects of shear strength of the reinforcement in determining the factor of safety against failure of the in-situ reinforced soil mass. The modes of failure evaluated in the design process and established for this project included:

* Slope failure outside the reinforced soil mass (Global Stability)
* Slope failure inside the reinforced soil mass
* Rupture of the reinforcement (tensile failure)
* Pullout of the reinforcement

External Stability

The external stability of an in-situ reinforced soil wall depends on the ability of the reinforced soil mass to withstand the external loads, including the horizontal earth pressure from the soil being retained behind the wall and loads applied to the top of the wall, without failure by one of the following mechanisms:

* Sliding along the base of the wall or any plane above the base of the wall
* Overturning about the toe of the wall
* Bearing capacity failure of the foundation soil

* General slope stability failure

The reinforced soil mass was examined for sliding, overturning and foundation failure using conventional geotechnical engineering methods developed primarily for gravity-type retaining walls. A minimum factor of safety of 1.5 was used for sliding stability and a minimum factor of safety of 2.0 was used for overturning and bearing capacity failures.

The overall (global) stability of the wall support system was evaluated to investigate the stability of the soil-nailed retaining wall, the slope above the wall, and the existing soil behind the reinforced soil mass against a deep seated failure mechanism. This global stability analysis was performed with the aid of the computer program, PCSTABL (version 5m) developed by Purdue University. Two assumptions were made for this analysis. The soil-nails, designed for a load of 10,000 pounds, were modeled in the stability analysis as equivalent line loads at the face of the wall. The second assumption was that the critical failure surface will not shear through the reinforcements but rather be forced behind the reinforced soil mass.

Figure 3 shows the geometry of one of the wall sections analyzed at this site. This particular case was for a wall height of 15 feet and a soil slope above the wall of 45 degrees. In order to satisfy the state-of-practice requirement of an overall factor of safety of between 1.3 and 1.5, the nail lengths had to be increased to an overall length of 27 feet. General stability requirements therefore resulted in an increased length of reinforcement of approximately five feet over the required length for stability against sliding, overturning, or bearing capacity.

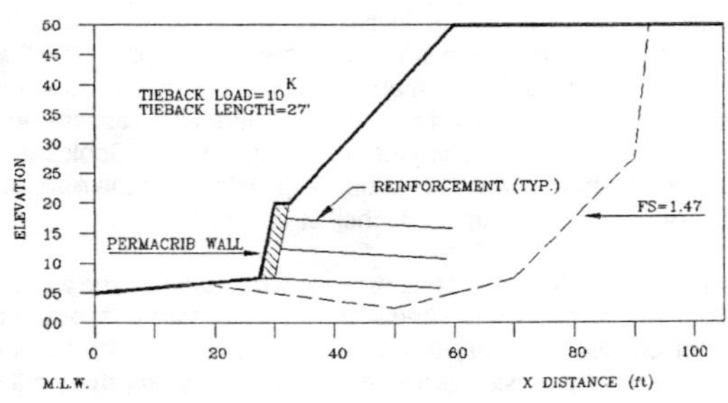

FIGURE 3
MINIMUM FACTOR OF SAFETY
GLOBAL STABILITY

Internal Stability

The reinforced soil mass was also analyzed to assure internal stability. This included an evaluation of the pullout capacity of the reinforcement as well as the ability of the reinforcement to withstand the tensile loads generated within this reinforced soil mass without rupture. The analytical procedures detailed in "Reinforcement of Earth Slopes and Embankments" (Mitchell and Villett, 1987) were used to assess the internal stability.

The connection between the reinforcement and the modular wood facing was designed for 100 percent of the design load in the reinforcement. Each connection was designed as a pin connection. The design life of the structure was assumed to be 50 years.

Construction Procedures

The wall was constructed in the fall of 1990. The following construction procedures were developed in order to minimize ground disturbance and movement both during and after construction.

The existing slope was excavated in four foot lifts. Each lift was nailed with helical screw anchors. The reinforcements did not require grouting to develop their pullout capacity. Therefore, once the anchor was installed, the positive effects of the reinforcement were realized. Two foot square timber anchor pads were installed over the ends of the anchors. Ten percent of the reinforcements were proof tested to assure development of the design load. After each lift was anchored and proof tested, the excavation proceeded. This top-down construction process was continued until subgrade was reached (Figure 4).

A stone base was installed as a leveling pad for the timber facing. The timber crib system was built in 4 to 5 foot lifts and connected to the reinforcement (Figure 5). Number 3 stone was placed inside the crib prior to backfilling between the soil-nailed excavated slope and the timber face. This process was repeated until the full height of the timber facing system was achieved.

FIGURE 4
WALL CONSTRUCTION

FIGURE 5
REINFORCEMENT WALL FACE CONNECTION

Conclusions

The Permacrib timber crib-faced soil-nailed retaining wall system provided an innovative and economical solution to a difficult grading problem that required both a temporary and permanent earth retention system. By combining soil nailing with an aesthetically pleasing timber-crib facing system, the designers were able to provide the owner of the marina with a retaining wall system that satisfied stability requirements and provided a cost savings of over $70,000 as compared to a conventional retaining wall system.

The wall was inspected in September of 1991, one year after construction. Both the appearance and performance of the wall at that time were considered excellent. This retaining wall system appears to provide an economical alternative to conventional walls in cut applications.

References

Gassler, G. and Gudehus, G. (1981). "Soil-Nailing - Some Soil Mechanical Aspects of In-Situ Reinforced Earth", Proceedings of 10th International Conference on Soil Mechanics and Foundation Engineering, Vol. 3, Session 12, Stockholm.

Mitchell, J.K. and Villet, W.C.B. (1987). "Reinforcement of Earth Slopes and Embankments", National Cooperative Highway Research Program Report 290.

Schlosser, F. (1983). "Analogies et differences dans Le Comportement et le Calcul des Ouvrages de Soutenement en Terre Armee et par Clouge du Sol", Annales de L'Institut Technique du Batiment et des Travaux Publics No. 418.

Shen, C.K., Herman, L.R., Romstad, K.M., Bang, S., Kim, Y.S. and Denatale, J.S. (1981). "An In-Situ Earth Reinforced Lateral Support System", Report #81-03, University of California, Davis, California, March.

DESIGN OF ANCHORED GEOSYNTHETIC SYSTEMS FOR SLOPE STABILIZATION

Roman D. Hryciw, A.M. ASCE[1] and Kamarudin Haji-Ahmad[2]

ABSTRACT

Design procedures for slope stabilization by anchored geosynthetic systems are presented. Charts are developed to assist in selection of optimum combinations of anchor length, spacing and orientation for a desired increase in stability. Final selection of the anchor parameters is also guided by the fabric strength, stress distribution, erosion susceptibility, stratigraphy and installation details. Two-dimensional infinite slopes, cohesionless soils and shallow failure surfaces are assumed and example problems demonstrate the design methodology.

INTRODUCTION

Hazardous slopes may be stabilized against shallow failure by anchored geosynthetics (Koerner and Robbins, 1986) or anchored nets (Koerner, 1984). In an anchored geosynthetic system, the geosynthetic is draped over the face of the slope and steel rods are driven into the soil mass through reinforced openings in the fabric. Prior to driving the last 1 to 2 ft., the geosynthetic is connected to the anchor. The final driving then tensions the fabric and imparts a curvature to the soil-fabric interface. The developed tension and fabric curvature combine to compress the soil and, if properly installed, to favorably increase the stresses on potential failure surfaces. Anchor pullout resistance is developed through friction between the anchors and soil.

The present paper analyzes the optimum orientation, length and spacing requirements for the anchors. Since

[1]Asst. Prof., Dept. of Civ. and Envir. Engrg., Univ. of Michigan, Ann Arbor, MI 48109-2125.
[2]Grad. Stdnt., Dept. of Civ. and Envir. Engrg., Univ. of Michigan, Ann Arbor, MI 48109-2125.

anchored geosynthetics are most attractive for stabilization of shallow failure surfaces to a depth of approximately 10 ft., the analyses assumed infinite slope conditions. Soils were assumed to be cohesionless.

Hryciw (1991) previously presented solutions based on two assumptions that will be improved upon herein. First, it was previously assumed that the increase in load applied to the failure surface was the same as that transferred from the fabric to the soil. This simplifying assumption neglected any reduction in load with depth due to load take-out along the length of the anchor. In the present work, it will be assumed that the increase in load on a failure surface will be limited to the anchorage loads developed beneath the failure surface. While it is reasonable to assume that the actual load applied to the failure surface will be intermediate between the former and present assumptions, the present analysis will provide a more conservative solution.

Previous analyses also assumed that optimum anchor orientations were those that produced a desired increase in factor of safety at minimum anchor length to anchor spacing ratios. The present analysis will consider additional factors including fabric strength, stress distribution, installation details and other practical considerations.

Two other assumptions are made. First, the bending stiffness of the anchors is conservatively neglected and the load resultant is assumed to act parallel to the axis of the anchor. Secondly, although the load transfer from fabric to soil is typically concentrated within approximately 1 ft. of the anchors (Vitton, 1991), due to the spreading out of stresses with depth and the coalescence of pressure bulbs from adjacent anchor points, the applied surface loads are assumed to be uniformly distributed on potential failure surfaces.

THEORETICAL CONSIDERATIONS

The factor of safety against sliding for infinite slopes is given by the ratio of resisting to driving forces on a potential sliding surface as shown in Figure 1a. For cohesionless, dry soils on infinite slopes the expression takes the form:

$$F_o = \frac{W \cos\beta \tan\phi}{W \sin\beta} \quad \text{.....(1)}$$

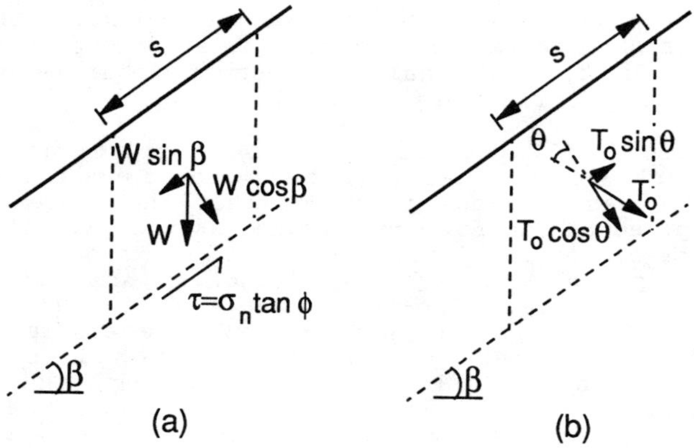

Figure 1. Loads on a Soil Slice in an Infinite Slope: (a) Soil Weight, (b) Load Applied From Anchored Fabric

where the subscript "o" indicates original conditions prior to stabilization, W=the weight of an s x s slice shown in Figure 1a, β=slope angle and ϕ=soil friction angle. If an external load of magnitude T_o is applied to each slice at an angle θ to the slope normal as shown in Figure 1b, (1) may be modified to:

$$F = \frac{(W\cos\beta + T_o\cos\theta)\tan\phi}{W\sin\beta - T_o\sin\theta} \quad \text{.............................(2)}$$

If the dimensionless ratio of T_o to W is defined as:

$$\xi = \frac{T_o}{W} \quad \text{.......................................(3)}$$

then (2) may also be written:

$$F = \frac{(\cos\beta + \xi\cos\theta)\tan\phi}{\sin\beta - \xi\sin\theta} \quad \text{.............................(4)}$$

The increase in slope stability by anchored geosynthetics may be conveniently quantified by a factor of safety ratio (FSR) defined as:

$$FSR = \frac{F}{F_o} \quad \text{.....................................(5)}$$

Replacing the expressions for F_o and F given by (1) and (4) in (5) yields:

$$FSR = \frac{\sin\beta + \xi\cos\theta\tan\beta}{\sin\beta - \xi\sin\theta} \qquad (6)$$

The maximum load that could be transferred from the fabric to the soil is limited by the pullout resistance of the anchor which is:

$$T_{max} = \int_0^L 2\pi r \tau_{max} dz \qquad (7)$$

where r=anchor radius, z=distance along the anchor with origin at the ground surface, L=anchor length and

$$\tau_{max} = \sigma_n \tan\delta \qquad (8)$$

where δ=the soil-anchor interface friction angle. Hryciw (1991) approximated an expression for σ_n, the normal stress on the anchor in infinite slopes. It is given by:

$$\sigma_n = K_n \gamma z \qquad (9)$$

where K_n = a stress factor given by:

$$K_n = \frac{K_X + 2K_Y + 1}{4}\cos(\theta+\beta) + \frac{K_X - 1}{4}\cos(\theta+\beta)\cos 2(\theta+\beta) +$$

$$\frac{1+2\nu}{2}\sin(\theta+\beta)\tan\beta + \frac{\theta+\beta}{\pi}\cos(\theta+\beta)\sin 2(\theta+\beta)\tan\beta \qquad (10)$$

where γ= unit weight of soil; ν=Poisson's ratio and K_X and K_Y = coefficients of lateral earth pressure in directions facing and parallel to the slope face. Hryciw (1991) conservatively suggested using K_X and K_Y = K_o = 1-sinϕ in the absence of measured K_X and K_Y.

As stated earlier, it was previously assumed that T_o=T_{max}. While this assumption simplified the problem for computational purposes, it unconservatively assumed that the loads applied at the surface would be fully transferred to the potential failure surface. A more conservative formulation assumes:

$$T_o = \int_{z_i}^L 2\pi r \, \tau_{max} dz \qquad (11)$$

where z_i= anchor distance at which the failure surface is intersected as shown in Figure 2.

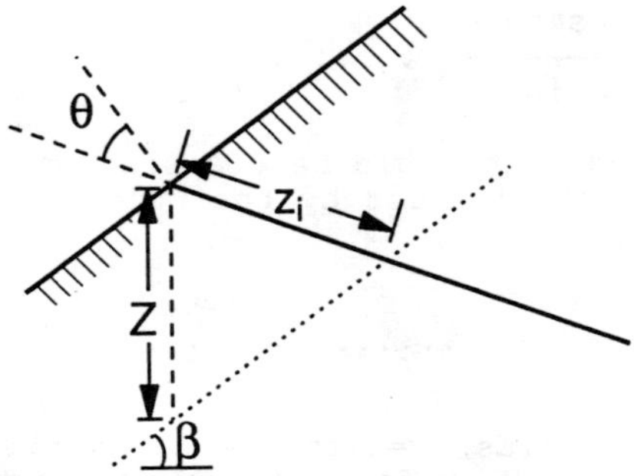

Figure 2. Slope and Anchor Geometry

Combining (8) and (9) with (11) and performing the integration yields:

$$T_o = \pi r K_n \gamma (L^2 - z_i^2) \tan\delta \qquad (12)$$

The weight of the slice shown in Figure 1 is:

$$W = Z\gamma s^2 \cos\beta \qquad (13)$$

where Z=vertical distance to the failure surface as shown in Figure 2. Replacing (12) and (13) in (3) produces:

$$\xi = \frac{T_o}{W} = \frac{\pi r K_n (L^2 - z_i^2) \tan\delta}{Z s^2 \cos\beta} \qquad (14)$$

It should be noted that ξ is inversely proportional to Z. Therefor, according to (6), *FSR* will decrease with Z and will asymptotically approach 1.0. This observation is particularly relevant to selection of a critical Z for design, as will be subsequently discussed.

The terms z_i and Z are related through:

$$z_i = Z \frac{\cos\beta}{\cos\theta} \qquad (15)$$

Equations (14) and (15) may be introduced into (6) which after some rearrangement and simplification becomes:

$$FSR = \frac{\sin\beta\cos^2\beta\cos^2\theta - C_1 K_n \cos^2\beta\sin\beta\cos\theta + C_2 K_n \sin\beta\cos^3\theta}{\sin\beta\cos^2\beta\cos^2\theta + C_1 K_n \cos^3\beta\sin\theta - C_2 K_n \cos\beta\cos^2\theta\sin\theta} \qquad (16)$$

where:

$$C_1 = \frac{Z\pi r\tan\delta}{s^2} \qquad (17)$$

and

$$C_2 = \frac{L^2}{Z^2} C_1 \qquad (18)$$

Equation (16) shows that for a given slope, the *FSR* is a function of only C_1K_n; C_2K_n and θ. Although the number of combinations of these three parameters that produce a desired *FSR* is infinite, for any given C_1K_n and C_2K_n pair there is a unique optimum angle (θ_{opt}) which yields the maximum factor of safety ratio, FSR_{max}.

Since (16) cannot be rearranged explicitly for θ, determination of θ_{opt} is difficult. Further complexity arises from K's dependency on θ. Unfortunately, differentiating (16) with respect to θ and setting it equal to zero does not help since the resulting equation also cannot be solved explicitly for θ_{opt}, and the differentiated expression is much lengthier. To avoid these problems in design, charts (Figures 3 and 4) have been developed to assist the user in optimally designing an anchored geosynthetic system. The figures show, for slope angles of 30°, 35°, 40° and 45° the optimum anchor angles and the corresponding FSR_{max} as functions of C_1K_n and C_2K_n.

ADDITIONAL FACTORS IN DESIGN

While equations (16) through (18) provide a theoretical framework for design, additional factors must be considered in selection of the final anchor spacing, length and orientation. The additional factors include fabric strength, stress increase distribution, erosion control, stratification and installation details.

Geosynthetic Strength

Koerner (1986) has suggested that a fabric strength of 200 lb/in. should be used but that the geosynthetic should also be reinforced around the anchor points to a strength of 500 lb/in. Noting that considerable stress relaxation and creep develop in fabrics at high stresses, Vitton (1991) recommended that the geosynthetics should only be loaded to approximately 60 to 66% of its tensile strength, or 300 to 330 lb/in. Assuming that the fabric is connected to a 2" diameter connector as suggested by Vitton (1991), the maximum T_o that can be applied to the slope becomes approximately 2000 lbs.

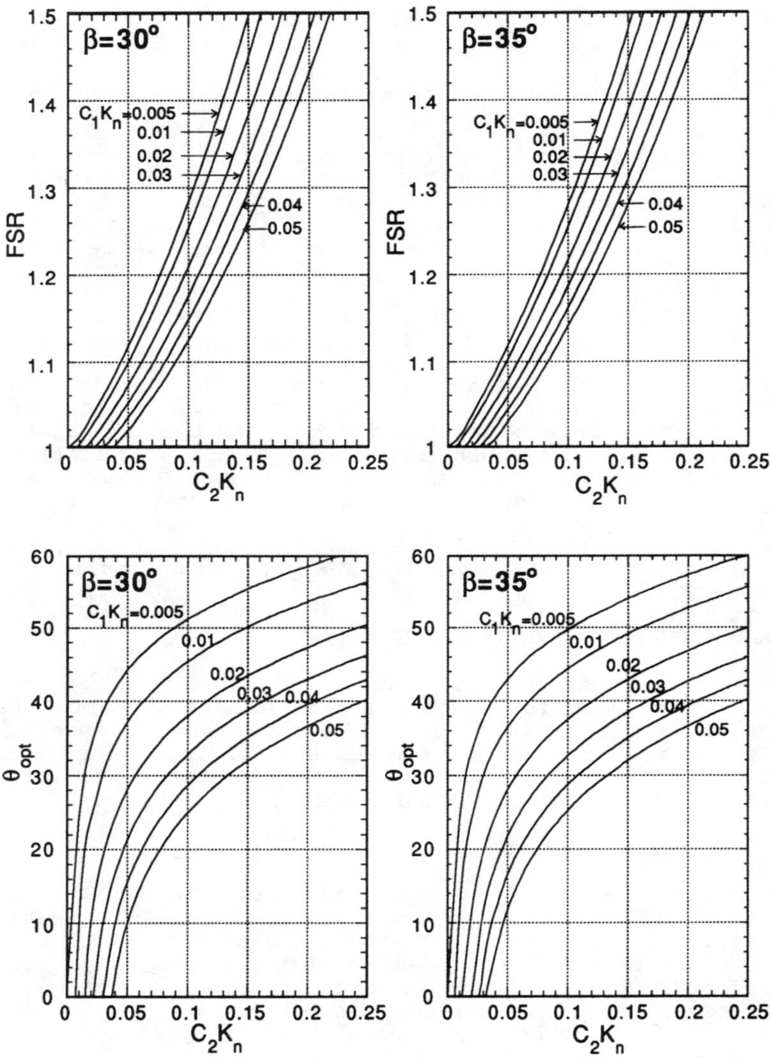

Figure 3. Charts for FSR_{max} and Corresponding θ_{opt} for 30° and 35° Slopes.

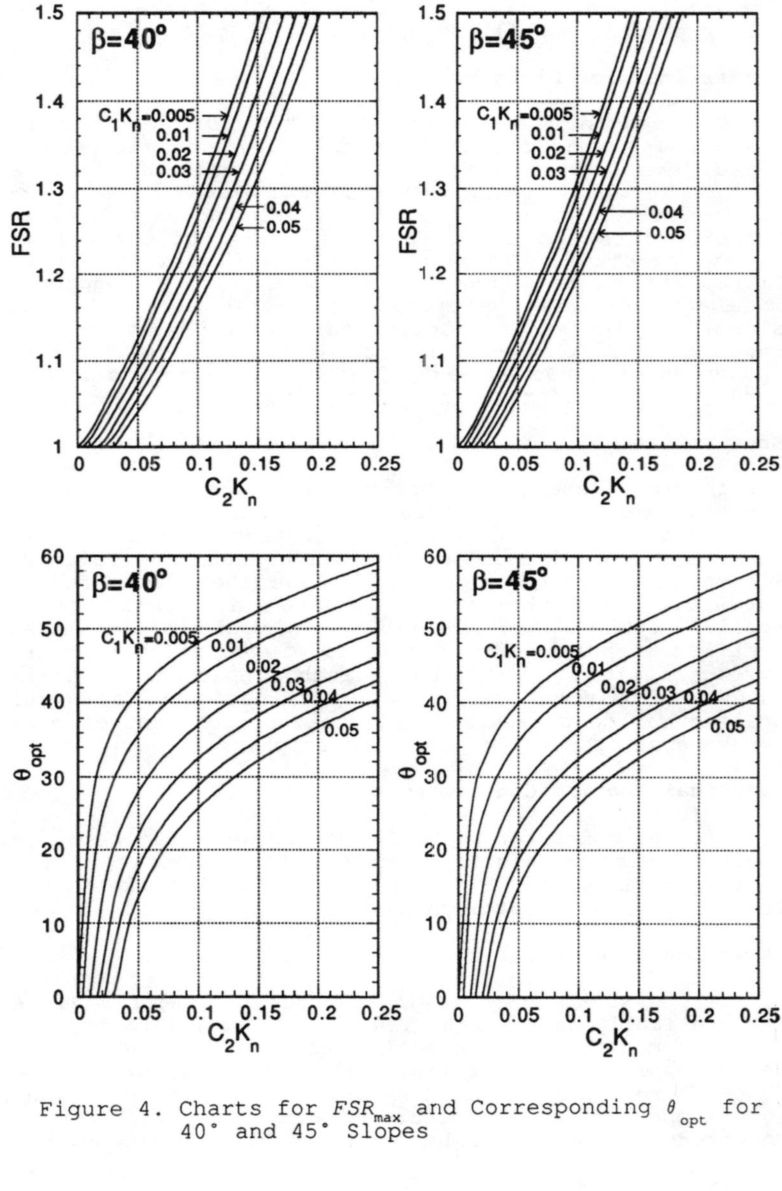

Figure 4. Charts for FSR_{max} and Corresponding θ_{opt} for 40° and 45° Slopes

The T_o required to achieve a desired *FSR* must increase if the anchor spacing is to increase (in order to maintain the same average applied pressure). Thus, the limit on T_o will essentially also be a limit on the anchor spacing as will be subsequently shown.

Stress Increase Distribution

The spacing cannot be too large for yet another reason. In order for a slope to be effectively stabilized, the increases in stress on a potential failure surface should be relatively uniform. Too large a spacing could result in lapses in stress increase between anchor points and local failures could still occur at these locations. However, based on elasticity solutions, it can be shown that below a depth of approximately 0.75 times the spacing (measured normal to the slope) the increase in stress varies by less than 10%. Based on this observation, it is recommended that the anchor spacing not exceed $1.3(Z\cos\beta)$ in order to insure a relatively uniform stress distribution.

Erosion Control

In many cases, a clear failure surface at depth cannot be identified, but rather, failure may be characterized by progressive surface sloughing or erosion. In such cases, a smaller anchor spacing should be favored. Vitton (1991) and Hryciw (1990) have shown that fabric curvature, and therefore increase in normal stress applied to the slope develops to a distance of less than 1 ft. from the anchor. Therefore, in cases where near-surface failure is a concern, anchor spacings as small as 2 ft. could be prescribed. Alternatively, a larger spacing may be prescribed if other erosion control measures, such as seeding, are also adopted.

Practical Spacing Considerations

Geosynthetics for anchoring would typically be manufactured with reinforced openings for anchors at even increments i.e. s=4, 5 or 6 ft. Therefore, unlike anchor lengths, variable spacings cannot be prescribed.

Selection of Critical *Z*

As previously discussed, *FSR* decreases with Z while asymptotically approaching 1.0. Therefore, in a hypothetical homogeneous slope with a uniform F_o, F will also decrease with Z and failure will still occur at the depth where $(FSR)F_o<1.0$. Fortunately, in most cohesionless slopes F_o tends to increase with Z as evidenced by the predominance of shallow failures in such slopes.

Selection of the design Z must be based on field investigations. Inferred friction angles may be used to compute F_o with depth. The desired F, which should be based on an acceptable level of risk, is then used to compute the required *FSR* with depth. At this point, there is no precise way to determine the critical Z due to the interdependence of many factors. Therefore, a trial Z is selected with preference given to higher required *FSR*s at depth. Once design of the anchored geosynthetic system for the assumed critical Z is performed, the developed *FSR*s are computed and compared to the required *FSR*s. Failure to meet or exceed the required *FSR* at any depth redefines the critical Z and the design procedure is repeated for the newly defined critical Z.

Anchor Angle Limitations

In most cases the anchors in anchored geosynthetic systems will be installed using hand held vibratory percussive tools. The anchors cannot and should not be installed with inclinations above the horizontal for two reasons. First, handling of the installation equipment would be difficult, but more importantly, the installation of anchors by vibratory means against gravity could cause local soil ravelling and slope destabilization.

Length Limitations

While there is no manufacturing constraint on anchor lengths, geologic constraints could limit the depth to which anchors could be driven. If loose materials are underlain by denser sediments or weak rock, penetration by vibratory driving may meet refusal. In such cases, shorter anchor lengths in conjunction with smaller spacing could be used.

Summary of Additional Factor Requirements

The foregoing discussion of additional factors leads to the following supplemental recommendations for anchor design:

$$T_o \leq 2000 \text{ lbs.} \qquad (19)$$

$$s \leq 1.3(Z\cos\beta) \qquad (20)$$

and

$$\theta \leq 90°-\beta \qquad (21)$$

In addition, *FSR* (developed) $\geq$ *FSR* (required) for all Z must be verified.

EXAMPLE PROBLEMS

Several example problems will illustrate how Figures 3 and 4 and equations (19) through (21) should be used. In Table 1 a summary is given of the available data for three different slopes to be stabilized by anchored geosynthetic systems. In all cases a fabric with a tensile strength of 500 lb/in at the anchor connection will be used. It will also be assumed that the given *Z* is a verified critical depth.

Table 1. Parameters for example problems

	Example problem		
	#1	#2	#3
β	30	35	40
Z (ft)	7	9	4
ϕ (deg)	31	33	40
γ (pcf)	100	105	110
K	0.49	0.46	0.36
ν	0.25	0.25	0.25
δ (deg)	35	35	40
r (in.)	0.50	0.50	0.25
F_o	1.04	0.93	1.00
F	1.30	1.25	1.20
FSR	1.25	1.35	1.20

Notice should be taken that different factors of safety reflect different acceptable risk levels. Also, it should be noted that the interface friction angle between soil and anchors may be higher than the soils angle of internal friction. This will occur if passive soil pressures develop against the ribs on reinforcing rods as shown by Irsyam and Hryciw (1991).

Having determined the required *FSR*s, acceptable combinations of C_1K_n and C_2K_n and the corresponding θ_{opt}s for each of the slopes may be obtained from Figures 3 and 4. These combinations are provided in Tables 2, 3 and 4. The remaining table entries are obtained as follows. K_n is computed for each θ_{opt} by (10). Values of C_1 and C_2 are computed by dividing C_1K_n and C_2K_n by K_n. The corresponding anchor spacings, *s*, and lengths, *L*, are computed by (17) and (18) respectively. The anchor distance to the failure surface, z_i, and the required anchor loads, T_o, are found from (15) and (12).

Table 2. Worksheet for Example Number 1.

C_1K_n	C_2K_n	θ_{opt}	K_n	C_1	C_2	s (ft)	L (ft)	z_i (ft)	T_o (lbs)
0.005	0.0900	50.0	0.568	0.009	0.158	8.6	29.7	9.5	4142
0.010	0.0990	45.0	0.637	0.016	0.156	6.4	22.0	8.6	2409
0.015	0.1030	42.5	0.669	0.022	0.154	5.4	18.3	8.2	1654
0.020	0.1120	39.5	0.706	0.028	0.159	4.8	16.6	7.9	1381
0.030	0.1230	36.0	0.745	0.040	0.165	4.0	14.2	7.5	992
0.040	0.1340	33.0	0.774	0.052	0.173	3.5	12.8	7.2	797
0.050	0.1470	31.5	0.787	0.064	0.187	3.2	12.0	7.1	677

Table 3. Worksheet for Example Number 2.

C_1K_n	C_2K_n	θ_{opt}	K_n	C_1	C_2	s (ft)	L (ft)	z_i (ft)	T_o (lbs)
0.005	0.1190	52.0	0.563	0.009	0.211	9.7	43.9	12.0	9702
0.010	0.1270	47.0	0.632	0.016	0.201	7.2	32.1	10.8	5563
0.015	0.1320	44.0	0.674	0.022	0.196	6.1	26.7	10.3	3954
0.020	0.1400	42.0	0.701	0.029	0.200	5.4	23.8	9.9	3173
0.030	0.1510	39.5	0.735	0.041	0.205	4.5	20.2	9.6	2246
0.040	0.1620	36.0	0.780	0.051	0.208	4.0	18.1	9.1	1845
0.050	0.1720	34.0	0.804	0.062	0.214	3.6	16.7	8.9	1549

Table 4. Worksheet for Example Number 3.

C_1K_n	C_2K_n	θ_{opt}	K_n	C_1	C_2	s (ft)	L (ft)	z_i (ft)	T_o (lbs)
0.005	0.0720	45.0	0.691	0.007	0.104	5.5	15.2	4.3	887
0.010	0.0800	41.0	0.745	0.013	0.107	4.1	11.3	4.1	503
0.015	0.0830	38.0	0.785	0.019	0.106	3.4	9.4	3.9	349
0.020	0.0890	35.5	0.817	0.024	0.109	3.0	8.4	3.8	283
0.030	0.0980	32.0	0.860	0.035	0.114	2.5	7.2	3.6	205
0.040	0.1060	30.0	0.883	0.045	0.120	2.2	6.5	3.5	160
0.050	0.1120	27.5	0.910	0.055	0.123	2.0	6.0	3.5	132

Tables 2, 3 and 4 reveal that any number of possible combinations of s, L and T_o, when installed at their respective θ_{opt} will yield the desired *FSR*. Figures 5 and 6 illustrate the acceptable combinations for the three example problems.

While any point on the theoretical curves would satisfy the desired *FSR* requirement, the additional factors given by (19) through (21) must still be considered. For slope #1, the fabric strength requirement (19) will limit the spacing of anchors to approximately 6 ft.

as shown in Figure 5. The corresponding required length is 20 ft. (Figure 6) and θ_{opt} = 43° to 44° (Table 2). The spacing and orientation criteria given by (20) and (21) are met.

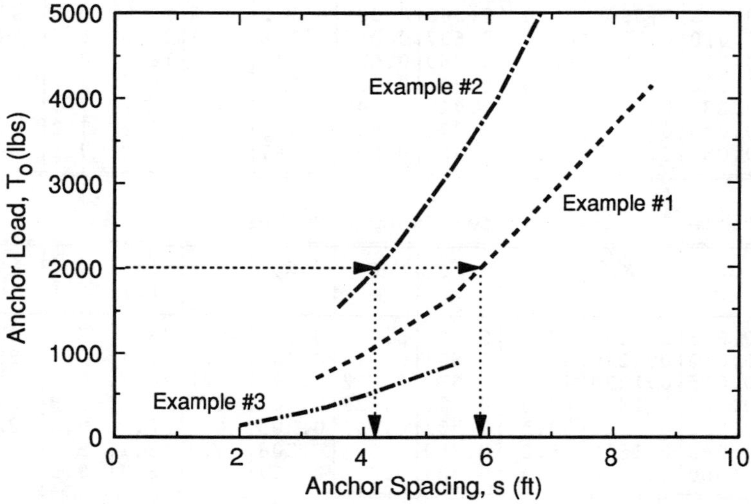

Figure 5. Optimum Combinations of Anchor Load and Spacing for the Example Problems

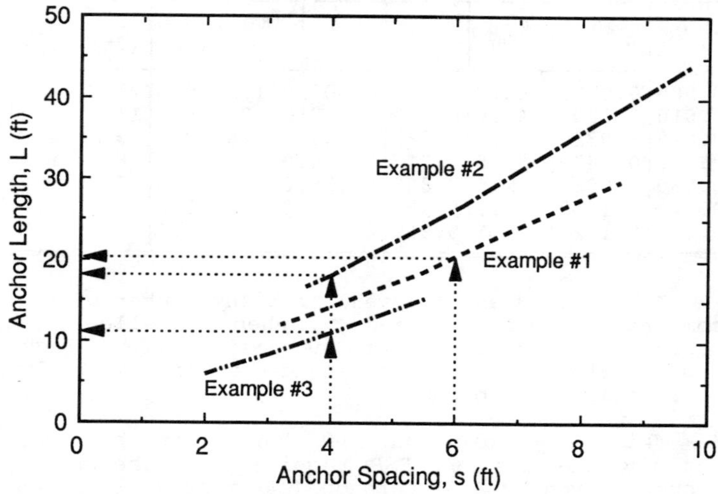

Figure 6. Optimum Combinations of Anchor Length and Spacing for the Example Problems

For slope #2, fabric strength again controls design. Figure 5 shows that an anchor spacing of 4 ft. can be used and Figure 6 shows that a length of 18 ft. will be required. The anchor orientation should be 36° (Table 3). Equations (20) and (21) are easily met.

Finally, for slope #3, the fabric strength requirement (19) is easily met. However, the spacing is limited by (20) to approximately 4 ft. The required anchor length will be 11 ft. (Figure 6) and θ_{opt} = 40° (Table 4). Equation (21) is met. For this slope, we see from Figure 5 that an anchor load of only about 500 lbs will be required for the 4 ft. spacing. Therefore, a lighter fabric could be used for this slope.

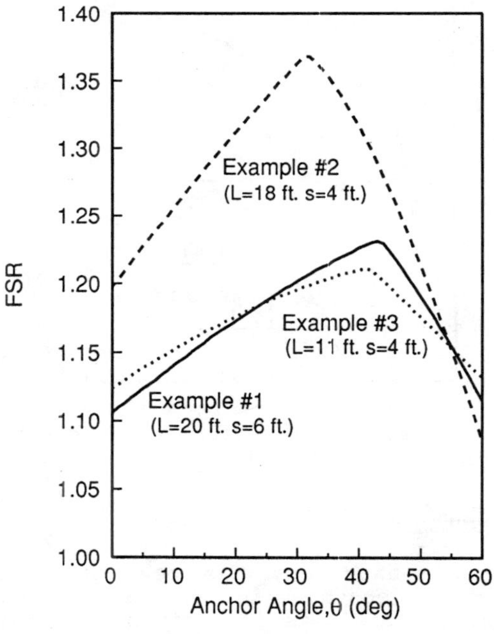

Figure 7. Factor of Safety Ratio Versus Anchor Angle for the Example Problems

To illustrate the importance of anchor orientation to design, Figure 7 shows the relationship between θ and *FSR* for the three example problems. The curves were developed holding the design anchor spacings and lengths constant. T_o was computed by (12) for each orientation but limited to 2000 lbs. as required by (19). Clearly, the orientation of anchors, in addition to the spacing and length is an important design parameter.

Finally, Figure 8 reveals the developed *FSR*'s with depth for the three problems. These curves would be compared to the required *FSR*'s to insure that the critical *Z* was correctly chosen.

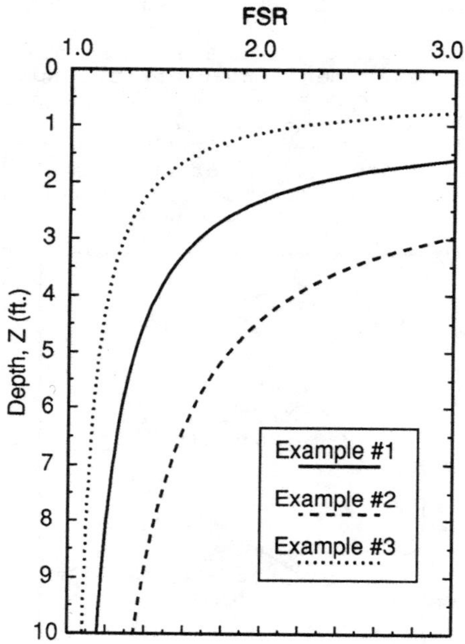

Figure 8. Developed *FSR* versus depth for the Example Problems.

SUMMARY AND CONCLUSIONS

An improved methodology has been presented for designing anchored geosynthetic systems for stabilization of slopes. The analysis is limited to cohesionless soils and shallow failures. An infinite slope geometry was assumed and a limiting equilibrium analysis was adopted. The primary stabilizing mechanism is the increase in stress on potential failure surfaces. A procedure was developed for establishing the most effective combinations of anchor spacing, length and orientation to achieve a desired increase in factor of safety for a given slope. However, practical considerations, including fabric strength, stress distribution, erosion control and installation details dictate the final design.

ACKNOWLEDGEMENTS

This study was partially funded by the Air Force Office of Scientific Research under Grant No. AFOSR-88-0166.

APPENDIX I. —— CONVERSION TO SI UNITS

To convert:	To:	Multiply by:
in.	cm	2.54
ft	m	0.305
lb	N	4.448
lb/in.^{3}	N/m^3	175.6
lb/ft^3	N/m^3	157.0

APPENDIX II. —— NOTATION

F = factor of safety
F_o = original factor of safety
FSR = F/F_o
K = coefficient of lateral earth pressure
K_n = coefficient of normal stress on the anchor
L = anchor length
L_{req} = required anchor length
r = anchor radius
s = anchor spacing
T_o = applied surface load per slice
X, Y, Z = cartesian coordinate system (Z=down)
x, y, z = cartesian coordinate system (z coincides with anchor)
W = weight of slice
β = slope angle
γ = soil unit weight
δ = soil-anchor friction angle
ϕ = angle of internal friction of soil
θ = angle of anchor to slope normal
θ_{opt} = optimum anchor angle
ξ = T_o/W
σ = stress
σ_n = average normal stress on the anchor
τ_{max} = maximum shearing resistance between soil and anchor

APPENDIX III. —— REFERENCES

Hryciw, R. D. (1990). "Load Transfer Mechanisms in Anchored Geosynthetic Systems," University of Michigan Research Report to the Air Force Office of Scientific Research, Grant No. 88-0166, 59 pp.

Hryciw, R. D. and Irsyam, M. (1991). "Friction and Passive Resistance in Soil Reinforced by Cylindrical Ribbed Inclusions," tentatively accepted by the International Journal for Numerical and Analytical Methods in Geomechanics.

Hryciw, R. D. (1991). "Anchor Design for Slope Stabilization by Surface Loading" Journal of the Geotechnical Engineering Division of ASCE, Vol. 117, No. 8, pp. 1260-1274.

Irsyam, M. and Hryciw, R. D. (1991). "Friction and Passive Resistance in Soil Reinforced by Plane Ribbed Inclusions," Geotechnique, Vol. 41, No. 4, pp. 485-498.

Koerner, R. M. (1984). "In-Situ Soil Slope Stabilization Using Anchored Nets," Low Cost & Energy Saving Construction Materials Conference Proceedings, Rio De Janiero, Brazil, 465-478.

Koerner, R. M. and Robins, J. C. (1986). "In-Situ Stabilization of Soil Slopes Using Nailed Geosynthetics," Proc. of the Third International Conference on Geotextiles, Vienna, Austria, 395-400.

Vitton, S. J. (1991). "Load Transfer Mechanisms in Anchored Geosynthetic Systems," thesis submitted to the University of Michigan in partial fulfillment of the requirements for the degree Doctor of Philosophy.

Vitton, S. J. and Hryciw, R. D. (1991). "Soil-Anchor Interaction in Anchored Geosynthetic Systems," Proceedings of the Geotechnical Engineering Congress, Boulder, CO, pp. 958-968.

A DESIGN METHOD FOR REINFORCED CLAY EMBANKMENTS ON SOFT FOUNDATIONS

Glen A. Roycroft,[1] Member ASCE

ABSTRACT: A rational design method and case study are presented for geosynthetic reinforced clay embankments constructed on soft foundation soils. The design method is based on traditional limit equilibrium slope stability analyses and accurate characterization of the engineering properties of embankment and foundation materials. Stability analyses are made using Spencer's procedure which allows for total moment and static force equilibrium. Tension crack behavior is evaluated and induced tensile forces in clay embankments are analyzed for undrained, end-of-construction conditions. Reinforcement schemes are designed to resist tensile forces and increase embankment stability. A case study using the design method is presented for reinforced embankments constructed on marshland deposits of soft Bay Mud in the San Francisco Bay Area.

INTRODUCTION

Considerable research has been devoted to the subject of reinforced soil structures. Holtz (1990) recently provided an excellent and broad summary on state-of-the-art procedures for the design and construction of geosynthetic reinforced embankments constructed on very soft foundation soils. Other researches, Mylleville and Rowe (1991) for instance, have used finite element methods (FEM) for analysis and design of reinforced embankments.

Practical methods, though, for design of geosynthetic slope reinforcement to stabilize clay embankments constructed on soft foundation soils are not widely available in geotechnical literature or promotional materials made available by the geosynthetic industry. This paper summarizes a practical analysis and design method for

1. Senior Engineer, Miller Pacific Engineering Group, 165 North Redwood Drive, Suite 120, San Rafael, CA 94904.

stabilizing such embankments using geosynthetic slope reinforcement. After discussing the concepts and analysis procedures, a case study is presented in which the method is applied to the design of about 4000 linear feet (1200 m) of reinforced embankment fills, 200 feet (60 m) in width, constructed on normally consolidated marshland deposits of soft Bay Muds in the San Francisco Bay Area.

ANALYSIS AND DESIGN METHOD

Embankment Stability

Instability of a clay embankment develops from two factors: (1) the weight of the embankment itself which causes unbalanced gravity loads and overturning, and (2) the tendency for the embankment to spread laterally. Both factors induce shear stresses within the embankment and foundation soils. If the induced shear stresses exceed the mobilized strength of the embankment or foundation soils, failure will occur.

Tensile Forces and Tension Cracks

The tendency of fills on soft soils to spread laterally induces tensile stresses in cohesive embankments. The tensile forces are highest at the crest of such embankments and their magnitude decreases with depth. Where tensile stresses exceed the tensile strength capacity of the soil, tension cracks are formed. While cohesion and undrained shear strength for a phi = 0 stability analysis implies a certain tensile strength, tensile strengths are not considered realistic for soils. The common approach in stability analyses, then, is to assume the formation of tension cracks.

Chirapuntu and Duncan (1975) provide procedures for estimating the height of an embankment constructed on soft soils at which cracking will begin. They explain that crack formation is dependent on the shear strength and stiffness of the embankment and foundation. If the embankment exceeds a certain height, then a tension crack is assumed. They provide a recommendation for the depth of the crack and strength reduction factors of embankment soils for analysis of stability.

Alternatively, Wright (1982) suggests that in conducting embankment stability analyses, the calculated tensile stresses be considered erroneous and eliminated by introducing a vertical "tension crack" having an appropriate depth below the crest of the embankment. He recommends the depth of the tension crack be estimated on the basis of active Rankine earth pressure theory.

Effects of Reinforcement

Using reinforcement within embankments has two effects of differing importance on stability. The first effect is that an internal stabilizing tensile force is applied to the embankment and foundation system. The tensile force acts on the shear surface of the critical sliding mass at the depth of the reinforcement. The stabilizing effect becomes greater as the reinforcement is placed deeper.

The second and more important effect to cohesive embankments is the lateral restraint imposed on the embankment by the reinforcement. This effect prevents formation of tension cracks and maintains the continuity and integrity of the cohesive embankment. Low and Duncan (1985) confirm that one of the major benefits in using reinforcement in these conditions is preventing formation of tension cracks and utilizing the mobilized strength of the cohesive fill. They conclude that in analyzing the stability of a reinforced cohesive embankment, only the uncracked condition need be considered. They reason that reinforcement in or beneath cohesive embankments on soft foundations restrains lateral spreading of the fill and induces higher internal horizontal forces within the embankment. Provided the reinforcement is not overstressed and has a sufficiently high tensile modulus at low strains, cracks will be prevented from developing in the embankment.

The problem then becomes one of determining the tensile forces induced and designing a reinforcement system to withstand lateral spreading forces. Of course, the engineering properties of the entire foundation, embankment, and reinforcement system must be accurately characterized and appropriately analyzed for stability. An appropriate level or factor of safety should also be selected based on the importance of the project, the consequences of failure, and the accuracy of both the engineering properties ascertained and analyses conducted.

Stability Analysis

The stability of reinforced embankment and foundation systems may be analyzed by limit equilibrium slope stability methods or finite element methods (FEM), among others. In consulting practice however, FEM analyses are still considered a rather esoteric design tool limited to larger projects and are not commonly or widely used. Additionally, Low and Duncan (1985) found FEM analyses for this application to be an unreliable and unconservative method, apparently due to inaccuracies that arise with conditions approaching failure.

Stability analyses by limit equilibrium methods, then, are preferred. Stability analyses using methods such Spencer's procedure which allow total moment and static force equilibrium are preferred and have been found to be the most accurate. Low and Duncan (1985) found Bishop's Modified Method to be conservative by about 18 percent in the evaluation of one test embankment and concluded that Spencer's procedure yielded the most accurate results.

Analysis Software

There are many computer programs to aid the engineer in analyzing embankment stability available on the geotechnical software market - some good and some not so. For analysis of reinforced embankments, the software should have the capability of utilizing Spencer's procedure for safety factor calculations. It also should allow for the detailed examination of normal stresses on the critical slip surface and forces between slices. An excellent program with these capabilities was written by Wright (1987) and designated "UTEXAS2." The theory and instructions for use of this software are documented by the Corps of Engineers (1986a, 1986b).

CASE STUDY APPLICATION

Project Description

The design method was applied to a project located in a former marshland area that was diked, drained, and farmed early in the 1900's. In the 1960's, embankments were constructed using materials from adjacent excavations to create lagoons. The foundation materials underlying the embankments consist of about 60 to 80 feet of soft, compressible Silty Clay (CL/CH) or Clayey Silt (ML/MH) known locally as soft Bay Mud. The embankments were raised by fill construction using both hydraulically placed Bay Mud and coarse grained soils excavated from adjacent hillside areas. Initially, a Bay Mud dike or levee was constructed after which the dredged Bay Mud was placed hydraulically, allowed to drain and dry. When a sufficient crust had formed, dry fill was spread over the surface to raise the elevation. The current elevation of the original fills is in the +6 to +8-foot (+1.8 to +2.4 m) range. The original embankments were not reinforced and during initial filling, slope failures occurred.

Additional new fill must now be placed on the original embankments to bring the final construction elevations high enough to protect future improvements from flooding. In addition, the fill elevations at the end of construction must be sufficiently high to compensate for future settlements that will occur due to the consolidation of the underlying Bay Mud. About 5 inches (13 cm) of new settlement will occur for each 1 foot (30 cm) thickness of new fill placed in addition to the 12 inches (30 cm) of settlement still expected due to the original fills. Thus, grading will consist of excavation of adjacent cohesive channel materials and placement of these materials as embankment fills to achieve a construction elevation of +12 feet (3.7 m). A 12-foot construction elevation will provide for a final design elevation of roughly +8.5 feet (2.6 m) after 50 years of settlement.

Characterization of Engineering Properties

The critical loading condition for an embankment constructed on soft foundation soils is the end-of-construction or short-term loading condition also known as the undrained, phi = 0 condition. Because the stability of the project embankments was marginal and primarily a strength problem, it was important to obtain an accurate strength profile, particularly in the soft Bay Mud foundation soils.

An extensive field exploration and laboratory testing program was conducted to determine the undrained strength characteristics of the fills and soft foundation soils. This necessitated high quality, nearly undisturbed subsurface sampling. Test borings were drilled through the existing fill and soft, compressible Bay Mud deposits to the top of firm, incompressible formations. Rotary wash type drilling equipment was used since this drilling technique leaves a clean hole for sampling and imposes minimal disturbance to soils in the vicinity of the drill bit. Samples of the fills were obtained by using a modified California split sampler lined with brass tubes. In soft soils where the highest quality, nearly undisturbed sample was desired, a thin-walled piston tube sampler activated by hydraulic pressure from the drilling fluid was used.

Strength tests consisting of unconfined compression tests were conducted to determine the undrained strength of the existing fill and soft Bay Mud deposits. Unconfined compression tests were conducted at the job site within a few days after sampling to minimize sample disturbance. Consolidation tests were conducted to determine the settlement and compression characteristics of the soft Bay Mud. The shapes of the consolidation curves, showing a sharp, well-defined break near the preconsolidation pressure, verified and provided confidence that the samples obtained were of very high quality and that the undrained strength profile had a high degree of accuracy. The usual complement of water content, density, Atterberg limits, and compaction tests were also conducted to define soil index properties and make appropriate correlations with other engineering properties.

Test results indicated that shear strength in the existing fills ranges from 500 to 2000 psf (24 to 96 kPa). The undrained shear strength of the underlying Bay Mud varies with depth and prior loading conditions. Bay Mud shear strengths beneath fills are slightly higher than in adjacent flat marshland locations due to consolidation from the increased load of the fills. The upper 10 feet (3 m) of Bay Mud directly below fills have relatively high shear strengths due to prior desiccation and fill surcharge influence. The weakest Bay Muds are in the -10 to -20-foot (-3 to -6 m) elevation range and have a shear strength of about 250 psf (12 kPa). Below elevation -20 feet (-6 m), the undrained shear strength increases linearly at a rate of about 7 psf per foot (1.1 kPa/m) of depth. This indicates a c/p ratio of 0.25 which is in the expected range for Bay Muds having a plasticity index of 50 percent. In the 75 to 80-foot (23 to 24 m) range of depth, the soils become stiff with undrained strengths exceeding 1000 psf (48 kPa). A summary of the results of the unconfined compression testing, and selected design shear strengths is presented in Figure 1, Undrained Shear Strength versus Elevation.

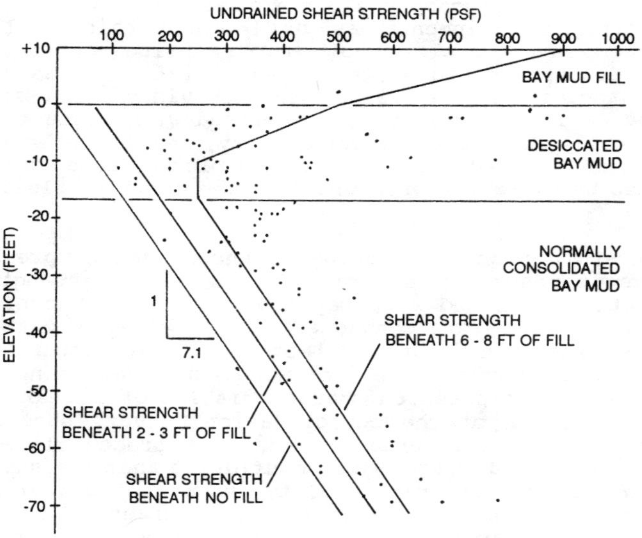

Figure 1. Undrained Shear Strength Versus Elevation

Embankment Geometry

The embankment geometry for this project is illustrated in Figure 2. Original embankments were placed as previously described. New construction will consist of dredging and compacted fill. Submerged slopes will be dredged to 6 horizontal to 1 vertical. The new fills and original embankment above the adjacent lagoon water level will to be constructed at 3 to 1 slopes. The embankment widths will be about 200 feet (60 m). The total embankment length will be about 4000 feet (1200 m).

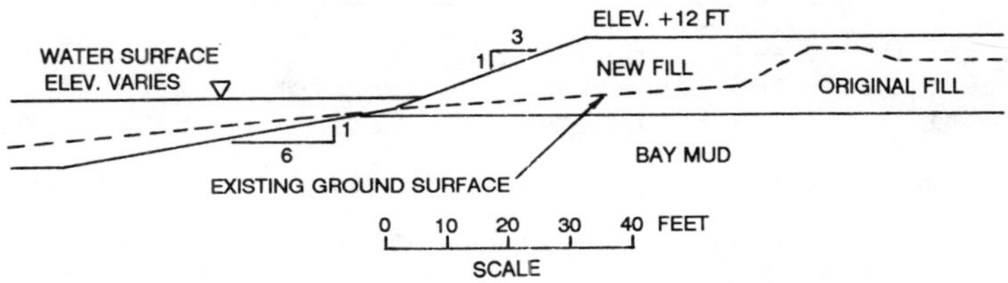

Figure 2. Embankment Geometry

End-of-Construction Stability

Critical loading conditions for clay embankments on soft soils occur during and immediately after construction. During this period, slope stability is affected by the adjacent lagoon water levels and the formation and depth of tension cracks in the embankment materials. Stability analyses for various combinations of water levels and tension crack depths were conducted for the assumed embankment slope geometry. The calculated safety factors range from a low of 0.99 for a water level at elevation -2 feet (-0.6 m) and tension crack depth of 6 feet (1.8 m) to 1.39 for a water level at elevation +3 feet (0.9 m) and no tension crack. Slope stability analyses for end-of-construction loading conditions are summarized in Table A. As indicated in Table A, low water elevations and deeper tension cracks severely affect embankment stability. Considering normal variations in soil materials, the calculated safety factor of 0.99 for a low-tide water level and deep tension crack indicated that slope failure would be likely. It therefore became apparent that lagoon water levels would have to be kept high during construction and some type of soil reinforcement would be required to maintain embankment stability.

Evaluation of Tensile Forces

To provide a rational basis for design of the geosynthetic reinforcement, it was necessary to conduct analyses to evaluate the tensile forces induced within the embankment. As previously mentioned, computer software using total force and moment equilibrium which allows the detailed examination of shear stresses and normal forces, etc., is needed to make an evaluation of the tensile forces within an embankment.

TABLE A

SUMMARY OF STABILITY SAFETY FACTORS
END-OF-CONSTRUCTION LOADING CONDITIONS

Water Elevation (feet)[1]	6-ft Tension Crack	3-ft Tension Crack	No Tension Crack	No Tension Crack with Reinforcement[2]
+3	1.23	1.30	1.39	1.44
0	1.06	1.13	1.20	--
-2	0.99	1.05	1.12	--

(1) 1 foot (0.3048 m)
(2) One layer of reinforcement with a design strength of 2000 lbs/ft (29 kN/m)

To do this, slope stability analyses were made to locate critical sliding masses passing through a series of points on the crest of the embankment. The points were spaced 5 feet (1.5 m) apart starting near the top of the 3 to 1 slope of the new fill construction. The critical potential sliding mass was determined for circular failure surfaces restricted to passing through each point. Tension cracks within the embankment were not allowed in order to make an estimate of the tensile forces induced. The safety factors for the corresponding series of critical shear surfaces were calculated. Additionally, computer software output, detailing normal shear stresses on each of the shear surfaces, forces between slices, iterative solutions for the safety factors, etc., was closely examined and evaluated.

As previously discussed, negative stresses on the shear surface indicate tensile forces. To obtain an estimate of the induced tensile forces at different sections within the embankment, the negative shear stresses occurring for slices near the top of the embankment were summed for each failure surface. Figure 3, shows the estimated tensile forces induced in new fills prior to the formation of tension cracks. The safety factors at various locations from the top of the finished 3 to 1 slope are also shown. The lowest or critical safety factor is 1.39 for the slope configuration and potential circular failure surface shown.

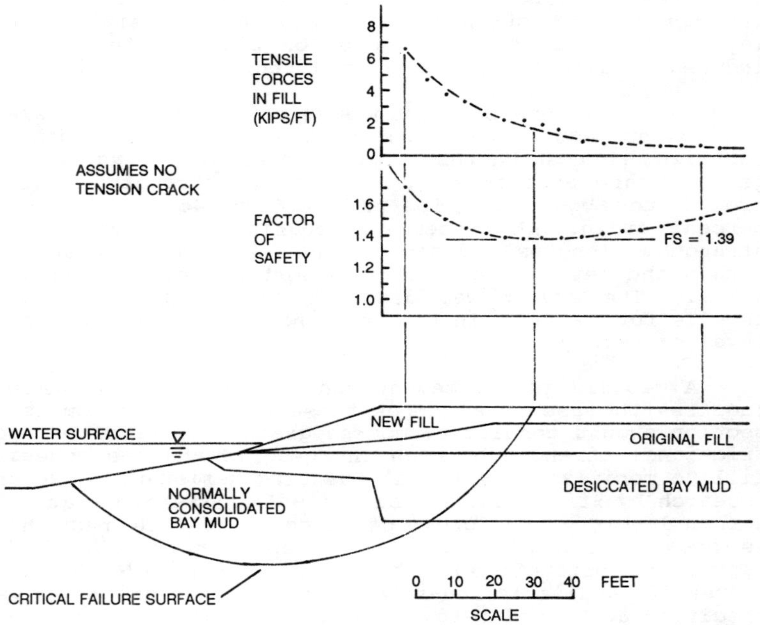

Figure 3. Induced Tensile Forces and Safety Factors

DESIGN OF REINFORCEMENT

Provided the reinforcement is not overstressed and has a sufficiently high tensile modulus at low strains, cracks will be prevented from developing in the embankment. Preventing the formation of tension cracks is the primary benefit to be realized from using slope reinforcement. This allows the full strength of the fill materials to be mobilized in resisting the gravity forces causing instability. The stabilizing moment provided by the reinforcement and applied to the sliding mass at the depth of reinforcement only adds a slight increase in safety factor. As shown in Table A, one layer of reinforcement increases the safety factor for the critical failure surface shown in Figure 3 from 1.39 to 1.44.

Figure 3 provides a rational basis for selecting the limits of horizontal geosynthetic reinforcement. Reinforcement should be placed in those areas of the embankment where tensile forces are high and safety factors are low. The number and strength of geosynthetic reinforcement layers should be selected based on the magnitude of the induced tensile forces to be resisted.

A geogrid-type material was selected for reinforcement of the project embankments because of its ability to mobilize high strengths at low strains. A variety of geogrids are available having tensile strengths ranging from 1 to about 3.2 kips/ft (14.6 to 46.7 kN/m) at 2 percent stain. A system of geogrid layers of various strengths, lengths and depths was selected for placement within the new fill and the top portions of the original fills. The system was designed to resist the induced tensile forces shown in Figure 3 and provide an acceptable level of safety.

A geogrid reinforcement scheme for clay embankments requires the use of a composite section. This composite section should consist of coarse grained material of high frictional strength between the geogrid and the cohesive fill to mobilize interfacial shearing resistance. Recent research by Sridharan, et. al., (1991) has shown that only a thin layer, in the order of 15 mm or less than an inch, is needed of high frictional strength material. As a matter of construction practicality, a 3-inch (76 mm) layer of readily available, on-site coarse grained materials was specified above and below each layer of geogrid.

CONCLUSIONS

Design Method

This paper has presented a practical and rational method of analysis for design of geosynthetic reinforced clay embankments constructed on soft soils. The key factors for the analysis and design of such embankments are summarized as follows:

(1) Obtain the highest quality and most accurate strength profile of the embankment/foundation system that is practical.

(2) Use appropriate computer software and total force and moment equilibrium procedures for stability analyses.

(3) Initially, analyze the stability of a clay embankment and foundation by assuming tension cracks. Estimate tension cracks extending to depths as given by Chirapuntu and Duncan (1975) or Wright (1982). If, analyses show stability to be marginal and other factors (such as water infilling tension cracks) indicate, use geosyn-

thetic reinforcement for the primary purpose of preventing tension cracks.

(4) Determine induced tensile forces within the embankment assuming no tension cracks and design a system of geosynthetic reinforcement to resist. Select a geosynthetic material of sufficient strength and compatible modulus.

(5) When selecting a particular geosynthetic reinforcement scheme for clay embankments, insure that the interface is not the weak point in the system. A composite system of reinforcement embedded in a thin layer of coarse grained soils is recommended for geogrid-type reinforcement.

Earthquake Effects

Although beyond the scope of this paper, geosynthetic slope reinforcement will also mitigate the potential damages caused by earthquake-induced embankment deformations and lateral spreading. Practical and reliable analytical methods as well as empirical data do not currently exist to quantify the magnitude of positive mitigation. However, qualitatively, the horizontal geosynthetic reinforcement will resist embankment deformations and lateral spreading caused by earthquake motions. The embankment and reinforced soil masses will act as stiffer, more cohesive, intact structural units.

Other Considerations

A number of other considerations may affect the design of embankments overlying soft foundations: overall bearing capacity, geosynthetic deformation, longitudinal strength, etc. Holtz (1990) provides an excellent and practical discussion of additional factors to consider in design.

Field Verification

As the project described in this paper has not yet been constructed, a unique opportunity exists to study on a large scale, the field behavior of reinforced embankments and accuracy of the applied method of analysis and design. Due to the quantity of reinforced embankment to be constructed, and the importance of avoiding unfavorable performance, test fills have been proposed to verify the adequacy of the design. The test fills will include both a non-reinforced and reinforced embankment as designed. The test sections will be instrumented and their movement and behavior will be monitored and evaluated. Appropriate modifications to the design will be made depending on the results of subsequent analysis and evaluation. The results of these field studies should provide useful information to add to the available geotechnical literature on reinforced soils.

APPENDIX.

REFERENCES

Chirapuntu, S. and Duncan, J.M., (1975), "The Role of Fill Strength in The Stability of Embankments of Soft Clay Foundations," Geotechnical Engineering Report No. TE 75-3, University of California, Berkeley, California, December 1975.

Holtz, R.D., (1990), "Design and Construction of Geosynthetically Reinforced Embankments on Very Soft Soils," State of the Art Paper, Session 5, International Reinforced Soil Conference, Glasgow, Scotland, 10-12 September 1990.

Low, B.K., and Duncan, J.M., (1985), "Analysis of the Behavior of Reinforced Embankments on Weak Foundations," Virginia Polytechnic Institute and State University, Department of Civil Engineering Report NO. VPI/CE-GT-85/11, October 1985.

Mylleville, Brian L.J., and Rowe, R. Kerry, (1991), "On the design of reinforced embankments on soft brittle clays," Geotechnical Fabrics Report, pp 12-24, Vol 9, No 4, May/June 1991.

Sridharan, A., Srinivasa Murthy, B.R., Bindumadhava, and Revanasiddappa, K., (1991), "Technique for Using Fine-grained Soil in Reinforced Earth," Journal of Geotechnical Engineering, ASCE, Vol 117, No. 8, August 1991.

U.S. Army Corps of Engineers, (1986), **User's Guide: UTEXAS2 Slope Stability Package; Volume I: User's Manual**, by CAGE, G-CASE Task Group on Slope Stability, U.S. Army Engineer Waterways Experiment Station, April 1986.

U.S. Army Corps of Engineers, (1986), **User's Guide: UTEXAS2 Slope Stability Package; Volume II: Theory Section**, by CAGE, G-CASE Task Group on Slope Stability, U.S. Army Engineer Waterways Experiment Station, June 1986.

Wright, Stephen G., (1987), Computer Program Designation - UTEXAS2, Version No. 1.209, College of Engineering, University of Texas at Austin, November 23, 1987.

Reinforced Soil-Cement Embankment

Safdar A. Gill, F.ASCE[1]
and Ted D. Bushell, M.ASCE[2]

Abstract

A soil-cement arch embankment was constructed over an existing tunnel. The soil-cement arch was reinforced with geogrid and underlain by expanded polystyrene panels to redistribute the embankment load and minimize load transfer to the tunnel. Finite element analysis was used to evaluate the effectiveness of the arch. Instrumentation was installed to evaluate stress changes in the tunnel structure and to confirm the analysis.

Introduction

To accommodate increased operating energy levels, an additional 12 feet (3.7 m) of shielding fill was required over an existing synchrotron tunnel structure. The analysis of stresses under the existing loads indicated that the design of the existing tunnel was marginal and could not sustain the additional shielding. The originally proposed concepts included a pile supported bridge across the tunnel to carry the new fill, corrugated metal pipe sections, as well as precast concrete arches. However, these various schemes were considered too costly, as well as time consuming and were abandoned.

The system developed for the new shielding fill consisted of a soil-cement embankment in the shape of an arch. The arch shape was selected since it would carry its own self weight, as well as the overlying embankment fill, and would minimize the stress increase on the existing tunnel. Figure 1 illustrates the soil-cement arch at Station 23+50 where the proposed embankment

---

[1] Senior Principal Engineer, STS Consultants, Ltd., 111 Pfingsten Road, Northbrook, IL 60062

[2] Principal Engineer, STS Consultants, Ltd., 111 Pfingsten Road, Northbrook, IL 60062

height will reach elevation 110.5 ft. (33.7 m). The arch consisted of an expanded polystyrene (EPS) base and soil-cement reinforced with geogrids. The haunches of the arch were angled away from the tunnel to transfer the load to the subgrade at the arch haunches and minimize load transfer to the tunnel walls.

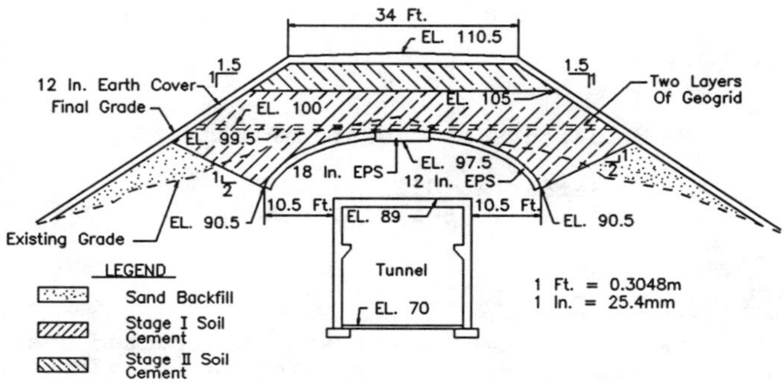

Figure 1. Soil–Cement Arch, Station 23+50

Both the stiffness of the arch, and the differential settlement across the arch were important factors in developing the desired arching action. Soil-cement provided a sufficiently stiff member to redistribute the weight of the fill and transfer the stresses from the fill to the outside haunches of the arch. In addition to the stiffness, the material below the crown of the arch had to deflect more than the haunches of the arch in order for the arching action to take place. To provide for the desired amount of deformation, compressible expanded polystyrene (EPS) panels were placed beneath the arch. In addition, loose granular fill was placed immediately below these EPS panels. The soil in the excavated haunches was compacted in order to provide a stiff reaction to the outside haunches of the arch. In this manner, the existing tunnel structure behaved similar to a negative projecting conduit.

Several methods of analysis were used to design and evaluate the effectiveness of the soil-cement arch. An arch member was intuitively conceived as being the most efficient means of constructing the embankment. The structural analysis of the soil-cement arch was performed using conventional structural mechanics. The soil-cement arch was modeled as a simply supported beam. This analysis provided the required strength for both the soil-cement, as well as the geogrid reinforcement. A finite element method (FEM) analysis was performed to evaluate the stresses on the existing tunnel, as well as to assist in the design of the soil-cement arch.

Analysis of the existing tunnel indicated that the tunnel structure could withstand the weight of the existing overburden pressure plus some limited live load surcharge. Soil-cement arch construction was broken into two stages, with the first stage being limited to a height which would not exceed the allowable tunnel loading. The analysis indicated that the soil-cement needed to achieve a strength of 250 psi (1.72 MPa) prior to proceeding with the remainder of the soil-cement arch construction.

Since the existing tunnel structure was not capable of sustaining substantial stress increases from the new embankment and the concept of the shielding upgrade relied upon the effectiveness of the soil-cement arch, it was considered essential that the tunnel be instrumented. Instrumentation, consisting of strain gages, earth pressure cells, and a convergence meter, was installed during construction.

This paper provides details of the design and construction process, along with a review of the performance as evaluated from the instrumentation program.

Material Properties

The materials used in the construction of the arch consisted of the natural soils which served as the bearing stratum for the arch, soil-cement which formed the arch itself, expanded polystyrene (EPS) panels and the geogrid reinforcement.

Natural Soils

The existing tunnel was built in 1958 by cut and cover construction methods. The on-site fine to coarse sand was used to backfill around the tunnel. The backfill around the tunnel was found to be in a loose to medium dense condition with an average Standard Penetration Resistance of 13 blows per foot (blows per 30 cm). The natural soil below the existing structure was found in a much denser condition with the average Standard Penetration Resistance value being 25 blows per foot (blows per 30 cm).

The loose soil conditions immediately surrounding the tunnel combined with the denser soils found below the tunnel were favorable for construction of an arch structure. Loose soils around the tunnel would permit deflection at the arch crown which would redistribute loads to the outer more dense soils found in the arch haunches.

Soil-Cement

Since sand was the prevalent material at the site, it was proposed that the arch embankment be constructed of soil-cement using this on-site sand. The soil-cement was mixed in an on-site batch plant and then transported to the arch location. Light weight equipment was used to place and compact the soil-cement in layers so as not to overstress the tunnel. The strength required for the soil-cement was dictated by the results of the structural analysis. A mix design was performed in the laboratory to determine the optimum cement content which would produce the desired strength. The results of the mix design indicated that the soil-cement should contain 10% cement by weight of sand to produce the desired strength. The mix design also indicated that the soil-cement would achieve strengths of 200 psi (1.38 MPa), 400 psi (2.76 MPA) and 500 psi (3.45 MPa) following a curing period of two days, three days, and five days, respectively. Based upon this data, three days of curing time was required to achieve the desired 250 psi (1.72 MPa) compressive strength in the field prior to placement of the second stage of the soil-cement arch.

During construction, the properties of the soil-cement were controlled at the batch plant and monitored in the field. The moisture content of the soil-cement typically ranged from 6 to 8% with the in-place dry unit weight ranging from 116 to 119 pounds per cubic foot (1.86 to 1.91 Mg/m^3). Little difficulty was experienced achieving the desired degree of compaction during construction of the arch. The soil-cement was cured by sprinkling with water and covering the soil-cement with visqueen to minimize evaporation.

EPS Panels

Low modulus expanded polystyrene panels (EPS) were placed on loose sand beneath the crown of the arch to allow required arch deflection and to transfer loads to the haunches of the arch. The properties of the EPS panels had to be selected to support construction activity without collapsing or significantly strain hardening. The properties of the EPS panels were determined by constructing a simple finite element model of a compression test specimen and matching the output to those obtained in a laboratory compression test. Based upon these analyses, the required deflection of a 6-inch-thick (152 mm) EPS panel was estimated to be on the order of 0.5 inches (13 mm), requiring a modulus of 60 psi (414 kPa). The specified minimum compressibility was 10% at 5 psi (35 kPa). The unit weight of the EPS panels was required to be less than 0.90 pounds per cubic foot (14 kg/m^3).

Expanded polystyrene appeared to be the ideal material since it had a high initial modulus to limit deflections under the initial Stage I fill, and had a much lower modulus at higher loads so as to deform the required amount under subsequent filling to create the arching effect. Expanded polystyrene panels conventionally used for construction were selected as they were readily available in sizes that were conducive to construction; based upon product literature, they had the potential to meet the design requirements; and, they were cost effective. The panels consisted of polystyrene spheres that were bonded together under pressure to produce a deformable panel. Even so, initial testing indicated that conventional panels were too stiff. The manufacturer modified the panels to meet the design requirements based upon laboratory test results. Laboratory compression tests indicated that the strain was affected by the thickness of the sample. One inch (25.4 mm) samples had strains ranging from 10.4 to 13.5%, while 6 inch (152 mm) samples showed smaller strains of 5.8 to 10.4%. The manufacturer modified the product so that the compressibility exceeded 8% at 5 psi (35 kPa) on a 6 inch (152 mm) thick sample.

Geogrid

Two layers of Tensar UX1600 HDPE grids were placed at the invert of the arch to provide lateral restraint during construction. The reinforcement was selected to limit strains to 2% to prevent cracking of the soil-cement.

The Tensar UX1600 was found to have a strength of 360 pounds per inch (6.4 kg/mm) at 5% strain, based upon laboratory testing. Compressive strength tests on the soil-cement specimens gave an average strength of 437 psi (3.01 MPa) at 3% strain. The modulus of these two materials provided close strain compatibility which was also a requirement for selection of the geogrid. The selected HDPE geogrid would also resist chemical degradation in the high alkaline environment of the cement stabilized soil, and, being inert, would resist corrosion better than steel reinforcement in the prevailing environment of stray currents.

Analytical Techniques

The design of the reinforced soil-cement embankment included determining the required soil-cement strength, maximum allowable fill height, as well as selecting the required reinforcement strength. This analysis was performed by modeling the soil-cement arch as a reinforced beam. A finite element analysis was performed to estimate stresses on the existing tunnel caused by the proposed increase in embankment height and to evaluate the effectiveness of the soil-cement arch.

Beam Analysis

The soil-cement arch was analyzed in stages to reflect the anticipated construction. The first stage arch was analyzed as a 7 foot (2.1 m) thick simply supported beam spanning 20 feet (6.1 m). Assuming the soil-cement to be an unreinforced beam, the tensile stresses within the beam were found to be less than 1/2 those typically assumed permissible for soil-cement based upon empirical correlations. Assuming that the first stage soil-cement consisted of a simply supported reinforced beam, the tensile stress within the reinforcing was found to be approximately 640 pounds per inch (11.4 kg/mm). Both of these analyses were then repeated for the full 12 foot (13.7 m) deep soil-cement arch with similar results.

Beam analyses indicated acceptable stresses within the soil-cement arch assuming that the soil-cement obtained an initial strength of 250 psi (1.72 MPa) with an ultimate strength of 500 psi (3.45 MPa). The mix design performed on the soil-cement indicated that this strength was achievable.

The reinforced beam analysis indicated a required strength at 5% strain of approximately 300 pounds per inch (5.4 kg/mm) for the geogrid to prevent total failure allowing for a design factor of safety of 2. The specified ultimate tensile strength of the geogrid reinforcing was 625 pounds per inch (11.2 kg/mm). Strength requirements were based upon the wide width strip method as defined by ASTM D-4595. Although a factor of safety of 3 would be used normally to include effects of creep and durability, the lower factor of safety was considered adequate as the arch itself would normally remain in compression to sustain the loading.

Finite Element Analysis

To estimate the stresses on the existing tunnel caused by the proposed 12 foot (3.7 m) increase in the soil embankment and to evaluate the effectiveness of the proposed soil-cement arch, a finite element method (FEM) analysis was performed. The FEM solution was obtained using an incremental finite element computer program for 2-dimensional plane-strain analyses, FEADAM 84 (Duncan, et al. 1984). This program calculates the stresses, strains and displacements in embankments by simulating the actual sequence of construction operations and post-construction loading conditions. The program approximates and accounts for non-linear and stress dependent stress-strain properties of the various materials included in the model, according to a hyperbolic soil model. Figure 2 shows the finite element mesh used for the model, along with the various soil materials used in the analysis. While a full size model was constructed to perform the analysis, only one-half of the model is shown for presentation purposes. Details regarding the model, as well as the soil parameters and boundary conditions used in the analyses, are presented in a companion paper describing the first phase of this project constructed in 1989 (Gnaedinger and Gill, 1991).

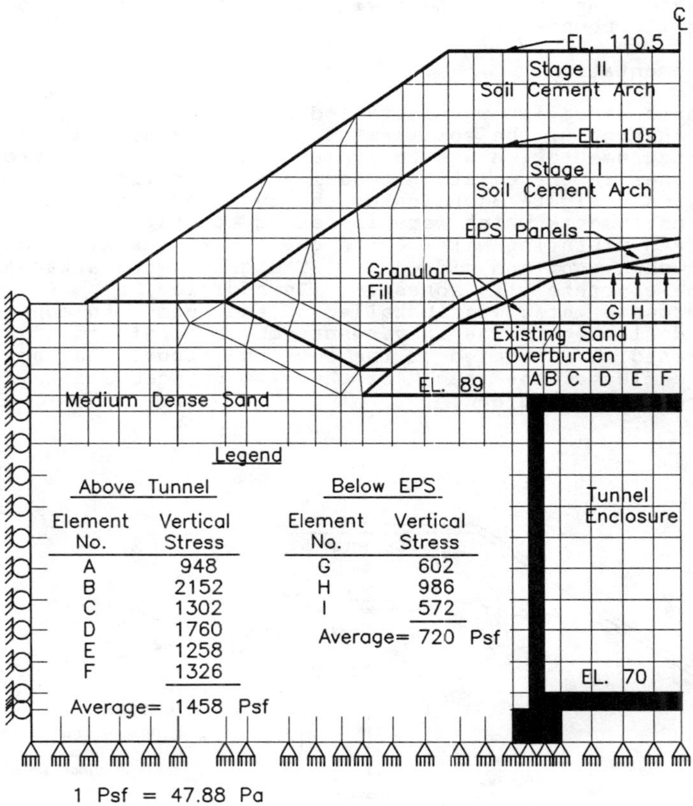

Figure 2. Finite Element Model

The results of the finite element analysis showed that the soil-cement acts effectively as an arch to substantially reduce the stress that would have been transferred to the tunnel crest and sidewalls due to embankment construction. The vertical stresses imposed on the tunnel roof as predicted by the FEM analysis are indicated in the elements across the tunnel roof as shown in Figure 2. The average total vertical stress change across the tunnel roof was found to be 1458 pounds per square foot (70 kPa) which represents only approximately 25% of the actual imposed load. By subtracting the existing overburden stress, the predicted net stress change over the tunnel roof was 738 pounds per square foot (35.3 kPa) due to construction of the soil-cement arch. The average predicted stress increase immediately underlying the EPS panels was found to be 720 pounds per square foot (34.5 kPa).

Instrumentation

Instrumentation was installed before and during construction of the soil-cement arch. The purpose of the instrumentation was to evaluate the potential stress increases in the existing tunnel and to confirm the analyses. Figure 3 indicates the type and location of the instruments which were installed at Station 23+50. Six IRAD vibrating wire strain gages were installed on the tunnel roof and sidewalls. The gages were attached to the concrete with adhesive. In addition, a convergence meter was installed in the tunnel to monitor roof deflection. The convergence meter consisted of a tensioned wire adhered to the roof and floor. As the roof deflected the length of the wire changed and this movement was recorded.

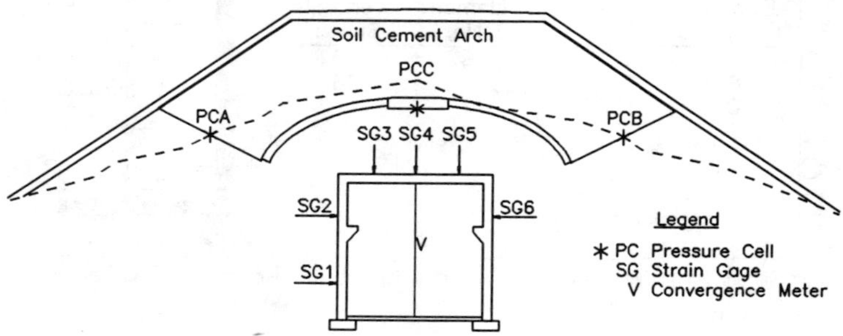

Figure 3. Instrumentation Cross-Section, Station 23+50

Pressure cells were also installed to monitor the change in stress caused by the arch construction. Pressure cells PCA and PCB were installed at the haunches of the arch, while pressure cell PCC was installed at the crown of the arch directly over the tunnel roof. All of these cells were installed immediately below the EPS panels within the soil subgrade overlying the tunnel.

Strain Gage Data

Figure 4 indicates a plot of the strain gage data obtained during the course of construction. In addition, the ground elevation overlying the tunnel is also indicated in this plot. The initial 100 to 150 microstrains measured early in construction at the center of the tunnel roof was due to inadvertent excavation and backfilling. Had the strain gages been installed prior to this excavation, there would have

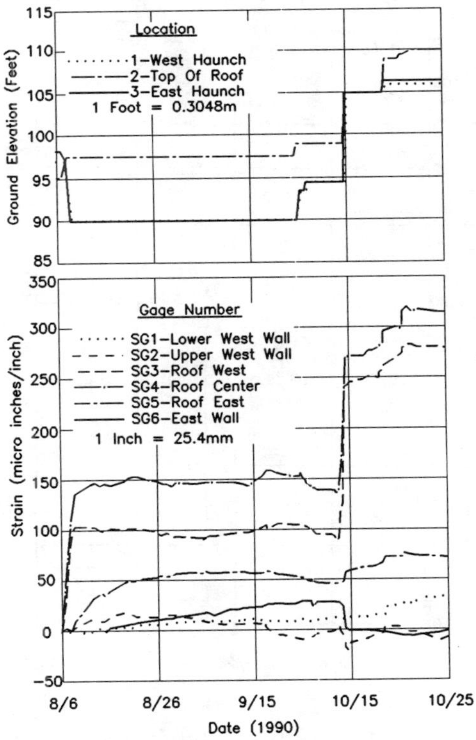

Figure 4. Strain Gage Data

been virtually no net strain due to this unloading and subsequent reloading. Therefore, only the net strain on the tunnel roof of 175 microstrains actually occurred due to the arch construction above the existing ground elevation of 99 ft. (30.2m).

Using the correlations established during the previous construction of 70 microstrains for a 330 pounds per square foot (15.8 kPa) stress change, the net change of 175 microstrains would correspond to a stress change of approximately 825 pounds per square foot (39.5 kPa) on the tunnel roof.

Earth Pressure Cell Data

A plot of the earth pressure cell data versus time, as well as fill elevation, is indicated in Figure 5. The

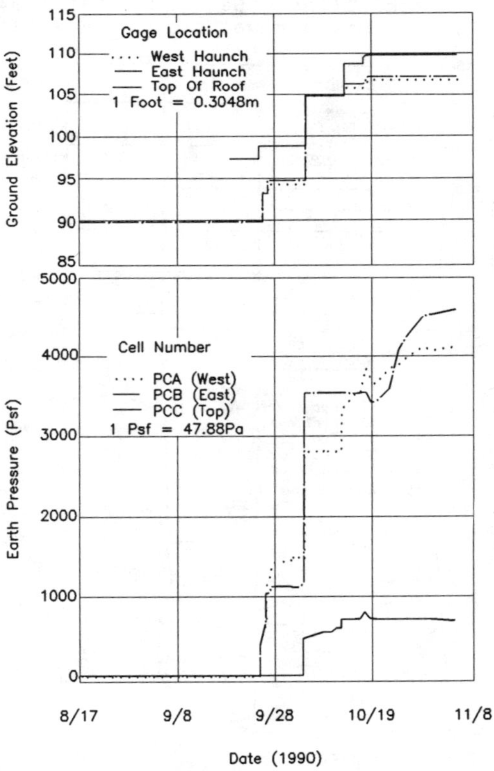

Figure 5. Earth Pressure Cell Data

plot of the earth pressure cell data clearly indicates the arching effect taking place. PCC located over the tunnel crown indicates a reading of 731 psf (35 kPa) which represents only approximately 45 to 55% of the actual overburden pressure of the newly constructed arch. Cells PCA and PCB located in the haunches of the arch indicated readings ranging from 3,000 to 5,000 psf (143.6 to 239.4 kPa) which are approximately 20 to 100% greater than the actual overburden stress due to the arch itself. The crown of the tunnel received less than the overburden stress, while the haunches of the arch attracted the load being shed at the center of the arch. Hence, this data clearly demonstrates the arching effect. Also, the 731 psf (35 kPa) reading obtained over the top of the tunnel compares favorably with the 720 psf (34.5 kPa) predicted by the FEM analysis and the 825 psf (39.5 kPa) stress change indicated by the strain gages.

Convergence Meter Data

Finally, the convergence meter indicated a downward deflection of the tunnel roof of 0.04 inches (1.0 mm) as indicated in Figure 6. Considering the roof slab of the existing tunnel as a fixed-end beam, a vertical deflection of 0.04 inches (1.0 mm) would correspond to a roof load of 829 pounds per square foot (39.7 kPa). This back calculated load is comparable to that indicated by both the strain gages and the pressure cells.

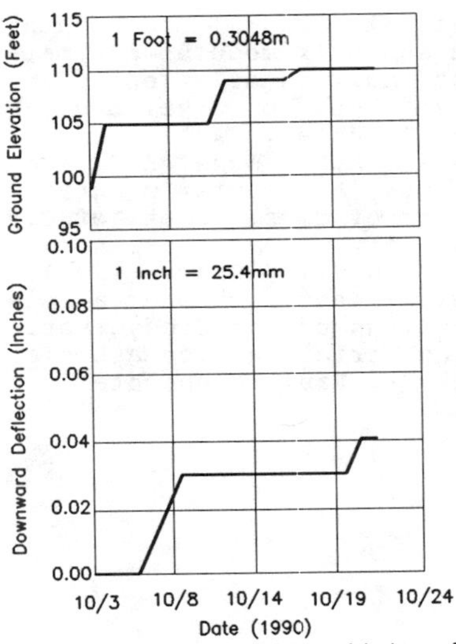

Figure 6. Convergence Meter Data

Conclusions

The soil-cement arch successfully minimized the additional stresses on the existing tunnel enclosure caused by the addition of 12 feet (3.7 m) of fill over the tunnel roof. FEM analysis of the soil-cement arch predicted an increase in vertical stress on the tunnel roof of 738 pounds per square foot (35.3 kPa). The various instruments measured a change in stress over the tunnel roof ranging from 731 to 829 pounds per square foot (35 to 39.7 kPa) which is within approximately 15% of that predicted by the FEM analysis. The measured increase in stress over the tunnel roof corresponds to only approximately 50% of the actual overburden loading which clearly demonstrates the arching action.

Geosynthetics assisted the soil-cement arch in stress transfer over the existing tunnel. The geogrids resisted lateral stretching of the soil-cement arch and minimized this movement which otherwise could have caused increased vertical stress on the tunnel. Geogrid reinforcement provided a strain-compatible material for the arch reinforcement at the anticipated stress levels in the soil-cement. Expanded polystyrene panels (EPS) placed at the invert of the arch assisted in load relief and stress transfer across the tunnel.

References

Duncan, J.M., et. al., (1980). "Strength, stress-strain and bulk modulus parameters for finite element analysis of stresses and movements in soil masses." Dept. of Civil Eng., University of California, Berkeley.

Duncan, J.M., Seed, R.B., Wong, K.S. and Ozawa, Y., (1984). "FEADAM84: A computer program for finite element analysis of dams." Dept. of Civil Eng., Stanford University.

Gnaedinger, J.P. and Gill, S.A., (1991). "Geogrid reinforced soil-cement arch over accelerator ring." Proceedings of the Geosynthetic '91 Conference, Industrial Fabrics Association International, St. Paul, Minnesota.

Two New Specialty Geotechnical Processes for Slope Stabilization

Dr. Donald A. Bruce,[1] Member, ASCE

Abstract

The paper describes two newer specialty geotechnical construction techniques employed in slope stabilization. Case histories are used to illustrate contemporary practice in in situ earth reinforcement (Type A Walls) and large diameter deep drains (RODREN). The paper also touches upon the use of various alternative contracting procedures to facilitate the introduction of these recent techniques to maximum mutual benefit.

1. Introduction

The solution of the problems inherent with slope instability remains a fundamental challenge to the geotechnical community. Indeed, the increasingly onerous demands placed by developers ensure that the community must continue to evolve technically despite the vast store of knowledge and expertise already accumulated. This evolution is reflected in the sophistication of methods of design and analysis now employed by designers. Equally dynamic, however, are the developments being made by specialty geotechnical contractors, in order to fulfill the developers' goals and the designers' concepts.

These developments are taking several forms but principally

- refinements of existing techniques, such as anchored earth retaining structures (e.g., ASCE, 1990);
- new applications for existing techniques, such as diaphragm walls (e.g., ASTM, 1991; Xanthakos, 1979);

---

1. Technical Director, Nicholson Construction of America, P.O. Box 308, Bridgeville, PA 15017

- resurrection of older techniques previously underexploited, such as certain types of in situ reinforcement; and
- importation of new techniques, such as RODREN.

This paper focuses on the latter two groups, which feature techniques which may be less well known to readers in North America than the former pair.

2. In Situ Earth Reinforcement: Type A Wall

Background

In the last decade or so in the United States, there has been increasing use made of small diameter cast-in-place bored inclusions. Most have been designed to act as conventional load bearing piles, commonly known as pinpiles (Bruce, 1988; 1989). However, these elements (100-250 mm in diameter) are finding growing popularity in the field of slope stabilization, where, installed in densely spaced patterns (Figure 1), they act as in situ reinforcement (Bruce and Jewell, 1986; 1987). The concept of their performance is that they form a composite structure with the included soil: this structure then constitutes an in situ barrier to arrest actual or potential slope movements.

General Features

Early applications of conventional, axially loaded pinpiles indicated, surprisingly, a positive "group effect", thought to be due to beneficial soil-structure interaction (Bruce, 1988a). This advantage was then exploited in slope stability applications in Western Europe and later - but infrequently -in the United States. In urban environments similar groups of pinpiles (or "INSERTS" in this context) can be used in cut and cover, as well as bored tunnel, construction. There the concept is to create protective structures in the ground to separate the foundation soil of the building from the zones that are potentially subject to disturbance (Figure 2). All these INSERT structures rely for their effectiveness on soil/pile interaction, and not on intergranular soil cementation. This composite structure - referred to as a "Type A Wall" because of its distinctive cross sectional appearance - is intended to stop loss of soil from behind it, and to prevent sliding along potential failure planes passing through it.

Design approaches continue to lag behind other aspects of the technology, but several instrumented field

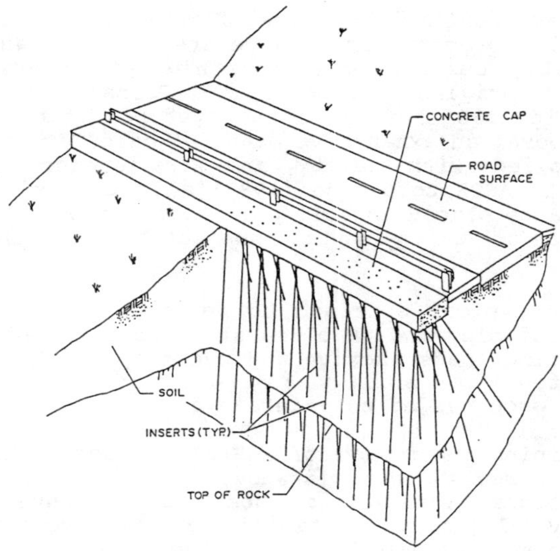

Figure 1. Typical arrangement of INSERTS (Boley and Crayne, 1984)

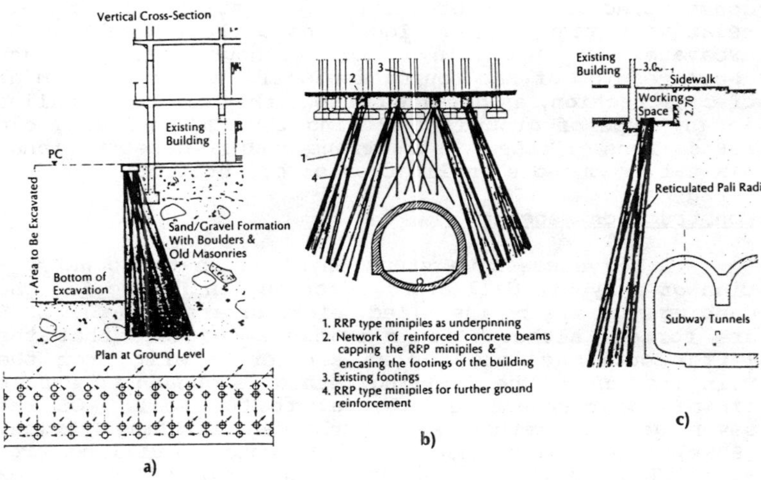

Figure 2. Applications of INSERTS a) for cut and cover and b) and c) around bored tunnels (Bruce, 1989a)

programs have confirmed that reinforcement stresses and overall wall movements in service are minimal, and that most probably designs have been highly conservative. Even their original proponent - Fernando Lizzi - confirmed in 1982 that "it is not yet possible to have at our disposal an exhaustive means of calculation ready to be applied with safety and completeness." In addition, an ASCE Committee (1987) also alluded to the great reliance placed on soil/pile interaction, the safe exploitation of "which is still subject to experience and intuition".

The typical approach to design is, of course, relatively simple, and involves standard basic steps:

- estimating loads (active and passive) on the wall;
- conducting a stability analysis to determine the shear force needed to maintain a required factor of safety;
- determining the number of INSERTS needed to provide the required shear resistance;
- calculatons (similar to those for a conventional gravity wall) to check stability against overturning, sliding and bearing failure at the base of the wall.

Usually the INSERTS are extended into bedrock where economically possible, but, in any event, always below the potential failure plane . INSERT Walls can be constructed in close proximity to existing buildings in relatively tight access locations without the need to excavate or underpin, and without causing any decompression of the foundation soil. Given their mode of construction, as detailed below, they can be installed in any type of ground, including through boulders, old foundations or other obstructions with no constructional limitation on hole inclination or orientation.

Construction Aspects

The successive steps involved in the construction of a Type A Wall are illustrated in Figure 3. The capping beam may be installed before or after the INSERTS are formed, although field evidence suggests that the latter option allows for an earlier benefit from the reinforcement. The drilling method is chosen to ensure minimal disturbance or upheaval to the soil. Of the seven generic methods of overburden drilling (Bruce, 1989b), the most common method is rotary drilling with water flush, either via a single casing or by the duplex method, depending on ground conditions. Once the casing has been advanced to target depth it is filled with a stable, high strength cementitious grout, and the permanent reinforcement is placed. This may be a solid

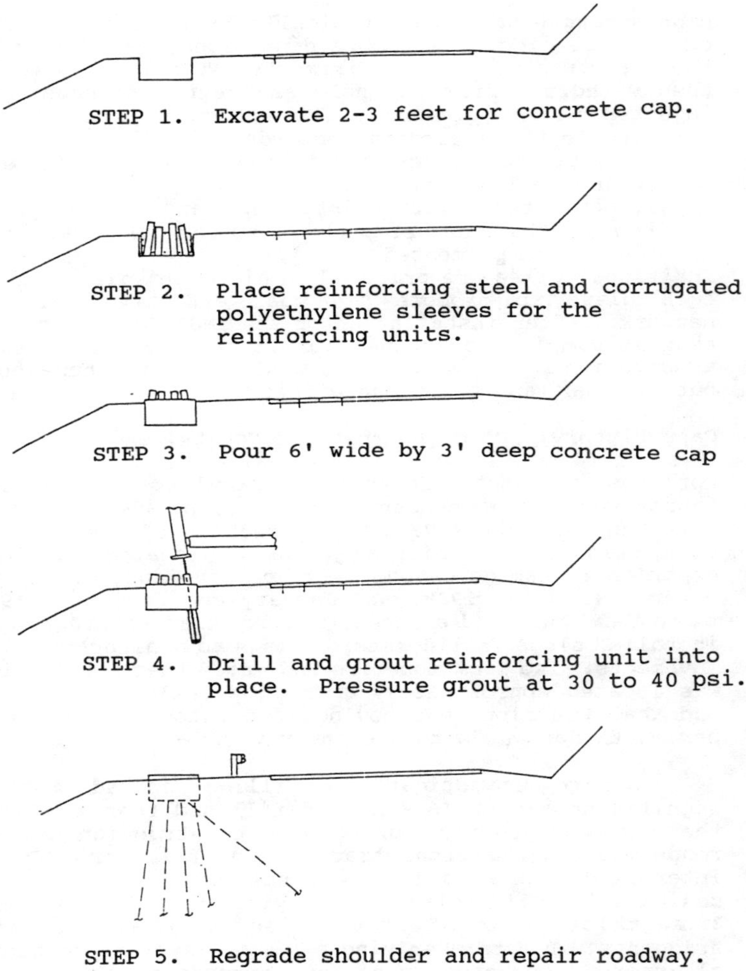

<u>Figure 3.</u> Typical steps in INSERT Wall construction (Boley and Crayne, 1984)

high strength steel bar, typically 25-50 mm in diameter, or a steel pipe of suitable dimensions, as dictated by the structural design requirements. The drill casing is then withdrawn from the hole as grout continues to be injected under pressure. The effect of the pressure grouting is three-fold in most conditions:

- it ensures all voids or drilling related disturbances to the soil are filled;
- it permeates a little into sands and gravels;
- it compacts somewhat soil around the pile that is too fine to be permeated.

Individual piles are oriented in different directions in each plane to promote the most effective soil/pile network. After installation of the INSERTS, the capping beam is simply graded over, or it can form the base of a guard rail or similar: the whole wall is thus wholly out of sight and maintenance free.

Case History: Road in Armstrong County, PA

Portions of State Route 4023 north of Kittanning, Pennsylvania, were constructed on a slope adjacent to the Allegheny River. A 73 m. long section of the two-lane road, and the railroad tracks located upslope, experienced damage caused by slope movements toward the river. In June, 1988, and January and February, 1989, the owner conducted a subsurface exploration program and installed slope inclinometer casings to monitor the slope movements. The inclinometers indicated that a slip-plane was located approximately 8 to 11 m. below the roadway and that the slope was moving at a rate of up to 18 mm per month downwards toward the river.

Site investigation drilling showed that a significant amount (6-9 m.) of fill had been placed at the site apparently during the construction of the roadway and/or railroad tracks. The fill consisted of intermixed loose to medium dense rock fragments and medium stiff silty clay. Underlying the fill was a 1.5-3 m. thick layer of stiff colluvial clay with rock fragments, in turn overlying a 1-6 m. layer of weathered claystone. Competent rock was encountered at about 15 m. below the roadway, and generally consisted of medium hard siltstones and sandstones.

The owner designed a repair of the failed section using an anchored caisson wall extending into competent rock. The earth pressures used for the design were based on the results of stability analyses, for which the soil along the slip-plane was assigned a residual friction angle of 17°. This design provided a minimum factor of safety with regard to the overall slope stability equal to 1.5 and 1.2 for the normal and rapid drawdown conditions, respectively. A row of 0.9 m. diameter caissons were foreseen at a center-to-center spacing of 1.4 m. and located immediately downhill of the roadway. The caissons were to be connected at the top by a cast-in-place reinforced concrete cap which was to have 27 m. long prestressed rock anchors extending underneath the roadway at 2.1 m. lateral intervals.

In 1989, the contractor proposed and the owner accepted an alternative design employing an INSERT Type "A" Wall. The wall consisted of four rows of pinpiles extending across the slip-plane and into competent bedrock. It comprised two equal length sections designated as Wall A and Wall B. Wall A contained a higher density of piles than Wall B, because the top of weathered rock dipped to a lower elevation in the area of Wall A which resulted in a larger volume of soil to be stabilized in this area. In general, Wall A contained 4 piles per lineal meter, and Wall B contained 3 piles per lineal meter. Besides providing a significant economic savings over the original design, the selection of the INSERT Wall allowed for one lane of roadway to remain open during construction (February to May, 1989). The wall was constructed as described above, with the cap poured after pinpile installation for practical reasons.

To monitor the INSERT Wall performance, two sections of the wall were instrumented with strain gauges, inclinometers, telltales, and survey pins. The inclinometers yielded the most useful information regarding the performance of the wall. The data for inclinometers located relatively close to and within the wall indicated that up to 38 mm of horizontal movement occurred during the 75-day construction period, but that a maximum of 8 mm of movement occurred in the 7-month period following the completion of the wall.

Overall, the inclinometer data indicated that the wall performed as expected, and had effectively stopped the slope movements at the site. These data also confirmed that some deflection of the relatively flexible INSERTS may be required to mobilize their lateral resistance.

Further details on the design, construction and performance of this project and several others of similar nature are provided by Pearlman et al. (1992).

Final Remarks

Two major restrictions have so far limited the

popularity of this type of in situ reinforcement in the United States. Firstly, there remains no rigorous design methodology and, secondly, often as a direct consequence, the high density of INSERTS required has often compromised the cost effectiveness of this novel method. However, fundamental research is underway to try to better understand the basic soil-structure phenomenon, while intensive field monitoring of installations is also clearly increasing knowledge. With a more precise design rationale, and more cost effective solutions, this technique offers a very potent option in combatting slope stability problems.

3. Large Diameter Deep Drainage Wells

Background

Drainage has long been recognized an effective method to stabilize landslides in saturated soil slopes. A valuable new technique introduced recently in Italy extends this potential. In essence, it features deep large diameter drainage wells, interconnected at their toes by a small diameter discharge pipe. The technique - registered as RODREN - had its first applications to stabilize several landslides affecting urban areas and other important structures, principally along the A1 Motorway, between Bologna and Florence, in the Tuscan-Emilian Apennines zone. The effectiveness of RODREN, and its favorable cost/benefit ratio, have encouraged several further uses in other regions with different geological conditions but similar problems.

General Features

RODREN consists of alignments of separate vertical drainage wells, 1200-1500 mm in diameter and 5-7 m. apart, connected near their toes by a pipe, 76-100 mm in diameter. This collector pipe transfers by gravity the groundwater intercepted by each well, and eventually exits at a convenient discharge or pump out location. The depth of the wells depends on the geological features of the slope, and it is important that the discharge pipe is located below the slip surface. The depth of RODREN is normally limited to about 30 m. although, in favorable geological conditions, wells of 2 m. diameter, 20 m. apart, have been drilled to depths of over 50 m. (Bruce and Bianco, 1991).

As the system is not structurally continuous, say like a drainage trench, it is particularly convenient when shallow obstructions are present, such as buried utilities in urban areas. Other advantages of the system are that it is:

- highly efficient and effective, especially in restricted areas;
- environmentally compatible and safe;
- cost effective, in comparison with equivalent drainage systems, such as continuous drainage trenches;
- possible to adapt the layout to the actual geometry and geology of the unstable slope, as revealed during construction.

Another important technical feature of RODREN relates to the flow rate conditions: the discharge rate responds to the actual natural hydraulic conditions, driven by external factors such as rainfall patterns, and the effectiveness of any pre-existing surface collector systems. It is not an active pumpout system in which the piezometric level is reduced by electric submersible pumps in deep, cased boreholes.

Construction Aspects

Shafts are drilled using the standard equipment commonly employed for bored piles requiring temporary steel casing. Before backfilling and withdrawal of the casing, each shaft is connected to the adjacent wells by a horizontal hole, drilled just above the shaft toe (<u>Figure 4</u>). The wells are then backfilled using a selected drainage filter material, graded according to the surrounding soil grain-size characteristics. Generally the grain size distribution of the filter is between 3 and 20 mm, with no more than 3% passing the 200 ASTM sieve.Several types of wells can be constructed:

- standard drainage wells, as just described;
- inspection drainage wells, having a permanent steel inner casing, for inspection and maintenance purposes (1500 mm outside diameter);
- structural drainage wells, having a reinforced concrete liner of minimum thickness 300 mm.

For these three types, the following steps are taken:

a) Vertical drilling operation: wells are drilled using a temporary casing. The thickness and the strength of the casing are chosen to guarantee adequate safety for the operators.
b) Horizontal drilling operation: the hole for the horizontal pipe is drilled by means of a special rig having a circular frame sized to operate inside the vertical steel liner and to safeguard the operator. All engines and power packs are placed outside the shaft.

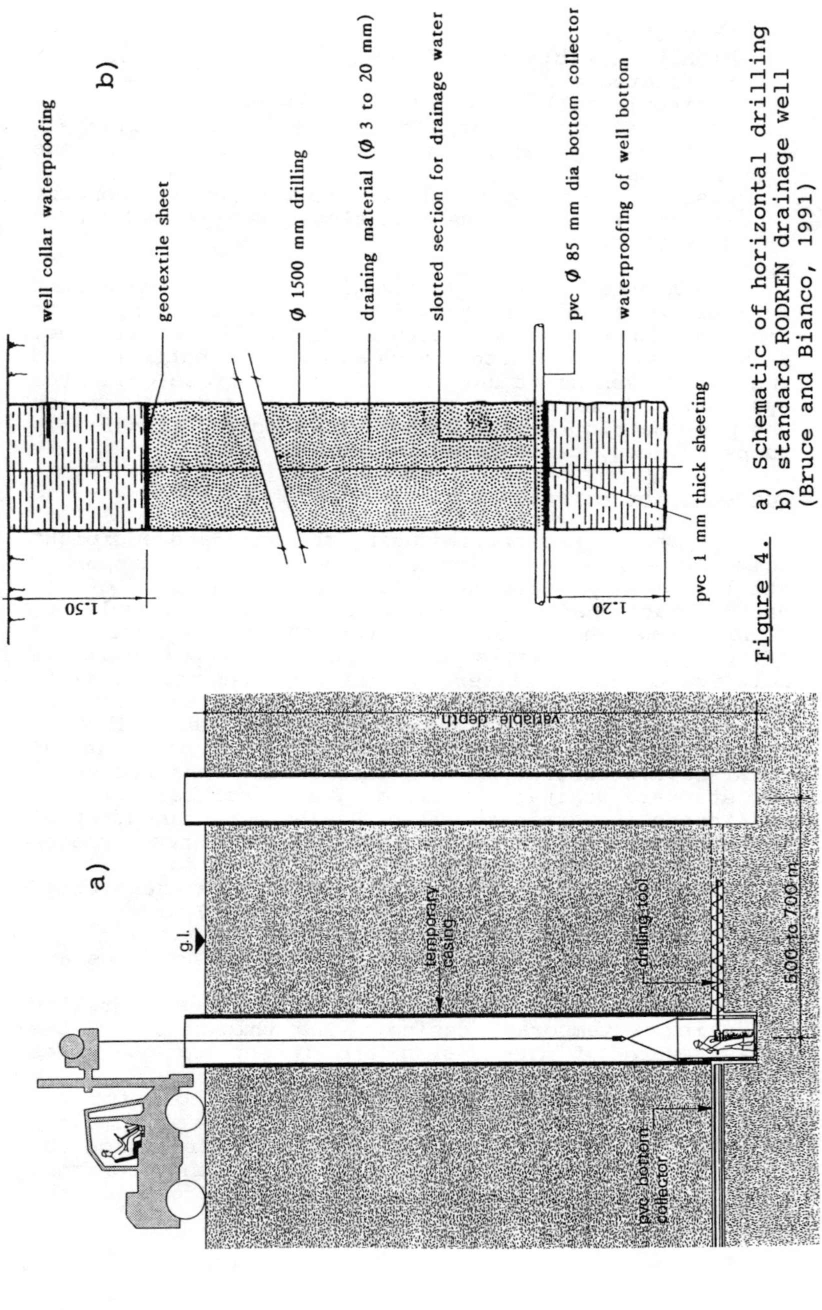

Figure 4. a) Schematic of horizontal drilling
b) standard RODREN drainage well
(Bruce and Bianco, 1991)

c) Waterproofing of the bottom of each well and the pipe connection with concrete.

d) Well completion: depending on the type of well, different procedures are applied:
   1) standard drainage wells: the drain material is poured into the well, and the temporary casing then removed.
   2) inspection wells: a permanent steel casing, having a smaller diameter (generally 1200 mm), is installed inside the temporary casing. The drain filter material is then placed around it in the annulus and the temporary drill casing is then removed as before.
   3) structural wells: in this case the diameter of the wells is increased to a minimum of 2 m. A drain filter is first placed, applying the same procedure as the inspection wells (permanent casing of 1800 mm diameter). Then a second steel liner, 1200 mm in diameter, is placed and the annular space concreted. If necessary, the concrete liner may be reinforced by means of a steel cage.

e) Waterproofing of the top and, for inspection and structural wells, closing of the wells by steel covers, completes the operation.

All construction steps, and particularly the horizontal drilling, are conducted under special safety measures, such as forced fresh air supply, safety belts and cables, phone, dewatering pipes, and remote TV survey of horizontal drilling operations.

Case History: Ancona Palombella, Italy

In December, 1982, a suburban area of Ancona, the District capital of the Marche region in central Italy, was affected by a catastrophic landslide involving more than 3,000,000 sq. m. of surface area (Hegg, 1985).

Geologically, this area features a Pliocene basement (grey overconsolidated clay), with several thin sandy layers, covered by a silty-clayey stratum, from 10 to 40 m. thick. In the upper part of the slope a large Pleistocene sandy layer is confined between these two principal formations. At the toe of the slope, the Adriatic Sea produces a progressive erosion. The unfavorable hydrogeological conditions of the slope, named the Balducci Landslide, were known for centuries, and so the expansion of the city had been limited to other areas. However, as the Balducci Landslide is close to Ancona Harbour, some important elements of the infrastructure, such as railways and highways, had to be located on the lower part of the slope, along the coast.

To permit improvements to the coastal highway, it was therefore necessary to execute a significant stabilization of the east side of the slipped area (Palombella). In Figure 5, one typical group of wells is shown. Due to the unfavorable stratigraphy, the wells were drilled to depths of over 50 m. This necessitated the solution of many technological problems, such as tolerance in vertical and horizontal drilling, the power required to drill very competent bedrock, hydraulic and earth pressures on permanent casing, backfilling operations, and so on. For these reasons, it was decided to increase the standard well dimensions of the RODREN system to a diameter of 2.0 m., and the spacing to 15-20 m. To check the accuracy of drilling operations, the verticality of the wells was measured combining displacements and rotations (polar coordinates). For the alignments completed to date, deviations of about 1% have been recorded.

The work is currently still in progress, but already very encouraging piezometric and inclinometric data have been recorded. As the well drilling has been carried out starting from the toe of the slope, the flow discharge of each alignment increases progressively. A significant change in flow rate is obtained when each well group intercepts the Pleistocene sandy formation present in the upper part of the slope. For Alignment B (Figure 5), the flow increased from 10 to 30 liters/min. at this stage.

Final Remarks

The RODREN technique of providing slope stability through water table lowering is clearly a proven success when applied in the appropriate set of hydrogeological conditions. Recent steps have been taken to guarantee the personal safety of the operators who must drill the horizontal connecting holes from well to well. Increasing experience with the method is reflected in a flexibility in the design-build process to optimize technical and economic return. One of the main attractions of the method is its environmental compatibility. It is not a forced drawdown dewatering system requiring active pumping: rather it is a maintenance free, gravity driven interceptor array, responding directly to the ambient conditions. The advantages of RODREN are being exploited in Italy and are currently favored for stabilizing very sensitive environmental and urban areas in Spain and Southern California. An expansion of the technique into other areas with similar problems and restraints would appear

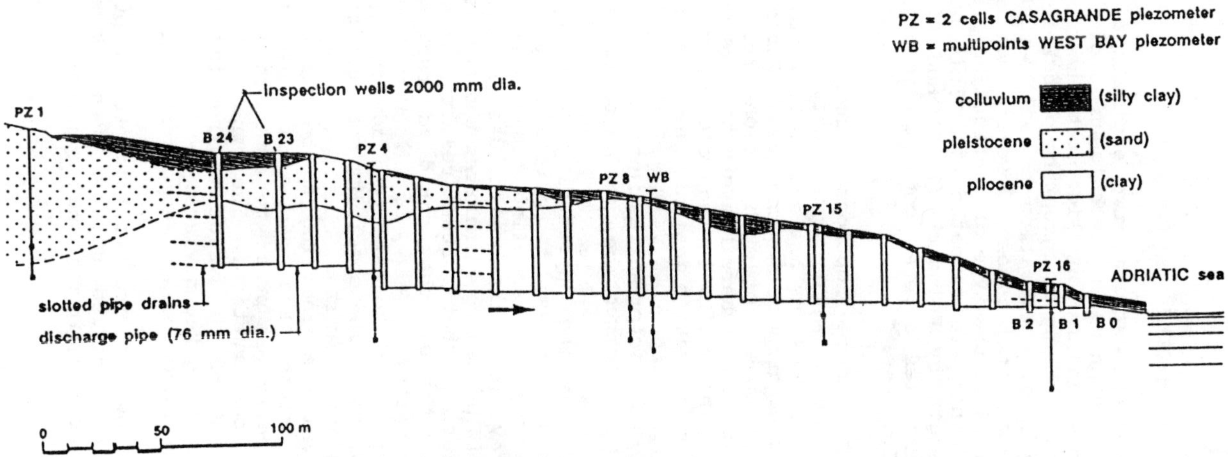

Figure 5. Cross section through RODREN Alignment B.
Ancona, Italy
(Bruce and Bianco, 1991)

most likely.

4. Contractual Frameworks

This paper illustrates two entirely different specialized geotechnical techniques of great technological potential in landslide stabilization. One owes its roots to drilling and grouting principles while the other is a novel application of well established groundwater drainage techniques. However, such new ideas can only be applied if the type of contracting practices adopted by the owner are sympathetic. This has been discussed at length by Wolosick (1989) and Nicholson (1990), who identified several contract specification options available.

Many of these options, such as open specifications, performance specifications, prebid designs and post-bid designs are excellent vehicles for introducing new cost effective techniques. For example, the advantages to owners of such options relative to "closed" specifications (wherein everything is designed by the owner from the onset) include the following:

- They allow for innovative designs, resulting in lower construction costs.
- The specialty contractor shares in the responsibility for the design, construction, and performance of the completed system.
- The engineer has the opportunity to review each contractor's design and to critique the various design elements, thus enabling him to incorporate necessary revisions into each design prior to bidding. This procedure eliminates the confusion or misinterpretation that often results in design changes after the bid that could adversely affect construction costs.
- They allow the owner to take full advantage of the experience of qualified specialty contractors.
- Design costs are minimized.
- They allow value engineering to be utilized more effectively since the value engineering is essentially performed prior to contract award and the start of construction. The owner retains one hundred percent of the value engineering savings rather than sharing the savings.
- They minimize change orders and contract disputes.

Assuming owners are willing to adopt such options, the novel techniques outlined in this paper have an excellent chance to gain national application.

References

A.S.C.E. (1987), "Soil Improvement - A Ten Year Update". Proc. Symp. at ASCE Convention, Atlantic City, NJ, April 28.

A.S.C.E. (1990), "Design and Performance of Earth Retaining Structures". Proc. Conf. at Cornell University, Ithaca, N.Y., June 18-21, 904 pp. Edited by P.C. Lambe and L.A. Hansen.

A.S.T.M. (1991), International Symposium on Slurry Walls, Atlantic City, NJ, June 27-28.

Boley, D.L. and Crayne, L.M. (1984), "INSERT Wall Saves Roadway". Highway Builder, Fall, pp. 14-16.

Bruce, D.A. and Jewell, R.A. (1986, 1987), "Soil Nailing: Application and Practice". Ground Engineering, 19 (8), pp. 10-15, and 20 (1), pp. 21-32.

Bruce, D.A. (1988, 1989), "Aspects of Minipiling Practice in the United States". Ground Engineering 21 (8), pp. 20-33 and 22 (1), pp. 35-39.

Bruce, D.A. (1988a), "Developments in Geotechnical Construction Processes for Urban Engineering". Civil Engineering Practice, 3 (1), Spring, pp. 49-97.

Bruce, D.A. (1989a), "American Developments in the Use of Small Diameter INSERTS as Piles and Insitu Reinforcement". DFI Int. Conf. on Piling and Deep Foundations, London, May 15-18, 12 pp.

Bruce, D.A. (1989b), "Methods of Overburden Drilling in Geotechnical Construction: A Generic Classification". Ground Engineering, 22 (7), pp. 25-32.

Bruce, D.A. and Bianco, B. (1991), "Large Landslide Stabilization by Deep Drainage Wells". Proc. Int. Conf. on Slope Stabilization Engineering, Shanklin, Isle of Wight, April 15-19, 8 pp.

Hegg, U. (1985), "Case History of a Large Coastal Landslide in Pliocene Clays". Proc. 4th Intl. Symp. on Landslides, Toronto.

Lizzi, F. (1982), "The Pali Rodice (Root Piles) Symp. on Rock and Soil Improvement, AIT, Bangkok, Nov. 29 - Dec. 3, Paper D-1, 21 pp.

Nicholson, A.J. (1990), "Contracting Practices for Earth Retaining Structures". As Reference 1, pp. 139-154.

Pearlman, S.L., Campbell, B.D., and Withiam, J.L. (1992), "Slope Stabilization Using In-Situ Earth Reinforcements". This Conference, 16 pp.

Wolosick, J.R. (1989), "Contract Specification Options for Retaining Wall Design and Construction: Discussion of Construction Alternates and Case Histories". Proc. 40th Highway Geology Symposium, Birmingham, AL, May 17-19, 16 pp.

Xanthakos, P. (1979), "Slurry Walls". McGraw-Hill Inc., ISBN 0-07-072215-3, 622 pp.

A Benchmark Slope Stability Study

Jose L.M. Clemente[1], M.ASCE

Abstract: A benchmark slope stability study was conducted during the fall of 1990 and spring of 1991. The purpose of the study was to assess how slope stability analysis results were affected by software users. The study was not intended to be a software calibration study, but it served as a software survey. A rather straightforward slope stability problem was devised and mailed to more than 60 people in the industry and academia. Responses were obtained from 13 people who ran a combined total of 20 analyses using 10 different software packages. This paper describes the problem submitted for analysis, and requested answers and a summary of the responses obtained. It was observed that all but one of the computed minimum safety factors fell within 5% of the average of all values reported. However, critical failure circles and the location of their respective centers disclosed a more significant scatter.

[1]Sr. Engr., Bechtel Corp., 9801 Washingtonian Blvd., Gaithersburg, MD 20878. Formerly Sr. Geotech. Engr., Law Engrg., 8940-A Route 108, Columbia, MD 21045

SEISMIC ANALYSIS AND DESIGN OF LINED WASTE FILLS: CURRENT PRACTICE

by

Raymond B. Seed[1] and Rudolph Bonaparte[2], Members, ASCE

ABSTRACT

Prior to the slippage which occurred in a waste fill at the Kettleman Hills Waste Repository on March 19, 1988, there was relatively little effort devoted to the general study of seismic stability and performance of solid waste landfills and repositories in the United States. Seismic stability issues were not consistently addressed, even in seismically active regions, and both standards and practices varied widely between geotechnical consulting organizations and even within many firms.

Over the past four years (1988 - present), major changes have occurred in the field of "seismic engineering" of solid waste landfills. This paper presents a survey and commentary on current U.S. practice in this area, addressing current major regulations as well as engineering practices, and also makes some comments regarding directions for further developments in the years ahead.

INTRODUCTION

In the wake of the Kettleman Hills Waste Repository (non-seismic) stability failure (discussed in the companion paper in these Proceedings by Byrne, et al., 1992), the past four years have seen a substantial increase in interest in the

---

[1]Professor, Department of Civil Engineering, 440 Davis Hall, University of California, Berkeley, CA 94720

[2]Vice President, GeoSyntec Consultants, 5950 Live Oak Parkway, Suite 330, Norcross, GA 30093

evaluation of both the static and seismic stability and performance of lined waste fills. Spurred jointly by interested clients, engineers, and regulatory agencies, the ongoing evolution of practice in areas related to seismic stability and performance of waste fills has progressed unusually rapidly. Engineers working in these areas, as well as clients and regulatory agencies, currently find it a challenge to identify what appear to be very rapidly evolving standards of practice. This situation is further complicated by the periodic promulgation of new regulatory requirements, and by uncertainties regarding interpretation of recent regulations as well as their application to some sites and conditions.

This paper presents the results of a survey of "practice" in the area of seismic stability and performance of lined waste fills. Five major geotechnical consulting firms, highly active nation-wide in this field, were surveyed with regard to their practices during the 1991 calendar year, and also with regard to the directions of evolution of their currently evolving practices at present. The results of the survey are discussed, along with major applicable Federal and State regulations, and serve to at least partially define "current" practice among the selected group of firms. This, in turn, is contrasted in several specific areas with the somewhat broader range of practices sometimes encountered by the authors within the field at large. Finally, several suggestions are made both for further research studies, and for modifications in some fairly common current practices in the field at large.

GENERAL ISSUES

A majority of the solid waste landfills in the U.S. can be divided, based on landfill and foundation configuration, into one of the three general categories illustrated schematically in Figure 1. Figure 2 shows examples of some of the principal potential modes of slippage, or large-displacement sliding, associated with these general landfill types. Slippage modes (a) through (d) are illustrated for a partially-filled "covered pit" type of landfill (the covered pit type is often inherently relatively stable with regard to deep-seated slippage after completion of fill placement), but these four modes are also applicable to "upslope" and "surface" waste fills. Another potentially significant slippage mode not illustrated is localized basal slippage, either on sliding surfaces within the base liner system, or on weak underlying foundation soils. Finally, Figure 3 illustrates two potential slippage modes associated with final cover systems and final fill slopes. These are by no means all of the potentially significant modes of slippage to consider in landfill design, but they represent an illustrative collection of many of the modes which commonly impact design and/or design studies. Additional examples are presented and discussed in the companion paper in these Proceedings by Mitchell and Mitchell (1992).

The slippage/displacement potential for many of these modes is controlled to large degree by the shear strengths developed at (or near to) interfaces within

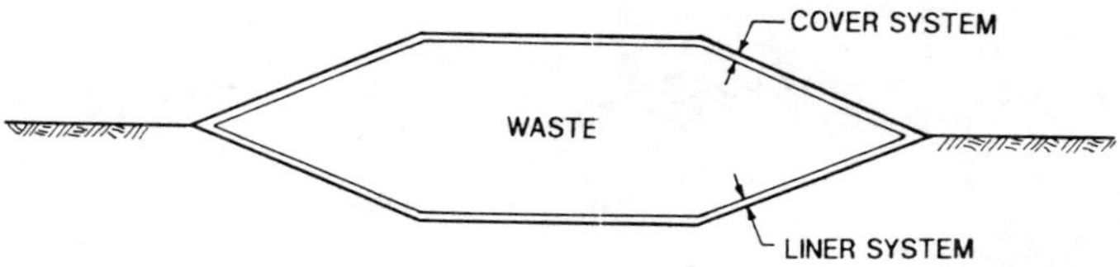

(a) "Covered Pit" Waste Fill - Typically Used for Relatively Level Sites with Firm Foundations (e.g.: Stiff Alluvium or Bedrock, and with Deep Water Tables.)

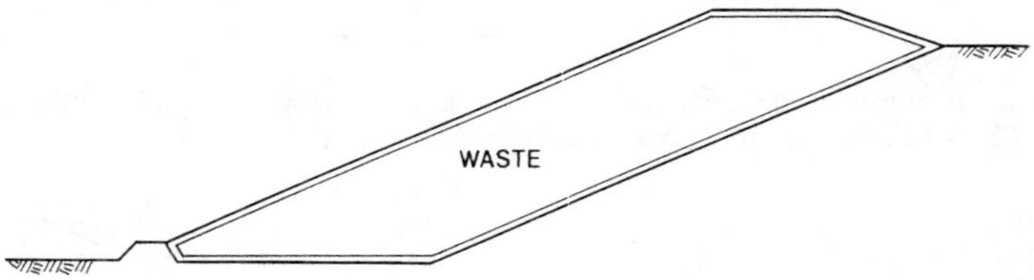

(b) "Upslope" Waste Fill - Typically Used in Sloping Canyons and Swales.

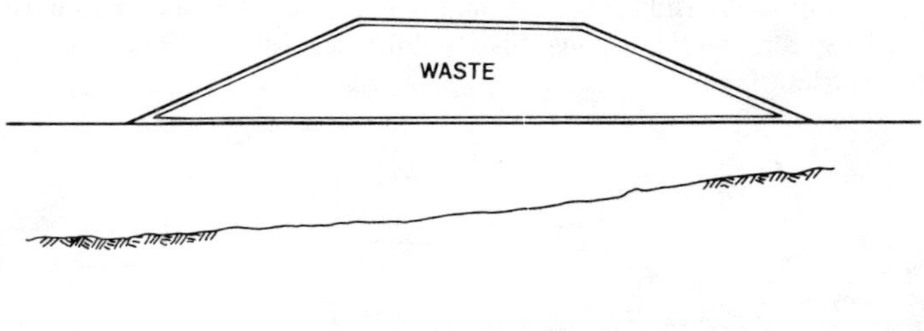

(c) "Surface" Waste Fill - Typically Founded on Alluvium (Often with Relatively Shallow Water Tables.)

Figure 1. Schematic Illustration of Three Common Solid Waste Fill Configurations

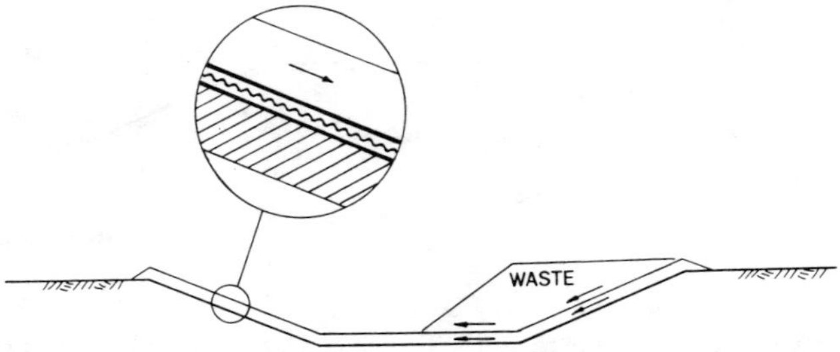

(a, b) Slippage and/or Fill Displacement Along Shear Surfaces Within the Liner System - Partially Completed Fill

(c, d) Side Slope Movement and/or Displacement of Fill and/or Impoundment Faces - Partially Completed Fill

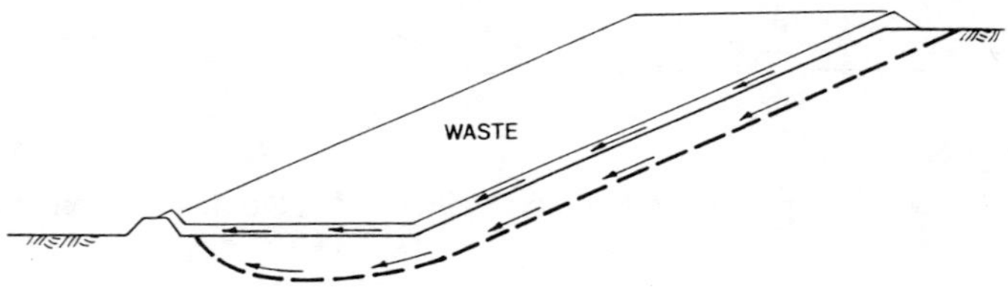

(e, f) Slippage/Displacement Primarily Along Shear Planes Within the Liner System and/or Along Sliding Surfaces within Underlying Geologic Materials

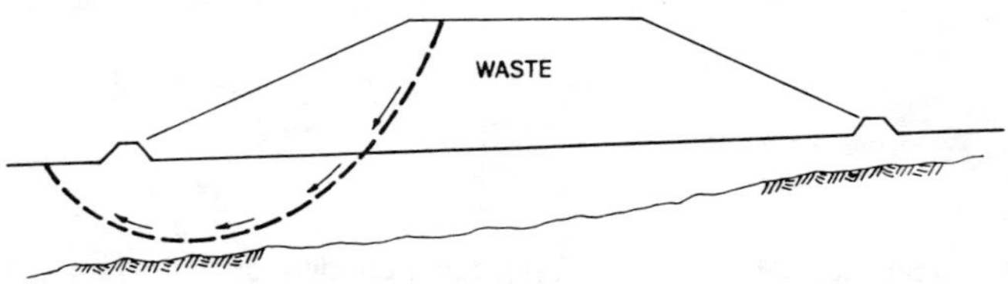

(g) Slippage/Displacement Through Deep Alluvial Foundation Soils

Figure 2. Schematic Illustrations of Examples of Potential Waste Fill and/or Foundation Failure/Slippage Modes

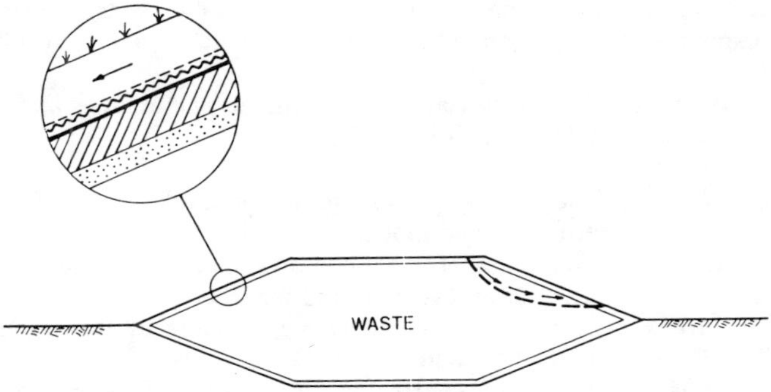

Figure 3. Schematic Illustration of Potential Modes of Slippage and/or Displacement of Final Fill Slopes and Surface Cover System

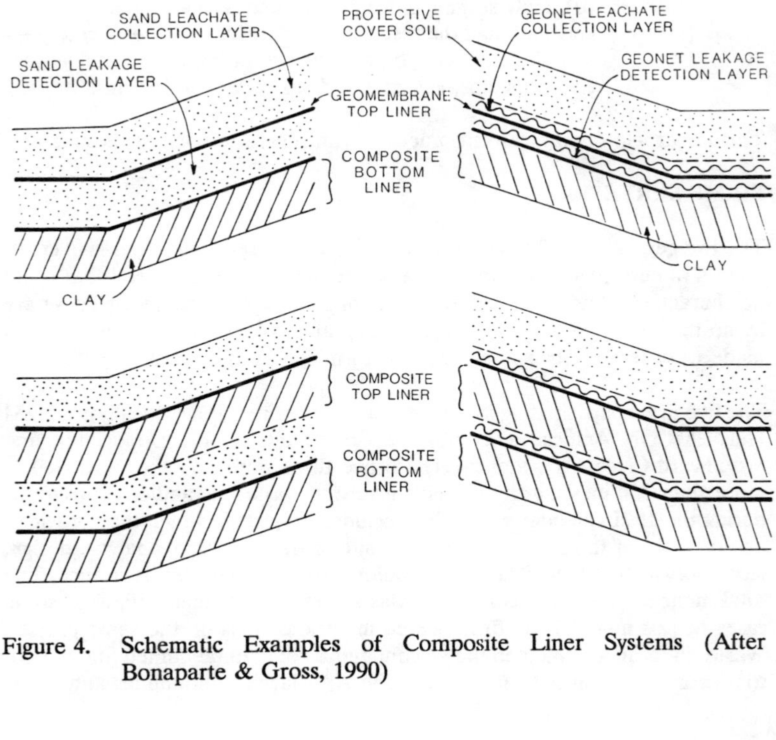

Figure 4. Schematic Examples of Composite Liner Systems (After Bonaparte & Gross, 1990)

the multi-layered soil and geosynthetic liner systems underlying the fills and providing surface coverage. Figure 4 shows several illustrative examples of multi-layered liner systems, and additional examples are shown in the companion papers by Byrne et al. (1992) and Mitchell and Mitchell (1992).

Some of the principal issues associated with seismic analysis and design of lined solid waste fills include the following:

(i) Selection of appropriate "design level" seismic events and/or strong ground motions.

(ii) Evaluation of the impact of local site conditions on the design level event ground motions.

(iii) Evaluation of the dynamic response of the waste fill and foundation system.

(iv) Evaluation of dynamic properties and shear strengths for: (a) the waste fill, (b) the base liner system(s), (c) the underlying foundation materials, and (d) the final surface cover system.

(v) Evaluation of either: (a) overall "seismic stability" using simplified analysis methods (e.g.: pseudo-static analysis), or (b) likely seismic deformations and displacements.

(vi) Evaluation of the resulting ramifications with respect to post-seismic integrity and performance of basal liner and surface cover systems, leachate and gas collection systems, etc.

This paper will discuss current practices with regard to each of these issues.

FIRMS SURVEYED

The five geotechnical firms surveyed were carefully selected to obtain a cross-section of current seismic waste fill engineering practice at a level that could be characterized as "significantly better than average" at present. These are by no means the only five firms practicing at this level, and no attempt was made to systematically survey firms performing at other levels.

The five firms participating in the formal survey were (alphabetically): CH2M Hill, EMCON Associates, GeoSyntec Consultants, Golder Associates, and Woodward-Clyde Consultants. These five firms were, together, involved in more than 350 engineering projects involving waste fills and/or repositories during the 1991 calendar year. This includes projects in which they provided one or more of the following services: site characterization, analysis, design, laboratory testing, technical review, and/or construction quality assurance. The total number of waste fills involved was somewhat less than 350, as in some cases two or more of the firms served in various roles on the same project. Waste fill locations spanned the 48 contiguous continental United States, with some concentration on states east of the Appalachian Mountains and states west of the Rocky Mountains, and also included several waste fill projects in Alaska. The firms surveyed will not be individually identified or specifically associated with individual responses in the remainder of this paper.

REGULATIONS

Regulations addressing the siting and design of solid waste landfills in seismically active regions of the country exist at the federal level, and, in a few cases, at the state level. Perhaps the most far-reaching regulations at the federal level are those for municipal solid waste landfills, promulgated by the U.S. Environmental Protective Agency (EPA) on 9 October 1991. These regulations, which will become effective on 9 October 1993, are contained in Section 258.14 of Title 40 of the Code of Federal Regulations (40 CFR 258.14) and they are summarized in Table 1. As seen in this table, these federal regulations require that containment systems for landfills constructed in seismic impact zones be designed to resist the maximum horizontal acceleration (MHA) in lithified earth material at the site. Containment systems covered by this requirement include liners, leachate collection systems, and surface-water control systems. Seismic impact zones are defined by EPA as areas with a ten percent or greater probability that the MHA in lithified earth material will exceed 0.1 g in 250 years. Based on the probabilistic estimates of maximum horizontal acceleration developed and presented as MHA "maps" by Algermissen et al. (1982), areas of the country that will be designated as seismic impact zones by this definition include, for example, most of California and parts of Washington, Oregon, Arizona, Nevada, New York, Massachusetts, South Carolina, Tennessee, Utah, and Idaho. Clearly, landfills in many areas of the country will be affected by the new EPA regulations.

California presently has the most comprehensive state regulations for the siting and design of landfills in seismically active regions. The California requirements include those of Title 23, Chapter 15 and Title 14, Chapter 3 of the California Code of Regulations (23 CCR Chapter 15 and 14 CCR Chapter 3). These, too, are very briefly summarized in Table 1. The regulations in Title 23 are primarily concerned with water quality protection and therefore address landfill siting and containment system design and construction. Title 23 requires that Class I (hazardous waste) and Class II (designated waste) landfills be designed to withstand the maximum credible earthquake (MCE), which is defined by the California Division of Mines and Geology (CDMG) as the "*maximum earthquake that appears capable of occurring under the presently known tectonic framework. It is a rationally believable event that is in accord with all known geologic and seismologic facts.*" Title 23 requires Class III (municipal solid waste) landfills to be designed to withstand the maximum probable earthquake (MPE), which is defined by the CDMG as "*the maximum earthquake that is likely to occur in a 100 year interval*". Title 23 requires that landfill containment systems, such as liners, leachate collection systems, gas

TABLE 1
SUMMARY OF SELECTED FEDERAL AND STATE REGULATIONS FOR SITING AND DESIGN OF LANDFILLS IN SEISMICALLY ACTIVE REGIONS

JURISDICTION/AGENCY	SITING REQUIREMENTS	DESIGN REQUIREMENTS
Federal/EPA (40 CFR Section 258; "Subtitle D")[1] (Effective Date: October 9, 1993)	• No siting within 200 ft of a fault having displacement in Holocene time[2].	• Facilities in seismic impact zones must have designs for all containment structures, including liners, leachate collection layers, and surface-water control systems, capable of resisting the maximum horizontal acceleration (MHA) in lithified earth material for the site. • "Seismic impact zone" is an area with ≥ 10% probability of MHA exceeding 0.1 g in lithified material in 250 years. • MHA is the acceleration for which the probability of exceedance in 250 years is 10%.
California/RWQCB (23 CCR Chapter 15; "Chapter 15")[1] (Effective Date: Now)	• No siting on a known Holocene fault.	• Facilities must be designed to withstand the maximum probable or maximum credible earthquake without damage to the foundation or to structures which control leachate, surface drainage, erosion, or gas.[3]
California/IWMB (14 CCR Chapter 3; "Title 14")[1] (Effective Date: Now)		• A foundation stability report must be prepared for landfills located: (a) in areas subject to liquefaction, or (b) unstable areas with poor foundations. • The minimum factor of safety for the critical final landfill slope, and the foundation that supports the slope, under dynamic loading conditions must be 1.5 (see alternate allowable design basis below). • In lieu of the above factor of safety, a more rigorous analytical method that provides a quantified estimate of the magnitude of movement of the final landfill slopes may be used; the more rigorous method must demonstrate that the final cover accommodates the calculated movement without jeopardizing the integrity of the cover or the other environmental control systems. • The evaluation must be based on the MHA associated with the maximum credible or maximum probable earthquake, depending on the classification of the landfill.[3]

Notes: [1] EPA = U.S. Environmental Protection Agency; RWQCB = California Regional Water Quality Control Board; IWMB = California Integrated Waste Management Board; CFR = Code of Federal Regulations; and CCR = California Code of Regulations. Common name of regulation is given in quotes.

[2] Variances from these restrictions are allowed under certain conditions.

[3] The "Maximum Credible Earthquake" (MCE) is to be used for Class I (hazardous waste) and Class II (designated waste) landfills. The "Maximum Probable Earthquake" (MPE) is to be used for Class III (municipal solid waste) landfills.

collection systems, and surface-water collection systems, be designed to resist the regulatory earthquake event without damage to the structure.

Title 14 addresses the closure of solid waste landfills (Classes I, II, and III). The Title 14 regulations include the requirement that the final slopes of the landfill, and the foundation beneath the slopes, have a minimum factor of safety of 1.5 under seismic loading conditions. The analysis of the slopes must account for the maximum credible or maximum probable earthquake event, depending on the classification of the landfill (as in Title 23 above). While not explicitly stated in the regulations, the analysis will normally be based on pseudo-static limit equilibrium slope stability methods. The seismic coefficient for the analysis must be based on the anticipated MHA in rock at the site during the regulatory earthquake event (MCE or MPE). The regulations allow the MHA in rock to be modified to account for the local soil conditions and for the presence of the waste fill. Obtaining a pseudo-static factor of safety of 1.5 in the manner described may be difficult, if not impossible, in many cases. In recognition of this fact, Title 14 regulations also allow an alternate design basis involving evaluation of seismic displacements and/or deformations as: "*In lieu of achieving a factor of safety of 1.5 under dynamic conditions, a more rigorous analytical method that provides a quanitified estimate of the magnitude of movement may be employed. In this case, the report shall demonstrate that this amount of movement can be accommodated without jeopardizing the integrity of the final cover or environmental control systems.*" Largely as a result of the difficulty and cost of achieving a factor of safety of 1.5 under pseudo-static conditions (with relatively high levels of MHA associated with either MCE events [Class I & II fills] or MPE events [Class III fills]), these alternate and more detailed methods of analysis of seismic movements are currently the most common basis for seismic design of waste landfills in California.

Most states other than California do not have extensive requirements for the siting and design of landfills in seismically active regions. A few states provide either regulations or guidance regarding seismic coefficients for pseudo-static slope stability analyses, but most other states do not directly address the issue of seismic design in any detail. This situation is likely to change, however, in response to the new EPA regulations as codified in 40 CFR 258.

REGIONAL VARIATIONS IN PRACTICE

All five of the engineering firms surveyed reported significant variations in practice based on project site locations. All five firms devote significant emphasis to seismic stability and performance issues for waste fills in the state of California, where seismicity is relatively high and state regulations mandating seismic design requirements are currently most rigorous. All five firms also devote similar emphasis, and employ similar analytical methods and design standards, to the seismic design of waste fills sited in the other "west coast" states of Oregon and Washington.

Four of the five firms employed similar or slightly relaxed levels of seismic analysis and design in selected additional areas including: the State of New York, the New Madrid Fault region, the Charleston (South Carolina) area, and the Wasatch Fault region in Utah. (The fifth firm had no projects in these particular regions during 1991.) Very few of the waste fill projects with which these firms were involved in 1991 in other areas had significant seismic study. Even in other localized regions of higher seismic hazard exposure, however, the surveyed firms reported becoming involved "after the fact" in a number of projects in which previous engineering efforts had not addressed seismic issues in detail.

These findings reflect the authors' own experience and contacts with additional projects and organizations. At present, seismic engineering is consistently implemented for essentially all waste fills in California, and similar standards and design are fairly commonly implemented in Washington and Oregon. Practice in other seismic impact areas is considerably "softer": the more technologically advanced geotechnical firms are increasingly applying relatively high standards of seismic analysis and design in localized "seismic impact areas", but it is not uncommon to find other examples in these same areas of lower levels of practice and even projects which are currently well-advanced without any apparent formal consideration of seismic issues. Indeed, several of the interviewed firms that had recommended seismic studies and design in some of these areas reported that these recommendations came as something of a surprise to clients, regulatory personnel and/or other engineering organizations involved in the project. Away from the major "seismic" zones, in areas comprising a majority of the U.S., relatively little formal seismic analysis and design presently occurs in connection with waste fills.

DESIGN MOTIONS

All five of the firms surveyed employ design motions <u>at least</u> consistent with the requirements of California's "Title 14" and "Title 23" (see Table 1) for solid waste landfills sited in California. As explained previously, these regulations require use of an MHA associated with the "maximum credible earthquake" (MCE) for Class I and Class II landfills, and the "maximum probable earthquake" (MPE) for Class III landfills. In practice, Class III landfills in California are sometimes designed for an MHA greater than that associated with the MPE event (a) in areas where a generally advanced level of overall seismic geotechnical practice (for projects other than waste landfills) has led to common usage of higher design levels, or (b) at sites where local seismicity appears to call for special treatment.

There is little regulation (or even guidance) for the selection of earthquake magnitudes (or durations) associated with the MPE or MCE design events. As a result, geotechnical firms are left largely on their own to select appropriate design levels. This (magnitude or duration) can have a significant impact on calculated seismic displacements/deformations. Interestingly, all five of the firms surveyed make similar assessments of faults and seismicity in the vicinity of a fill site, and all develop similar values of "magnitudes" associated with design-level MHA's. These magnitudes are generally in the range of 6 ≤ Design Magnitude ≤ 8, and depend primarily on local and regional fault characteristics. These important magnitude "values" are not simply "assumed".

The Federal criteria of "Subtitle D" (see Table 1) mandate the use of an MHA associated with a 10% probability of exceedance in 250 years (or, very approximately, the 2,500-year return-interval acceleration) in "seismic impact zones". This is an intermediate seismic event, being somewhat more severe than the MPE design basis required for Class III landfills in California, and somewhat less severe than the MCE design required for California's Class I and II landfills. By way of illustration, Figure 5 presents a plot of MHA (on "rock") for a site in central California, based on probabilistic seismic hazard studies. The MPE, "Subtitle D", and MCE values of MHA for this site are approximately 0.2 g, 0.28 g and 0.32 g, respectively. The "spread" between these three levels of MHA can become much greater in cases where a fault of relatively low activity (low annual probability of producing a major event) exists in very close proximity to a waste fill site. In such cases, the "MCE" level of MHA can be very much higher than either the MPE level or the "Subtitle D" level. The Federal criteria of "Subtitle D" do not address well the issue of selecting a design event magnitude (or duration) associated with the MHA.

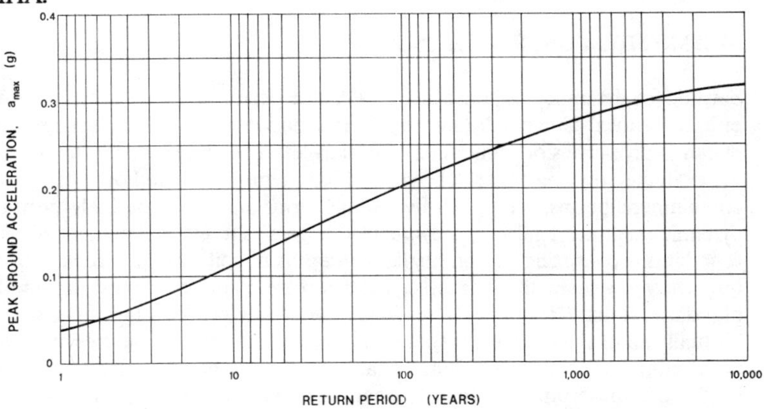

Figure 5. Peak Horizontal Ground Acceleration (or "MHA") vs. Return Period Based on Probabilistic Seismic Hazard Studies for a Site in Central California (After Seed & Marachi, 1987)

All five of the firms surveyed used "California-like" levels of MHA (MCE for Class I & II, MPE or greater for Class III) for waste fills in Oregon and Washington in 1991, and all are now also considering the "Subtitle D" criteria as well. Away from the "west coast states", in other seismic impact zones, the use of MCE levels (and magnitudes or durations) was highly unusual, and the five firms more typically employed levels roughly analogous to those of the now-pending "Subtitle D" criteria. It should be noted that these 10% probability of exceedance over 250 years values can readily be obtained from maps prepared in support of national seismic building code efforts (Algermissen et al., 1982).

Interestingly, none of the five firms surveyed used probabilistic seismicity assessments or "seismic hazard studies" to develop design level motions for "final" waste fill designs during the 1991 calendar year. In the very few cases wherein such detailed probabilistic seismicity assessments were performed, they were done either: (a) as part of environmental impact studies (EIR/EIS studies), (b) for purposes of clarifying seismicity levels (usually in response to issues and/or questions raised at regulatory hearings), or (c) to provide a basis for risk analyses to select design levels of seismic shaking for use in analysis and design of "interim" fill stages. In several cases, at sites where: (a) the annual probability of occurrence of MCE-level accelerations was very low, (b) the environmental consequences were minimal (complete remediation of any potential failure was judged to be technically and economically feasible), and (c) the unlikely MCE-level MHA was very high (representing a high-cost design basis), the use of lower levels for interim stages ("during" fill placement) was considered. None of the five firms reported actually adopting such "lowered" motions for interim conditions for design studies performed during the 1991 calendar year, however.

SEISMIC RESPONSE ANALYSES

Seismic (or dynamic) response analyses can be performed: (a) to evaluate the effects of local site conditions (e.g.: foundation stratigraphy, topography, dynamic properties of foundation materials, etc.) on site response, and/or (b) to evaluate the response of waste fills themselves (and/or engineered containment berms, etc.). All five of the firms surveyed routinely perform dynamic response analyses to evaluate the "local site effects" associated with foundations consisting of non-negligible depths of soil over "bedrock". In all but a very few cases, these analyses were performed during 1991 using simple, one-dimensional ("columnar") analyses of vertically propagating shear waves. Virtually all of the dynamic response analyses performed by the five surveyed firms employed the "equivalent linear" method (Schnabel, et al., 1972) to model the important nonlinear (shear strain-dependent) soil modulus and damping characteristics. Most of these analyses were performed using either standard or slightly modified versions of the well-known program SHAKE (Schnabel, et al., 1972). Appropriate evaluation of dynamic soil properties for these types of analyses is a fairly straightforward issue, and will not be discussed herein.

Only one of the five firms surveyed routinely performs dynamic response analyses of the actual waste fills, while the other four firms virtually never perform response analyses for the actual (or proposed) waste fill masses. This represents an interesting divergence in otherwise very similar practices among the five firms, and merits some discussion.

The four firms not currently performing response analyses of the waste fill masses use the previously described response calculations for the foundations ("site conditions") as a basis for development of conservative estimates of peak acceleration and/or acceleration time history at the base of the waste fill mass. They then employ either pseudo-static stability analyses, and/or "Newmark-type" seismic displacement analyses, in conjunction with these calculated "base" accelerations. The degree of conservatism involved will be discussed in the next section ("Seismic Stability and Displacement Analyses").

These four firms do not currently attempt a more rigorous approach to analyses of waste fill response due to the very large uncertainties associated with attempting to evaluate and model the dynamic properties of waste fill materials. Major uncertainties in this regard are due to several factors unique to waste fills, and these include:

1. A relative lack of data regarding in-situ measurements of seismic wave velocities.
2. The strongly heterogeneous nature of many waste fills.
3. The large compressibility of many waste fill materials, and the associated likely changes in properties (with compression) over time.
4. The practical impossibility of performing any type of representative laboratory tests for evaluation of dynamic waste fill properties at sufficient scale as to be meaningful.
5. A pronounced lack of field data regarding actual seismic response of full-scale waste fills in earthquakes causing strong levels of shaking.
6. The unpredictable nature of many waste streams.

The last of these, the unpredictable nature of many waste streams, is arguably the most difficult of these very major problems. As an example, current state-level plans in some states for increasing requirements related to mandatory recycling are expected to have a pronounced impact on the nature(s) of the waste stream(s) entering waste landfills in the coming decade: detailed quantitative response analyses will be difficult, as these wastes do not yet exist.

For waste fills already in-place, field (in-situ) measurements of shear wave velocity (v_s) can provide useful data regarding dynamic waste fill moduli (G_{max}) at small shear strains, but these do not provide insight regarding either: (a) shear modulus variation, or (b) damping characteristics at the larger strain levels of interest for typical design studies of "strong motion" response. There are, in fact, very few useful data of any type currently available regarding large-scale damping characteristics of waste fills.

All five of the firms surveyed have "in-house" programs for collection of available data (both published data and proprietary, project-specific data) regarding v_s measurements in waste fills, (in anticipation of future need for response analyses) but this will not (by itself) remedy problems #2 through 6 in the above list. The best data regarding dynamic moduli and damping characteristics, at various strain levels, in waste fills must eventually come from back-analyses of field recordings of full-scale fill response to various levels of seismic excitation. The companion paper by Anderson et al. (1992) in these Proceedings (see Session 7: "Case Histories") provides a start, but there is a need for much more data. The requirement by regulatory agencies of appropriate instrumentation of completed waste fills would appear to be the most effective means of increasing the likelihood of acquisition of useful response data in the years ahead. Such instrumentation should include placement of strong motion recording stations on both waste fills and on their abutments and/or foundations (ideally on "rock".)

One of the five firms surveyed routinely performs response analyses of waste fill masses, in order to evaluate the effects of the waste fills (and in some cases the two- and even three-dimensional geometries of the waste fill and foundation configurations) on acceleration and shear stress time-histories within the waste fill/foundation systems. Due to the present uncertainties regarding dynamic waste fill properties, this approach currently requires very broad enveloping of projected dynamic waste fill properties. The net result of this modelling of a broad range of properties appears likely to be a conservative assessment of response not dissimilar to the end results obtained by the other four surveyed firms (who conservatively apply "site" response evaluations in a manner not requiring dynamic analyses of the fill itself).

SEISMIC STABILITY AND DISPLACEMENT ANALYSES

All five of the firms surveyed currently employ pseudo-static stability analyses for waste fills located in areas of sufficiently low seismicity (low MHA) that, with favorable fill and foundation geometries, satisfactory factors of safety can be achieved. In some areas of moderate to high seismicity (e.g., many "seismic impact zones"), however, the "pseudo-static" factors of safety typically called for by regulatory agencies are often difficult and/or very expensive to achieve for design-level MHA values. Accordingly, evaluation of seismic performance in these areas is frequently based on more detailed analyses of seismic deformations or displacements.

All five of the surveyed firms currently employ some form of "Newmark-type" analysis of seismic displacement for waste fills in regions of significant seismicity (Newmark, 1965). The "Newmark-type" analysis is illustrated schematically in Figure 6, and generally consists of the following basic steps:

1. Evaluation of the "yield" acceleration a_y (or pseudo-static horizontal yield coefficient, k_y) necessary to reduce the factor of safety against slippage of a potential slide mass (e.g.: the block on the inclined plane in Figure 6) to F.S. = 1.0.
2. Evaluation of the effective acceleration time history of the potential slide mass. (Variations on this approach include evaluation of either the acceleration time history or the shear stress time history at the base of the slide mass.)
3. Based on the assumption that sliding (or slippage) would occur when the accelerations are larger than those necessary to reduce the factor of safety to 1.0, the excess accelerations (above a_y or k_y) and their durations are then integrated twice with respect to time to obtain the net displacements that would accrue.

Four of the five firms surveyed routinely employ a variation of the original "Newmark" method developed by Makdisi and Seed (Makdisi & Seed; 1977, 1978). The Makdisi/Seed procedure allows use of chart solutions representing enveloped summaries of detailed displacement calculations performed as part of the original studies. Using this approach, a bounded estimate of expected resulting net displacements can be quickly developed as a function of:

1. The pseudo-static horizontal yield coefficient (k_y) for a potential slide mass.
2. The peak effective horizontal acceleration within the slide mass ($a_{h,max}$).
3. The duration of strong shaking (or number of cycles): this is empirically accounted for as a function of earthquake magnitude (longer durations or increased numbers of cycles being associated with larger causitive magnitude earthquake events).

The four firms surveyed use a variation on the actual Makdisi/Seed procedure which, while still permitting use of these enveloped chart solutions, replaces

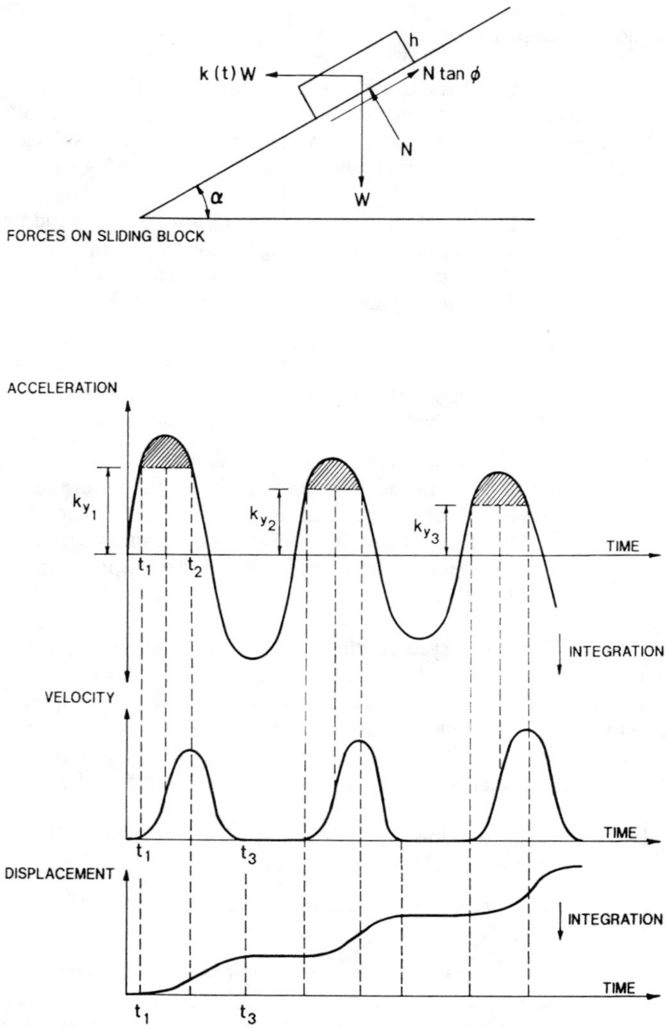

Figure 6. Schematic Illustration of the "Newmark-Type" Approach to Estimation of Cumulative Seismic Displacement.

the need for dynamic analyses of the response of the actual waste fill itself with a conservative set of assumptions. As noted previously, all four of the firms employing this methodology routinely use the peak horizontal acceleration at the base of the potential slide mass (in most cases at or near the base of the waste fill) as a conservative basis for evaluation of $a_{h,max}$. This approach may neglect some potential for amplification of absolute peak horizontal accelerations at any location within the waste fill, but this is offset because the counterbalancing effects of shear wave reversal within the fill are also neglected. (The "average" or "effective" maximum net horizontal acceleration within a large potential slide mass is generally much lower than the instantaneous peak horizontal acceleration at any single point within the mass.)

The fifth firm, which routinely performs broadly-banded response analyses of the waste fill itself, uses these response analyses to develop broadly-banded (conservative) estimates of the shear stress time-history at the location of a potential sliding surface. These time-histories are transposed to equivalent acceleration time histories, and these are then integrated twice with respect to time to develop estimates of resulting net displacements. Based on the information provided by the responding firms, it appears that this use of broadly-banded waste fill response analyses produces conservative results roughly comparable with the banded displacement estimates developed using the slightly modified "Makdisi/Seed" methodology employed by the other four firms surveyed. Both approaches appear to represent conservative implementations of the general "Newmark-type" of analysis.

A key element of all of these analyses is the evaluation of the yield acceleration, or horizontal pseudo-static yield coefficient (k_y), necessary to initiate slippage of a potential slide mass. This involves evaluation of dynamic shear strengths in one or more of the following three general zones:

(a) Shear Strengths within Foundation Materials:

Evaluation of dynamic shear strengths within foundation soils (and/or potentially unstable foundation "rock" units) is a generally well-established area of practice and should ideally require little discussion. All five of the firms surveyed appear to be using appropriate methods and procedures, and all appear to develop suitably conservative strength estimates for design.

Unfortunately, the selection of appropriate dynamic strength properties for foundation materials remains a somewhat "soft" area overall in geotechnical practice, and the authors sometimes encounter less appropriate practices in dealing with some other firms and/or projects. One potential problem area is the difficulty in determining whether (and when) unconsolidated-undrained (U-U), or consolidated-undrained (C-U) conditions are a more appropriate basis for strength assessment, and in evaluating the transitions between these conditions for projects routinely involving either incremental or continuous fill placement over periods of many years. Another common problem area is difficulty in dealing with post-peak strain softening in sensitive alluvial clays and/or residual weathered clay soils (and these types of foundation materials are attractive with regard to natural containment of potential leakage). A third difficult area not always well-addressed is the potential for dynamic shear strength increases due to rate effects; it is not unusual for undrained shear strengths in cohesive soils sheared very rapidly (under seismic rates of loading) to be on the order of 1.1 to 1.5 times higher than "static" shear strengths measured by conventional means (under less rapid rates of shearing) for use in static (non-seismic) analyses. These higher "dynamic" shear strengths are, however, often offset (and sometimes more than offset) by cyclic degredation due to repeated cycles of earthquake loading, and this can be particularly important in strain-softening soils. There is some excellent practice in dealing with these issues, particularly in geotechnical engineering firms with significant prior experience with these issues in conjunction with seismic analyses of major dams and other critical structures, but these difficult issues are not yet consistently well-addressed by all geo-consultants at this time.

(b) Shear Strengths within Waste Fill Masses:

The characterization of shear strengths within waste fills themselves is a difficult problem, largely as a result of many of the same types of difficulties associated with assessment of dynamic modulus and damping characteristics (as discussed previously). Important among these are difficulties associated with: (a) scale (the difficulty of performing laboratory tests at sufficiently large scale as to provide meaningful and representative results), (b) the strongly heterogeneous nature of many waste fills, and (c) the difficulty in predicting, accurately and reliably, the actual nature of the wastes (or "waste stream") likely to arrive at some future date(s).

Fortunately, the practical significance of the uncertainties listed above is limited because, even with very conservative selection of waste fill shear strength characteristics for analysis and design, the principal potential slippage surfaces controlling most design considerations usually tend to involve slippage primarily along surfaces either within liner systems and/or within weak underlying geologic foundation materials. Shear strengths of the waste fill itself tend to have relatively little impact on these potential slippage modes. As a result, waste fill shear strengths can often be very conservatively evaluated without significantly impacting overall waste fill design.

All five of the firms surveyed use conservative estimates of waste fill shear strength characteristics, based on conservative interpretations of "lowest possible" shear strengths (combinations of c' and ϕ') back-calculated from statically stable field fill geometries. All five firms use combinations of published data and their own private case studies for these purposes. As waste fills are not generally permitted to become saturated, effective stress analyses are routinely employed. As conservatism in selection of design strength values seldom significantly affects design (which is usually controlled by modes of slippage within liner systems or foundation materials), all five firms also conservatively neglect any potential increase in waste fill strengths associated with very "rapid" (seismic) loading, and use the same waste fill shear strengths for both static and seismic stability (and seismic deformation/displacement) studies. These practices all appear to be representative of typical current "common practice" based on the authors' experiences with other firms and agencies.

(c) Shear Strengths within Multi-Layered Liner Systems and Cover Systems:

Potential slippage along surfaces within multi-layered liner systems often represents "critical" conditions for analysis and design. Evaluation of shear strengths at interfaces between various combinations (pairs) of geosynthetics is usually relatively straightforward. While some geosynthetic/geosynthetic interface shear strengths can be quite low, the degree of uncertainty associated with the strength is also quite low. Accordingly, the strength can be reliably accounted for during the design process. Shear strengths at or near the interface between compacted clay layers and geomembrane liners can also be quite low. In constant to geosynthetic/geosynthetic interfaces, however, there remains substantial uncertainty in selecting compacted clay/geomembrane interface shear strengths for design. Furthermore, the lowest possible compacted clay/geomembrane interface strength will frequently be lower than the lowest possible geosynthetic/geosynthetic interface strength. For these reasons, selection of compacted clay/geomembrane interface strengths becomes one of the most crucial steps in the stability evaluation (either static or seismic) of lined waste fills.

Shear strengths of compacted clay/geomembrane interfaces are strongly affected by the "as-compacted" water content (w_c) and density (γ_d) of the compacted clay. These shear strengths can change significantly even over the very small ranges of γ_d usually allowed by very strict clay liner compaction specifications. An example is presented in Figure 7, which shows compacted clay/HDPE geomembrane interface shear strengths varying by a factor of approximately two over a small range of compaction water contents and densities in the range(s) routinely required to assure minimum permeability of the clay liner; high degrees of compaction, at water contents a few percent "wet" of the modified AASHTO [ASTM-D1557] optimum water content.

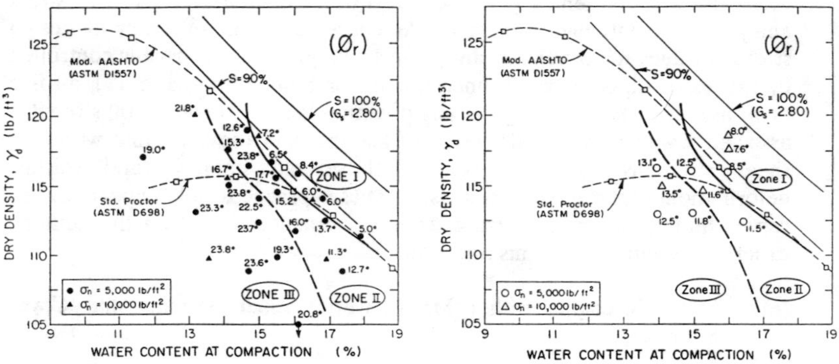

Figure 7. Effects of Compaction Water Content and Density on Compacted Clay/HDPE Geomembrane Interface Shear Strengths (After Seed & Boulanger, 1991).

Another factor that can significantly affect shear strengths at this potentially critical interface is "wetting" of the interface after placement and compaction of the clay. Such "wetting" can occur by a number of means, including:

1. Migration of leachate to the base of the waste fill.

2. Consolidation of the compacted liner clay (compacted "wet of optimum"), which can result in expulsion of significant volumes of water (e.g.: Anderson, et al., 1990; Bonaparte and Gross, 1990).

3. Periodic application (spraying) of water onto the surface of the completed compacted clay liner to prevent drying and cracking for the time period(s) required to cover the clay with an overlying geomembrane. The need for inspection, testing, and approval of the clay liner, as well as the care required for placement, seaming and inspection of the geomembrane can cause prolonged exposure, and uncontrolled application of surface water ("spraying") for periods of several days to as much as several weeks is a not uncommon field practice.

4. "Thermal" effects associated with cyclic heating and cooling of geomembranes placed over compacted liner clays (due to exposure to sunshine during daytime, until the geomembrane

is subsequently covered with soil) will draw water from the clay and cause this water to condense on the underside of the geomembrane. In some circumstances, this will result in "wetting" of the interface.

It is not generally feasible to prevent, with certainty, any (and all) of these and other potential "wetting" mechanisms. Accordingly, potential "wetting" should be considered in evaluation of interface shear strengths for design.

All five of the firms surveyed are fully aware of the potential significance of minor variations in "as-compacted" water content and density, and employ testing methods and conservative enveloping of resulting strengths to account for this. Their approach routinely requires a significantly higher number of compacted clay/geomembrane interface shear strength tests, performed over a range of as-compacted clay water contents and densities, than the authors' often encounter in dealing with some other firms and agencies. Similarly, all five firms surveyed are aware of the potential effects of various types/modes of "wetting" (and under surcharges varying from very small to very large). All five firms also include performance of compacted clay/geomembrane interface shear strength tests on "wetted" samples in their testing programs. "Wetting" details (usually soaking under some specified light surcharge load) vary from firm to firm. Similarly, several of the firms perform some of their tests on samples allowed to consolidate under a specified overburden stress prior to rapid, undrained shearing (C-U conditions) for some long-term seismic conditions, whereas the remainder of the reported testing was more typically performed under unconsolidated, undrained shearing (U-U) conditions.

The practices of these five firms represent a considerably more advanced level of understanding, and accounting for, the effects of both compaction conditions and potential "wetting" than the authors sometimes encounter at present in dealing with some other firms and agencies. Significantly, this approach generally results in a more conservative final design than is sometimes encountered. As it is only over the past two to four years that the potential significance of these factors has come to be understood, it may be expected that the treatment of these issues by the profession at-large will continue to improve rapidly.

Another potential area of concern is the lack of communication regarding both clay compaction conditions and "wetting" between the various engineers who routinely share responsibility for a major waste fill project. It is not unusual to have different engineers, and often even different firms, take primary responsibility for different engineering tasks including preliminary design, laboratory shear strength testing, preparation of specifications, interim stage design and design modification, and field inspection ("construction quality assurance"). As a result, the engineers performing stability analyses often have little or no direct contact with engineers responsible for actual field placement, and the potentially significant effects of deviations in field conditions from those upon which design strengths were based can often fail to be fully accounted for. It would represent a significant advancement of current practice if steps could be taken to ensure that the primary design engineer maintains sustained and meaningful oversight of all subsequent activities, including testing and field placement operations, that could potentially affect liner system stability.

ASSESSMENT OF PERFORMANCE

The assessment of the seismic performance of solid waste landfills, including their containment and other environmental protection systems, is a complex issue requiring considerable engineering judgement. Seismic performance assessments are currently expressed in many cases in terms of "acceptable" worst-case seismic displacements and/or deformations. Current practice does not yet include rigorous calculations of the effects of both transient and residual stresses and deformations on the performance of the principal environmental protection systems. Instead, these issues are typically dealt with on a partially judgemental basis. Issues that should ideally be addressed in this regard include:

1. The potential for rupture or tearing of geomembrane liners due to transient or residual stresses and deformations. Considerable amounts of data exist regarding both tensile stresses and tensile strains which can be sustained by these materials under slow (static) loading conditions. There is presently, however, much less data regarding the tensile strength and ductility of geomembranes (and their seams) under very rapid (seismic) rates of loading.

2. The potential for failure of leachate collection systems and final gas collection systems. Seismic deformations and displacements of the overall waste fill and containment systems can have potential ramifications for the piping/conduits associated with leachate and gas collection systems. Issues to consider include potential piping failures in bending, shear, tension and/or compression, as well as potential "punching" of nearby geomembranes.

All five of the firms surveyed currently employ the approach of limiting the calculated potential seismic displacements to a sufficiently small level as to ensure, in their judgement, acceptable performance of the environmental protection systems. This, of course, raises the interesting question as to the magnitudes of seismic displacement considered "acceptable".

Interestingly, all five of the firms surveyed have independently evolved over the past few years towards very similar conclusions regarding acceptable seismic displacements/deformations. Three of the five firms currently consider likely maximum seismic displacements (calculated using the conservative procedures described previously) of on the order of six inches (15 cm) or less to correspond to acceptable performance for most well-designed liner systems and geometries. The other two firms surveyed also routinely consider conservatively estimated maximum seismic displacements of on the order of six inches or less to be acceptable in most cases, but also occasionally consider slightly larger (based on conservative calculations) displacements of up to 12 inches (30 cm) or less to be acceptable when the liner systems and geometries are judged to be able to sustain these levels of potential movement without failure. All five of the firms surveyed employed similar criteria for both "final" fill configurations and interim fill stages. Several of the firms reported considering use of slightly relaxed criteria for evaluation of final cover liners and gas collection systems, based on the understanding that these would be relatively easily accessible and thus quickly repairable in the event of damage by a major seismic occurrence. Such relaxed criteria apparently were not, however, actually implemented by any of the five firms during 1991.

All of the foregoing is compatible with the authors' experiences in dealing with other firms and agencies over the past several years. As increased attention has come to be focussed on seismic performance of waste fills, the issue of "acceptable" levels of seismic displacement has drawn particular attention. Current practice spans a somewhat broader range in this regard than the fairly well-defined (and fairly uniform) range of criteria employed by the five firms surveyed, but the authors have noted a general evolution within the profession at-large towards the use of criteria corresponding, at least approximately, to consideration of conservatively-estimated maximum calculated seismic displacements of on the order of either six or twelve inches (or less) as "acceptable" for most conditions, with minor modifications of these criteria being applied to special cases.

SUMMARY AND CONCLUSIONS

Over the past four years there has been a significant increase in interest in the evaluation and design of lined waste landfills with regard to seismic performance. Although the full spectrum of current practice in this rapidly evolving area is somewhat broader, the survey of five major firms presented herein has shown a convergence towards a fairly well-defined set of analytical and design approaches within this selected group of firms. This coincides with the authors' more general observations, based on interactions with other firms and agencies, regarding a progressive evolution towards generally similar methods and procedures.

There are some considerable uncertainties with respect to material properties and dynamic response characteristics associated with both: (a) waste fill masses, and (b) base liner systems and final cover systems. As a result, current analysis and design methods are generally based on a sequence of conservative assumptions at various stages of the analyses, with the resulting cumulative level of conservatism generally selected so as to offset the possible impact(s) of uncertainties at each stage of analysis and design. In the opinion of the authors, the compounding of conservative assumptions, both in terms of property/parameter selection and analysis/evaluation methodology, at multiple stages during the overall analysis and design process should provide a conservative final result when carefully implemented.

There is significant potential for refinement of current analysis methods, and thus for increasingly efficient design, based on the implementation of less conservative assumptions at the various stages of analysis and design. All five of the firms surveyed are currently collecting the types of full-scale field data necessary to develop and/or refine and verify these types of methods, but none judge the current data base to be sufficient at this time to fully justify the use of less conservative (and generally more complex) modelling and analysis approaches.

ACKNOWLEDGEMENTS

The authors wish to thank the five firms surveyed, and their representatives, both for sharing their data and experiences as well as for their constructive comments and suggestions on this text. The authors are particularly grateful to Larry Roth and Rick Mitchell of CH2M Hill, Dr. Hari Sharma of EMCON Associates, Dr. Gary Schmertmann of GeoSyntec Consultants, John Byrne of Golder Associates, and Dr. Pat Lucia and Linda Locke of Woodward-Clyde Consultants.

REFERENCES

Algermissen, S. T., Perkins, D. M., Thenhaus, P. C., Hanson, S. L., and Bender, B. L. (1982). "Probabilistic Estimates of Maximum Acceleration and Velocity in Rock in the Contiguous United States", United States Department of the Interior, Geologic Survey, Open-File Report 82-1033, 99 p.

Anderson, D. G., Hushmand, B. and Martin, G. R. (1992). "Seismic Response of Landfill Slopes", Proc., ASCE Specialty Conference on Stability and Performance of Slopes and Embankments-II, Berkeley, Calif., June 28-July 1.

Byrne, R. J., Kendall, J. and Brown, S. (1992). "Cause and Mechanism of Failure, Kettleman Hills Landfill B-19, Unit IA", Proc., ASCE Specialty Conference on Stability and Performance of Slopes and Embankments-II, Berkeley, Calif., June 28-July 1.

Bonaparte, R. and Gross, B. A. (1990). "Field Behavior of Double-Liner Systems", Proceedings, Symposium on Waste Containment Systems, ASCE Geotechnical Special Publication No. 26, pp. 52-83.

California Code of Regulations, Title 23. California Water Regulations, Chapter 15. Discharges of Waste to Land (1984, as modified through 1992).

California Code of Regulations, Title 14. California Waste Management Board, Chapter 3. Minimum Standards for Solid Waste Handling and Disposal (1989).

Mitchell, R. A. and Mitchell, J. K. (1992). "Stability Evaluation of Waste Landfills", Proc., ASCE Specialty Conference on Stability and Performance of Slopes and Embankments-II, Berkeley, Calif, June 28-July 1.

Seed, R. B. and Boulanger, R. W. (1991). "Smooth HDPE/Clay Liner Interface Shear Strengths: Compaction Effects", J. of Geotechnical Engineering, ASCE, Vol. 117, No. 4, pp. 686-693.

Schnabel, P., Lysmer, J. and Seed, H. B. (1972). "SHAKE: A Computer Program for Earthquake Response Analysis of Horizontally Layered Sites", Report No. EERC 72-2, Earthquake Engineering Research Center, Univ. of California at Berkeley.

United States Code of Federal Regulations, Title 40. Protection of the Environment, Part 258. Solid Waste Disposal Facility Criteria (1991).

SUBJECT INDEX

Page number refers to first page of paper.

AUTHOR INDEX

Page number refers to first page of paper.

Please may I have a copy of:
pp. 267–291

Many thanks
Sunil.